PLANT BIOCHEMISTRY

PLANT BIOCHEMISTRY

Caroline Bowsher, Martin Steer, and Alyson Tobin

Garland Science
Taylor & Francis Group

Vice President:	Denise Schanck
Editor:	Elizabeth Owen
Editoral Assistants:	Sarah Holland and Kirsty Lyons
Production Editor and Layout:	Georgina Lucas
Copyeditor:	Penny Bowers
Proofreader:	Sally Livitt
Illustration:	Phoenix Photosetting and Matthew McClements, Blink Studio, Ltd
Cover Design:	Andrew Magee

Front cover images courtesy of: Alex Tobin; Brian E. S. Gunning, Australian National University; Professor Karl Oparka; Samfoto Photo Agency, Oslo, Norway; Ronald Gaubert, Prairieville, Louisiana; Christopher Sheehan/Ark Gallery; Alyson Tobin; David A. Anderson, Centre College, Danville, Kentucky. Back cover images courtesy of: Brian E. S. Gunning, Australian National University; Professor Karl Oparka; Ronald Gaubert, Prairieville, Louisiana; Samfoto Photo Agency, Oslo, Norway; Alex Tobin.

10-digit ISBN 0-8153-4121-0
13-digit ISBN 978-0-8153-4121-5

1005621180

Library of Congress Cataloging-in-Publication Data

Bowsher, Caroline.
  Plant biochemistry / Caroline Bowsher, Martin Steer, and Alyson Tobin. -- 1st ed.
    p. cm.
  ISBN-13: 978-0-8153-4121-5
  1. Botanical chemistry. I. Steer, Martin W. II. Tobin, Alyson K. III. Title.
  QK861.B69 2008
  572'.2--dc22

                    2008003086

Published by Garland Science, Taylor & Francis Group, LLC, an informa business
270 Madison Avenue, New York, NY 10016, US, and
2 Park Square, Milton Park, Abingdon, OX14 4RN, UK

Printed in the United States of America
15 14 13 12 11 10 9 8 7 6 5 4 3 2 1

Taylor & Francis Group, an informa business

Visit our web site at http://www.garlandscience.com

Dedication

For:
Dave
Brian, Phil, Robin, Ros and Sarah
Tim, Alex and George

Preface

This book is written for anyone studying plant biochemistry at an advanced under-graduate or graduate level, whether they are taking a course in plant biochemistry or studying metabolism, cell biology, general biochemistry, or more ecological fields including physiology, ecology, and plant–animal interactions. It will also be of value to experienced research scientists who might be entering the field of plant biochemistry for the first time, perhaps with early career training in animal or biomedical sciences. In all cases we expect the reader to have taken some introductory courses in biology and to have a working knowledge of basic biochemistry and cell biology. Consequently, we have chosen to avoid covering topics such as protein and DNA structure, the mechanisms of gene expression and nucleic acid biosynthesis, for example, which are widely covered in introductory level textbooks. Instead, we have taken a research-led approach to each topic in order to bring the reader to a stage where they can then delve more deeply into the scientific literature. With this aim in mind, we have included a list of further reading at the end of each chapter, along with some brief comments that should help to point you towards the most suitable material with which to start your research into the topic.

In writing this book, we have each brought our own perspective on the subject. Between us, we have carried out research (in cell biology, biochemistry, and physi-ology) and teaching (in biology, biochemistry, cell biology, plant sciences, medicine, and ecology) at universities across Europe, North America, and Asia. This variety of experience and expertise has allowed us to take a wide-ranging approach to each topic with the aim of setting the biochemistry firmly within the context of the whole plant and its surrounding environment. We feel that this scene setting is an impor-tant aspect of teaching, so that a biochemical pathway is not simply learned for its own sake but is placed within the framework of what it does, and why it is needed. In many cases we have extended this approach to include a consideration of how our lives might benefit from the exploitation of certain plant biochemical pathways; for example, by the production of pharmaceuticals, food products, and fuels. In many cases these topics are embedded within the main body of the chapter. However, we have also used the boxes as a means of delving more deeply into a topic and of highlighting an area that merits a particular focus in itself. The boxes may be particularly useful as a means of guiding students towards a special topic that could be extended through their own further reading; for example, crop improvement, poisons and feeding deterrents, and novel plant products.

As discussed in Chapter 1, the biochemistry of plants is sufficiently unique that it merits its own textbook. Nevertheless, while some pathways are unique to plants, others, for example the respiratory pathways, are of fundamental importance to most organisms. Consequently, we have taken a comparative approach, using infor-mation from animal, bacterial, fungal, and plant studies, where we feel it is relevant to do so. This approach is particularly useful when we have explored the evolution of plant biochemical pathways, with many plastid-located processes; for example, having originated in the prokaryotic ancestor to the plastid. Hence, while we hope that this textbook will serve to stimulate and to reinforce an interest in plants and their biochemical processes, we also urge students to read widely and to learn from studies of other organisms.

While the three of us have each spent many hours working alone in our offices and studies to produce each chapter, we have also benefited enormously from the support of our colleagues and families. We are particularly grateful to Kirsty Lyons and Liz Owen, of Taylor & Francis, for striking the right balance between gentle and firm reminders of impending deadlines and for providing encouragement to keep us on the right track. We must also thank our families for their tolerance and understanding while we spent yet another evening or weekend immersed in the book. We also thank the following reviewers who gave us careful and constructive feedback on one or more of the chapters: Owen Atkin (Australian National University), Professor Tony Bacic (University of Melbourne), John Barnett (University of Reading), Donald Briskin (University of Illinois at Urbana – Champaign), Lailiang Cheng (Cornell University), Dean DellaPenna (Michigan State University), Jim Dunwell (University of Reading), Robert Edwards (University of Durham), Peter Hedden (Rothamsted Research), Norman Huner (University of Western Ontario), Brian Larkins (University of Arizona), Peter Lea (University of Lancaster), Richard Leegood (University of Sheffield), Enrico Martinoia (Universität Zürich), Denis Murphy (University of Glamorgan), Nicholas Smirnoff (University of Exeter), Chung-Jui Tsai (Michigan Technological University), Astrid Wingler (University College London), and Andreas Weber (Heinrich-Heine-Universität, Düsseldorf). Finally, we thank Curtis Givan (University of New Hampshire) for reading and re-reading each chapter and for passing on his own enthusiasm for teaching and learning that has inspired us to try and do the same in return.

Caroline Bowsher

Martin Steer

Alyson Tobin

Contents

Preface   vii

Abbreviations   xv

1 Introduction to Plant Biochemistry   1

2 Approaches to Understanding Metabolic Pathways   5

What we need to understand a metabolic pathway   5

Chromatography   7

Electrophoresis   11

The use of isotopes   14

Current research techniques use a range of molecular biology approaches   16

Unique aspects of plant metabolism and their impact on metabolic flux   26

Metabolic control analysis theory   27

Coarse and fine metabolic control   29

Compartmentation: keeping competitive reactions apart   33

Understanding plant metabolism in the individual cell   33

The isolation of organelles   34

Summary   36

Further Reading   36

3 Plant Cell Structure   39

Cell structure is defined by membranes   40

The plasma membrane   46

Vacuoles and the tonoplast membrane   48

The endomembrane system   49

Cell walls serve to limit osmotic swelling of the enclosed protoplast   54

The nucleus contains the cell's chromatin within a highly specialized structure, the nuclear envelope   56

Mitochondria are ubiquitous organelles, which are the site of cellular respiration   57

Peroxisomes house vital biochemical pathways for many plant cell processes   58

Plastids   59

Summary   63

Further Reading   64

4 Light Reactions of Photosynthesis   65

Bacteria evolved the basic photochemical pathways found in plants today   65

Chlorophyll captures light energy and converts it to a flow of electrons   71

Carotenoids extend the spectral range of light that can be utilized in photosynthesis   74

Photosystem II splits water to form protons and oxygen, and reduces plastoquinone   75

The Q cycle uses plastoquinol to reduce plastocyanin and transport protons into the lumen   78

Photosystem I takes electrons from plastocyanin and reduces ferredoxin, which is used to make NADPH and other reduced compounds   80

ATP synthase utilizes the proton motive force to generate ATP   83

Cyclic photophosphorylation generates ATP independently of water oxidation and NADPH formation   85

Regulation of electron flow pathways in response to fluctuating light levels   86

Scavenging and removal of superoxides, peroxides, and other radicals by dismutases and antioxidants   87

Mechanisms for safely returning the levels of trapped high energy states to the ground state   88

Nonphotochemical quenching and the xanthophyll cycle   89

Summary   90

Further Reading   91

5 Photosynthetic Carbon Assimilation   93

Photosynthetic carbon assimilation produces most of the biomass on Earth   93

Carbon dioxide enters the leaf through stomata but water is also lost in the process   94

Carbon dioxide is converted to carbohydrates using energy derived from sunlight   94

The Calvin cycle is used by all photosynthetic eukaryotes to convert carbon dioxide to carbohydrate   96

Discovery of the Calvin cycle   96

There are three phases to the Calvin cycle   97

Calvin cycle intermediates may be used to make other photosynthetic products   108

The Calvin cycle is autocatalytic and produces more substrate than it consumes   108

Calvin cycle activity and regulation   109

Rubisco is a highly regulated enzyme   111

Rubisco oxygenase: the starting point for the photorespiratory pathway   113

The photorespiratory pathway: enzymes in the chloroplast, peroxisome, and mitochondria   113

The isolation and analysis of mutants and the photorespiratory pathway   117

Photorespiration may provide essential amino acids and protect against environmental stress   117

Photorespiration and the loss of photosynthetically fixed carbon   118

Photorespiration uses ATP and reductant   119

Decreasing global carbon dioxide concentrations caused a rapid evolution of C_4 photosynthesis   119

C_4 photosynthesis concentrates carbon dioxide at the active site of Rubisco   120

Spacial separation of the two carboxylases in C_4 leaves   120

Stages of C_4 photosynthesis and variations to the basic pathway   122

C_3–C_4 intermediate species may represent an evolutionary stage between C_3 and C_4 plants   126

The C_4 pathway can exist in single cells of some species   128

Some of the C_4 pathway enzymes are light-regulated   129

Crassulacean acid metabolism as a feature of desert plants   130

Temporal separation of the carboxylases in CAM   130

Crassulacean acid metabolism as a flexible pathway   130

Phospho*enol*pyruvate carboxylase in crassulacean acid metabolism plants is regulated by protein phosphorylation   133

Crassulacean acid metabolism is thought to have evolved independently on several occasions   133

C_3, C_4, and CAM photosynthetic pathways: advantages and disadvantages   134

C_3, C_4, and CAM plants differ in their facility to discriminate between different isotopes of carbon   138

Summary   140

Further Reading   141

6 Respiration   143

Overview of respiration   143

The main components of plant respiration   144

Plants need energy and precursors for subsequent biosynthesis   144

Glycolysis is the major pathway that fuels respiration   145

Hexose sugars enter into glycolysis and are converted into fructose 1,6-bisphosphate   148

Fructose 1,6-bisphosphate is converted to pyruvate   148

Alternative reactions provide flexibility to plant glycolysis   149

Plant glycolysis is regulated by a bottom-up process   151

Glycolysis supplies energy and reducing power for biosynthetic reactions   151

The availability of oxygen determines the fate of pyruvate   151

The oxidative pentose phosphate pathway is an alternative catabolic route for glucose metabolism   153

The irreversible oxidative decarboxylation of glucose 6-phosphate generates NADPH   153

The second stage of the oxidative pentose phosphate pathway returns any excess pentose phosphates to glycolysis   153

All or part of the OPPP is duplicated in the plastids and cytosol   155

The tricarboxylic acid cycle is located in the mitochondria   155

Pyruvate oxidation marks the link between glycolysis and the tricarboxylic acid cycle   155

The product of pyruvate oxidation, acetyl CoA, enters the tricarboxylic acid cycle via the citrate synthase reaction   164

Substrates for the tricarboxylic acid cycle are derived mainly from carbohydrates   167

The tricarboxylic acid cycle serves a biosynthetic function in plants   167

Anaplerotic reactions are needed to enable intermediates to be withdrawn from the tricarboxylic acid cycle   169

The tricarboxylic acid cycle is regulated at several steps   170

Recent research into a thioredoxin/NADPH redox system for regulating tricarboxylic acid cycle enzymes and other mitochondrial proteins   172

The mitochondrial electron transport chain oxidizes reducing equivalents produced in respiratory substrate oxidation and produces ATP   172

Main protein complexes of the electron transport chain   173

Plant mitochondria possess additional respiratory proteins that provide a branched electron transport chain   175

Plant mitochondria contain four additional NAD(P)H dehydrogenases   176

The physiological function of the alternative NAD(P)H dehydrogenases remains the subject of some speculation   176

Plant mitochondria contain an alternative oxidase that transfers electrons from QH_2 to oxygen and provides a bypass of the cytochrome oxidase branch   177

The alternative oxidase is a dimer of two identical polypeptides with a nonheme iron center   178

Alternative oxidase isoforms in plants are encoded by discrete gene families   178

Alternative oxidase activity is regulated by 2-oxo acids and by reduction and oxidation   180

The alternative oxidase adds flexibility to the operation of the mitochondrial electron transport chain   181

The alternative oxidase may prevent the formation of damaging reactive oxygen species within the mitochondria   182

Alternative oxidase appears to play a role in the response of plants to environmental stresses   182

Alternative oxidase and NADH oxidation   183

Plant mitochondria and uncoupling proteins   183

ATP synthesis in plant mitochondria is coupled to the proton electrochemical gradient that forms during electron transport   183

ATP synthase uses the proton motive force to generate ATP   184

Mitochondrial respiration interacts with photosynthesis and photorespiration in the light   187

Emerging research area into supercomplexes and metabolons   191

Summary   191

Further Reading   192

7 Synthesis and Mobilization of Storage and Structural Carbohydrates   195

Role of carbohydrate metabolism in higher plants   195

Sucrose is the major form of carbohydrate transported from source to sink tissue   197

Sucrose phosphate synthase is an important control point in the sucrose biosynthetic pathway in plants   198

Sensing, signaling, and regulation of carbon metabolism by fructose 2,6-bisphosphate   200

Fructose 2,6-bisphosphate enables the cell to regulate the operation of multiple pathways of plant carbohydrate metabolism   200

Fructose 2,6-bisphosphate as a regulatory link between the chloroplast and the cytosol   204

Sucrose breakdown occurs via sucrose synthase and invertase   205

Starch is the principal storage carbohydrate in plants   209

Starch synthesis occurs in plastids of both source and sink tissues   209

Starch formation occurs in water-insoluble starch granules in the plastids   213

The composition and structure of starch affects the properties and functions of starches   215

Starch degradation is different in different plant organs   216

The nature and regulation of starch degradation is poorly understood   216

Transitory starch is remobilized initially by a starch modifying process that takes place at the granule surface during the dark period   218

The regulation of starch degradation is unclear   219

Fructans are probably the most abundant storage carbohydrates in plants after starch and sucrose   220

A model has been proposed for the biosynthesis of the different fructan molecules found in plants   220

Fructan-accumulating plants are abundant in temperate climate zones with seasonal drought or frost   222

Trehalose biosynthesis is not just limited to resurrection plants   222

Trehalose biosynthesis in higher plants and its role in the regulation of carbon metabolism   223

Plant cell wall polysaccharides   224

Synthesis of cell wall sugars and polysaccharides   225

Cellulose   225

Matrix components consist of branched polysaccharides   228

Expansins and extensins, proteins that play both enzymatic and structural roles in cell expansion   234

Lignin   234

Summary   235

Further Reading   235

8 Nitrogen and Sulfur Metabolism   237

Nitrogen and sulfur must be assimilated in the plant   237

Apart from oxygen, carbon, and hydrogen, nitrogen is the most abundant element in plants   238

Nitrogen fixation: some plants obtain nitrogen from the atmosphere via a symbiotic association with bacteria   239

Symbiotic nitrogen fixation involves a complex interaction between host plant and microorganism   242

Nodule-forming bacteria (Rhizobiaceae) are composed of the three genera *Rhizobium*, *Bradyrhizobium*, and *Azorhizobium*   242

The nodule environment is generated by interaction between legume plant host and rhizobia   244

Nitrogen fixation is energy expensive, consuming up to 20% of total photosynthates generated   245

Mycorrhizae are associations between soil fungi and plant roots that can enhance the nitrogen nutrition of the plant   246

Most higher plants obtain nitrogen from the soil in the form of nitrate   248

In higher plants there are multiple nitrate carriers with distinct properties and regulation   249

Nitrate reductase catalyzes the reduction of nitrate to nitrite in the cytosol of root and shoot cells   250

The production of nitrite is rigidly controlled by the expression, catalytic activity, and degradation of NR   251

Nitrite reductase, localized in the plastids, catalyzes the reduction of nitrite to ammonium   253

Plant cells have the capacity for the transport of ammonium ions   255

Ammonium is assimilated into amino acids   258

Ammonium originates from both primary and secondary sources   258

Ammonium is assimilated by glutamine synthetase and glutamate synthase, which combine together in the glutamine synthetase/glutamate synthase cycle   259

GS is an octameric protein with two isoforms, localized in the cytosol and plastid   259

The GS genes and proteins show discrete cellular localization and different responses to light and nutrients   260

Glutamine synthetase activity is regulated by metabolites and effectors, and may be modified by phosphorylation and 14-3-3 binding   261

Further evidence of the functions of glutamine synthetase isoenzymes has come from studies of mutants and transgenic plants   261

Higher plants contain two forms of GOGAT, one is ferredoxin-dependent and the other is NADH-dependent   263

Both Fd- and NADH-GOGAT are located in the plastid and exist as monomeric proteins in most species   263

The tissue and cellular localization of Fd- and NADH-GOGAT provides a clue to their function in higher plants   264

Further evidence of the function of Fd- and NADH-GOGAT has come from the analysis of mutants and transgenic plants   264

Sulfur is an essential macronutrient but it represents only 0.1% of plant dry matter   265

Sulfate is relatively abundant in the environment and serves as a primary sulfur source for plants   266

The assimilation of sulfate   267

Adenosine 5′-phosphosulfate reductase is composed of two distinct domains   268

Sulfite reductase is similar in structure to nitrite reductase   269

Sulfation is an alternative minor assimilation pathway incorporating sulfate into organic compounds   269

Amino acids biosynthesis is essential for plant growth and development   270

Carbon flow is essential to maintain amino acid production   270

There are species differences in the form of nitrogen transported through the xylem   272

Aminotransferase reactions are central to amino acid metabolism by distributing nitrogen from glutamate to other amino acids   273

Asparagine, aspartate, and alanine biosynthesis   275

Glycine and serine biosynthesis   276

The aspartate family of amino acids: lysine, threonine, isoleucine, and methionine   276

The branched chain amino acids valine and leucine   279

Sulfur-containing amino acids cysteine and methionine   280

Glutamine, arginine, and proline biosynthesis — 282

The biosynthesis of the aromatic amino acids: phenylalanine, tyrosine, and tryptophan — 284

Histidine biosynthesis — 285

Large amounts of nitrogen can be present in nonprotein amino acids — 285

Plant storage proteins: why do plants store proteins and what sort of proteins do they store? — 286

Vicilins and legumins are the main storage proteins in many dicotyledonous plants — 288

Prolamins are major storage proteins in cereals and grasses — 290

2S albumins are important but minor components of seed proteins — 292

Where are seed proteins synthesized and how do they reach their storage compartment? — 292

Protein stores are degraded and mobilized during seed germination — 296

Vegetative organs store proteins, which are very different from seed proteins — 297

Despite their diversity, storage proteins share common characteristics — 299

Summary — 299

Further Reading — 300

9 Lipid Biosynthesis   303

Overview of lipids — 303

Fatty acid biosynthesis occurs through the sequential addition of two carbon units — 307

The condensation of nine two-carbon units is necessary for the assembly of an 18C fatty acid — 307

For the assembly of an 18C fatty acid from acetyl CoA using type II fatty acid synthase, 48 reactions are necessary and at least 12 different proteins involved — 312

Acyl-ACP utilization in the plastid — 314

Regulation of fatty acid formation — 314

Source of NADPH and ATP to support fatty acid biosynthesis — 315

Glycerolipids are formed from the incorporation of fatty acids to the glycerol backbone — 315

Phosphatidic acid, produced in the plastids or endoplasmic reticulum, is a central intermediate in glycerolipid biosynthesis — 316

Lipids function in signaling and defense — 318

The products of the oxidation of lipids and the resulting metabolites are collectively known as oxylipins — 320

A waxy cuticle coats all land plants — 322

Role of suberin as a hydrophobic layer — 324

Storage lipids are primarily a storage form of carbon and chemical energy — 325

Release of fatty acids from acyl lipids — 328

The breakdown of fatty acids occurs via oxidation at the β carbon and subsequent removal of two carbon units — 329

Summary — 333

Further Reading — 334

10 Alkaloids   335

Plants produce a vast array of chemicals that deter or attract other organisms — 335

Alkaloids, a chemically diverse group that all contain nitrogen along with a number of carbon rings — 336

Functions of alkaloids in plants and animals — 336

The challenges and complexity of alkaloid biosynthetic pathways — 336

Amino acids as precursors in the biosynthesis of alkaloids — 338

Terpenoid indole alkaloids are made from tryptamine and the terpenoid secologanin — 338

Isoquinoline alkaloids are produced from tyrosine and include many valuable drugs such as morphine and codeine — 344

Tropane alkaloids and nicotine are found mainly in the Solanaceae — 349

Pyrollizidine alkaloids are found in four main families — 354

Purine alkaloids as popular stimulants in beverages, and as poisons and feeding deterrents against herbivores — 355

The diversity of alkaloids has arisen through evolution driven by herbivore pressure — 356

Summary — 360

Further Reading — 361

11 Phenolics   363

Plant phenolic compounds are a diverse group with a common aromatic ring structure and a range of biological functions — 363

The simple phenolics — 364

The more complex phenolics include the flavonoids, which have a characteristic three-membered A, B, C ring structure — 367

Lignin is a complex polymer formed mainly from monolignol units — 369

The tannins are phenolic polymers that form complexes with proteins   370

Most plant phenolics are synthesized from phenylpropanoids   370

The shikimic acid pathway provides the aromatic amino acid, phenylalanine, from which the phenylpropanoids are all derived   371

The core phenylpropanoid pathway provides the basic phenylpropanoid units that are used to make most of the phenolic compounds in plants   375

Flavonoids are produced from chalcones, formed from the condensation of p-coumaroyl CoA and malonyl CoA   379

Simple phenolics from the basic phenylpropanoid pathway are used in the biosynthesis of the hydrolyzable tannins   391

Lignin is formed from monolignol subunits in a complex series of reactions that are still being unraveled   392

Summary   397

Further Reading   398

12 Terpenoids   399

Terpenoids are a diverse group of essential oils that are formed from the fusion of five-carbon isoprene units   399

Terpenoids serve a wide range of biological functions   402

The biosynthesis of terpenoids   411

Subcellular compartmentation is important in the regulation of terpenoid biosynthesis   426

Summary   428

Further Reading   428

Index   431

The colour plate section appears between pages 352 and 353.

Abbreviations

1-FFT	fructan 1-fructosyl transferase		F2,6BPase	fructose 2,6-bisphosphatase
3-PGA	3-phosphoglycerate		F6P	fructose 6-phosphate
4CL	4-coumarate coenzyme A ligase		F6P2K	fructose 6-phosphate 2 kinase
AACT	acetoacetyl coenzyme A thiolase		FAD	flavin adenine dinucleotide
ACCase	acetyl coenzyme A carboxylase		Fd-GOGAT	ferredoxin-dependent glutamine-oxoglutarate aminotransferase
ACP	acyl carrier protein		Fd_{ox}	oxidized ferredoxin
ADP	adenosine diphosphate		Fd_{red}	reduced ferredoxin
AOX	alternative oxidase		G6P	glucose 6-phosphate
APS	adenosine 5′-phosphosulfate		GABA	γ-aminobutyric acid
ATP	adenosine triphosphate		GAPDH	glyceraldehyde 3-phosphate dehydrogenase
C4H	cinnamate 4-hydroxylase		GDC	glycine decarboxylase complex
CA1P	2-carboxyarabinitol 1-phosphate		GOGAT	glutamine-oxoglutarate aminotransferase or glutamate synthase
CAM	crassulacean acid metabolism			
CDP	cytidine diphosphate		GPP	geranyl diphosphate
CHI	chalcone isomerase		GS	glutamine synthetase
CHS	chalcone synthase		H6H	hyoscyamine 6β-hydroxylase
cMEPP	2-*C*-methyl-D-erythritol-2,4-cyclodiphosphate		HDR	HMBPP; 1-hydroxy-2-methyl-2-(*E*)-butenyl 4-phosphate reductase
CO_2	carbon dioxide		HMBPP	1-hydroxy-2-methyl-2-(*E*)-butenyl 4-phosphate
CoA	coenzyme A			
DAG	diacylglycerol		HMG	3-hydroxy-3-methylglutaryl
DAHP	3-deoxy-D-arabino-heptulosonate 7-phosphate		HMGR	hydroxy-methylglutaryl coenzyme A reductase
DHAP	dihydroxyacetone phosphate		HMGS	hydroxy-methylglutaryl coenzyme A synthase
DHS	deoxyhypusine synthase			
DMAPP	dimethylallyl diphosphate		HMW	high molecular weight
DXP	1-deoxy-D-xylulose 5-phosphate		HPLC	high performance liquid chromatography
E3BP	E_3-binding protein		HSS	homospermidine synthase
EPSP	5-*enol*pyruvylshikimic acid 3-phosphate		IDH	isocitrate dehydrogenase
ER	endoplasmic reticulum		IEF	isoelectric focusing
F1,6BP	fructose 1,6-bisphosphate		IFS	isoflavone synthase
F2,6BP	fructose 2,6-bisphosphate		IP_3	inositol 1,4,5-triphosphate

IPG	immobilized pH gradient	PFK	phosphofructokinase
IPP	isopentenyl diphosphate	PFP	pyrophosphate-dependent fructose 6-phosphate 1-phosphotransferase
LATS	low affinity transport system		
LHC	light harvesting chlorophyll	P_i (in circle)	inorganic phosphate
LOX	lipoxygenase	PI	phosphatidylinositol
LPA	lysophosphatidic acid	PMT	putrescine N-methyltransferase
LPAAT	lysophosphatidic acid acyltransferase	PPDK	pyruvate orthophosphate dikinase
MDH	malate dehydrogenase	PP_i (in circle)	inorganic pyrophosphate
ME	malic enzyme	PPI	poly-phosphoinositide
MEP	2-C-methyl-D-erythritol phosphate	PQ	plastoquinone
MVA	mevalonic acid	PQH_2	plastoquinol
NAD	nicotinamide adenine dinucleotide	PS	Photosystem
NAD^+	nicotinamide adenine dinucleotide	PSV	protein storage vacuole
NADH	nicotinamide adenine dinucleotide (reduced form)	PSY	phytoene synthase
		RG	rhamnogalacturonan
NADPH	nicotinamide adenine dinucleotide phosphate (reduced form)	RNAi	RNA interference
		ROS	reactive oxygen species
NDin(NADH)	inner membrane dehydrogenases	RuBP	ribulose 1,5-bisphosphate
NH_3	ammonia	S^{2-}	sulfide
NH_4^+	ammonium	SDS	sodium dodecyl sulfate
NIP	nitrate reductase inhibitor protein	SHMT	serine hydroxymethyltransferase
NiR	nitrite reductase	SiR	sulfite reductase
NO_2^-	nitrite	SMT	scoulerine 9-O-methyltransferase
NO_3^-	nitrate	SO_3^{2-}	sulfite
NR	nitrite reductase	SPDS	spermidine synthase
OAA	oxaloacetate	SPS	sucrose phosphate synthase
OPPP	oxidative pentose phosphate pathway	STS	stilbene synthase
P (in circle)	phosphate (PO_4^{2-})	TAG	triacylglycerol
PA	phosphatidic acid	TCA	tricarboxylic acid
PAGE	polyacrylamide gel electrophoresis	TDC	tryptophan decarboxylase
PAL	phenylalanine ammonia lyase	TPP	trehalose 6-phosphate phosphatase
PDC	pyruvate dehydrogenase complex	TPS	trehalose 6-phosphate synthase
PDH	pyruvate dehydrogenase	UCP	uncoupling protein
PDK	pyruvate dehydrogenase kinase	VLCFA	very-long-chain fatty acid
PDP	phospho-pyruvate dehydrogenase kinase	WUE	water use efficiency
		XET	xyloglucan endotransglycosylase
PEP	phospho*enol*pyruvate		
PEPCK	phosphoenolpyruvate carboxykinase		

Introduction to Plant Biochemistry

Plants are present in all aspects of our everyday life. They provide our food, produce the oxygen that we breathe, and serve as raw materials in many of our clothes and buildings, and in the manufacture of biofuels, drugs, dyes, perfumes, and pesticides. Land plants have existed for over 400 million years, and the first flowering plants (angiosperms) evolved about 135–190 million years ago. Through their continued evolution they have acquired an enormous diversity of form and physiology that has enabled them to colonize successfully a wide range of habitats. There are currently thought to be between 250,000 and 400,000 angiosperm species and they survive on all continents, from the Arctic to the Antarctic, and in all manner of environments from deserts to flooded plains. Such breadth of environments means that plants also survive a range of temperatures.

Despite the diversity and prevalence of plants, modern agriculture relies on just a few species to provide much of the human diet. Indeed, just six crop plants—wheat, rice, corn, potatoes, sweet potatoes, and manioc—provide directly or indirectly (i.e. after having been fed to animals) over 80% of the total calories consumed by humans. Although these plants are rich in carbohydrates, to ensure a balanced diet proteins can be provided by legumes, including the common beans, peas, lentils, peanuts, and soybeans, while additional vitamins and minerals are provided by leafy vegetables such as lettuce, cabbage, and spinach. Finally, fats are provided by seeds and fruits of plants such as sunflower, peanuts, canola, and olives. Although other crops are of great importance to humans, including sugar cane, sugar beet, barley, sorghum, coconut, and bananas, it is clear that we currently use only a tiny proportion of the available plant species in our daily diet. As the world population continues to grow, there is an increasing demand not only for food but also the raw materials that plants supply. To meet this demand we need to be able to maximize the production of plant products, by improving existing crops and by developing novel crops. Consequently, an important aim of plant biochemical research is to discover how plants produce, and control the production of these valuable food and non-food products. In reading this book, we hope that you will gain some insight into our current knowledge of plant biochemistry.

The biochemistry of plants is sufficiently distinctive to merit separate treatment from that of other organisms. This is for two main reasons. First, the sessile plant cannot avoid environmental stress or predation by moving away, and second, plants are autotrophic and their resources are relatively simple (e.g. inorganic nutrients, light, water, and CO_2) and frequently in short supply. Consequently, plant biochemical pathways tend to be particularly flexible and to be responsive to changes in the plant's environment, both as a survival mechanism and as a means of making optimum use of limiting resources. Throughout this book, examples of plants' responses to their environment and the challenges the environment presents will be considered. As we shall see, a recurring feature of plant biochemical pathways is the existence of isoenzymes—enzymes that carry out essentially identical reactions and yet differ in their regulatory properties and subcellular location. For example, many enzymes are duplicated between plastids and cytoplasm [e.g. in ammonium assimilation (Chapter 8), glycolytic reactions (Chapter 6), and reactions that produce precursors for terpenoid biosynthesis (Chapter 12)]. One function of these isoenzymes is to provide an added flexibility to the biochemical pathway. It is almost as if this duplicity of function is providing a safety net whereby a metabolic bypass may occur should stress produce an untoward effect on one compartment. The second major feature that marks out plant biochemistry is the autotrophic nature of plants—hence, even respiratory pathways such as the tricarboxylic acid cycle, which are primarily catabolic in non-photosynthetic organisms, serve an anabolic function in plants, providing carbon skeletons for biosynthesis (Chapter 6). A good example where both of these key features coincide to enable plants to survive is that of the secondary metabolic pathways leading to the biosynthesis of alkaloids (Chapter 10), phenolics (Chapter 11), and terpenoids (Chapter 12) where the environmental stress imposed by herbivory, in particular, has led to the evolution of increasingly complex biochemical pathways that result in a chemical arms race, with plants producing a massive array of defensive chemicals as a means of deterring potential herbivores. This, of course, presents a significant drain on the photosynthetic productivity of a plant and yet biochemical pathways are sustained as a survival mechanism.

As you will find in any biochemistry textbook, we have presented the metabolic pathways rather like a route map to show how one reaction follows another. However, it is important to see these pathways as dynamic and fluid processes that do not operate in isolation and are responsive to considerably fluctuating conditions, both in their immediate vicinity (i.e. cellular or subcellular) and in the wider (i.e. whole plant–environment interactions) environment. To understand specifically not just the route map but the flux and regulation, i.e. the dynamic nature of the pathways, what is being produced/degraded and how this is balanced according to the needs of the plant, biochemists have developed a wealth of methods. In recent years there has been a move away from the reductionist approach where the biochemical reactions were only ever investigated at the level of single proteins in isolation. Current progress is taking place through a more holisitic, or systems-based approach, where there is a realistic hope of understanding the interlinking processes of biochemical and gene regulation in relation to the function and metabolism of the whole plant. We have introduced and explained some current methodologies that are providing a greater understanding of the biochemistry of the whole plant (e.g. metabolomics, Chapter 2). Indeed, a theme of this book is to set the biochemistry into the context of the function and response of the whole plant, where we have been able to do so. In several chapters you may find the boxes to be particularly useful in this respect.

Many biochemical pathways in plants, as in other eukaryotes, show cellular and subcellular compartmentation, with enzymes localized in distinct compartments (Chapter 2). It is important to understand plant cell structure in order to understand fully this compartmentation, and in Chapter 3, therefore, we introduce the structural aspects of plant cells. This topic serves as a reminder of the significant contribution that compartmentation makes to the regulation and control of biochemical pathways. Compartmentation is often essential as a means of preventing unwanted reactions from taking place, i.e. by isolating enzymes from potential substrates that might otherwise be converted to toxic products. Nevertheless, many biochemical pathways are distributed among a number of different subcellular compartments. Photorespiration, which involves reactions that take place in chloroplasts, peroxisomes, cytosol, and mitochondria, is a good example of a highly compartmentalized pathway. In pathways such as this, there is the need for substrates and products to be moved across membranes, hence providing additional control points that need to be considered when investigating the regulation of a pathway. Specific metabolite translocator proteins, which serve to carry metabolites across cell membranes, are therefore introduced in Chapter 2, and are further discussed in the context of photosynthesis (Chapter 5), respiration (Chapter 6), carbohydrate metabolism (Chapter 7), nitrogen and sulfur assimilation (Chapter 8), and fatty acid biosynthesis (Chapter 9).

The autotrophic nature of plant biochemistry is introduced in the two photosynthesis chapters (Chapter 4, Light reactions of photosynthesis, and Chapter 5, Photosynthetic carbon assimilation). Chapter 5 also illustrates, once again, the environmental influences on plant biochemical pathways, where the more complex carbon concentrating pathways of C_4 photosynthesis and of Crassulacean acid metabolism are thought to have evolved in response to limited CO_2 concentrations and to a reduction in water availability, respectively.

Autotrophy is not just about gaining carbon; it also involves the acquisition of other major minerals that are needed for plant growth and development. Hence nitrogen and sulfur assimilation (Chapter 8) are just as essential a part of autotrophy as is carbon assimilation. The way in which plants acquire and reduce inorganic nitrogen is another good example of how environmental conditions influence plant biochemistry. As explained in Chapter 8, the form of inorganic nitrogen that is assimilated by a plant depends not only on the plant species, but also on the conditions of the soil, with anaerobic and acidic soils being primarily colonized by plants that are adapted to using ammonium as their nitrogen source. In contrast, most agricultural soils contain the bulk of their nitrogen as nitrates; hence, in most crop species (with the noted exception of rice and legumes) nitrogen assimilation begins with the uptake and reduction of nitrate, while plants living in acid bogs and shrubby heathlands are generally better adapted to the uptake and assimilation of ammonium. Other plants, such as the legumes, have co-evolved with symbiotic bacteria so as to acquire their nitrogen from free atmospheric nitrogen, with the bacteria releasing ammonium for the plant, which, in turn, supplies the bacteria with organic acids for respiration.

In plants, respiration involves the combined operation of glycolysis, the oxidative pentose phosphate pathway, tricarboxylic acid cycle, and the mitochondrial electron transport chain, and respiration serves both a catabolic and a biosynthetic function (Chapter 6). Respiration is a highly flexible process allowing plants to respond to varying environmental conditions and to changes in photosynthetic and photorespiratory activity. For example, plant glycolysis is unique in having dual locations, in the cytosol and the plastids. Plant glycolysis also uses alternative enzymes to supplement or even replace those conventionally present, ensuring metabolic flexibility and allowing glycolysis to proceed and the plant to survive under varying and potentially stressful environmental and/or nutritional conditions. Furthermore, plant mitochondria possess a branched electron transport chain that provides a mechanism for varying the ATP production in response to metabolic demands, in some cases supplementing the chloroplast to support carbon assimilation. For example, the possession of an alternative respiratory pathway, in addition to the conventional cytochrome pathway, enables plants to respire at high rates without the usual respiratory control by ADP that would otherwise limit respiration. This pathway has been particularly successful, as we shall see, in some plants where high rates of carbohydrate oxidation result in heat production and volatilization of scents to attract pollinating flies (e.g. Arum lilies, Chapter 6).

Carbohydrates are the main respiratory substrate in most plant tissues (Chapter 7). They are also an important source of carbon skeletons for the biosynthesis of many organic molecules in a plant, e.g. amino acids, lipids, and structural carbohydrates. Plant carbohydrate biosynthesis is highly compartmentalized, with the two major storage carbohydrates of sucrose and starch being synthesized in the cytosol and plastids, respectively. Starch is generally used for relatively long-term storage of carbohydrates, while sucrose is soluble and readily transported from tissue to tissue and cell to cell. The balance between the biosynthesis, transport and utilization of carbohydrates has to be regulated to ensure that there is an adequate supply of carbohydrate for both photosynthetic and non-photosynthetic tissue. Consequently, carbohydrate metabolism is a highly regulated process with a number of feedback controls that serve to control and coordinate the supply and demand of source and sink tissue. The presence of isoenzymes adds further flexibility to the pathways for sucrose and starch metabolism, as explained in Chapter 7.

Lipids are not only essential to the structure of cells but they also serve as important respiratory substrates in certain plant tissues (e.g. oily seeds). Fats, oils, some hormones and pigments, and most non-protein membrane components are lipids. A saponifiable lipid is made of a glycerol backbone and fatty acids. Fatty acids are carboxylic acids with long-chain hydrocarbon side groups. The type of fatty acid that is made, in terms of chain length and degree of saturation, will determine the type of lipid produced. Similarly, the selection of fatty acid incorporated into the glycerol backbone influences the type of lipid. Finally the location where the fatty acids are built into lipids will influence the combination of fatty acids and the type of lipid produced. In Chapter 9 we focus on fatty acid biosynthesis and the lipids arising as a result of this. We further discuss the subsequent release of energy from lipids when they are metabolized. Fatty acid and lipid metabolism in plants have a number of features in common with other organisms. Again in plants, fatty acid and lipid metabolism is more complex than in other organisms because of its cellular compartmentation. By necessity lipid pools must be moved around to meet metabolic demands (Chapter 9).

This book illustrates some of the ways in which plants have exploited the full range of possibilities offered by carbon chemistry to flourish on this planet. The study of the numerous synthetic pathways present in plants, unraveling the structures of intermediates and end products and characterizing the enzymes involved, is a challenge for the modern plant biochemist. With the toolkits of today's chemists, it is at last possible to explore the fascinating range of processes that underlie the success of plants.

The topics covered in this book clearly demonstrate that plants and their biochemistry have a direct impact on human activity and success, both in terms of their importance as a food supply and as raw materials for fibers and other industrial and pharmaceutical products. With the growing world population currently estimated to reach 9–10 billion by 2050 it is predicted that food will be in increasingly limiting supply. There is an abundance of unused land on the earth, and much of it is nutrient-poor; the challenge is to develop effective strategies to exploit these areas. Both increased productivity on current areas of agricultural land, and the successful cultivation of crops in more stressful environments (e.g. saline or mineral-deficient soils) will be needed to meet the growing demand for crops. Knowledge of the biochemistry of plants will provide us with the opportunity to try to develop the means to at least address some of these issues. By understanding the limits, controls, and potential sites for the improvement of existing biochemical pathways it will be possible to significantly improve the productivity and versatility of plants. We will also be able to modify or to introduce novel pathways to produce new plant products.

Approaches to Understanding Metabolic Pathways

2

Key concepts

- A range of techniques including chromatography, electrophoresis, and the use of radioisotopes, are available to study metabolic pathways.

- Individual genes and enzymes can be studied through the analysis of mutant and transgenic plants.

- The availability of entire genome sequences and developments in bioinformatics means that multiple genes can be studied at the same time.

- Transcriptomics, metabolomics, and proteomics can give an understanding of entire biological systems.

- Plant metabolism is diverse, flexible, and robust because plants can't move away from environmental stress or predators.

- Metabolic Control Analysis is more realistic than studying only rate limiting steps to understand metabolic flux.

- There are common mechanisms for regulating enzyme activity in plants.

- Long-term control is achieved by altering the rate of transcription, mRNA processing, translation, and proteolysis and is energetically expensive.

- Short-term control is achieved by altering substrate or cofactor concentration, varying pH, allosteric effects, covalent modification, subunit association-dissociation, and reversible enzyme associations and is energetically inexpensive.

- Compartmentation keeps competitive reactions apart and is vital for plant metabolism.

- Characterizing organelle metabolism has given much information and understanding of plant metabolism.

What we need to understand a metabolic pathway

The focus of this chapter is to broadly consider the methods available to enable a plant biochemist to characterize the metabolites and enzymes involved in a particular metabolic pathway. The approaches considered provide a broad overview of ones that will have been used many times throughout the chapters of this book and, where appropriate, the reader is referred to examples discussed in specific chapters. In addition readers are directed to the Further Reading section at the end of this chapter for more in depth details of the application of particular approaches and the information this generates.

Understanding the composition of metabolic pathways and how they operate and interact in a plant has been the aim of plant biochemists for many years. During this time the level of understanding and the approaches used reflect the methods that were available to investigate these pathways. With time, the array of techniques available has become increasingly sophisticated. This has led to a dramatic shift in the amount of information generated and to potentially enhance our level of understanding. However, although these new techniques are providing the opportunity to gather more information, the interpretation of this information, as with all experimental approaches, remains the challenge.

With respect to plant nutrition, for example, a hundred years ago, scientists were only able to identify a particular process at an observational level. For instance, the effect of an individual nutrient supplied or withheld from a plant had an effect on the physical appearance of that plant. This was seen as indicative of a specific need by the plant for that nutrient. However, this then raised the question as to how that nutrient was used. Was it used directly in the form it was supplied, or was it altered by a series of biochemical conversions? If the nutrient was the start of a metabolic process, then it was acting as the substrate, but the specific nutrient might then be converted into a series of metabolic intermediates before giving the final end product required by the plant (Figure 2.1). Over time, a range of analytical techniques have been developed to isolate specific components of a metabolic pathway, allowing a greater understanding of that pathway. Although the experimental approaches are varied, there are some key ways in which an extensive amount of information has been generated.

To understand the need for a particular nutrient and its impact on the plant it was necessary to be able to identify the start and end products and any intermediates. Metabolites present in a plant were initially determined via classical chromatography and then using more sensitive techniques including high performance liquid chromatography (HPLC), enzyme-linked assays, and so on.

By knowing what happens to the substrate in terms of what it is converted to, one is then in a position to be able to follow that conversion. The ability to do

Figure 2.1 Understanding a metabolic pathway. Techniques in bold are discussed in more detail in the text.

• Identify the starting substrate, metabolic intermediates, and end product. Methods available include: **chromatography, HPLC**, enzyme-linked assays, **metabolomics**.

Substrate
↓ Enzyme a
Metabolic intermediate A
↓ Enzyme b
Metabolic intermediate B
↓ Enzyme c
Metabolic intermediate C
↓ Enzyme d
End product

• Identify the enzymes involved in catalyzing the individual conversions of substrate to metabolic intermediate to end product. Methods available include: **chromatography**, enzyme assays, **electrophoresis, radioactive feeding and labeling studies, proteomics, metabolomics**.

• Characterize the regulatory properties of individual enzymes. Methods available include: enzyme assays, **electrophoresis, radioactive feeding and labeling studies, proteomics**.

• Identify genes of interest and their regulation, localization, and expression. Methods available include: isolation of genes, **generation of transgenic and mutant plant lines, transcriptomics**.

this also provides a starting point to identify the component enzymes that are involved. A key step in being able to do this is the ability to measure the activity of the enzyme using an enzyme assay. This then provides a means to monitor for the presence of that particular enzyme.

A focus on identifying the enzyme involved and understanding how it works has generally been undertaken by protein purification from a particular plant tissue source, for example, leaves, roots, or fruits. Purifying a specific enzyme involved in a metabolic pathway has made it possible to work out its regulatory properties and whether there is sufficient activity present in the tissue to meet the needs of the plant.

With the arrival of molecular biology techniques it then became possible to clone the genes encoding particular proteins and identify their regulation and localization and expression using a variety of approaches, including the production and analysis of transgenic plants.

Chromatography

A method that has proved extremely useful and continues to have a marked impact on biochemical research is chromatography. Chromatography can be used to separate, identify, and quantify components of mixtures, whether they are solid, liquid, or gas. The underlying principles of chromatography involve the separation of components of a heterogeneous mixture by differential migration followed by detection and/or visualization. This technique has been of immense value as a means of separating and analyzing metabolites, enzymes, and proteins. By successive applications of different chromatographic approaches, a metabolite or protein of interest can be preferentially purified, ultimately leading to a homogeneous sample. Once purified the sample of interest can be further characterized in terms of molecular structure, determination of enzyme reaction products, and enzyme substrate specificity. There are three main types of chromatography: adsorption, partition, and gel permeation.

In adsorption chromatography, compounds adsorbed on to the surface of a solid adsorbent are desorbed by using an eluting solvent (Figure 2.2). Solid adsorbents include calcium phosphate gels, charcoal, cellulose, starch, and silica gels. Solvents commonly used include benzene, chloroform, diethyl ether, petroleum ether, and various alcohols. A range of aqueous buffers and salts are also used, sometimes in combination with organic solvents. The ability of the compound to bind to the adsorbent and subsequently be eluted reflects the chemical properties (for example, ionic charge) of the substance, and the nature of the solvent and the adsorbent. Separation of compounds is achieved based on the balance between the affinity of the compounds for the adsorbent and their solubility in the solvent.

Generally adsorption chromatography, which includes ion-exchange chromatography, is performed in columns. The sample is applied to the top of the column. Compounds move down the column according to their solubility and affinity for the adsorbent. For example, a sample that is more soluble and has less affinity for the adsorbent, moves down the column more quickly than a sample with higher affinity for the adsorbent (Figure 2.2).

The substances adsorbed on the column can be eluted in a number of ways. The simplest approach is to run a single solvent continuously through the column until the compounds of interest have been separated and eluted from the column. Alternatively two or more different solvents of fixed volumes can be

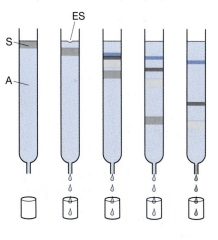

Figure 2.2 Adsorption chromatography. Solid adsorbent (A) is packed into a column and the sample (S) from which the metabolite or protein is to be purified is applied to the top of the column. An eluting solvent (ES) is then applied to the top of the column. The eluting solvent continues to be run through the column allowing the sample to run down the column. As the sample runs down the column, different components run at different speeds and are separated according to their ability to bind to the adsorbent and be eluted by the chemical properties of ES. Eluted fractions from the column are collected as run-off from the column and available for subsequent analysis.

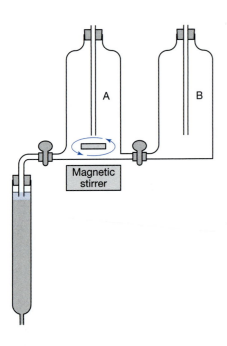

Figure 2.3 Gradient elution. Flow of solvent B into solvent A with mixing continuously changes the composition of solvent A as it flows into the column.

added sequentially to elute particular compounds using a stepwise elution. Finally, a gradient elution can be used where the composition of the solvent is continuously changing (Figure 2.3). Such an approach overcomes the problem that when using a single solvent the eluting components tend to form a broad trailing band. A simple linear gradient with a starting solvent (A) and a final solvent (B) is initiated once solvent A starts flowing on to the column. That is, the final solvent B is allowed to flow into the solvent A (Figure 2.3) resulting in the composition of solvent A continuously changing as solvent B is added. Where appropriate variations can be made by introducing a third vessel and varying the composition of the solvents in the vessels to make, for example, exponential, concave, or convex gradients. All of these column elution methods can be used with the other chromatographic methods described.

Partition chromatography is performed on paper sheets (paper chromatography) or a thin layer of silica gel on a glass plate (thin-layer chromatography). It is based on a thin film formed on the surface of the solid support by a liquid stationary phase. Solute equilibrates between the two liquid phases, a stationary phase bound to the inert support and a mobile phase that passes over the stationary phase. Depending on solubility, the mobile phase distributes the substances to be separated between the two liquids.

Gas–liquid chromatography is a type of partition chromatography with the stationary phase a high-temperature boiling liquid and the mobile phase an inert gas (Figure 2.4). Any samples to be separated are, due to different solubilities, partitioned between the stationary and mobile phases. Where a volatile substance is being separated it is introduced at one end of the column (filled with a liquid phase) and an inert carrier gas, for example, helium or nitrogen, is passed through. Normally the column is placed in an oven and the separation is carried out at either fixed or variable temperatures. The sample components leave the column at the opposite end and pass through either a flame ionization or a thermal conductivity detector.

Gel filtration or gel permeation chromatography separates molecules according to their molecular size (Figure 2.5). A sample of interest may contain solutes with a range of molecular weights from less than 100 to several million Daltons. Gel filtration chromatography purifies a wide range of biomolecules by using a gel that is usually an inert, cross-linked polymer (bead)

Figure 2.4 The main components of a gas chromatography system. (From Boyer, Rodney F., *Modern Experimental Biochemistry*, 2nd Edition, © 1993, P. 66. Reprinted with permission by Pearson Education Inc., Upper Saddle River, New Jersey, USA.)

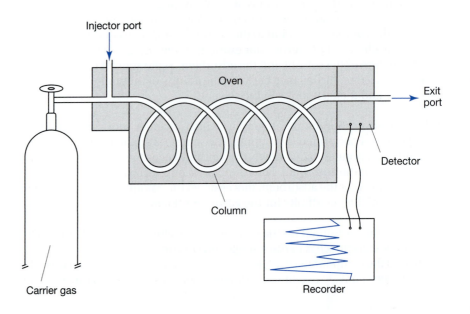

containing small pores of a specific size. The amount of cross-linking of the polymer determines the average pore size of the gel beads and affects the separation range of the column. Solute molecules that are larger than the pores are unable to enter the gel beads meaning they can only move through the space between the beads. This reduces the volume of the gel matrix that is available and these molecules progress very quickly through the column and elute rapidly (Figure 2.5). Smaller molecules are able to diffuse in and out of the gel beads and therefore move more slowly. All molecules above the exclusion limit, the molecular weight of the smallest molecule that cannot diffuse into the inner volume of the gel matrix, elute rapidly in a single zone. For example, the exclusion limit of a typical gel, Sephadex G-50, is 30 kDa. All solute molecules with a molecular size greater than this value move quickly straight through the column bed without entering the gel pores. Depending on the size of the molecules being fractionated different sized gel beads are available. For example, Sephadex G-50 has a fractionation range of 1.5–30 kDa. Molecules within this size range are separated in a linear fashion. Adaptations of this type of chromatography allow the separation of mixtures of compounds for analytical or preparative processes.

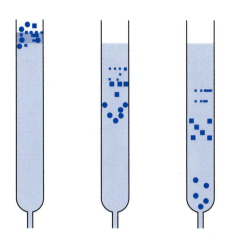

Figure 2.5 Gel filtration or gel permeation chromatography.

High performance liquid chromatography

When samples are applied to a gel matrix and separated during the chromatographic process, components are eluted as a series of zones, which may be composed of either multiple or individual components with similar properties (Figure 2.6). The chromatographic peak resolution is related to the column particle size; i.e. small column particles give higher resolution. It is necessary to use high pressure technology to get reasonable solvent elution through a column containing very small particles. An HPLC system operates at pressures of up to 5000 psi due to the development of support materials able to withstand such high pressure (Figure 2.7). To achieve this the column support medium tends to be made using rigid core particles with a shallow pore layer where separation takes place, or fully porous particles of very small diameter (≤5 μm). Initially glass beads were coated with a layer of ion-exchange resin or a thin layer of silica gel and used as the support material. With further technological developments smaller particles of 5 μm diameter and alternative chemical bonded phases on the particle surface have become available. The three main types of stationary phases are absorbants (with microparticulate silica and alumina being the most common), size exclusion gels (with silica and rigid styrene-divinyl benzene copolymers that are able to withstand high pressure and a range of both aqueous and organic solvents), and chemically bonded stationary phases (a large variety of substitutents are chemically bonded to the

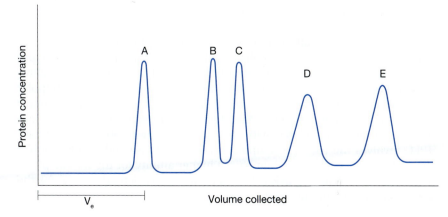

Figure 2.6 Elution curve for proteins from a chromatographic procedure. Different protein peaks are labeled A–E. The elution volume, V_e, for each protein peak can be estimated as shown in the figure for protein A. (From Boyer, Rodney F., *Modern Experimental Biochemistry*, 2nd Edition, © 1993, P. 88. Reprinted with permission by Pearson Education Inc., Upper Saddle River, New Jersey, USA.)

Figure 2.7 A schematic diagram of a high performance liquid chromatography system. (From Boyer, Rodney F., *Modern Experimental Biochemistry*, 2nd Edition, © 1993, P. 92. Reprinted with permission by Pearson Education Inc., Upper Saddle River, New Jersey, USA.)

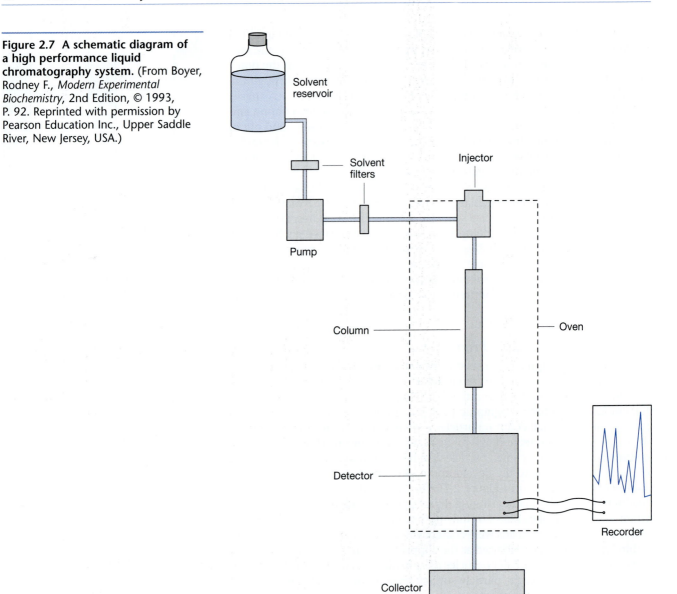

silanol groups, Si-OH) on silica including long-chain hydrocarbon groups and ion-exchange groups. HPLC has become an increasingly important technique because it facilitates the separation of a wide variety of complex mixtures in a very short time (5–30 minutes) and is suitable for thermally labile compounds, strongly polar compounds and biopolymers.

HPLC can be used for all the types of chromatographic separations already discussed. In addition, reversed phase and paired-ion partition are two types of chromatography developed specifically for HPLC. Normal phase chromatography uses polar solid supports and an organic, relatively nonpolar mobile phase solvent. However, as highly polar molecules are strongly attracted to the stationary phase then there are problems of retention times and peak trailing associated with this type of system. So reversed-phase HPLC involves the use of a nonpolar stationary phase and a polar mobile phase. The polar molecules then have a higher affinity for the mobile phase and tend to elute quite quickly. This approach is particularly useful for the separation of polar biomolecules.

With paired-ion chromatography charged polar substances pair with their counter-ion and therefore create a less polar species:

$$R_3NH^+ + ClO^-_4 \longrightarrow R_3NHClO_4$$

The ion pairs partition into the nonpolar phase, while the free ion R_3NH^+ is attracted to the polar phase.

An alternative fast chromatographic approach to HPLC is fast protein liquid chromatography. This approach uses experimental conditions intermediate between those of traditional column chromatography and HPLC. Fast protein liquid chromatography does not involve separation under such high pressure as HPLC. However, the chromatographic materials developed to use with fast protein liquid chromatography mean that separation can be achieved very rapidly and with similar efficiency and reproducibility to that seen with HPLC.

Chromatography involves the separation of chemical substances by partitioning them between two media. One medium, either solid or liquid, is stationary and the other, either liquid or gas, is moving. Because the solvent and solid can be varied so much and because chromatography can be used to separate all sorts of things, from small metabolites through to proteins, chromatography is a very powerful method, providing a means of separating substances that may be very similar. It is the availability of all of these chromatographic approaches that provides a firm platform for a greater understanding of the biochemistry of plants. The information generated is fundamental to understanding metabolic pathways and forms the basis of the biochemical information presented in all of the chapters of this book.

Electrophoresis

Electrophoresis is an important biochemical method that allows the separation of charged biological molecules through a support medium such as agarose, starch, or polyacrylamide gels in an electric field towards a positive or negative electrode. Polyacrylamide gel electrophoresis (PAGE) is the most commonly used technique to separate proteins. Depending on the conditions, molecules are separated according to size and charge. For instance proteins or protein complexes are often electrophoresed under dissociating conditions (that is in the presence of agents that disrupt the native structure). Sodium dodecyl sulfate (SDS) is a detergent with both polar and nonpolar properties. The nonpolar hydrophobic portion of SDS binds to the nonpolar region of most proteins, while the negatively charged sulfate portion is exposed to the solvent. The binding of SDS first interferes with the native hydrophobic and ionic interaction, resulting in the dissociation of most oligomeric proteins into their monomer subunit and disruption of their secondary structure. Where polypeptide chains are joined by covalent disulfide linkages, heating the protein in the presence of a disulfide reducing agent, such as β-mercaptoethanol, is also needed to ensure the protein dissociates into its smallest subunits. A second effect of SDS is that as proteins become saturated with negatively charged SDS molecules, electrophoretic separation is entirely dependent on molecular weight, with low molecular weight monomers moving fastest through the gel. SDS binds to protein in a ratio of approximately 1.4 g SDS per 1.0 g of protein (although the binding ratios can vary from 1.1 to 2.2 g SDS/g protein). SDS binding in this way eliminates any differences in charge density for proteins so that the main resolution factor during migration is sieving through the SDS gel. Electrophoresis under these dissociating conditions (SDS–PAGE) allows the estimation of the molecular weight of proteins to be made (Figure 2.8). Alternatively a protein

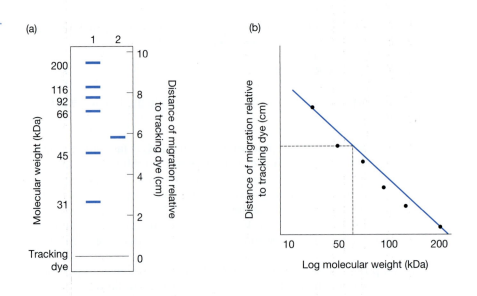

Figure 2.8 The use of sodium dodecyl sulfate–polyacrylamide gel electrophoresis to determine the molecular weight of a protein.
(a) Separation of molecular weight standards (1) and sample (2).
(b) Calibration curve for molecular weight estimation of sample (---) by measuring mobility of protein and plotting as a function of the log of the molecular weight.

may be run under native conditions leading to movement through the gel based on its charge and size. In this situation neither the gel nor sample contain detergent and the protein remains in its native and active state. For either denaturing or native PAGE separation of proteins with different properties can be improved by altering the percentage of the gel matrix.

Proteins can also be separated on the basis of their net charge and irrespective of their size by isoelectric focusing (IEF). Here electrophoresis is carried out in a pH gradient that allows each protein to migrate to the place in the pH gradient at which the protein has no net charge. The pH where the molecule has no net charge is the isoelectric point or pI (Figure 2.9). In standard electrophoresis the buffer used has a uniform pH meaning that there is no pH gradient. As such the charge density of each protein stays the same throughout electrophoresis with each protein eventually reaching either the cathode or anode. In contrast during IEF the charge density of each protein decreases as it moves towards its pI along the pH gradient. Once it reaches its pI the protein charge density is zero and it stops migrating. Although diffusion still acts against the tendency of a protein to focus to a single point on the gel, a protein diffusing away from its pI becomes charged and therefore moves back towards its focus. At the start of IEF there is a sieving effect that separates proteins based on size. However, by running the gels for a sufficiently long period then proteins with different pI values will focus at different points in the pH gradient (and independent of size). The pH gradient in an IEF gel can be generated by either using ampholytes or immobilized pH gradient (IPG) gels. In the case of synthetic carrier ampholytes these are collections of small amphoteric molecules

Figure 2.9 The principle of isoelectric focusing. A sample (protein mixture) is loaded at the basic end of a gel that has a pH gradient. An electric field is applied and the proteins separate according to their charge, focusing at positions where the isoelectric point (pI) value is equivalent to the surrounding pH. Larger proteins move more slowly through the gel but with time catch up with small proteins of equivalent charge. The circles represent proteins, with shading indicating protein pI value and diameters representing the molecular weight of the proteins. (From R. Twyman, Principles of Proteomics, Taylor and Francis, 2004.)

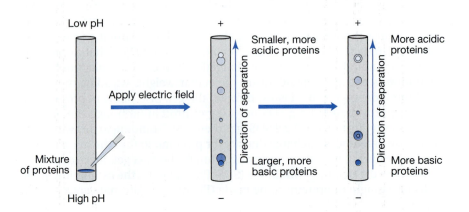

with pI values equivalent to a given pH range (Figure 2.10a). In the first instance the ampholytes are evenly distributed throughout the gel meaning that there is no pH gradient. Once the electric field is applied the ampholytes are subjected to electrophoresis. According to their pI the acidic ampholytes move towards the anode and the basic ampholytes move towards the cathode, with all other ampholytes establishing intermediate zones. Upon completion of this stacking process the system reaches equilibrium and is characterized by a continuous pH gradient. The proteins, which migrate more slowly than the ampholytes, then move towards their isoelectric points in the gel.

IPG gels have buffering groups attached to the polyacrylamide gel matrix (Figure 2.10b). As this is a highly reproducible technique then this has become the method of choice for proteomic studies (see Box 2.7). Immobilines, a group of nonamphoteric molecules, contain weak acid- or base-buffering groups at one end, and an acrylic double bond to facilitate the immobilization reaction at the other. The gel is run as for other electrophoretic procedures; however, the pH gradient is present before the electric field is applied and remains stable even when the gel has been run for a long period of time. Once the protein sample is loaded on to the IPG gel the proteins then migrate to their isoelectric point.

Although PAGE can be used to separate a protein according to its charge and mass by exploiting both principles in the same direction, this leads to low resolution separation. For better resolution of proteins alternative methods of gel electrophoresis are now routinely used where proteins are separated independently on the basis of charge or size in a combined approach of two-dimensional gel electrophoresis (Figure 2.11). This approach is widely used for proteomic studies (see Box 2.7).

Running a protein sample on a gel provides a wide range of information, depending on how the proteins in the gel are examined. Once separated by electrophoresis, a variety of methods are available to identify compounds. Gels may be stained generally for proteins with specific protein dyes such as Coomassie blue or the more sensitive silver stain. This will provide information relating to the number of proteins present in a sample in terms of size

Figure 2.10 Different ways of forming a pH gradient for isoelectric focusing. (a) With ampholytes, the buffering molecules are free to diffuse and initially are distributed evenly so there is no pH gradient. When an electric field is applied, the ampholytes establish a pH gradient and become charge neutral. This results in the separation of proteins based on their pI values. (b) Where there is an immobilized pH gradient, the buffering molecules are attached to the polyacrylamide gel matrix. No movement occurs when the electric field is applied but proteins are separated. Dotted arrows indicate direction of separation and shading indicates protein pI values. (From R. Twyman, Principles of Proteomics, Taylor and Francis, 2004.)

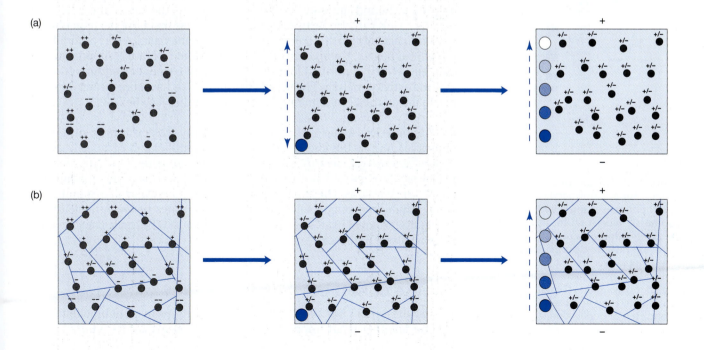

Figure 2.11 Two-dimensional gel electrophoresis using a tube gel for isoelectric focusing and a slab gel for sodium dodecyl sulfate (SDS)–polyacrylamide gel electrophoresis. The proteins are separated in the first dimension on the basis of charge and in the second dimension on the basis of molecular mass. The circles represent proteins, with shading to indicate protein pI values and diameters representing molecular weight. The dotted lines represent the direction of separation. (From R. Twyman, Principles of Proteomics, Taylor and Francis, 2004.)

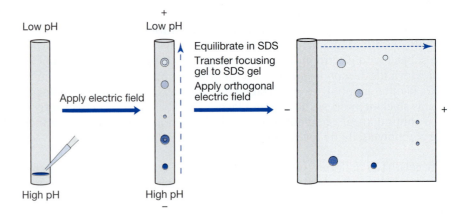

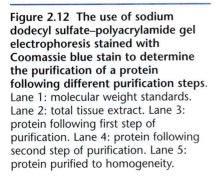

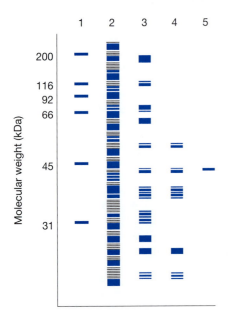

Figure 2.12 The use of sodium dodecyl sulfate–polyacrylamide gel electrophoresis stained with Coomassie blue stain to determine the purification of a protein following different purification steps. Lane 1: molecular weight standards. Lane 2: total tissue extract. Lane 3: protein following first step of purification. Lane 4: protein following second step of purification. Lane 5: protein purified to homogeneity.

(SDS–PAGE), charge and size (native PAGE), or charge (IEF). If the aim of the electrophoresis is to identify the success of a particular approach to purifying a protein of interest, then it may be appropriate to run out samples taken during the purification procedure to see if a particular protein is being preferentially purified (Figure 2.12). This approach may lead to the identification of the protein of interest in terms of its molecular weight. However, it would still be necessary to confirm that this is the protein involved in a particular metabolic process. There are a number of possible ways to try and identify a band on a gel as being a protein of interest. When running a native gel the protein remains in its active form. This means the gel can be incubated with substrates for the enzyme of interest and, where possible, visualized with a colored product. Using this approach it can be seen as to whether a particular band on a gel matches the enzyme activity. If specific antibodies are available that recognize the protein of interest, then these can be used to identify its location within a gel. In this case, protein detection is usually performed by a process described as immunoblotting or Western blotting. The proteins from the gel are transferred to and immobilized on a membrane that is then incubated with a solution of the antibody. The antibody binds to any proteins it recognizes and can then be visualized using a secondary antibody conjugated to either an enzyme or fluorescent tag. The addition of appropriate reagents allows the bound antibody to be visualized and the target protein to be detected. Gels can also be used to run out samples that have been incubated with radioactively labeled compounds and proteins visualized by autoradiography (see below). Compounds may also be eluted from the support media and recovered in a purified form for subsequent sequence determination or enzyme activity analysis. The use of electrophoresis has more recently proved invaluable in the development of proteomic studies (see Box 2.7).

Electrophoresis also forms the backbone of many molecular biological techniques leading to the characterization and identification of nucleic acids. That is, agarose gel electrophoresis is the standard method used to separate nucleic acids (Figure 2.13). This simple method allows the rapid separation of a wide range of sizes. The bands of nucleic acids are visualized by staining the gel in ethidium bromide solution, which intercalates between nucleic acid base pairs. When activated with ultraviolet radiation the nucleic acid can be visualized as it fluoresces.

The use of isotopes

The use of isotopes has a long history of applications in plant biochemistry. Isotope labeling is an important method for observing the operation of biochemical pathways. It may involve the use of radioisotopes such as ^{14}C, ^{3}H, ^{121}I, ^{32}P, and ^{35}S, or stable isotopes such as ^{2}H, ^{13}C, ^{15}N, and ^{18}O.

Radioisotopes are atoms with unstable nuclei that undergo transformation into other atoms with more stable nuclei. It is this transformation that can be quantitatively measured as energetic particles or radiant energy are released. Radiotracers can be detected by measuring the β-particles or γ-rays released. β-particles are typically detected by light emission in liquid scintillation detectors and γ-rays are detected by γ counters, which are similar to a Geiger counter. This can be used in combination with an approach such as HPLC to allow the detection of radioisotopes in specific compounds. The ability of radioisotopes to participate in chemical or biological reactions, just like their nonradioactive counterpart, is why they have been used so widely. Radioisotopes have become indispensable tools in the quantitative determination of compounds in complex mixtures. Often radioisotopes have provided a means of measuring metabolic substances present in tissues or cells at very low concentrations that are not accessible by alternative chemical methods of analysis. Radioisotopes have been used time and again in plant biochemistry to locate compounds in complex systems and for following the fate of a particular atom or compound in a dynamic process (Box 2.1). Without the use of radioisotopes our understanding of how biological processes occur in the living cell and isolated enzyme systems would have been more difficult to achieve. For example, the use of radioisotopes has enabled the elucidation of carbon reactions of photosynthesis (see Chapter 5), the tricarboxylic acid (TCA) cycle (see Chapter 5), protein biosynthesis, and the genetic code.

In a pulse–chase experiment the progression of a radiolabeled molecule is tracked through a cell. The pulse phase begins when a radioactive substrate or compound is added to cells in a culture medium or fed directly to whole plants or organs. The reaction mixture is then chased by the addition of unlabeled substrate. The pulse–chase allows for the movement of radiolabel to be traced as the reactions progress through the pathway. Product analysis determines the amount of labeled residues incorporated into other molecules. In addition, where in the molecule the label occurs gives important positional information and a means of tracing the mechanism of the reaction.

Autoradiography allows the determination of the location of radioisotopes in tissue, tissue sections, chromatograms, and gels. It is a nondestructive method where the location of radioactive compounds can be visualized by placing the radioactive sample in close contact with X-ray film. Emissions from the radioactive sample hit and reduce the silver bromide in the film emulsion. This reduces the silver ions to metallic silver. Following a period of exposure in a light-tight container, the film is developed and the location of the radioactivity in the sample determined by darkening of the film corresponding to metallic silver. A number of radioactive isotopes can be detected using autoradiography including ^{14}C, ^{3}H, ^{32}P, and ^{35}S. For instance, to work out metabolic pathways ^{14}C-labeled substrates can be fed to plant tissue during growth and the leaf exposed to film.

Stable isotopes can be used instead of radioisotopes, for example ^{15}N and ^{13}C can be traced by their different atomic mass from the more abundant ^{14}N and ^{12}C. ^{15}N labeling can be used to trace nitrogen in the field while analysis of ^{12}C:^{13}C ratios is particularly useful for analyzing carbon fluxes in photosynthesis (see Chapter 5). The detection of stable isotopes by either mass spectrometry or by nuclear magnetic resonance spectroscopy provides a more direct approach (when compared with radioisotope labeling) for identifying labeled metabolites.

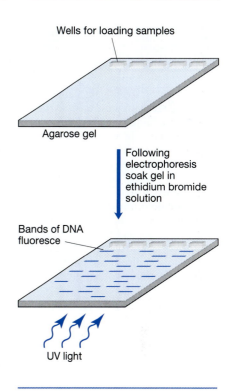

Figure 2.13 Nucleic acid gel electrophoresis.

Box 2.1 The use of radioactive tracers in the study of enzyme reaction mechanisms and metabolic pathways

The use of specifically labeled substrates is an effective way of studying how an enzyme catalyzes its reaction. For instance, aldolase cleaves fructose 1,6-bisphosphate into glyceraldehyde 3-phosphate and dihydroxyacetone phosphate (Figure 1). Using radioactivity specifically labeling the carbon atom in position 1 [1-^{14}C] fructose 1,6-bisphosphate, allows the reaction to be followed and which part of the fructose molecule gives rise to each of the products can be determined (Figure 1). The labeled C in position 1 would give ^{14}C labeled dihydroxyacetone phosphate, indicating that this three-carbon product comes from carbon atoms 1, 2, and 3 of the fructose. When fructose 1,6-bisphosphate specifically has carbon atom 6 radioactively labeled, the reaction would give ^{14}C-labeled glyceraldehyde

3-phosphate, indicating that this carbon product comes from carbon atoms 4, 5, and 6 of the fructose.

Another example is provided by the study of the series of the reactions of the TCA cycle in intact cells. The pathway and mechanism of the pathway were originally established using compounds such as acetate labeled in different positions and pyruvate. The use of these labeled precursors resulted in labeled cycle intermediates with the radioactive label located in specific positions. The labeled isotope location was determined by chemical or enzymic degradation of the intermediates and provided important information relating to the mechanism of the reactions occurring in the cycle.

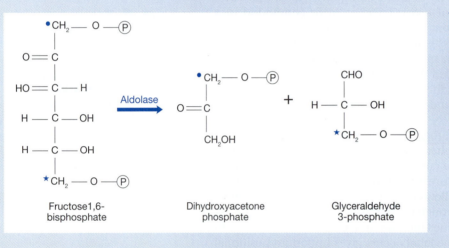

Figure 1 The use of radioisotopes in the study of enzyme reaction mechanisms and metabolic pathways.
● indicates specific labeling of the carbon atom in position 1 and
★ indicates specific labeling of the carbon atom in position 6 of the fructose 1,6-bisphosphate.

Current research techniques use a range of molecular biology approaches

Current research techniques using a range of molecular biology approaches provide a means of determining the function of genes and their products. Traditional approaches have characterized pathways via a systematic method of identifying intermediates, component enzymes, individual genes, etc. This information has facilitated the generation and analysis of mutant and transgenic plants (Box 2.2). By altering levels of specific components the plant can be further analyzed to try to understand functionality of particular components (Box 2.3). This reductionistic method tends to be laborious due to the single-enzyme approach that is needed to reassess the system from the bottom up. Furthermore, a major limitation of this piece-by-piece analysis of component parts is that the isolation and identification of enzymes is dependent on a prior knowledge of their properties—what reaction they catalyze, what cofactors they will require. Such a trial and error approach may lead to an incomplete knowledge of a pathway. In contrast, in the last decade there has been a notable change in the acquisition of information and its impact on our understanding of, among other things, plant metabolism. More recently scientists have tended to move away from studying a single gene or gene product and instead consider its analysis in the context of multiple genes and their protein

Box 2.2 Plant transformation techniques

Plant transformation provides an important approach for studying the role of genes in an organism. It is also routinely used as an approach to improve crop plant characteristics. Plant transformation is dependent upon the stable introduction of transgene(s) into the plant genome. Different methods have been used to achieve this and a wide number of plant species have now been successfully transformed. A number of transformation methods are dependent on using a pathogen of broad-leafed dicots *Agrobacterium tumefasciens*. This pathogen has proved to be an extremely efficient system for transforming dicots, but unfortunately it has proved more problematic with monocots. As such a number of more direct gene transfer methods using, for example, biolistic gun approaches have proved to be the method of choice for transforming monocots (see Further Reading for more details).

Transformation can take a number of forms, including reinserting cloned genes into mutants to complement the mutant and confirm function, or to make mutant lines through antisense and RNA interference techniques (Box 2.3). Overexpression can be achieved by integrating multiple copies of the gene into the plant genome and/or the use of a strong promoter plus translational enhancer to drive expression of the gene. To successfully manipulate a plant the introduction and stable integration of transgenes into the genome of a host plant is only the first step. The transgene must then be expressed, both spatially and temporally, in the correct manner. The transcript must be properly processed and the protein product modified correctly and targeted to the correct cellular location. As such much effort has to be made by researchers to ensure the transgene is designed appropriately before it is introduced into the plant.

Routine transformation of an increasing number of different crop species has resulted from continual developments and improvements of available methodology. Throughout this book there will be examples presented where the use of plant transformation has provided invaluable information towards our understanding of plant metabolism and/or been the basis of advances in plant biotechnology.

products. This shift was first in response to the technological breakthroughs that meant that it was possible to rapidly carry out large-scale DNA sequencing. Progress has been further aided by enormous advances in instrumentation allowing high throughput analysis of samples and, perhaps more importantly, parallel developments in bioinformatics. The important thing about these approaches is that they are system-based and allow for a whole system analysis. Such information has provided scientists with knowledge that allows a detached understanding of how biological systems work.

Box 2.3 Gene silencing or the suppression or complete absence (null, knockout) of gene expression

Gene silencing can be achieved at the DNA level by insertional mutagenesis where a T-DNA is inserted in a gene to prevent transcription. However, this approach cannot efficiently target a specific gene. At the RNA level gene silencing can be achieved by antisense where expression of DNA encoding antisense RNA is used to eliminate or reduce transcription of a target gene. Co-suppression involves the expression of additional copies of DNA encoding sense RNA resulting in reduction of the expression of the endogenous target gene. RNA interference (RNAi) is the mechanism of gene silencing mediated by double-stranded RNA (also called post-transcriptional gene silencing).

Large-scale mutagenesis

Mutation of a gene and observing its effect on a phenotype is a good way to establish the function of a gene (Box 2.4). Although mutations have been used since the start of the twentieth century, more recently it has been possible to generate mutant libraries. Insertion mutagenesis inserts a known piece of DNA into a gene to modify the expression of that gene (Figure 1). Random T-DNA insertions have been used in plants and T-DNA-mutagenized populations of arabidopsis are available through arabidopsis stock centers. Also endogenous transposons such as *Mu*, *Ac/Ds*, *Em/Spm*, and *Tam* have been used in maize and antirrhinum (snapdragon) and to a lesser degree in arabidopsis, flax, and tomato. Usually a transposon insertion disrupts the open reading frame of the gene resulting in null alleles. However, where there are insertions into regulatory elements then this can alter expression levels. An advantage of insertional DNA mutagenesis is that the interrupted gene is tagged with a DNA sequence that may be isolated by hybridization or PCR, allowing the mutated gene to be mapped and identified (Box 2.4). A disadvantage of this approach is that where

continued

Box 2.3 Gene silencing or the suppression or complete absence (null, knockout) of gene expression (continued)

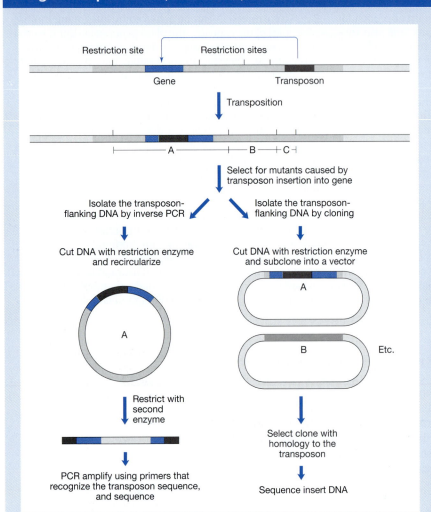

Figure 1 Insertion mutagenesis by transposon tagging. Transposon insertion inactivates the gene resulting in a mutant phenotype. Genomic DNA is then isolated, restriction enzyme digested, and the DNA flanking the transposon can be isolated by inverse polymerase chain reaction (PCR) or subcloning as shown. (From M. Dickinson and J. Beynon, editors, Molecular Plant Pathology, London: CRC Press, 2000.)

transposase-expressing element can be segregated out resulting in a stabilized mutation.

Insertion elements have also been developed that identify insertions into or near regulatory elements in plant genomes (gene traps). If a reporter gene (e.g. β-glucuronidase) is transformed into plants in the absence of a promoter, then it will only be expressed when it inserts adjacent to a functional promoter. The resulting expression pattern of the reporter gene would then reflect the expression pattern of the gene, which is normally controlled by the promoter. As most insertions are not into regulatory elements then this is a fairly inefficient approach generating only a small number of transgenic lines with any detectable expression of the gene in question.

RNA interference (RNAi) is a cellular defense mechanism evolved to protect cells from viruses

Plant viruses have an RNA genome and plants use a gene silencing mechanism to specifically target viral RNA and gain resistance. In the early 1990s studies showed that expression of viral sense or antisense RNA in transgenic plants could cause viral resistance. This phenomenon was called virus-induced gene silencing or post-transcriptional gene silencing. It was later shown that expression of viral-derived double-stranded RNA in plants caused more efficient gene silencing and resulted in a gain of

the gene mutation is lethal then no mutant line will be recovered. Also transposable-element generated mutations are generally unstable as the transposable element can transpose out of the gene as well as into it. A method for overcoming this and creating stable insertions is to use a two-element system where one element contains the transposable enzyme but lacks the target excision sequences, while the second element has the necessary excision sequences but no transposase. Only the second element is capable of transposition but needs the presence of the transposase-expressing element. To generate mutants the two elements are combined in a single plant by crossing the two lines that initiates transposition. Once a plant with the desired phenotype is identified then the

resistance to viral infection. Upon analysis of RNAi in plants in more detail small (21–25 nucleotide) RNA was observed in plants exhibiting gene silencing. It was suggested that this small RNA is a marker for RNAi mediated gene silencing. These small (or short) interfering RNA or simply siRNA are generated by cleavage of double-stranded RNA by an RNase type-III endonuclease called dicer (Figure 2). siRNAs form a complex that targets complementary mRNA for silencing. That is, these siRNAs are incorporated into the RNA induced silencing complex (RiSC). This is extremely active, reducing the mRNA of most genes to undetectable levels. Because RNAi is a systemic phenomenon, moving between cells, it can cause silencing throughout.

Box 2.3   Gene silencing or the suppression or complete absence (null, knockout) of gene expression (continued)

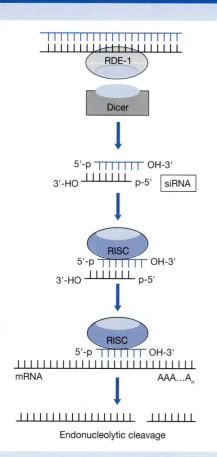

Figure 2   The mechanism of RNA interference. Double-stranded RNA is recognized by the protein RDE-1, which uses the nuclease dicer. Dicer cleaves dsRNA into short 21–23 bp fragments with two base overhangs called short interfering RNAs (siRNAs). The siRNA is incorporated into the RNA induced silencing complex (RISC). siRNA acts as a guide for RISC, upon perfect pairing the target mRNA is cleaved in the middle of the duplex formed with the siRNA. (From R. Twyman, Principles of Proteomics, Taylor and Francis, 2004.)

RNAi has the ability to silence specific genes more efficiently than antisense technology. For example, reduction of lysine catabolism specifically during seed development by an RNAi approach improves seed germination. RNAi technology has also been used in a number of plants to improve their nutritional quality. For example, the caffeine content in coffee plants has been reduced by RNAi-mediated suppression of the caffeine synthase gene (see Chapter 10).

Cells contain another class of small RNAs called microRNAs (miRNAs) which regulate gene silencing in a similar manner to siRNAs

miRNAs are 21–24 nucleotide non-coding single-stranded RNA derived from endogenous non-protein coding genes that unlike siRNAs are not the target gene. Arabidopsis has hundreds of endogenous miRNAs. Generally their function is unknown although they are thought to have gene regulatory roles. However, manipulation of miRNA has been suggested to have potential biotechnological applications. This is an active area of research and it is envisaged that a combination of approaches using knowledge from both RNAi and miRNA may be the future for the development of highly nutritional designer food crops (Figure 3).

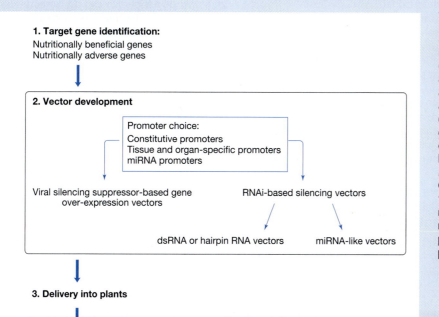

Figure 3   RNAi-based approaches for improving the nutritional qualities of plants. (1) Target genes are identified and cloned, (2) beneficial genes are coordinately overexpressed and any adverse genes silenced, (3) transgenic plants are generated and (4) screened to confirm the properties are beneficial. (From *Trends in Biotechnology*, Volume 22, Guillard Tang and Gad Galili, "Using RNAi to improve plant nutritional value: from mechanism to application", page 7, 2004. Reprinted with permission from Elsevier.)

With the completion of genome sequences of model organisms including arabidopsis (*Arabidopsis thaliana*) and rice (*Oryza sativa*), functional identification of unknown genes has become a major biological challenge. The availability of the entire genome sequences of arabidopsis and rice has led to a revolution in how plant scientists can interpret molecular information. There are now a wealth of approaches available to identify the specific function of a gene by using forward genetics (Box 2.4) or reverse genetics (Box 2.5). The genome is static information that remains essentially the same, regardless of cell type or environmental conditions. The central dogma of molecular biology is that a gene (DNA) is transcribed into RNA and then translated into protein. The availability of the genomic data means there is now a need to be able to substantiate this information with what is happening at the transcriptional and translational level. Postgenomic science as a result of technological advances mean that the central dogma of molecular biology can now be rewritten as the genome (all the genes in the organism) gives rise to the transcriptome (the total mRNAs present in the cell at any point in time; Box 2.6), which is translated to produce the proteome (the total proteins present in the cell at any point in time; Box 2.7). More recently plant scientists have also been able to move into the area of metabolomics (the analysis of the metabolites present in the cell at any point in time; Box 2.8).

Box 2.4 Forward genetics for obtaining mutants with desired phenotypes

Forward genetics is the classical way to identify the specific function of a gene. It starts with a phenotype and moves towards identifying the gene. Forward genetics screens allow mutants to be identified in all the genes involved in a specific process. Detailed phenotypic analysis of these mutants and the characterization of the corresponding genes allows the biochemical mechanisms underlying the process to be determined. Mutagenesis can be introduced physically (ultraviolet, X-rays, γ-irradiation), chemically (ethyl methane sulfonate) or biologically (transposons). Mutants with altered phenotypes are identified by screening these mutagenized populations. By carefully choosing the mutant phenotype and designing appropriate screens, mutations can be targeted either to a specific pathway or a specific enzyme step of a pathway. The most common methods used to clone genes in forward genetics are insertion mutagenesis (Box 2.3), map-based cloning or by PCR using primers designed from heterologous genes in other organisms.

Mutants generated by chemical mutagens or irradiation map-based cloning can be used to identify the mutant gene

Map-based cloning identifies the co-segregation of the mutant phenotype and known mapped markers (Figure 1). The resolution of mapping depends on the size of the segregating population, which is scored in the mapping cross, together with the number of available markers.

Chromosome maps have been developed using molecular markers by, for example, restriction fragment length polymorphisms. Also random amplified polymorphic DNA sequences is a PCR-based technique using random primers, which was developed to rapidly screen smaller samples of genomic DNA. Markers usually represent base differences in the priming site meaning there is a PCR product in one parent and not in the other. As random amplified polymorphic DNA sequences usually result in the presence/absence of polymorphisms, then heterozygous individuals cannot be distinguished from one of the homozygotes in diploid organisms without testing the next generation, making them less useful in haploid organisms. Therefore, random amplified polymorphic DNA sequences have been replaced by co-dominant markers such as cleaved amplified polymorphic sequences in diploid organisms, where the basis of allelic variation is the presence of a restriction site in one parent but not the other. PCR primers are used to generate a genomic fragment encompassing the polymorphic restriction site, and the PCR products are digested with the diagnostic restriction enzyme and the digests are run on a gel for scoring. Generally both of the parental genotypes and the heterozygote are all clearly distinguishable.

Box 2.4 Forward genetics for obtaining mutants with desired phenotypes (continued)

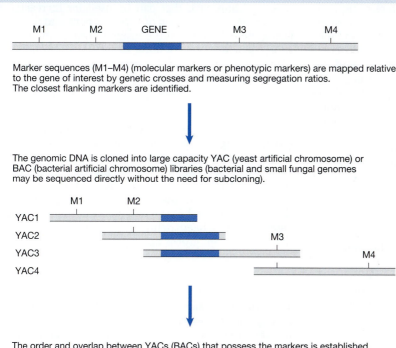

Marker sequences (M1–M4) (molecular markers or phenotypic markers) are mapped relative to the gene of interest by genetic crosses and measuring segregation ratios. The closest flanking markers are identified.

The genomic DNA is cloned into large capacity YAC (yeast artificial chromosome) or BAC (bacterial artificial chromosome) libraries (bacterial and small fungal genomes may be sequenced directly without the need for subcloning).

The order and overlap between YACs (BACs) that possess the markers is established and the DNA in the relevant YAC (BAC) clones can then be sequenced.

Figure 1 Map-based cloning. The gene to be cloned is flanked by markers (M2 and M3) and an approach for cloning the gene is based on the information presented. (From M. Dickinson and J. Beynon, editors, Molecular Plant Pathology, London: CRC Press, 2000.)

Simple sequence length polymorphisms and microsatellites have been developed as molecular markers in plants. The polymorphisms are based on differences in lengths of short repeat sequences, and the PCR products are separated on a gel to check for differences in length. Markers based on single nucleotide polymorphisms have been developed and offer great potential for the use in mapping, marker-assisted breeding, and molecular diagnostics. Real-time PCR techniques, used for the detection of single nucleotide polymorphisms and for quantitative PCR, are based on ways of monitoring for the accumulation of PCR over time through the labeling of primers, probes or amplification products with fluorogenic compounds. Another method used in map-based cloning is amplified fragment length polymorphism. This is a PCR-based technique where genomic DNA is cut with two restriction enzymes and adaptors specific to each type of cohesive end are ligated on to the fragments. Primers complementary to the two adaptors are used to amplify the restriction fragments. Two to three extra bases are added at the 3′ end of the primers so that only a subset of the restriction fragments serve as templates for the PCR reaction. This decreases the complexity of the mixture of PCR products to about 50–100 products per reaction, so that when run on a polyacrylamide gel, the majority of bands represent a single PCR product and differences in the presence or absence of bands can be detected. The use of various combinations of primers with different extensions means that large numbers of genomic fragments can be screened for polymorphisms. This is a very effective strategy to isolate the gene corresponding to a chemically or radiation derived mutant. Two resources are needed to obtain the gene of interest, molecular or genetic markers and a mapping population. In the case of molecular or genetic markers these identify chromosomal map positions. A mapping population segregates for the mutation of interest. It is generated by crossing the mutant plant in a specific ecotype to wild-type plants of a different ecotype. The resulting F_1 progeny are grown to maturity and seeds harvested. The recessive mutant characteristic reappears in the F_2 population, and the segregation of the molecular markers in this population allows the identification of markers that are closely linked to the mutation of interest.

Box 2.5 Reverse genetics

This approach starts with the gene of interest and ends with identifying the relevant mutant. It allows gene families to be characterized where functional redundancy among closely related gene members generally obscures their phenotypes. Most techniques replace the endogenous wild-type gene by inserting a modified copy and allowing homologous exchange. Reverse genetics allows mutations in all members of a gene family to be identified. Simple crosses can be undertaken to combine mutations in closely related genes. This approach allows the function of genes to be identified. Screening of a gene of interest can now be performed *in silico*. Alternatively, the gene of interest can be down-regulated by antisense or co-suppression or RNAi (Box 2.3).

Reverse genetics involves producing populations of plants, by, for example, chemical mutagenesis, with random mutations. DNA is prepared from the individual plants and mixed in pools of, for example, 10 individuals in each pool, and the gene of interest is then PCR amplified from each pool. If a mutation is present in one of the plants in the pool, the PCR products will be of two types, wild type and mutant, and when this DNA is heated and cooled, it will form a heteroduplex that can be detected by denaturing HPLC. The individual plant within the pool can then be identified and, following appropriate control experiments, the phenotype associated with the mutation in the gene of interest determined. This technique has been widely used in the arabidopsis community where mutant lines can be held as a community resource and automation of the PCR and denaturing HPLC techniques facilitates rapid screening of these populations.

A second approach that can be used for increasing or decreasing the expression of a specific gene and then analyzing the phenotype of the resultant organism is that of gene silencing such as sense or antisense suppression, which works through post-transcriptional gene silencing mechanisms (Box 2.3).

T-DNA tagging

The use of T-DNA as an insertional mutagen provides a molecular tag, which facilitates the isolation of the corresponding gene. Over 225,000 independent insertion events have been made, with the precise location determined for T-DNA insertions in about 90,000 lines at the Salk Institute. T-DNA insertion databases are available to the public and can be searched on the available websites. Once lines have been identified they can be ordered from the stock center, grown up and the plants genotyped to identify plants homozygous for the mutation.

Transposon tagging

This utilizes the fact that transposable elements mutagenize genes by insertion into coding and regulatory regions. There are a number of public collections, including gene trap lines and enhancer trap lines.

Target-induced local lesion in genomics (TILLING)

This is a general reverse genetics tool, combining random chemical mutagenesis with PCR-based screening to identify point mutations in a genomic region of interest.

Box 2.6 Transcriptome studies mRNA expression profiles on a global scale

A lot can be revealed about the role of a gene in a cell and also its functional links to other genes based on its expression profile. This may include information relating to cell or tissue localization, expression in response to particular environmental conditions, signals, etc. Often an understanding of gene function can be deduced by determining genes with similar expression profiles. This may allow functions to be predicted in a 'guilt by association' manner. Mutation of one gene may affect the expression profile of another and thereby lead to linking genes into functional pathways and networks.

The two major technologies emerging for large-scale expression analysis are large-scale DNA sequence sampling, based on standard DNA sequencing methods and the use of DNA assays for expression analysis by hybridization.

Sequence sampling is the most direct way to study transcriptomics

Randomly picked clones from cDNA libraries are partially sequenced to give 200–300 bp information. This is sufficient information to allow identification through comparison with the sequence databases. By determining the number of times each clone appears in the sample, the abundance of the corresponding transcript in the transcriptome of the original biological sample can be estimated. Comparisons of samples from different sources/treatments can be made and used to identify differentially expressed genes. However, this approach is laborious and expensive. A number of alternative high throughput approaches have been developed including EST databases, differential display PCR and serial analysis of gene expression (Table 1).

Box 2.6 Transcriptome studies mRNA expression profiles on a global scale (continued)

DNA microarrays are now the method of choice in transcriptomics

DNA microassays are miniature devices on to which many different DNA sequences are immobilized in the form of a grid. They are generally prepared by either mechanically spotting DNA molecules on to a coated glass slide or by *in situ* oligonucleotide synthesis. Expression analysis is based on multiple hybridization using a complex population of labeled DNA or RNA molecules (Plate 2.1). In either case, a representative complex cDNA population is formed by reverse transcription en masse of a given mRNA population from a particular source.

For spotted microarrays a fluorophore-conjugated nucleotide is included in the reaction mix to ensure the cDNA population is universally labeled. While with oligonucleotide chips the unlabeled cDNA is converted to a labeled complementary RNA (cRNA) population by the incorporation of biotin. This is later detected by fluorophore-conjugated avidin, which binds to biotin. The complex population of labeled nucleic acid is then applied to the assay and left to hybridize. As each array contains 10^6–10^9 copies of the same DNA sequence, it is unlikely to be saturated completely during this hybridization reaction. The intensity of the hybridizing signal is proportional to the relative abundance of the particular cDNA or cRNA in the mixture and reflects the abundance of the corresponding mRNA in the original source population and the particular conditions used. As such this approach allows thousands of transcripts to be monitored in one experiment. To compare samples cDNA or cRNA isolated from each of the samples can be hybridized to identical microarrays. Alternatively, the samples can be labeled with different fluorophores and both labeled populations hybridized to the same array. By scanning the array at different wavelengths it is possible to compare the relative levels of mRNAs between samples (Plate 2.1).

There is a complete range of bioinformatics tools and databases that can be used to compare microarray data obtained by different labs (for further details see Rhee et al 2006 in Further reading). These tools allow researchers to perform *in silico* analysis of gene expression.

Table 1 Sequence sampling methods for the global analysis of gene expression

Method	Key steps
Random sampling of cDNA libraries	Clones are randomly picked and searched against databases to identify the corresponding genes. The frequency each sequence is represented gives a guide as to the relative abundance of different mRNAs in the original sample. • Very labor intensive
Analysis of EST databases	ESTs are generated by single-pass sequencing of random cDNA clones to generate a clone signature. The representation of each sequence in the database gives an estimation of abundance. • Rapid • Carried out *in silico* • Dependent on representative EST samples
Differential display PCR	Allows the rapid identification of cDNA sequences differentially expressed between two or more samples. Populations of labeled cDNA fragments are generated by RT–PCR using an oligo-dT primer and an arbitrary primer, to give pools of cDNA fragments representing subfractions of the transcriptome. By running equivalent amplification products differentially expressed cDNAs can be determined. • Identifies differentially expressed genes • False positives are common
Serial analysis of gene expression (SAGE)	Very short sequence tags (signatures) are collected from cDNAs and linked to form long concatemers. These are sequenced and the representation of each transcript determined by the number of times a specific tag is counted. • Technically demanding • Very efficient

ESTs, expressed sequence tag; RT–PCR, reverse transcription–polymerase chain reaction.

Box 2.7 Proteomics uses a range of technological approaches for the large-scale characterization of proteins

The proteome is dynamic and will reflect the immediate environment in which it is studied. It has been described as a snapshot of the protein environment at any given time. Proteins respond to internal and external cues by undergoing post-translational modification, synthesis, or degradation. Proteomics allows an integrated view of biology by studying all proteins in a cell rather than one individually. Many different studies and/or experimental approaches contribute to proteomic studies (Table 1 and Figure 1). In addition to providing information about the function of individual proteins, proteomics also provides information about how proteins function in pathways, networks, and complexes.

Functional genomic technologies in the twenty-first century are based on high throughput clone generation and sequencing. This information is of little value if it is not analyzed and interpreted with a view to the functional

biological system. A need to determine protein function in the light of this information is crucial. It is for this reason that the area of proteomics has developed so rapidly in recent years. Two-dimensional gel electrophoresis forms the backbone of many proteomic approaches (see Figure 2.11 in text for further details). Although first used in 1975 and shown to be an efficient means of separating proteins, the potential of two-dimensional electrophoresis in terms of proteomics could not be achieved until it was possible to identify the separated proteins. The identification of proteins, first, by determination of the amino acid sequence using Edman degradation, and subsequently by microsequencing were important developments. However, the widespread application of proteomics today is a direct result of advances in large-scale nucleotide sequencing of genomic DNA. That is, the availability of information of the DNA sequence in databases allows any protein's sequence identified by

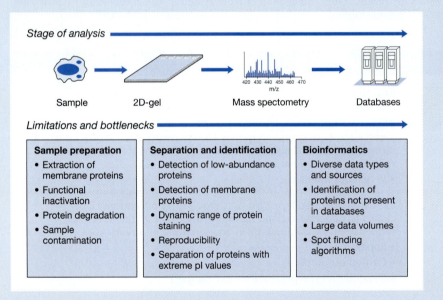

Stage of analysis

Sample → 2D-gel → Mass spectometry → Databases

Limitations and bottlenecks

Sample preparation	Separation and identification	Bioinformatics
• Extraction of membrane proteins • Functional inactivation • Protein degradation • Sample contamination	• Detection of low-abundance proteins • Detection of membrane proteins • Dynamic range of protein staining • Reproducibility • Separation of proteins with extreme pI values	• Diverse data types and sources • Identification of proteins not present in databases • Large data volumes • Spot finding algorithms

Figure 1 Proteomic approaches currently available.
(From R. Twyman, Principles of Proteomics, Taylor and Francis, 2004.)

Table 1 Proteomic approaches available

Type of proteomics	Information generated
Protein expression	By comparing protein expression of the entire proteome or a subsystem, novel proteins involved in particular responses can be identified.
Structural (cell map)	Identify all the proteins within a protein complex or organelle, determine where they are localized and characterize all protein–protein interactions. This will explain the overall architecture of cells and how expression of certain proteins gives unique characteristics.
Functional	Broad term for a range of directed proteomic approaches that allow a selected group of proteins to be studied and characterized.

Box 2.7 Proteomics uses a range of technological approaches for the large-scale characterization of proteins (continued)

mass spectroscopy or Edman degradation to be assigned an identity with a reasonable degree of confidence.

Mass spectrometry in proteomics

A mass spectrometer is an instrument that measures the mass:charge ratio of ions in a vacuum. Mass spectrometers consist of a source of ions, a mass analyzer, and an ion detector. The ion source converts the analyte into gas phase ions in a vacuum. These ions are then accelerated in an electric field towards the analyzer where they are separated according to their mass:charge ratios before the data for the individual ions are recorded in the detector. Such data mean that the molecular mass can be accurately determined so that the molecular composition of a particular sample or analyte can be determined. With proteomic studies the analyte usually consists of a number of peptides derived from a particular protein sample by digestion with a proteolytic enzyme, such as trypsin. From such an approach the masses of the intact peptides can be calculated and this can be used in a correlative database searching approach to identify the proteins in the sample. This sequence information can also be used to obtain *de novo* sequences.

Box 2.8 Metabolomics means it is possible to monitor and compare changes in metabolite levels under a range of different environmental conditions

The metabolome cannot be predicted from the genome, 25–30% of the arabidopsis genome has been estimated to encode metabolic enzymes. When characterizing the plant metabolome, its complex nature and diversity of chemical compounds creates a number of problems. It is thought that plants produce of the order of 200,000 metabolites. Many of these will play specific metabolic roles. The different chemical properties of metabolites have an impact on a number of factors, including extractability in different solvents, pH requirements, and sensitivity to extraction conditions (temperature, pressure, time). The metabolic profile obtained reflects the success of the extraction system in providing a picture resembling the *in vivo* state. Where metabolites remain in the plant matrix they cannot be profiled. As such, metabolic profiling involves a balance between accuracy and coverage of metabolite measurements. The range of chemical properties of metabolites present in a plant means that it is not straightforward to apply a particular approach to identify them. A way to address this is to use conditions and approaches to identify particular groups of metabolites. More recently advances in instrumentation developments for metabolite analyses are allowing increased metabolite identification within a single analysis. Alternatively it is possible to apply a range of analytical approaches, ideally operating in parallel, to profile multiple metabolites and to allow quantitative analysis of particular ones. The recent developments in metabolomics and its integration into systems biology are primarily due to combined expertise from biology, chemistry instrumentation, computer science, and mathematics (Figure 1). Separation of component metabolites is perhaps the most important part of the metabolomics approach and this has benefited from the development of equipment with

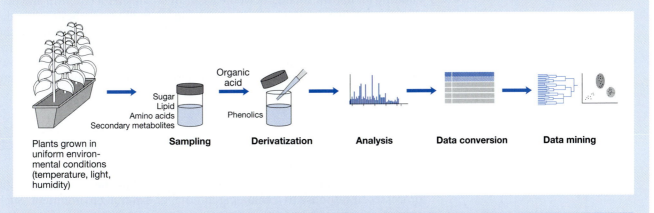

Figure 1 Metabolomics approach to analyzing plants grown under specific growth conditions. (From *Journal of Bioscience and Bioengineering*, Volume 100, Eiichiro Fukusaki and Akio Kobayashi, "Plant Metabolomics: potential for practical operation.", pages 347–354, 2005. Reprinted with permission from Elsevier.)

continued

Box 2.8 Metabolomics means it is possible to monitor and compare changes in metabolite levels under a range of different environmental conditions (continued)

increased sensitivity allowing not only the detection, but also the quantification of many compounds in a mixed sample. A wide range of separation methods is used including gas and liquid chromatography or electrophoresis coupled to mass spectroscopy. A complete range of mass spectroscopy approaches is now amenable to determining a wide range of metabolites from different samples and, as would be expected, different approaches are used for the identification of particular classes of compounds (for further information on specific approaches see the Further Reading section). Alternatively, nuclear magnetic resonance spectroscopy (NMR) can be used in combination with isotope labeling and metabolic flux analysis to determine metabolic fluxes in different pathways.

The ability to develop methodologies to use equipment in sequence has increased the number of metabolites detectable in crude extracts. The overall goal of metabolomics is to identify and quantify in an unbiased way all the metabolites present in a sample from an organism grown under particular conditions. This allows a complete or global picture of the metabolome to be determined as a function of particular environmental conditions, and also to study the effects of genetic modification on plant metabolism. The virtue of unbiased metabolite analysis is that a global picture of the way metabolites change in response to specific conditions can be determined. Such information can provide an important insight into previously ignored interactions between different metabolic pathways. An important area in metabolomic analysis (and indeed all omic technology) is that of analyzing and processing the data. Developments in these technologies have meant that it is possible to monitor and compare changes. These types of approaches should contribute to a greater understanding of plant metabolism. Not surprisingly it will be necessary to determine not only the full metabolic profile for a whole organ, tissue or cell, but to do so for each of the major metabolic compartments at their subcellular level. Such detailed profiles could then presumably be interpreted in relation to, for example, independent determinations of subcellular enzyme localization.

A combination of transcriptomics, proteomics, and metabolomics has facilitated a new understanding of biological systems and given a more complete picture in terms of interactions between genes, proteins, and metabolites. It is this interaction within and between samples that can be analyzed using available computational technologies (see Rhee *et al.* 2006 under Further reading). In the coming decades further advancements will mean that understanding complete pathways and their interactions at a number of levels will be a realistic proposition. No longer is a biochemist restricted to studying their favorite class of compounds. This analytical approach can further be combined with the availability of modeling-based predictions in a systems biology approach. Perhaps an interesting point to note here is that systems biology provides a better insight into the interconnectivity of pathways— allowing for holistic models of whole tissue or organism metabolism and predictive modeling; for example, of what happens to photosynthesis when the TCA cycle slows down. Likewise, when considering, as discussed earlier, the example of plant nutrition, it is clear from combined transcriptomic, proteomic, and metabolomic analysis that the global impact of specific nutrients on plant metabolism reaches out across biochemical pathways much more widely than was ever imagined.

Unique aspects of plant metabolism and their impact on metabolic flux

A key feature of all organisms is their ability to regulate the rate of metabolic processes in response to a changing internal or external environment. Such regulation ensures homeostasis, by conserving the stability of the intracellular environment, this in turn maintains an organism in an efficient state. Plant metabolism is, by necessity, a complex process. It has to be flexible as well as robust, given the sessile nature of plants and their exposure to fluctuating and frequently unpredictable environmental conditions. Anatomical, biochemical, and physiological adaptations are all needed to enable plants to cope with environmental stresses including heat, cold, drought, salinity, nutrient limitations, and anoxia. Furthermore, a green plant exhibits a complex metabolism highlighting its autotrophic nature, biosynthetic capacity, metabolic redundancy and duplicity, and unique cellular and subcellular compartmentation (Chapter 3). The major types of metabolic regulatory mechanisms found in plants are generally similar to those of other organisms. What sets plants apart, however, is often the diversity of metabolism—as exemplified by the array of defense chemicals that they produce (Chapters 10–12) and the flexibility of metabolism that enables them to optimize their use of resources. Furthermore, the precise mechanisms of regulation of a particular pathway or enzyme may not be the same between species or even between locations within the same plant or under different environmental conditions. Hence it is important to understand metabolic regulation in plants if we are to understand how plants perform. It is not enough to assume that we can extrapolate information gained from non-plant studies. Nevertheless, despite this variation there are a number of common features relating to metabolism that can be applied across all Kingdoms.

In any metabolic pathway, any flux or movement of metabolites must be closely coordinated with the needs of the cell, tissue, or organism for the end product(s) of the pathway. The focus of research in the past has tended to be towards characterizing any pacemaker enzymes perceived to be catalyzing a rate-limiting step of a particular metabolic pathway. Indeed these pacemaker enzymes were often described as being the most important enzymes *in vivo* for controlling metabolic flow. Generally a rate-determining step was defined as occurring at the first committed step of a pathway and representing an irreversible reaction with a high negative free energy change *in vivo*, together with a low overall activity. A rate-limiting step would also be identified as likely to occur after a major branchpoint and at the last step of a multi-input pathway. It was these so-called rate-limiting steps of metabolic processes that researchers had a tendency to focus on with respect to identifying, extensively purifying, and characterizing the properties of an enzyme. Such studies in the past led to researchers generating plants with altered levels of these enzymes with a view that this would have a marked impact on plant metabolism. However, it is clear that what is seen *in vitro* does not necessarily reflect the *in vivo* situation. For example, many attempts to increase different aspects of carbon metabolism in plants using genetic engineering have failed. This reflects the need to move away from the concept of rate-limiting steps in a pathway. Instead, metabolic control analysis is a more realistic tool to identify the enzymes that participate in controlling the metabolic flux through a pathway.

Metabolic control analysis theory

This theory was developed by Kacser in 1973. The development of the control analysis theory was an important step in recognizing the complexity of a metabolic system and interpreting the contribution of multiple enzyme steps. The essence of Kacser and Burns' control analysis theory is that metabolic control is shared among all steps in a pathway. More detailed reviews examining metabolic control analysis are listed under Further Reading.

Broadly speaking the flux control coefficient (C^J_E) value specifies the change in metabolic flux (δJ) in response to small changes in the activity of any enzyme (δE) in the metabolic system as follows:

$$C^J_E = \frac{(\delta J)}{(\delta E)}$$

A number of factors can influence the activity of the enzyme, for example, gene expression and protein degradation, regulatory changes, and environmental factors. Where the flux control coefficient is very small, any change in the activity of that particular enzyme will have little effect on the flux, in other words that enzyme has little control on the overall pathway flux. In contrast, where the flux control coefficient of a particular enzyme is nearer to 1.0, any change in the enzyme activity will result in an almost proportional change in flux through the pathway. This enzyme will be an important enzyme for regulating the overall rate of metabolite movement or flux through that pathway. That is, flux through a pathway will be particularly sensitive to any small change in the activity of that particular enzyme. The flux control coefficient of an enzyme only relates to the set of conditions under which it has been measured. Changing the conditions will usually affect the flux control coefficient and the influence that a particular enzyme has on the overall flux of metabolites through that pathway. Normally, control is shared between several enzymes, with similar control coefficients in a molecular democracy as first proposed by Kacser and Burns (Box 2.9). This is an important point, because it tells us that the control of a pathway rarely lies with a single, rate-limiting enzyme. More frequently, control is shared and in order to alter flux by selective breeding or genetic modification, it will be necessary to increase the concentration of all enzymes that have significant flux control coefficients.

Box 2.9 Metabolic control analysis

This is a tool to identify the enzymes (E1, E2 etc.) that contribute to the control of metabolite flux through a pathway:

$$\begin{array}{ccc} E1 & E2 & E3 \\ S0 \leftrightarrow S1 \longrightarrow S2 \leftrightarrow S3 \end{array}$$

The elasticity coefficient is a property of an individual enzyme in a pathway. It describes how the enzyme rate responds to a change in the concentration of any metabolite (e.g. substrate, S1, S2 etc.) that affects it.

$$\begin{array}{ccc} E1 & E2 & E3 \\ S0 \longrightarrow \left[S1 \longrightarrow S2 \right] \longrightarrow S3 \end{array}$$

Change in substrate S1 $\Delta S1/S1$
Results in change in rate of enzyme E2 $\Delta V_{E2}/V_{E2}$

Elasticity coefficient is $\sum_{S1}^{V_{E2}} = \dfrac{\Delta V_{E2}/V_{E2}}{\Delta S1/S1} = \dfrac{1\%}{2\%} = 0.5$

The connectivity theory describes how a change in the activity of an enzyme leads to a change in the activity of others.

$$\begin{array}{ccc} E1 & E2 & E3 \\ S0 \longrightarrow S1 \longrightarrow S2 \longrightarrow S3 \end{array}$$

Change in the activity of E2: $\Delta E2/E2$
Corresponds to a change in flux (J), e.g. S3 mole/h: $\Delta J/J$

A control coefficient is the system property of an enzyme that describes how for instance the flux or metabolite concentration depend on the activity of the enzyme. This involves all the elasticity coefficients.

(flux) Control coefficient for E2: $C_{E2}^{J} = \dfrac{\Delta J/J}{\Delta E2/E2} = \dfrac{2\%}{5\%} = 0.4$

Finally, the summation theory is that the sum of all the control coefficients for the pathway is equal to 1.

$$\begin{array}{ccc} 0.3 & 0.5 & 0.2 \\ S0 \longrightarrow S1 \longrightarrow S2 \longrightarrow S3 \end{array}$$

Can change under different conditions

$$\begin{array}{ccc} 0.4 & 0.3 & 0.3 \\ S0 \longrightarrow S1 \longrightarrow S2 \longrightarrow S3 \end{array}$$

Some values may be very low (0.01) and where an increase of an enzyme decreases the flux, the coefficient is negative

In summary, the higher the value of the flux control coefficient of an enzyme, the greater control that enzyme has over the pathway.

The availability of transgenic plants with altered levels of particular enzymes has been an important means of examining further the contribution of particular enzymes to metabolic control. An example of the application of metabolic control analysis and the effect of changing conditions on the control of a pathway can be seen in the analysis of a series of transgenic tobacco plants in which a progressive decrease in the amount of Rubisco levels was present as a result of antisense technology. Figure 2.14(a,b) compares the impact of decreased levels of Rubisco in transgenic tobacco plants with wild-type plants that contain normal levels of Rubisco. Calculations of the control coefficient of Rubisco with respect to photosynthetic flux show how this changes with differing environmental conditions (Figure 2.14c). Rubisco only exhibited a relatively high flux control coefficient at high light. In the wild-type plants the control coefficient was 0.7 at high light and 0.1 at low light. While in transgenic plants with markedly reduced levels of Rubisco the value was close to 1. That is, Rubisco levels had to be reduced by a lot before there was any effect on photosynthesis, confirming its low coefficient under these conditions.

Coarse and fine metabolic control

Coarse and fine metabolic control contribute to the regulation of an individual enzyme. Throughout the chapters of this book examples will be presented where metabolic pathways involve the interaction of a number of different components involving enzymes, metabolites, energy, and regulatory effectors. Generally, the information obtained is in response to intensive studies, often focusing on individual components of a particular pathway. The information for the individual components can then be collated and interpreted to understand the whole biochemical process more completely. In these studies it is clear that there are a number of common mechanisms relating to the regulation and control of enzyme activity. This can have a marked impact on the overall operation of a metabolic pathway and its interaction with other processes. These regulatory processes will be briefly considered in this section.

Long-term coarse metabolic control

Coarse metabolic control occurs over a long period of hours to days. This is a long-term control that is energetically expensive and is achieved through changes in the total cellular population of enzyme molecules, for example,

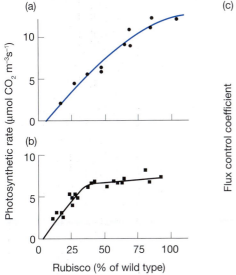

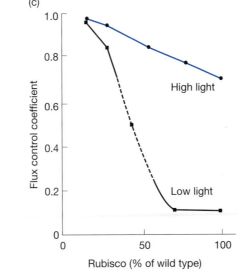

Figure 2.14 Impact of decreased Rubisco on photosynthesis in leaves of wild-type (100% Rubisco) and antisense tobacco grown under (a) high light and (b) low light, (c) data calculated as flux control coefficient. (From *Planta*, Volume 183, number 4. Reprinted with kind permission from Springer Science and Business Media.)

due to protein synthesis. It often occurs in response to tissue differentiation or long-term environmental changes. The relative rates of synthesis and degradation determine the amount of enzyme present at any given time. As such, coarse control may be caused by alterations in the rates of transcription or translation, mRNA processing or degradation, or proteolysis, as occurs, for example, in the case of nitrate reductase (see Chapter 8).

Fine metabolic control

Fine metabolic control is a rapid response taking place in seconds to minutes. This is generally energetically inexpensive and involves regulatory mechanisms that modulate the activity of the preexisting enzyme molecule. There are a wide range of mechanisms of fine control: alteration in substrate or cofactor concentration, pH, allosteric effectors, covalent modification, subunit association–disassociation, and reversible associations of metabolically sequential enzymes.

Alteration of substrate or cofactor concentration

Michaelis and Menten discovered that the rate or velocity (V) of an enzyme reaction varies with the substrate concentration. The rate of reaction increases up to a maximum velocity (V_{max}) and then remains constant (Figure 2.15a). Enzymes do not generally operate at their V_{max}, with substrate concentrations being subsaturating *in vivo*. The substrate concentration, when sub-saturating, affects the rate of an enzyme-catalyzed reaction (Figure 2.15). Hence changes in concentration of a substrate or cofactor can result in a direct increase in activity at that enzyme-catalyzed step in a pathway. While the Michaelis–Menten type plot (Figure 2.15a) is often observed for single subunit active site enzymes or multi-subunit enzymes with active sites operating independently, more complex relationships, for example, that of allosteric enzymes (see below), can also be observed as in a sigmoidal plot (Figure 2.15b). In this latter case at low concentrations of substrate there is a fairly low velocity of reaction, but as the substrate concentration increases there is a point when the velocity rapidly increases.

Variation in pH

Enzyme activity is affected by pH, with different enzymes having a different pH at which their activity is optimal. Above and below the pH optimum the enzyme activity normally declines. However, the response of the enzyme to pH varies from a broad to a narrow pH profile (Figure 2.16). Enzymes found in the same location do not necessarily exhibit the same pH optima. Furthermore, the pH optimum of a particular enzyme may be different to the pH of the intracellular location. Such differences are an additional means of regulation. For example, in the chloroplast the photosynthetic electron transport is linked to H^+ uptake into the thylakoid lumen, which leads to a proton gradient being established between the lumen and stroma. The proton gradient results in an increase of H^+ in the lumen and a light-dependent increase in stromal pH from 7.0 to 8.0. In the dark, H^+ leak back into the stroma and the pH falls back towards 7.0. A number of Calvin cycle enzymes exhibit a sharp pH optimum of between 7.8 and 8.2. This means these enzymes are most active when the stromal pH increases (see Chapter 5). Similar regulation by pH can be seen, for example, in germinating castor oil seeds under anaerobiosis. Here an anoxia-induced reduction in cytosolic pH leads to the activation of the cytosolic pyruvate kinase involved in glycolysis (see Chapter 6).

Allosteric effects

Allosteric effectors bind to sites other than the catalytic site leading to a precise change in the structural conformation of the enzyme. The conformational

(a)

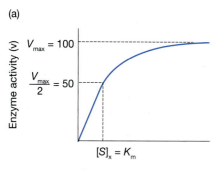

(b)

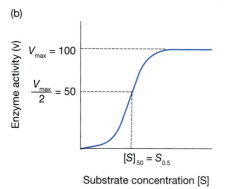

Figure 2.15 Relationship between the substrate concentration and reaction rate for (a) an enzyme showing hyperbolic substrate saturation kinetics and (b) showing sigmoidal substrate saturation kinetics. The V_{max} represents the maximum enzyme activity and the Michaelis–Menten constant of K_m is defined as the substrate concentration that gives rise to an enzyme activity that is equal to half the maximum activity.

change may activate or inhibit the enzyme substrate interactions, markedly altering the velocity of a particular enzyme reaction and therefore the flux of metabolites through a particular pathway (Figure 2.17). Allosteric effects are a means by which metabolites that are distinct from a particular pathway reaction may act as either a feedforward or feedback signal to that pathway. This is an important mechanism for coordinating distinct, but related, metabolic processes. For example, inorganic phosphate (P_i) is an allosteric activator of ATP phosphofructokinase and inhibitor of pyrophosphate dependent fructose 6-phosphate 1-phosphotransferase, PFP (as discussed in Chapter 6).

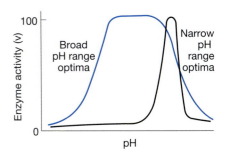

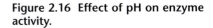

Figure 2.16 Effect of pH on enzyme activity.

Reversible covalent modification

Reversible covalent modification leads to an interconversion of an enzyme between an active and less active form. Covalent modification results from the formation of a new stable covalent bond as a result of thermodynamically favorable enzyme-catalyzed reactions. The reaction in either direction can be very fast, occurring in minutes, and may occur completely in a particular direction. The resulting enzyme conformation leads to an alteration in enzyme substrate interactions that affect (by elevating or reducing) V_{max} and K_m. Enzyme regulation by reversible covalent modification is an important mechanism allowing a fast response to external environmental stimuli such as hormones, light, or environmental stress. It provides an important means of coordinating intracellular pathways. In higher plants there are two major ways that reversible covalent modification regulates enzyme activity, by dithiol–disulfide interconversion and by phosphorylation–dephosphorylation.

Dithiol–disulfide interconversion is an important regulatory mechanism for photosynthetic organisms. This process links photosynthetic electron flow to light regulation of a number of chloroplastic enzymes involved in carbon dioxide fixation (discussed further in Chapter 5). Electrons generated from water by the photosynthetic electron transport chain are shuttled via ferredoxin, ferredoxin thioredoxin reductase, and thioredoxin to specific enzymes (Figure 2.18a). In chloroplasts the ferredoxin–thioredoxin system may be composed of different forms of thioredoxin including thioredoxin *f* or thioredoxin *m* (see Chapter 5). Thioredoxin *f* selectively activates a number of enzymes involved in carbohydrate biosynthesis. Thioredoxin *m* activates NADP-malate dehydrogenase and inhibits glucose 6-phosphate dehydrogenase. The dithiol–disulfide exchange leads to covalent modification, which either activates or inactivates the enzyme involved. The light-dependent alkalization of the stroma, together with increased concentrations of a number of allosteric effectors, enhances the thioredoxin-mediated disulfide–dithiol interconversion process of some chloroplastic enzymes. Enzymes reduced by thioredoxin in the light are oxidized to disulfide forms in the dark. This may be catalyzed by the action of either oxidized thioredoxin or possibly low molecular weight oxidants, such as oxidized glutathione, dehydroascorbate, or hydrogen peroxide.

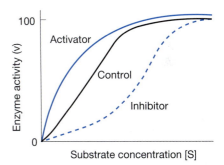

Figure 2.17 Effect on the substrate saturation kinetics of an allosteric enzyme of adding an activator (———) or inhibitor (--------).

Phosphorylation–dephosphorylation is the reversible covalent incorporation of phosphate occurring on either serine, threonine, or tyrosine residues. A protein kinase catalyzes the phosphorylation reaction and a phosphoprotein phosphatase catalyzes the reverse dephosphorylation reaction (Figure 2.18b). There are a range of classes of protein kinases and phosphoroprotein phosphatases, often these are regulated by particular metabolic signals and inhibitor proteins, respectively. They are usually specific for a particular enzyme. A well characterized example in plant biochemistry is the regulation of the mitochondrial pyruvate dehydrogenase complex (discussed further in Chapter 6). This was the first plant enzyme shown to be regulated by protein phosphorylation. Its conversion from an inactivated phosphorylated form to an activated dephosphorylated form is a key mechanism for regulating carbon flow between glycolysis, the TCA cycle, and fatty acid metabolism. Since

Figure 2.18 Regulation of enzyme activity by reversible covalent modification. (a) Dithiol–disulfide interconversions. (b) Phosphorylation–dephosphorylation.

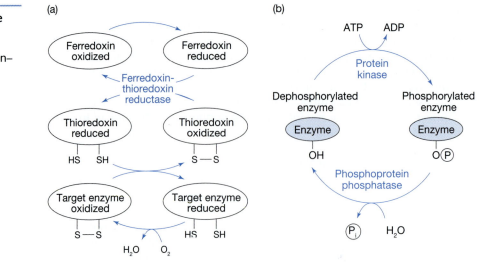

this initial discovery a number of other plant enzymes in key areas of metabolism have now been identified as being regulated by protein phosphorylation. These include sucrose phosphate synthase (Chapter 7) and nitrate reductase (Chapter 8). Understanding how protein phosphorylation can contribute to controlling and coordinating plant metabolic pathways is an ongoing area of plant research.

Subunit association–disassociation

The dissociation of the constituent subunits of a multimeric enzyme generally results in the enzyme becoming less active or totally inactive. The state of aggregation or dissociation can be influenced by metabolites binding to specific subunits. Pyrophosphate-dependent phosphofructokinase can be interconverted between an active 260 kDa tetramer and a less active 130 kDa dimer. The reversible association and dissociation of the pyrophosphate-dependent phosphofructokinase subunits is determined by the relative concentration of allosteric activators fructose 2,6-bisphosphate and P_i. With the tetrameric form enhanced by the presence of fructose 2,6-bisphosphate and the dimeric form by P_i. The association–disassociation is seen as potentially representing a glycolytic/gluconeogenic regulatory mechanism (discussed further in Chapter 7).

Zymogen activation

This process regulates enzyme activity through an irreversible activation involving the proteolytic cleavage of a domain of an inactive enzyme protein precursor. For example, during starch breakdown in monocot seed germination β-amylase is activated by the proteolytic cleavage of a small peptide from the C-terminus of the enzyme. This regulation is brought about by gibberellic acid promoting the expression of the necessary protease, which is then released from the seed aleurone layer (Chapter 7).

Reversible associations of metabolically sequential enzymes

The organization of multi-enzyme complexes of enzymes (metabolons) involved in particular metabolic processes has been proposed as an efficient means of metabolite channeling. In contrast to the more classic pathways involving separate enzymes with pools of intermediates and diffusion of products from one enzyme to become the substrate for the next, metabolic channeling allows the direct transfer of biosynthetic intermediates between catalytic sites. By regulating the movement of metabolites to particular enzymes or complexes to promote specific reactions and to minimize other, less desirable reactions, metabolic channeling through the so-called metabolons provides a system for enhancing and regulating metabolism. Examples of metabolons have been described for the Calvin cycle (Chapter 5), the cysteine synthase complex (Chapter 8), and the phenylpropanoid pathway (Chapter 11). All of these systems involve high-affinity, stable enzyme associations that have made it possible to study them in detail. However, it is becoming increasingly clear that even pathways once thought to be composed primarily of soluble enzymes may be subject to some sort of subcellular associations. Short-lived dynamic complexes involving weak or transient interactions between enzyme components may be formed and dissociate in response to metabolic status. Such a system would provide an efficient means of regulating metabolism. Having a pathway proceed in a metabolon rather than as individual enzymes has a host of implications for pathway flux control, regulation, and attempts at metabolic engineering. Moreover metabolons allow for a sub-compartmentation of the pathways with the enzymes themselves eliminating competition for common metabolic intermediates and non release of potentially cytotoxic intermediates.

Compartmentation: keeping competitive reactions apart

As described elsewhere the organization and structure of a plant cell leads to compartmentation of metabolism (see Chapter 3). Many metabolic reactions of a pathway may be located in either a specific location within the cell, or in particular organelles. Plant metabolism is dependent on compartmentation and this has long been established to be important for a number of different reasons. Throughout this book examples will be presented emphasizing the importance of compartmentation. By concentrating a metabolite, cofactor, or effector within a particular compartment, a more favorable environment for enzymes to operate can be established. Compartmentation also separates intermediates from enzymes and in doing so can avoid futile cycling or by-products inhibiting particular processes. The localization of a number of enzymes and metabolites within the same compartment can lead to a more efficient operation of a metabolic pathway. Similarly it can allow pathways to interact more efficiently. However, compartmentation also means that the different components of the plant cell have to interact at a number of levels. Substrates often must move between compartments from the site of synthesis to the site where they are consumed (Box 2.10). This movement often involves translocators and has to be coordinated and regulated to maintain an appropriate flux through that step of the pathway. It is not just metabolites that move. Most proteins are nuclear-encoded and synthesized in the cytoplasm, meaning that if they are located in an organelle then there is a necessity to ensure targeting of the protein to the correct organelle and this usually involves that the protein has to be transported across one or several membranes. There are a range of bioinformatics programs available that identify N-terminal transit or signal peptide sequences, which target proteins to particular locations based on sequence information (for further information see Rhee et al., 2006 and Lunn 2007 in Further Reading).

Understanding plant metabolism in the individual cell

In its simplest form a plant may be seen as an organ system consisting of roots, stems, and leaves; however, a plant is composed of many organs each contributing to the overall metabolism. Even at a cellular level each cell could be differently programmed and an understanding of the contribution

Box 2.10 Characteristics of plastid transporters

Plant cells contain specific compartments, for example, vacuole, endoplasmic reticulum, mitochondria, microbodies, and plastids, and communication with the cytosol and each other is vital in maintaining cell coordination. For example, communication between the plastid and cytosol is via the plastid inner and outer envelope. The inner membrane of the double membrane envelope contains a variety of transporters and channels for the exchange of metabolites between the cytosol and plastid.

Photosynthetic end products of carbon fixation and other secondary compounds are exported from the plastids into the cytosol for various metabolic processes, such as sucrose formation, sucrose is the dominant carbohydrate moved around the plant (Figure 1). In the 1970s the first plastidial translocator, the triose phosphate/phosphate translocator was identified in spinach chloroplasts inner envelope. The translocator was capable of exporting newly synthesized triose phosphate, and to a lesser extent 3-phosphoglycerate, into the cytosol from the chloroplast. Photosynthates are then partitioned to various metabolic pathways in the same or different sinks. The triose phosphate/phosphate translocator is an obligatory antiporter or counter-exchange system, expelling triose phosphate and 3-phosphoglycerate for P_i, the substrates are transported in a strict 1:1 counter-exchange. The reaction has been described as proceeding via a ping pong reaction. The substrate once transported across the membrane is then required to leave before the transport site can bind a second substrate, suggesting that the transport sites face alternating directions of the membrane envelope. Other translocators located in the plastid envelope are present and function in moving a variety of metabolites and nucleotides in and out of plastids, e.g. glucose, glucose phosphate, 2-oxoglutarate, phospho*enol*pyruvate, and adenine nucleotides.

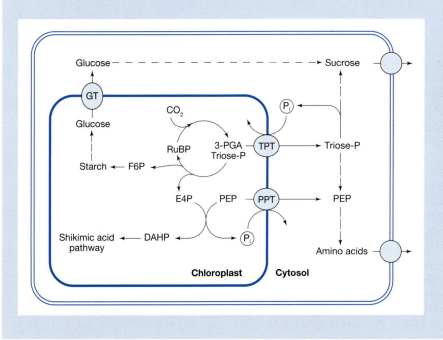

Figure 1 Processes involved in the transport of photoassimilates in leaves. TPT, triose phosphate/phosphate translocator; DAHP, 3-deoxy-D-arabino-heptulosonate 7-phosphate; F6P, fructose 6-phosphate; GT, glucose translocator; PEP, phospho*enol*pyruvate; PPT, phospho*enol*pyruvate/phosphate translocator; 3-PGA, 3-phosphoglycerate; E4P, erythrose 4-phosphate; Triose-P, triose phosphate; RuBP, ribulose 1,5-bisphosphate.

of individual cells within particular tissues and organs will have an important effect on our quantitative understanding of metabolic interactions. For instance in wheat leaf tissue it has been determined that 51% of the cells are mesophyll, 27% epidermal, and 4% vascular. This heterogeneity of cell types has an implication in terms of the metabolic function and protein composition. There is an increasing array of approaches available to resolve metabolism at the level of an individual cell (Table 2.1).

The isolation of organelles

The isolation of organelles separately and intact provides a system without interference from other cellular components. The understanding of where an

Table 2.1 Methods available to understand plant metabolism at the level of the individual cell	
Method	**Application**
Immunocytochemistry	Use antibodies to locate specific proteins in a cell and then visualize using microscopy
In situ hybridization	Usually used to localize (m)RNA
Ion-sensitive microelectrodes	Small probes that can be inserted into plant tissue to measure electrical potential difference of a number of ions
Reporter genes	The use of visible genes that code for products that can be visualized by fluorescence or staining. Useful for determining the cellular location of specific gene expression in tissues and organs
Single cell dissection and analysis	Manual dissection of cells for subsequent biochemical analysis
Single cell sampling and analysis	Microcapillary inserted into single cells and biophysical and inorganic measurements made on expressed contents

enzyme is specifically located provides an indication as to how it fits into a particular metabolic pathway. Such information allows an assumption to be made regarding how different parts of the pathway may be coordinately regulated and may also provide information on regulatory networks. Specific organelles perform specific functions. Such functional specification reflects the unique complement of proteins found in a particular location. A great deal of information and understanding of plant metabolism has been obtained by the ability to isolate specific organelles and characterize their metabolism *in vitro*. Cell fractionation has traditionally been used to determine the particular location(s) of specific enzymes.

Methods used for isolating organelles are dependent upon the type of organelle being isolated and its particular properties. In addition, consideration of the purpose of isolating the organelle is important. When studying the metabolism or composition of any organelle it is essential to try and achieve purity and structural integrity. There are three major problems encountered when trying to separate plant cell organelles. First, the shear forces required to disrupt the high tensile strength of the cellulose walls surrounding the plant cells are often sufficient to rupture the organelles. Depending on the tissue and robustness of the organelle, the cells should be broken under the mildest conditions possible, while still achieving good cell breakage and organelle yield. Extensive pulverization, osmotic shock, ultrasonic disruption, and rapid pressure changes are generally much too violent and disruptive for plant tissue. Rapid slicing is an effective means of cutting open a proportion of the cells so that their contents flow rapidly into the isolation medium. Generally, complete cell breakage is not achieved, as extensive homogenization may prove counter-productive by damaging any released organelles. Secondly, a number of harmful substances including tannins, hydrolases, and organic acids from the vacuoles, phenol oxidases, and lipolytic enzymes are released during extraction. As at pH 8 most lipolytic enzymes are inactive and the vacuole acids are neutralized, this is the pH usually used for the extraction buffer. A variety of protectants to maintain functional integrity are also included. These might include a chelating agent such as EDTA, to bind heavy metal contaminants, sulfydryl reagents such as dithiothreitol, to protect sulfydryl groups from being oxidized (which might otherwise result in enzyme inactivation), and magnesium salts to provide an ionic environment suitable for retaining the structural integrity of some proteins. The necessity for an osmoticum to maintain organelle structure results in the inclusion of, most commonly, sucrose in the buffer. Thirdly, rapid purification of the organelle to minimize its contact with deleterious components of the cell is a priority in order to retain the metabolic competence of the isolated organelles as close as possible to their *in vivo* capacity. To fractionate particular organelles differential centrifugation and/or density centrifugation of samples through a sucrose or percoll gradient is usually the method of choice.

For any quantitative studies, certain methods for the assessment of contamination and structural and functional damage to the isolated organelle must be used. The use of organelle-specific marker enzymes provide an estimate of both organelle recovery and cross-contamination by other organelles. However, enzymically unidentifiable membrane fractions may also be isolated with the organelle of interest or some damage could occur to it during extraction or centrifugation and as such affect its biochemical abilities in isolation. For complete identification of both the organelle and potential contaminants, it is advisable to examine the collected samples under an electron microscope.

Summary

Plant biochemists have used a wide range of experimental approaches to understand how metabolic processes operate. Traditionally this led to the generation of a static route map of a biochemical pathway. With increasing information the characterization of a diverse range of regulatory mechanisms has given further insight into the way a particular biochemical pathway operates and interacts with other processes. In more recent years technological developments have meant that biochemists no longer need to look at an enzyme in isolation, but instead can consider it as a component of an integrated biological system. This wealth of experimental information means plant biochemistry is now able to investigate the dynamics of a pathway including fluctuations in concentration, competition and movement between cellular and subcellular compartments. It is this level of information that will contribute to our understanding of the functioning of the biochemical processes that combine to make up a whole plant.

Further Reading

Protein purification

Roe S (2001) Protein Purification Techniques: A Practical Approach, 2nd ed. Oxford University Press.

Roe S (2001) Protein Purification Applications: A Practical Approach, 2nd ed. Oxford University Press.

Textbooks describing strategies and considerations for purifying proteins from a range of sources, including plant tissues.

Hames BD (2002) Gel Electrophoresis of Proteins: A Practical Approach, 3rd ed. Oxford University Press.

Textbook full of the wide range of electrophoresis methodologies available and their applications.

Transcriptomics

Clarke JD & Zhu T (2006) Microarray analysis of the transcriptome as a stepping stone towards understanding biological systems: practical considerations and perspectives. *Pl. J.* 45, 630–650.

This review tutorial considers methods to undertake and analyze microarray data to characterize plant transcriptomes to answer biological questions.

Proteomics

Twyman RM (2004) Principles of Proteomics. Taylor and Francis.

Textbook describing the different stages of proteomic analysis.

Chen S & Harmon AC (2006) Advances in plant proteomics. *Proteomics* 6, 5504–5516.

Excellent review considering up-to-date approaches in plant proteomics. Also considers the integration of "omic" technology into systems biology.

Metabolomics

Fukusaki E & Kobayashi A (2005) Plant metabolomics: potential for practical operation. J. *Biosci. Bioengin.* 100, 347–354.

Weckwerth W (2003). Metabolomics in systems biology. *Annu. Rev. Pl. Biol.* 54, 669–689.

Sweetlove LJ, Fell D & Fernie AR (2008) Getting to grips with the plant metabolic network. *Biochemical Journal* 409, 27–41.

These reviews provide an in-depth overview about metabolomic technology and how metabolomic networks can be linked to the reaction pathway structure.

Bioinformatics

Rhee YY, Dickerson J & Xu D (2006) Bioinformatics and its application in plant biology. *Annu. Rev. Pl. Biol.* 57, 335–360.

As the title suggests this excellent review covers the important role that bioinformatics now plays in plant biology from key concepts through to methodology, software packages and databases.

Plant biotechnology

Slater A, Scott N & Fowler M (2003) Plant Biotechnology. Oxford University Press.

General textbook offering an overview of the approaches used for producing transgenic crops.

Britt AB & May GD (2003) Re-engineering plant gene targeting. *Trends Pl. Sci.* 8, 90–95.

Gelvin SB (2003) Improving plant genetic engineering by manipulating the host. *Trends Biotechnol.* 21, 95–98.

Tzfira T & Citovsky V (2002) Partners-in infection: host proteins involved in the transformation of plant cells by *Agrobacterium. Trends Cell Biol.* 12, 121–129.

Tzfira T & Citovsky V (2006) *Agrobacterium*-mediated genetic transformation of plants: biology and biotechnology. *Curr. Opin. Biotechnol.* 17, 147–154.

Selection of up-to-date references covering different aspects of plant transformation.

Tzfira T & White C (2005) Towards targeted mutagenesis and gene replacement in plants. *Trends Biotechnol.* 23, 567–569.

Recent review discussing the available mutagenesis technologies.

RNA interference

Kusaba M (2004) RNA interference in crops. *Curr. Opin. Biotechnol.* 15, 139–143.

Tang G & Galili G (2004) Using RNAi to improve plant nutritional value: from mechanism to application. *Trends Biotechnol.* 22, 463–469.

Voinnet O (2002) RNA silencing: small RNAs as ubiquitous regulators of gene expression. *Curr. Opin. Pl. Biol.* 5, 444–451.

Reviews covering different aspects of the technology, applications and potential of RNA interference.

Metabolic control analysis

Kacser H & Porteous JW (1987) Control of metabolism: what do we have to measure? *Trends Biochem.* 12, 5–14.

Metabolism Control Analysis FAQ: http://dbkgroup.org/mca_home.htm

Poolman MG, Fell DA & Thomas S (2000) Modelling photosynthesis and its control. *J. Exp. Bot.* 51, 319–328.

Stitt M, Quick WP, Schurr U, Schulze E-D, Rodermel SR & Bogorad L (1991) Decreased Rubisco in transgenic tobacco transformed with antisense rbcs: flux control coefficients for photosynthesis in varying light, CO_2 and air humidity. *Planta* 183, 555–566.

A range of useful references considering metabolic control analysis.

Enzyme regulation

Cannon DM, Winograd N & Ewing AG (2000) Quantitative chemical analysis of single cells. *Annu. Rev. Biophys. Biomol. Struct.* 29, 239–263.

Huber, SC (2007) Exploring the role of protein phosphorylation in plants: from signalling to metabolism. *Biochem. Soc. Trans.* 35, 28–32.

Lunn J (2007) Compartmentation in plant biology. *J. Exp. Bot.* 58, 35–47.

Meyer Y, Reichheld JP & Vignols F (2005) Thioredoxins in Arabidopsis and other plants. *Photosynthesis Res.* 86, 419–433.

Winkel BSJ (2004) Metabolic channelling in plants. *Annu. Rev. Pl. Biol.* 55, 85–107.

General reviews considering enzyme regulatory mechanisms.

Plant Cell Structure

3

Key concepts

- The internal organization of plant cells is dependent on membranes.

- Membranes, which consist of lipids and proteins, depend on a water phase for their organization into two dimensional sheets.

- Proteins are targeted to specific membranes and confer specific properties on the membrane, including its permeability to solutes.

- The plasma membrane defines the outer boundary of the cell, regulates solute exchange, and participates in cell wall formation.

- Vacuoles occupy a very large proportion of cell space and store a variety of compounds, including pigments, proteins, and osmotically active solutes.

- The endomembrane system consists of two distinct compartments, the endoplasmic reticulum and Golgi. It provides a complex internal space bounded by a membrane that is involved in the synthesis of proteins, polysaccharides, and lipids, mostly destined for export from the cell.

- Cell walls act as mechanical supports and as osmoregulators and define the shape of each cell, so ultimately defining the shape of the whole plant.

- Cell walls regulate the flux of molecules between cells and provide transport pathways within the plant.

- In common with other eukaryotic cells, plant cells contain nuclei, concerned with the storage and expression of genetic information, mitochondria, mostly involved in respiration, and peroxisomes, which have diverse roles in plants.

- Plastids are a defining feature of every plant cell and are specialized for a variety of functions, notably photosynthesis in all green tissues.

Plants are communities of cells. This is immediately apparent when a slice is taken from any plant tissue, leaf, stem, or root, and viewed under a light microscope. Such slices, or sections, show that plant tissues are composed of internal box-like structures that were famously compared by Robert Hooke in 1665 with the small cells occupied by monks in a monastery. As we shall see, it is important to note that these boxes are the exoskeleton of the enclosed cells. Details of the cells can be revealed by preserving the tissue before slicing, and then staining the sections. Most mature plant cells have a deceptively empty appearance due to the presence of a large central vacuole that limits the stained cytoplasm to a thin layer beneath the cell surface. Nuclei are conspicuous, but with only one per cell they can be missing from the thin slice of a cell included in the section. Leaf cells have numerous large organelles, the chloroplasts, readily visible in this cytoplasmic layer. They may also contain prominent starch grains, but these are much more conspicuous as amyloplasts in sections of seeds and storage tubers. Examination of suitable sections at a higher level of resolution, in the light microscope or in an electron microscope, reveals much smaller structures, including mitochondria, microbodies, Golgi bodies, endoplasmic reticulum (ER) and ribosomes (Figure 3.1).

While sections give a static view of these structures at a fairly high level of resolution they do not allow us to see the behavior of cell components in the living cell. Over the past 10 years the availability of improved microscope systems and labels that can be applied to specific components without harming the cell has now revealed the extraordinary level of dynamic activity within plant cells. These can be seen as videoclips on numerous websites and a superb collection is available on DVD (Gunning 2007).

Cell structure is defined by membranes

Electron micrographs of plant cells show that the cell and the organelles within are all defined by membranes (Figure 3.1). These form crucial boundaries in all living systems. They separate internal spaces from the outside, and permit the regulated movement of defined compounds (usually small molecules and ions) across the membrane barrier. Differential staining shows that internal cell compartments, or organelles, are chemically different from one another, reflecting specialization for different biochemical activities. The modern eukaryotic cell is the end product of over 2 billion years of evolution and selection; therefore, it will be no surprise to find that the present day structures are extremely efficient. Here we will consider first the general structure and function of membranes, and then turn our attention to the different organelle types.

Membrane structure

Membrane structure depends on the interaction of lipid molecules with water. Membranes are composed of lipids and proteins. The lipids are organized into bilayers (two layers), whereas the protein polypeptide chains are embedded in the bilayer, sometimes protruding above the membrane surface

Figure 3.1 Electron micrograph of part of a plant cell. The cell is surrounded by a plasma membrane and membranes divide the protoplast into the nucleus and the cytoplasmic organelles: mitochondria, plastids, endoplasmic reticulum, and Golgi bodies. Clear areas represent small cell vacuoles, each surrounded by a membrane, the tonoplast. (Reprinted with permission from Gunning BES, Plant Cell Biology on DVD, 2007.)

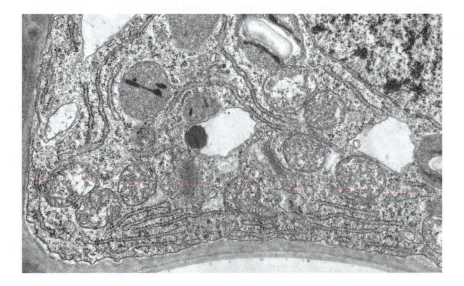

on both sides of the membrane, or lying predominantly in one or other of the two layers of lipid (Figure 3.2).

The organization of lipids and proteins in membranes is driven by the interaction of these molecules with water. In fact, the basic lipid bilayer structure can self-assemble in a test tube to form membrane-enclosed spaces, called liposomes. This, however, does not occur in cells. Instead specific lipids are made and inserted into existing membranes, so in living organisms all membranes are formed by expansion of existing membranes.

Most membrane lipids consist of fatty acid chains attached to a glycerol head. Glycerol is a three-carbon alcohol, with a hydroxyl group on each carbon atom. The hydroxyl groups interact with the carboxylic acid at the end of the fatty acid molecules to form an ester (Figure 3.3). This type of lipid is termed a saponifiable lipid; when boiled with strong alkali, free fatty acids are released (see Chapter 9). The fatty acids can have chain lengths from 14 to 18 or more carbons, and the C–C bonds can be saturated or unsaturated. Typically, plant lipids have more unsaturated bonds in each fatty acid chain than animal lipids, with plants having either predominantly 16:3 or 18:3 fatty acids (i.e. each having three unsaturated double bonds on the hydrocarbon chain). Peas are an example of 18:3 plants with most of the lipid being made in the ER, whereas spinach would be an example of 16:3 plants, with the plastids being involved in the lipid synthesis. Animals are unable to synthesize these unsaturated fatty acids. Consequently, humans regard plant lipids as essential because of their content of linoleic (18:2) and α-linolenic (18:3) fatty acids.

The unsaturated fatty acids play a crucial part in membrane fluidity because they introduce a kink in the fatty acid chain (Figure 3.3). This means that the lipids in the membrane layer cannot pack so closely together, increasing fluidity and making the membrane more resistant to low temperatures (see section Temperature-dependent lipid phase transitions determine cold tolerance in plants).

Each lipid molecule can be composed of two or three different fatty acids attached to the glycerol backbone. If there are only two, the third hydroxyl group on the glycerol binds a phosphate ester group to form a phospholipid (Figure 3.3). Several different polar groups can be linked by the phosphate to the glycerol backbone (Figure 3.4). Photosynthetic membranes, which are the major membrane component of green tissues, contain high levels of galactolipid, in which the third position on the glycerol backbone is occupied by a monosaccharide or disaccharide of galactose (Figure 3.4). Galactolipids are present in the largest amount of all lipids in the biosphere.

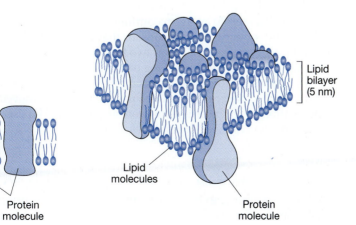

Figure 3.2 Lipid bilayer model of cell membranes. Lipid molecules in the two layers have their hydrophilic heads on the outside of the membrane, in contact with the water phase on each side of the membrane; their hydrophobic tails directed to the inside of the membrane. Membrane proteins are shown spanning the entire width of the membrane, but some proteins can be restricted to either one of the two layers. (From Alberts et al., Molecular Biology of the Cell, 4th edition, New York, NY: Garland Science, 2002.)

Figure 3.3 Phospholipids, like all saponifiable lipids, are based on a glycerol molecule. Fatty acids are attached to carbon atoms 1 and 2, while a phosphate ester, choline, is attached to carbon 3. One of the fatty acids is unsaturated with a single double bond, which introduces a kink into the tail. (From Alberts et al., Molecular Biology of the Cell, 4th edition, New York, NY: Garland Science, 2002.)

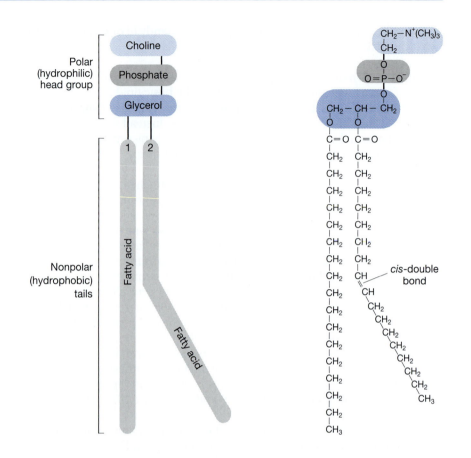

This structure renders the lipid molecules bipolar, i.e. different parts of the molecule have different affinities for water. The glycerol head (an alcohol) is readily soluble in water, as are galactose (a sugar), the phosphate ion, and the attached polar groups. The fatty acid tails, as their name implies, are much less soluble in water. In other words, the heads are hydrophilic and the tails are hydrophobic. A small drop of lipid placed on the surface of a sheet of water will spread out to form a monolayer, with the heads in the water and the tails sticking out at the surface. Pushing this monolayer down into the water results in the formation of a sheet with the typical bilayer structure: the heads are in the water phase on each side of the sheet, and the tails are in the relatively water-free (hydrophobic) interior of the membrane. This is the energetically most stable configuration. Its stability is crucially dependent on the presence of water on both faces of the bilayer, a fact that is of considerable importance in biology.

Membrane bilayers always form continuous surfaces in a water phase. These can be simple spheres, like balloons, or they can be flat sheets (equivalent to squashing a soft balloon between two flat surfaces), or they can form tubes, which can have complex branching patterns. In all cases the membrane surface is continuous, there are never any free edges exposed to the aqueous environment. In general lipid molecules are not easily moved from one side of the membrane to the other, but there are enzymes (flipases) that can facilitate this process for specific lipids.

Membrane proteins are stitched into the lipid bilayers

Membrane proteins can be inserted into lipid bilayers during synthesis (co-translation) or after translation on free ribosomes (post-translation). During

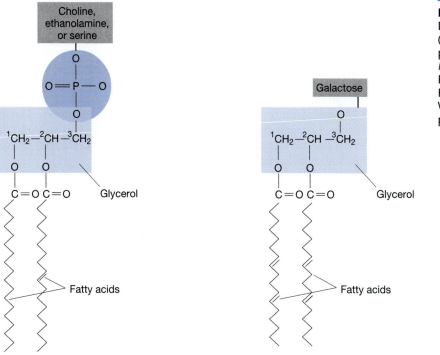

Figure 3.4 Membrane lipids.
Phospholipids (left) and galactolipids (right) are the major lipids found in plant cells. (From *Biochemistry and Molecular Biology of Plants*, Bob Buchanan, Wilhelm Gruissem, and Russell Jones, 2000. Copyright John Wiley & Sons Limited. Reproduced with permission.)

co-translation, ribosomes are attached to the surface of ER cisternae (rough ER) and the emerging polypeptides are inserted into the membrane and stitched into the bilayer (Figure 3.5). Specific sequences (transmembrane sequences) along the polypeptide chain consist of amino acids with hydrophobic side chains that interact with the interior hydrophobic lipid tail environment of the membrane. Frequently each coils up to form an α-helix that spans the membrane. The number of transmembrane sequences determines the number of stitches. Each type of membrane protein will have at least one of these transmembrane sequences; some have as many as six or eight or up to 12, so their polypeptide chains are folded back and forth through the membrane.

Membrane proteins appear as particles on freeze-etched membrane surfaces

Membrane proteins can be visualized by a process termed freeze etching. A very small tissue block is deep frozen very rapidly and cracked apart while still held at a very low temperature. The strength of the frozen block is due to the frozen water, ice. However, the hydrophobic interior of the membranes lacks water and is correspondingly weak. So when the frozen block is cracked open it preferentially splits along the planes of the membranes, cleaving them into the two separate leaflets. Exposing the frozen split block of tissue to a high vacuum causes water to be sublimed off from the frozen face, leaving cellular macromolecules protruding from the surface; this is the etching step in the freeze etching process. Suitable preparation can then reveal these inner faces of the leaflets.

Freeze etching of artificial membranes, consisting only of lipid bilayers, reveals inner faces that are smooth. In contrast, cell membranes have these faces covered with arrays of membrane particles (Figure 3.6). These membrane particles represent transmembrane proteins that have been preferentially retained in

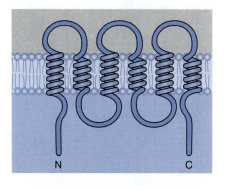

Figure 3.5 Membrane proteins have a number of transmembrane sequences along the length of the polypeptide chain. This one has six, so the C-terminus ends up on the same side as the N-terminus. If this were an endomembrane, then this N-terminus would belong to the amino acid that emerged first from the large ribosome subunit. (From *Biochemistry and Molecular Biology of Plants*, Bob Buchanan, Wilhelm Gruissem, and Russell Jones, 2000. Copyright John Wiley & Sons Limited. Reproduced with permission.)

one or other lipid leaflet during the freeze splitting process. This was one of the significant observations that led to the abandonment of earlier membrane models based on a lipid bilayer with the protein components held only on the surfaces of the lipid. Another problem with earlier models was the poor transport properties of lipid bilayers when compared with real cell membranes. Transmembrane proteins provide hydrophilic routes for transport across cell membranes. Comparison of the two faces split apart from the same membrane often reveals differences in the size distribution of membrane particles on each face. This confirms our belief that membranes do have different protein complements on different faces, a fact that has considerable practical significance for the cell.

The distinctly different leaflets of cell membranes are usually referred to by reference to the protoplasmic side of the membrane (P). Thus freeze fracture of the plasma membrane will reveal particles on the fracture face (F) of the bilayer lying against the protoplast (hence it is the PF face); this is a view from outside of the cell to the inside. The other face is called external (E) and the view of the EF face is from the inside of the cell looking out towards the lipid layer that is adjacent to the cell wall. Similarly, all internal membranes have a membrane layer adjacent to the protoplasm (P) and a layer that is external (E) in the sense that if that membrane were to fuse with the plasma membrane (as in the case of secretory vesicles), then it would be on the outside of the cell.

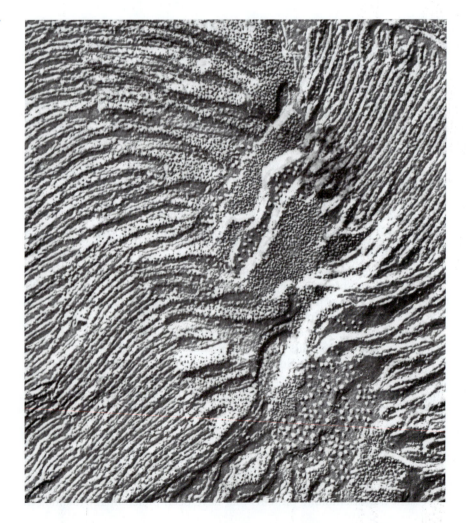

Figure 3.6 Freeze-etch view of membranes in a chloroplast, showing membrane particles on the exposed fractured face of the membranes. Large particles are readily visible on one of these faces (the EF face). The other face (PF) is covered with more numerous, and smaller, membrane particles. EF, fracture face of external leaflet; PF, fracture face of protoplasmic leaflet. (Reprinted with permission from Gunning BES, Plant Cell Biology on DVD, 2007. Originally published in *Oxygenic Photosynthesis: The Light Reactions*, Advances in Photosynthesis, Volume 4, 1996, Kluwer Academic Publishers, pages 11–30, figure 5. Permission granted by the co-author, Dr A. Staehlin and with kind permission of Springer Science and Business Media.)

Lipids and proteins can diffuse in the plane of the membrane

Each membrane consists of several different lipids and numerous proteins. All these molecules are able to diffuse within their layer, in the plane of the membrane. The lipids, which are smaller, move more rapidly than the proteins, which often span both layers. This gave rise to the name for the classic model for the general structure of all cell membranes, the Fluid Mosaic Model, proposed by Singer and Nicolson in 1972.

In animal cell plasma membranes, certain proteins and lipids can associate together to form discrete rafts that move about in the fluid phase of the bilayer. These are rich in another component of cell membranes, sterols, notably cholesterol. Similar rafts have recently been demonstrated to occur in plant cells. In plants the main sterol found is sitosterol, with cholesterol being only a minor component.

Membranes define compartments in the cell and act as enzymically active surfaces

The picture of cell membranes that emerges from these structural considerations is that of a boundary layer that encloses an inner space, forming a compartment. This can be the cell itself enclosed by the plasma membrane, it can be one of the membranes surrounding an organelle such as a chloroplast, or it can be the complex compartment of the ER. As the membrane has two faces, and as different proteins (enzymes) can be associated with either of the two faces, it is easy to see that a membrane can be involved in one set of biochemical reactions inside a compartment and a different set on the outside. Also, membrane proteins can be arranged in a membrane to perform sequential reactions, with the products of one reaction forming the substrate for the next. So the membrane surface is analogous to a factory floor supporting machinery, and an increase in productivity in both cases requires an increased surface area for more machines. Very large membrane surface areas for biochemical reactions are generated within the confines of a single cell.

The restriction of certain proteins (enzymes) to specific membranes provides a basis for understanding the evolution of the varied extents and arrangements of the different membrane systems within a cell. In general these arrangements maximize the functional activity of the membrane. So large areas of membrane, formed by folding, stacking or just sheer planar extent, provide a correspondingly large surface for the particular membrane proteins found in that membrane. An analogy can be made with the available floor area of a factory advertised for sale or rent. In both cases the maximum productive capacity is fixed by the area available for installing machinery, at either the molecular level or the engineering level.

Temperature-dependent lipid phase transitions determine cold tolerance in plants

Different membrane systems in a cell have different lipid and protein compositions. The lipid composition shows variation both between species and within different organs of the same species. One of the factors influenced by lipid composition is the ability of the plant to withstand low temperatures. Lipids exhibit temperature-dependent phase transitions (T_m) from gel, at low temperatures, to fluid at higher temperatures. In the gel state, membrane integrity and function are compromised. The exact mixture of lipids present in a membrane determines the T_m for that membrane, by determining the density of packing of the lipids. Membranes of cold-tolerant or cold-acclimated plants have a higher proportion of fatty acids with short chain lengths and unsaturated double bonds than cold-sensitive plants. This leads to more fluid membrane structures, because the fatty acid chains of these lipids cannot pack closely together as in fully saturated lipids, which have straight hydrocarbon chains (Figure 3.3). Plants acclimatizing to low temperatures shift the lipid composition of their membranes to lower their T_m value. This emphasizes that membranes are constantly turning over, with all the components being renewed on a regular basis.

One might expect that membranes having similar functions would have similar lipid and protein compositions. While this is true for proteins, there is a surprising degree of variation in lipid composition. In general it is believed that the protein function is unaffected by the lipid composition, as long as the lipids are in the fluid state, above the T_m transition. This statement may need some qualification in the future, as a membrane protein in one *in vitro* system does show altered function according to the lipids present, above the T_m temperature

Membranes serve a variety of roles in the cell

Membranes always enclose a space, e.g. the cytoplasm separated from the outside world, or an inner lumen segregated within the cytoplasm. Such arrangements may allow particular biochemical pathways to operate in a defined environment, as with mitochondria and chloroplasts, or simply allow segregation of materials by creating a storage receptacle, as in the vacuole. Membrane compartments also provide a means of keeping track of specific products, as in the ER and Golgi systems. By segregating two different environments, membranes in chloroplasts and mitochondria provide a high energy barrier that is discharged through specific turbines to synthesize the high-energy intermediate, adenosine triphosphate (ATP), used in cell metabolism.

The shape of the enclosed compartment is critical. Tubular systems, highly branched and interconnected, provide the maximum contact between the cytoplasm and the space enclosed by the membrane. Spherical structures provide efficient compartments for transport or storage of specific molecules; efficient in that a minimum of membrane is required to enclose the volume of the contents. The following sections demonstrate these principles when considering the main compartments encountered in plant cells.

The plasma membrane

The plasma membrane barrier controls the cell's import/export activities and has synthetic capabilities. The plasma membrane serves as the boundary to the plant cell, separating the inner living cell from the external, non-living cell wall continuum (Figure 3.7). However, it is far from being an inert barrier, as it has roles in transport in and out of the cell and in the synthesis of the cell wall. Ions, and other solutes, and water are transported across the membrane. The concentration of solutes inside the cell is relatively high (for example, K^+ is present at 100–150 mM) compared with that outside, in the cell wall. The lower osmotic potential of the cell solution inside the plasma membrane would drive, by osmosis, a continuous inward flow of water if it were not resisted by the cell wall. Stretching of the wall, as the cell swells, creates a hydrostatic pressure within the cell that raises the internal water potential from negative values to near zero. Because of constraints on water flows from roots to aerial shoots and leaves, many cells in these structures are normally found to have negative water potentials, i.e. they could take in more water.

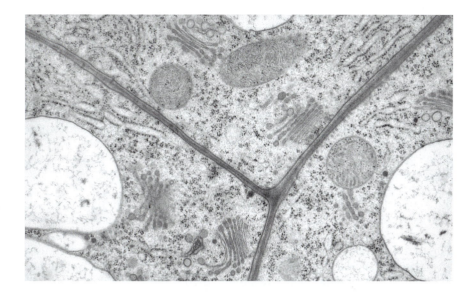

Figure 3.7 Parts of three cells seen in a section from tissue prepared by high pressure rapid freezing followed by freeze fixation and substitution of the ice phase for liquid embedding resin. The plasma membranes are seen as lying smoothly over the surface of the cell in close contact with the cell wall. Other organelles and the vacuoles are seen to have smooth membrane surfaces; this contrasts with the appearance after chemical fixation (see Figure 3.1), which is much slower and gives rise to artifacts, such as shrinkage. (Reprinted with permission from Gunning BES, Plant Cell Biology on DVD, 2007. Original image courtesy of Professor Adrienne Hardham, RSBS, ANU Canberra.)

The problem of transporting larger molecules (macromolecules) across the plasma membrane is circumvented by placing them in membrane-bound vesicles. Secreted proteins and polysaccharides arrive at the cell surface in secretory vesicles that fuse with the plasma membrane, releasing the contents to the outside. In the reverse situation, endocytotic vesicles are invaginated from the cell surface, carrying in molecules that were close to, or attached to the surface of the plasma membrane.

Plasma membranes are also important sites of macromolecule synthesis. Cellulose synthases embedded in the outer membrane bilayer make cellulose microfibrils for the cell wall (see Chapter 7).

Plasma membranes are continuously synthesized and degraded

New plasma membrane, including membrane proteins, is synthesized in the ER and Golgi apparatus and delivered by fusion of secretory vesicle membranes with the cell surface. The reverse process, endocytosis, pulls the plasma membrane in thus forming a vesicle from the cell surface, which migrates into the cytoplasm. The result is a continuous flow of membrane from the cytoplasm to the surface and back into the cytoplasm. This recycling can account for the replacement of the whole plasma membrane in as little as 10 minutes in active secretory cells, while even in rapidly growing cells the surface is still turned over once every 3–4 hours. Therefore, the plasma membrane is far from being a static boundary to the cell; it is part of a continuum of membrane constantly cycling to and from the cell interior.

Plasma membrane transport proteins

Several different transport proteins facilitate the movement of ions and solutes across the plasma membrane. Transporters are membrane proteins that have a high specificity for a particular ion or metabolite and have a binding site that opens alternately to either one side or the other of the membrane. Each transporter will only pump one ion in one direction; the flow through the pore is not reversible.

Antiporters are transporters that pump protons in one direction, while simultaneously a particular cation is moved in the opposite direction. For example, the proton pump transports H^+ out from the cytoplasm and pumps in K^+. This type of transport may be electro-neutral or electrogenic, depending on the ratios of the ions moved in opposite directions. Symporters pump protons and an anion, or other substrate molecule, in the same direction across the membrane.

The transport direction is coupled to an electron gradient, which establishes a membrane potential of about 70 millivolts across the cell surface, with the inside of the cell being negative (i.e. due to a net surplus of H^+ transported out by the proton pumps). This gradient is essential for many transport processes.

Ion channels allow passage of specific ions across the membrane, and the activity of these depends on the concentration of ions on each side. These channels may be gated, i.e. are closed, but will open upon receipt of a specific signal. Ion pumps transport specific ions against their concentration gradients, e.g. the calcium pump maintains the cytoplasmic $[Ca^{2+}]$ concentration at about 10^{-7} M, about 1000 times lower than the external concentration. Gated channels and pumps can work together to bring about specific signaling from outside the cell to inside it. For example, calcium pumps maintain a low internal calcium level, until the gated calcium channels receive a specific signal that causes them to open briefly and flood the cytoplasm with calcium ions that trigger internal responses. The pumps then restore the previous concentration differences.

Plasma membranes are attached to the cell wall and to the cytoskeleton

There are structural attachments to the plasma membrane; e.g. on the outside these are revealed by gentle plasmolysis of the cell, when it is placed in an external solution of sugar. The plasma membrane shrinks away from the wall interface, but remains attached at numerous points, so that it is drawn out into numerous fine strands, Hechtian strands. This process also activates ion pumps that are sensitive to the stretching of the plasma membrane.

Internally the plasma membrane is attached to the cytoskeleton, notably the microtubules, which somehow exert an influence across the plasma membrane that results in the ordered deposition of cellulose in the cell wall (Chapter 7).

Vacuoles and the tonoplast membrane

Most plant cells contain a large vacuole that occupies >90% of the cell volume and is separated from the cytoplasm by a membrane, the tonoplast. Meristematic cells possess a number of small vacuoles that coalesce as the cell enlarges (Figures 3.1 and 3.7). In some ways the tonoplast resembles the plasma membrane; it separates a space with low metabolic activity, the internal vacuole, from the cell cytoplasm. Like the plasma membrane it carries ion pumps, though they are often subtly different in their properties to those on the outside of the cell. Like the plasma membrane, the tonoplast can accept appropriately targeted secretory vesicles from the Golgi apparatus.

Vacuoles store a variety of compounds that serve different functions

The vacuole serves as a storage compartment for a wide range of materials and metabolites. It serves as a repository for anions and cations, e.g. potassium, calcium, chloride, and phosphate, and organic acids, such as malate and oxalate. Sometimes these are present at a sufficiently high concentration to initiate the formation of inorganic and organic crystals. These are often characteristic of the tissues and the plant species, and they may persist in sediments, allowing identification of the plant species from which they came. Some crystals are fine needles that readily penetrate the soft mucosa lining the mouths of herbivorous animals. They also act as mild poisons, discouraging further consumption of that species. Dumb cane (*Diffenbachia seguine*) is aptly named, as the swollen tongue can fill the whole mouth.

Vacuoles store pigment molecules, as seen in many blue/purple flower colors (anthocyanin pigments, see Box 11.2). They also store protective agents, inhibitors of insect digestive enzymes such as proteases, which render leaves useless, and therefore unattractive, to herbivores. Proteins are laid down in vacuoles as reserves of nitrogen compounds in many seeds (see Chapter 8). The solutes stored in vacuoles contribute to the osmotic load of the cell, which is offset by the cell wall. The wall mechanically resists the swelling forces generated by an inflow of water.

Vacuoles allow plant cells to grow to a relatively large size

Indirectly, the vacuole allows plant cells to grow very much larger than animal cells. Plant cell growth is largely driven by an increase in cell vacuole size and volume, so that the cytoplasm is confined to a thin layer pressed against the cell wall. This maintains an approximately constant ratio of volume of cytoplasm to surface area of plasma membrane. In contrast, some animal cells are spherical or cuboidal in shape. As they increase in size the volume increases as a cube function, whereas the surface area only increases as a square function. This limits growth, as the cell surface is the main barrier for nutrient and gas exchange, required for metabolic processes within. Thus plant cells are generally able to grow larger than animal cells, and with little additional expenditure of energy, as the solutes required to inflate the vacuoles can be inorganic ions rather than organic molecules, such as glucose. Further the vacuoles provide a greatly increased surface area for the distribution of chloroplasts and the interception of solar radiation.

The endomembrane system

The endomembrane system consists of two distinct compartments, the endoplasmic reticulum and the Golgi apparatus. The ER is a very extensive membrane system ramifying throughout the cytoplasm (Figure 3.8) that serves to separate an internal space, the lumen, from the rest of the cytoplasm. It also provides a very extensive surface area, up to 10 or even 20 μm^2 per μm^3 cytoplasm, on which biochemical reactions can take place. The membrane and its enclosed space is called a cisterna (Figure 3.9). Many synthetic and transport functions are performed by the ER, some in specialized regional localities that are recognizable by their characteristic morphologies. As far as we can tell, all the ER in a cell is interconnected into a single system. Two basic categories of ER can be distinguished, rough and smooth. There are also a range of subtypes in each category depending on morphology: flat sheets or tubular, and narrow lumen or wide lumen.

Rough endoplasmic reticulum

Rough ER synthesizes membrane and secretory proteins. It is studded with polysomes (ribosomes + mRNA) that are targeted to the outer surface by signal sequences in the nascent polypeptide chains emerging from the ribosome

Figure 3.8 The netlike arrays of endoplasmic reticulum in a leaf cell revealed by transforming the cell with a Green Fluorescent Protein (GFP) sequence attached to an endoplasmic reticulum-specific protein. Expression of the gene leads to accumulation of the GFP-labeled protein in the cisternae, which can then be visualized with ultraviolet light in a fluorescence microscope. (Reprinted with permission from Gunning BES, Plant Cell Biology on DVD, 2007. Original image courtesy of Professor Karl Oparka.)

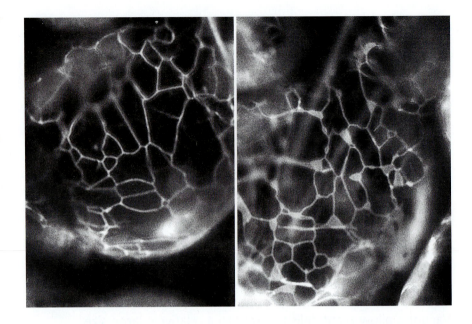

tunnels. These bound ribosomes are synthesizing proteins that are destined either for the ER lumen (soluble proteins) or are stitched into the ER membrane (membrane proteins). All the proteins start off in the ER, some stay there, others go to the Golgi apparatus. Secretory proteins undergo specific glycosylation modifications to form N-glycans within the ER. Membrane proteins are stitched into the ER membrane via specific transmembrane (TM) amino acid sequences in the emerging polypeptides. These membrane

Figure 3.9 Micrograph showing the endoplasmic reticulum cisternae, with a membrane-bound lumen. Note that the cisternae are continuous from one cell to the next through the plasmodesmata. (Reprinted with permission from Gunning BES, Plant Cell Biology on DVD, 2007.)

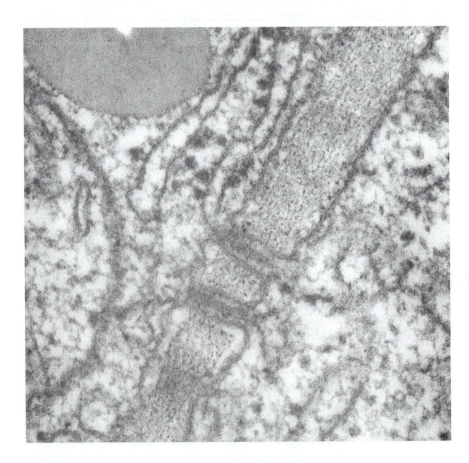

proteins might be enzymes retained in the ER for ER functions, or exported to perform functions elsewhere. For example, cellulose synthase molecules destined for the plasma membrane are made on the rough ER (see Chapter 7) or the membrane proteins are exported to function elsewhere, such as the Golgi apparatus and plasma membrane.

Rough ER is usually present as a series of interconnected, flattened cisternae that lie parallel to each other in the cytoplasm. Frequently protein products can be stained within its lumen (Figure 3.10). Rough ER is also modified at telophase to form the new nuclear envelope. In this capacity it loses its ribosomes on the inner face of the nucleus and the two ER membranes fuse at numerous discrete points to form nuclear pores, in conjunction with a support scaffolding of proteins.

Rough ER is found in all cell types, but is especially extensive in cells that are synthesizing either storage proteins, as in many seeds, or secretory proteins. It is also extensive in cells of leaf trichomes and the tapetum in anthers. In living cells it can be visualized in a confocal microscope with fluorescent tracers (Figure 3.8). In marked contrast to the static electron microscope images, the ER system of living cells is seen to be highly dynamic, with sheets and tubes constantly moving, coalescing, breaking apart and streaming around in the cytoplasm. We have no way of knowing, at the present time, what these activities represent in terms of ER function.

Smooth endoplasmic reticulum

Smooth ER synthesizes waxes and oils (see Chapter 9). It generally lacks ribosomes on its cytoplasmic surface and is often present as a meshwork of interconnected tubules or flattened cisternae. This type predominates in cells undergoing long chain carbon synthesis, such as in the formation of waxes and oils. Such cells are often found in the epidermis of leaves and flower petals, which form surface secretions of epicuticular wax crystals, and in developing seeds that store oils, such as castor bean and sunflower.

Some smooth regions of ER are specifically associated with the outer membranes of plastids and mitochondria, and are assumed to be involved in exchanges with these organelles (Figure 3.11). Other smooth faces of cisternae

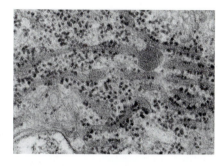

Figure 3.10 Products of the rough endoplasmic reticulum are readily stained within the lumen of the cisternae in this tapetal cell. (Reprinted with permission from Gunning BES, Plant Cell Biology on DVD, 2007.)

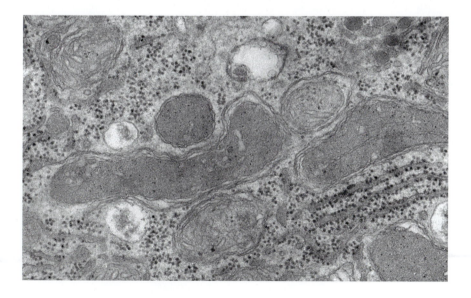

Figure 3.11 Smooth elements of endoplasmic reticulum, with narrow lumens, encase each of the plastids and mitochondria in the cytoplasm of a tapetal cell. (Reprinted with permission from Gunning BES, Plant Cell Biology on DVD, 2007.)

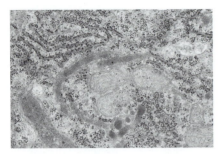

Figure 3.12 Three morphologically distinct types of endoplasmic reticulum are found in tapetal cells. Narrow smooth cisternae are wrapped around mitochondria and plastids and join on to rough cisternae containing stained product and bearing ribosomes (polysomes) on their outer faces. These rough elements are in turn continuous with distended smooth cisternae that contain larger accumulations of the unknown product. (Reprinted with permission from Gunning BES, Plant Cell Biology on DVD, 2007.)

are appressed to the plasma membrane and also pass through the cell wall as narrow tubes within plasmodesmata (Figures 3.9 and 3.14).

This functional diversity of ER within each cell (Figure 3.12) makes it difficult to investigate its biochemical activity. Cell fractionation leads to the ER fragmenting into a series of vesicles, which can then be separated from other components by differential centrifugation. Recently, fluorescent labeling techniques have been developed that enable some functions and activities to be observed in living cells by confocal light microscopy.

The Golgi apparatus

The Golgi apparatus makes polysaccharides and glycoproteins and packages them in secretory vesicles. The Golgi apparatus consists of a stack of three or more (can be up to 20) flattened circular cisternae about 1–3 μm in diameter (Figure 3.13, see also Figures 3.1 and 3.7). There can be several hundred such stacks in the cytoplasm of a higher plant cell, in which case each is referred to as a dictyosome. Some cell biologists would then regard the Golgi apparatus as the sum of all dictyosomes in a cell, rather than calling each individual unit a Golgi apparatus.

The cisternae in each stack are not identical. Morphologically there is a gradation in staining pattern and of membrane thickness from one cisterna to the next across the stack. The cisterna at the side, or face, of the dictyosome with the lightest staining and thinnest membranes is called *cis-*, whereas the cisterna at the opposite face, with the heaviest staining and thickest membranes is called *trans-*. Cisternae in the middle may be called *medial*. This transition from one face to the other reflects a biochemical transition in properties, e.g. in the multiple steps leading to glycosylation of a secreted protein, or the synthesis of a cell wall polysaccharide. Proteins are imported from rough ER in transitional vesicles that bud off the ER and travel to the *cis-*face and fuse with it. These transitional vesicles carry secretory proteins from the lumen of the ER and newly synthesized membrane proteins from the ER membranes that have contributed to the formation of the transitional vesicle membrane.

The Golgi cisternae are the site of complex carbohydrate synthesis. ER-derived proteins destined for the cell surface or cell vacuole have their glycan groups modified in the Golgi by trimming and extension of the N-linked

Figure 3.13 Golgi bodies lying near a small vacuole. The one on the right is sectioned vertically from top (*trans*-face) to bottom (*cis*-face). Transitional vesicles lie beneath the *cis*-face, above the endoplasmic reticulum cisterna. Large secretory vesicles, containing stained polysaccharide molecules, are seen near the *trans*-face. (Reprinted with permission from Gunning BES, Plant Cell Biology on DVD, 2007. Original micrograph courtesy of Dr. Tom Fraser.)

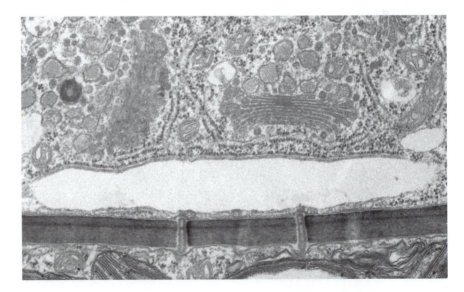

glycans to form complex glycoproteins. Further, O-linked glycans can be added to the polypeptide to form proteoglycans.

The Golgi also assembles very large carbohydrate molecules from sugars. These syntheses are very precise, despite the large number of possible combinations of different sugars and different linkages, each macromolecule is precisely tailored according to the plant species, cell type and function. In some unicellular algae the structures are scales, precisely patterned, that are up to 1 μm across. Some specific enzymes, synthases and transferases, have been isolated from dictyosomes, and it is known that many large carbohydrates are initially synthesized on membrane protein cores. In general little is known about how the specific assembly sequence of sugars and linkages is determined. Some seem to be assembled in stages starting at the *cis*-cisternae, and progressing to *medial* and *trans*, while others are made entirely within the *trans*-cisternae and the post-Golgi compartments of the trans-Golgi network (TGN).

Observation of plant Golgi stacks in living cells using specific fluorescent dyes has provided remarkable and unexpected insights into the dynamic activities of these organelles. Individual stacks move around the cell. This may mean that they move with circulating ER elements, remaining associated with specific regions of ER that gave rise to transitional vesicles, or that the Golgi stacks actively patrol the ER surface and harvest the transitional vesicles as they are formed. Observations using ER-specific fluorescent markers suggest that it is the ER that is moving and carrying the Golgi stacks along with it.

The products of the Golgi stacks leave from the *trans*-cisternae in large secretory vesicles. These pinch off from the *trans*-cisternae, incorporating both cisternal contents, including soluble proteins, and cisternal membrane, bearing membrane proteins made on the rough ER. The extent to which the *trans*-cisternae are entirely used up in the formation of secretory vesicles has been the subject of controversy among research workers for many years. One theory is that they are entirely consumed in the process, and replaced by the maturation of *medial* cisternae. So the whole Golgi stack can be viewed as a sequential formation of new *cis*-cisternae at the base and loss of complete cisternae at the top. Alternatively, the cisternae can be viewed as static compartments with specialized functions, receiving and passing on products sequentially. Studies of unicellular scale-forming algae have revealed the release of scales at the cell surface. Each scale was formed in, and filled, a single Golgi cisterna; the process could be traced from at least the *medial* position to final transformation of the whole *trans*-cisterna into a large secretory vesicle. This supported the cisternal maturation model.

There is some evidence that different products of the same Golgi stack are packaged into different secretory vesicles that travel to different destinations, some going to the plasma membrane and others to the tonoplast. Therefore, materials can be deposited simultaneously into the cell wall and into the internal cell vacuole. Some seeds may even send specific secretory vesicles from the Golgi, containing different products, to two different classes of vacuoles (Chapter 8). This is clear evidence for specific sorting and targeting systems at the *trans*-face of the Golgi.

Fusion of secretory vesicles at the cell surface not only delivers cargo to the cell wall space, but also incorporates the vesicle membrane into the plasma membrane. Rates of secretion in many cell types are high, such that appreciable areas of new membrane are delivered to the cell surface. This never leads to folding or pleating of the plasma membrane because excess membrane is simultaneously retrieved from the cell surface by endocytosis (i.e.

the invagination of part of the plasma membrane to form a small vesicle that is pinched off and internalized). Hence turnover of the plasma membrane is a continuous process, even in growing and expanding cells, and can lead to recycling of the whole cell plasma membrane on a time scale of minutes to a few hours.

Cell walls serve to limit osmotic swelling of the enclosed protoplast

The plasma membrane, as detailed above, is the boundary between a high solute compartment, the cytoplasm, and the low solute exterior. This poses a physical problem for all cells, plant and animal, because water will move into the high solute compartment by osmosis (due to the more negative water potential within the cell), causing swelling. Evolution has favored cells that are able to circumvent this potentially lethal process. Animals have systems for controlling the level of solutes in the body fluids so that the cells are bathed in an isotonic medium (similar solute levels in the fluids to those in the cells), while plant cells are surrounded by a cell wall that mechanically limits the swelling process. This has the added benefit for the plant as a whole, because such turgid (inflated) cells support aerial structures that are otherwise mechanically weak. This is readily, and regularly, seen when houseplants are deprived of water so that they wilt—the cells are no longer turgid, but flaccid. Cell wall structure and synthesis will be considered in detail in Chapter 7; here we will provide a general introduction to their roles in plant cells.

As stated previously all cells are surrounded by walls (with some notable exceptions in reproductive structures). New walls are generated at cell division during cytokinesis. The new cross-wall forms at the equator of the mitotic spindle and arises by fusion of secretory vesicles derived from the Golgi bodies. These form a flat disc of wall material (the cell plate, or phragmoplast in the older literature), derived from the vesicle contents, surrounded by a new plasma membrane, which is formed from the fused vesicle membranes. The cell plate expands by accretion of further vesicles at its edges until it reaches and fuses with the side walls of the parent cell. Each daughter cell then builds a new cell wall on each face of this disc. The preponderance of acidic carboxylic acid residues in the cell plate imparts distinctive staining properties to this middle part of the cell wall, so that it is readily recognized in stained sections as the middle lamella.

Other cell wall components, synthesized in the Golgi apparatus, are delivered in secretory vesicles that have membranes that recognize and fuse with the inner surface of the plasma membrane, the vesicle membrane becoming an integral part of the plasma membrane.

Cell growth and cell shape are governed by the cell wall

Cell growth and expansion is driven, at least in part, by solute uptake. Water enters by osmosis, increasing the turgor pressure and stretching the cell wall. This cell growth is accompanied by deposition of wall material that maintains wall thickness in the face of continual stretching and thinning. Such walls are called primary walls. Once cell expansion has ceased, wall development may continue according to the cell type forming secondary walls.

The organization of the primary wall is crucial to the shaping of the expanding cell. Anisotropic deposition of wall materials leads to areas of wall that are relatively weak, and so these areas expand preferentially under increased turgor pressure. The most common cell shape in stems and roots is cylindrical. These cells are formed at meristems as short cylinders that expand in length creating the familiar roots and stems of the plant. This is due to weakness at the ends of the cylinders compared with the sides, so there is relatively little increase in diameter, but the organ elongates considerably. Plant cells form a great variety of complex asymmetric shapes by simply altering the location of the weak and strong areas of the wall faces. So cortical cells of *Juncus* and other rushes form aerenchyma, a spongy mass of stellar-shaped cells surrounded by extensive air spaces. Some cells, known collectively as motor cells, have walls that are selectively strengthened so that when they go through cycles of high turgidity (induced by high internal solutes) and low turgidity (induced by lowering internal solutes) the cells expand and contract in specific directions, much like hydraulic rams. The simplest examples are guard cells, whose expansion springs open the stomata on a leaf surface. More complex pistons serve to power nastic movements of plant leaves (leaf folding at night, unfolding at dawn) while others serve to spring traps in carnivorous plants.

Primary walls contain microfibrils of cellulose embedded in a matrix of hemicellulose and pectins (Chapter 7). The matrix contains a highly hydrated gel, so that the water phase is a significant fraction of the cell wall's fresh weight. The microfibrils provide the strength and resist wall stretching along their length, so the wall expands at right angles to the preferred microfibril orientation. In cylindrical cells microfibrils run around the cell in hoops or spirals in the side walls, resisting increases in cell diameter, so the cells grow mainly in length.

Secondary walls are similar, but they contain a higher density of microfibrils and hence a higher cellulose content and a lower matrix water content than primary walls. The microfibrils may be laid down in a series of distinct layers that can be recognized by microscopy. The matrix may be modified, for example, by the deposition of suberin, which renders it impermeable to water flows. In xylem and some other cells, lignin is deposited in the matrix and becomes highly cross-linked to the microfibrils, adding considerably to the mechanical strength of the wall (see Chapter 7 for details of wall carbohydrates, microfibril synthesis and lignin). Twigs, branches and tree trunks owe their mechanical strength to these thickened and modified walls that have very elaborate spiral or reticulate patterns. These cells die at maturity forming continuous pipes that conduct water to transpiring leaves, sometimes high in the vegetation canopy. The strong negative pressures experienced in these tubes are resisted by the thickened walls. In gymnosperms, lateral connections between files of tubes are made by specialized bordered pits, which have a disc of wall material that acts as a safety valve. If a column of cells is damaged, for example, by insect attack or wind damage, the valves seal off and protect the adjacent columns.

Plasmodesmata connect adjacent cells across the cell wall

Communication between adjacent cells is essential for the functioning of the whole plant. The cell wall places a physical barrier between adjacent cells. Although solutes secreted from the surface of one cell can diffuse in the water phase of the wall (the apoplast) across to the plasma membrane surface of an adjacent cell, this appears to be too slow to support the needs of the whole plant. The process also lacks specific direction, once released into the apoplast molecules could diffuse in any direction and may not be taken up by the intended target cell. Plasmodesmata provide a transport route that is faster, directional and regulated. Plasmodesma transform the separate cells of the plant body into a symplast of interconnected cells, separated from the apoplast of the wall by the plasma membrane. Cells with thick primary and secondary walls often have pits, which are places with thinner walls where plasmodesmata are concentrated, facilitating cell–cell communication and transport.

Plasmodesmata are plasma membrane-lined tubes that cross the cell wall from one cell to the next (Figure 3.14). They can be very numerous, and can be branched within the wall structure. Each plasmodesma also contains a tubule of ER running across from one cell to the next, providing continuity between the endomembrane systems of adjacent cells. The lumen of this desmotubule is very narrow, only 2–3 nm, so it is not immediately clear to what extent this provides a pathway for free movement of solutes and other molecules. Fluorescent lipid markers that bind to the ER are unable to diffuse from cell to cell, so it seems that the lipid bilayer of the desmotubule does not have the fluid properties typical of other membranes. Microscope images support the suggestion that it is dominated by proteins.

The maximum size of molecules able to move through plasmodesmata has been established by carefully injecting fluorescently labeled dextrans into living tissue cells. The dextrans are inert; they are not metabolized. Small labeled dextrans (up to about 800 Da) are able to move into adjacent cells, while larger ones are unable to do so. This technique has been used to explore many cells and tissue boundaries in plants. Stomatal guard cells and mature root hair cells have no plasmodesmatal connections, and neither does the root cap have connections with the root tip. Some tissue boundaries (epidermis/cortex) and organ connections (petiole/stem connections) have plasmodesmata with reduced permeability. Plants under stress can have plasmodesmata with increased permeability, dextrans up to 5–10 kDa can pass through them.

Invading viruses travel through the plant body from cell to cell via plasmodesmata. They synthesize movement proteins that cause the plasmodesmata to enlarge in diameter, enabling viral nucleic acids to invade the adjacent cells. Plants transformed to express these virus movement proteins in all cells have plasmodesmata that can support transport of dextrans up to about 20 kDa in size.

At certain tissue interfaces plasmodesmata are not formed, for example, between different cell generations during plant reproduction (thereby protecting the next generation from viral attack), between phloem parenchyma cells and sieve tubes and to xylem tracheids and vessels, which are dead cells lacking plasma membranes. Nevertheless, there may still be a requirement for a high level of solute transfer across such boundaries. This is achieved in some plants by the development of a highly enlarged plasma membrane surface, which thus increases the capacity for secretion or uptake of solutes by the cells at these interfaces. These are then termed transfer cells. The increased surface is formed by finger-like ingrowths of cell surface that penetrate into the cytoplasm. These form a complex wall-membrane apparatus that consists of branching and anastomosing ingrowths, often associated with mitochondria in the adjacent cytoplasm.

The nucleus contains the cell's chromatin within a highly specialized structure, the nuclear envelope

The nucleus is a specialized compartment derived from the ER. It is formed at the end of each mitosis event, by the specific association of ER cisternae

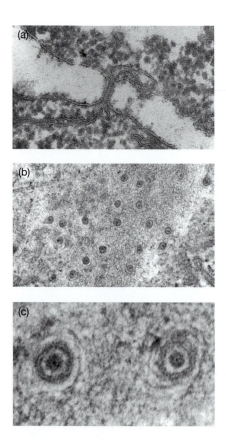

Figure 3.14 Sections through plasmodesmata. (a) Longitudinal section, showing the continuity of plasma membrane and endoplasmic reticulum through the plasmodesma from one cell to the next. (b) Transverse section in the plane of the cell wall, showing the distribution and frequency of plasmodesmata in the wall. Each circular profile is a cross section through the tube of plasma membrane that lines the plasmodesma. (c) As (b) but at a greater resolution, revealing the inner desmotubule of endoplasmic reticulum. The 'free space' between the two membranes severely limits the size of molecules that can pass from cell to cell. (Reprinted with permission from Gunning BES, Plant Cell Biology on DVD, 2007. (a) courtesy of Robyn Overall.)

with the coalescing telophase chromosomes to form a new nuclear envelope. This association leads to modification of the ER to form a highly specialized envelope that serves to separate the nucleoplasm from the cytoplasm, while maintaining a high level of continuity between the two (Figure 3.1). Morphologically, these modifications are most readily seen as a loss of polyribosomes from the face of the cisterna on the nucleoplasm side of the envelope and the formation of nuclear pores by local fusion of the ER membranes. The pores are guarded by protein complexes that are actively involved in the transport of proteins and nucleic acids both in and out of the nucleus. Less obvious is the development of specific membrane proteins on the inner face that bind chromatin, imparting order to the great lengths (many meters in some cases) of nucleosome-beaded DNA contained in each nucleus.

The structure/function relationships of the envelope are reflected in the link between activity within the nucleus and structure of the envelope. Less active nuclei are spherical, and have relatively fewer pores per unit area than highly active nuclei. The latter may also be characterized by the formation of lobes and infoldings, so increasing the area of envelope relative to the volume of nucleus that it encloses.

The roles of chromatin in providing the genetic material, and the means for its selective expression, are outside the bounds of this book. However, it should be clear that the great majority of the enzymes that form the basis of biochemical reactions described in these chapters are made from mRNA molecules specifically transcribed within the nucleus, or within the DNA-containing organelles, mitochondria, and plastids.

Mitochondria are ubiquitous organelles, which are the site of cellular respiration

Observation of living cells by differential contrast microscopy reveals hundreds of small ellipsoidal and cylindrical organelles moving in the cytoplasmic stream (see, for example, the DVD by Gunning, 2007). These are mitochondria, and they are remarkable for their constant change in size (length) and behavior: splitting, fusing, and dividing. They are said to be pleiomorphic. When thin sections are viewed by transmission electron microscopy, the outer surface of the enclosing envelope is seen to consist of a smooth outer membrane, while the inner membrane is thrown into a series of folds (the cristae), which thrust towards the interior of the mitochondrion (the matrix) (Figure 3.15). The outer membrane contains specialized transmembrane proteins termed porins that make the membrane highly permeable. The cristae bear elements of the electron transport pathway involved in respiration and ATP synthases (Chapter 6). The space between the inner and outer membranes is continuous with the space in the folds of the cristae. The matrix contains soluble enzymes of the TCA cycle, involved in carbon catabolism (Chapter 6), a nucleosome containing DNA arranged as closed loops without histones, ribosomes, and other factors required for protein synthesis.

The organization of the mitochondrial DNA and the characteristics of its protein synthesis machinery are very similar to those of eubacteria, which is in sharp contrast to the cell's nucleus and cytoplasm. These features prompted the suggestion that mitochondria have evolved from symbiotic associations between early eukaryotic cells and aerobic bacteria. Further work has confirmed, in so far as it is possible, that this is indeed the case. The mitochondrial envelope bears strong similarities to bacterial envelopes, including the relatively permeable outer layer with similar porins.

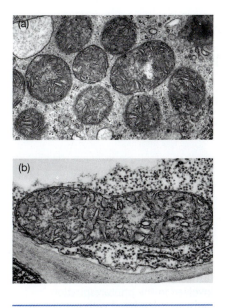

Figure 3.15 **Plant mitochondria typically have complex internal cristae with continuities to the inner membrane of the mitochondrial envelope.** The matrix contains densely staining ribosomes and clear patches with fibers, which represent the nucleoid containing DNA. (Reprinted with permission from Gunning BES, Plant Cell Biology on DVD, 2007.)

The central role played by mitochondria in cell metabolism is detailed in Chapter 6. They are the sites of cellular respiration, oxidizing pyruvate to carbon dioxide and water. The energy released is used to synthesize ATP, and the reducing power obtained from the oxidation process is in part used to synthesize NADH.

The link to energy production is reflected in the distribution of mitochondria in the cell. Most cells have energy requirements uniformly distributed throughout the cytoplasm, and the mitochondria are also uniformly distributed. Some cells have asymmetric functions, e.g. the active uptake or secretion of ions at one face of the cell. In these cells mitochondria are found clustered at this face, in close proximity to ATP-utilizing ion pumps in the plasma membrane.

Peroxisomes house vital biochemical pathways for many plant cell processes

All plant cells contain peroxisomes, also known as microbodies. These are single membrane-bound organelles with a dense matrix, which frequently contain a crystal lattice structure. The crystals are catalase protein, which promotes the rapid breakdown of peroxides to water and oxygen; therefore, they are oxygen-producing organelles. Peroxides are extremely harmful by-products of a number of biochemical pathways in plants; during the course of evolution, the pathways that produce them have become located in these discrete bodies where they can be contained and destroyed. Three of the main types are described here.

Peroxisomes in green leaves (Figure 3.16) are, by virtue of their positioning, associated with the chloroplasts and mitochondria and are an integral part of the photorespiration pathway (Chapter 6). Mutant plants lacking catalase are killed by exposure to normal light levels, they can only survive in very low light, or under conditions where photorespiration is avoided (e.g. by exposing leaves to an elevated ratio of $CO_2:O_2$).

Peroxisomes are involved in the mobilization of lipid reserves in seeds, as part of the glyoxylate cycle (see Chapter 9), in which case they are often termed glyoxysomes. In cotyledons of epigeous seeds that store lipids, whose cotyledons come above ground and turn green, the glyoxysomes function

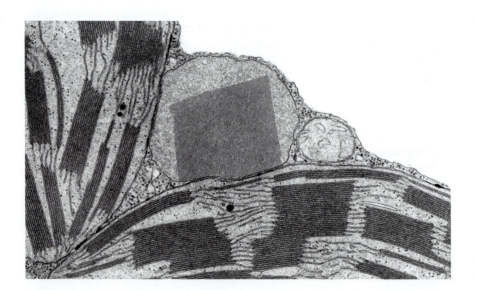

Figure 3.16 A peroxisome in a tobacco leaf cell, with a prominent catalase crystal. The peroxisome is physically associated with two chloroplasts and a mitochondrion, facilitating exchange of metabolites involved in the photorespiration pathway. (Reprinted with permission from Gunning BES, Plant Cell Biology on DVD, 2007. Original image reprinted with permission of the author Dr. S.E. Frederick, "Cytochemical Localization of Catalase in Leaf Microbodies (Peroxisomes)"; *Journal of Cell Biology*, Volume 43: 343, Figure 6, Rockefeller University Press. Reprinted with kind permission from Rockefeller University Press.)

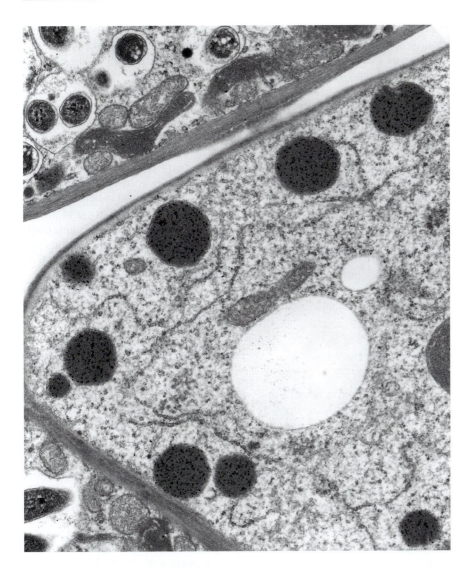

Figure 3.17 Section of a soybean root nodule. The cell to the left contains bacteria (circular dark profiles), which release ammonia fixed from atmospheric nitrogen. This is converted to uric acid by the mitochondria and plastids lying near the wall. The uric acid moves into the adjacent uninfected cell (main part of this picture) where it is converted to ureide by uricase in the peroxisomes. This section has been immunolabeled with an anti-uricase antibody labeled with gold particles. The black dots of gold are located exclusively over the peroxisomes. (Reprinted with permission from Gunning BES, Plant Cell Biology on DVD, 2007. Original image reprinted with permission from Springer Science and Business Media. *Planta* 167: 425, Figure 1; 1986; K.A. Bosch and E.H. Newcomb, "Immunogold localization of nodule-specific uricase in developing soybean root nodules.")

switches at the end of lipid mobilization to that of photorespiration and they become peroxisomes.

In root nodules of nitrogen-fixing plants, peroxisomes are part of the pathway that transfers nitrogen compounds from the endosymbiotic bacteria to the host plant. Uric acid produced in the infected cells (Figure 3.17) is converted to ureides by uricase enzymes housed in the peroxisomes of the adjacent uninfected cells.

Plastids

Plastids serve a variety of functions in plants depending on their location and tissue. A distinguishing feature of all plant cells is that they contain plastids. These organelles may be large, conspicuous and pigmented green or other colors, or they may be small, colorless and inconspicuous. They are surrounded by a plastid envelope consisting of a smooth outer membrane and an inner membrane, which has some folds projecting into the inner space or stroma. Internally, there are further stroma membranes, which may be very extensive, a nucleosome containing DNA arranged as closed loops without histones, ribosomes and other factors required for protein synthesis. This organization is clearly similar to that given above for mitochondria; the same arguments also apply for believing that plastids evolved from symbiotic associations of eukaryotic cells, containing mitochondria, with cyanobacteria capable of photosynthesis.

Plastids have a variety of functions in plant cells, depending on the cell type and tissue. Meristematic cells contain plastids that lack distinctive morphological features (Figures 3.1 and 3.7). This has led to them being referred to as proplastids, the juvenile progenitors of all plastids seen in mature tissues, with more apparent specialized internal structures. However, there is evidence that these plastids serve an important role in lipid synthesis; from a purely morphological examination they may well be more specialized than we suspect. Proplastids in roots play a crucial part in the reduction of nitrate to ammonia, and in the assimilation of ammonia to organic nitrogen compounds (see Chapter 8). As meristematic cells differentiate, the proplastids undergo a transition in their structure and properties as the cells mature into various tissues. Such transitions are well documented between other plastid types, so there is no good reason for regarding these morphologically simpler plastids as being especially juvenile.

Chloroplasts, the visually dominant plastids that play a critical role in life

Vegetation is green because of the presence of the green pigment chlorophyll in the chloroplasts of photosynthetic tissues. These plastids trap sunlight energy, which forms the energy basis for nearly all the life forms on this planet. Sunlight energy is used to synthesize ATP, and the reducing power obtained from splitting water is used to synthesize NADPH (Chapter 4). Both these compounds are used to support carbon fixation reactions in the stroma, known as the Calvin cycle (Chapter 5), which results in the conversion of carbon dioxide into carbohydrates.

The morphology of chloroplasts has attracted most attention, owing to their role in photosynthesis (Figure 3.18). The green color of leaves, stems, unripe fruits, and other plant organs is due to the presence of chlorophyll pigments in chloroplasts. The pigments are responsible for trapping light energy, and they are located within membranes that are organized to maximize the surface area available for energy trapping. Morphologically and functionally there are two classes of internal chloroplast membranes. The stroma lamellae are organized as a series of continuous almost flat sheets lying parallel to

Figure 3.18 Section of a chloroplast with a small starch grain near the center. The system of stroma lamellae and grana stacks is evident. The stroma contains numerous darkly stained ribosomes. (Reprinted with permission from Gunning BES, Plant Cell Biology on DVD, 2007.)

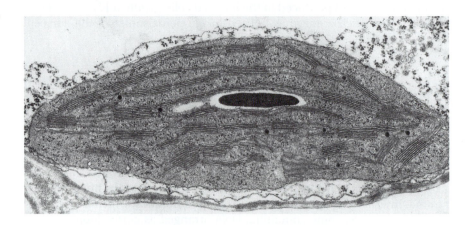

each other and to the base of the chloroplast. The grana consist of a stack of membrane discs, the thylakoids, which adhere to each other (Figures 3.18 and 3.20). The number of thylakoids in a stack is variable between tissues, species and plant growth conditions. The minimum number is two, and there can be as many as 20 or more discs in a stack. The internal lumen of each thylakoid is in continuity with the lumen of at least one of the stroma lamellae, hence all the internal membranes have a single continuous inner lumen space.

Chloroplast formation can involve an intermediate step, etioplast formation, if it occurs in the dark

The development of chloroplasts in meristematic tissues and differentiating cells of the stem apex has been followed in detail. Development in daytime (light) differs considerably from that at night (dark). In daytime the final steps of chlorophyll biosynthesis are catalyzed by light, which converts protochlorophyll, a yellow pigment, to chlorophyll. At the same time the internal membrane system expands, with flat disc-shaped structures, the thylakoids, growing out on to the surface of the stroma lamellae. These discs proliferate successively one on top of the other to form stacks of thylakoids, called grana.

In the dark, protochlorophyll accumulates and the resultant plastid is yellow and is termed an etioplast (remember that plants grown in the dark have excessive stem growth but limited expansion of leaves, they are etiolated). The stroma membrane grows as a succession of branched tubular structures interconnected with each other in three dimensions to form large semi-crystalline bodies, the prolamellar bodies (Figure 3.19). The name indicates what happens to these structures as daylight returns. The protochlorophyll is converted to chlorophyll and the tubules in the prolamellar bodies coalesce to form extensive flat sheets of stroma lamellae and thylakoids. Prolamellar bodies seem to be a mechanism for permitting the continued growth and storage of certain membrane components in the dark. In the light, these membranes are populated by chlorophyll protein complexes.

Successive blebbing of the stroma lamella to form thylakoids ensures that the space enclosed by this membrane system is continuous throughout the chloroplast. That is, the lumen of the stroma lamellae and the lumen of all the thylakoids in all the grana is one interconnected space. As we shall see in Chapter 4, this membrane and the enclosed space are critical structural elements in the process of photosynthesis. Individual thylakoids make connections to several stroma lamellae, as may be seen in favorable thin sections viewed by transmission electron microscopy or in freeze-etch preparations.

Chloroplast membranes are regionally differentiated into stacked and unstacked regions

Freeze-etch preparations of chloroplasts reveal a complex pattern of membrane particles of various size categories and various distributions across different membrane faces. Membranes of the stroma lamellae are referred to as unstacked, while those of the grana discs are termed stacked; therefore there are four possible freeze-etch views of chloroplast inner membranes, PF-unstacked and PF-stacked and EF-unstacked and EF-stacked. Correlation with the biochemical analysis of membrane fractions supports an identity between specific membrane particles and particular components of the photosynthetic machinery. Large photosystem II particles are located mainly on

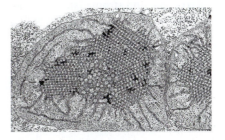

Figure 3.19 Section showing parts of two etioplasts in a leaf that was grown in the dark. Each contains a large pseudocrystalline membrane structure, the prolamellar body. (Reprinted with permission from Gunning BES, Plant Cell Biology on DVD, 2007.)

the EF-stacked face of the thylakoid discs while smaller particles indicative of the presence of photosystem I and the coupling factor protein are found on the PF-unstacked face of the adjacent stroma lamellae (see Figure 3.6). The photosystem II particles are so large that they project through and on to the inner surface of the lumen membrane as a series of bumps. In some plants, and under certain conditions, these bumps can be arranged in a regular pattern, which gave rise to the former term quantasome.

Considerable variation is found in the extent of grana formation in different chloroplasts

The extent of grana development is very variable between species, and in the same species grown under different lighting conditions (Figure 3.20). In shade plants the number of thylakoids in each granum stack can be very numerous, in comparison with those in the same plant grown in the open and exposed to higher light intensities. There are also differences between tissues of the same leaf. This is particularly noticeable in plants with Kranz leaf anatomy, which have the C_4 mechanism of photosynthetic carbon fixation (see Chapter 5). The cells of the bundle sheath have chloroplasts that tend to lack grana, they just have parallel sets of stroma lamellae. In contrast, the cells of the mesophyll have chloroplasts with the arrangement of grana stacks and stroma lamellae usually found in C_3 plants (see Figure 5.16). The significance of these different arrangements reflects the different roles of chloroplasts of C_4 plants in these two tissues. Lack of stacked membranes implies a reduced level of photosystem II components.

Chromoplasts impart colors to flowers, fruits, and some roots

Plastids can accumulate a variety of water-insoluble pigments that impart color to certain flowers, fruits and some roots, notably carrot, which gave its name to the pigment carotenoid. Chromoplasts provide flower colors that are typically yellow (e.g. daffodil and members of Cucurbitaceae: marrows, cucumbers, zucchini, melons), or red/orange (narcissi). Fruit ripening often

Figure 3.20 Mature chloroplast with prominent grana and interconnecting stroma lamellae. The small dark circles are sections through stained oil droplets. (Reprinted with permission from Gunning BES, Plant Cell Biology on DVD, 2007. Original image from the work of the late A.D. Greenwood, courtesy of Professor J. Barber, Imperial College, London.)

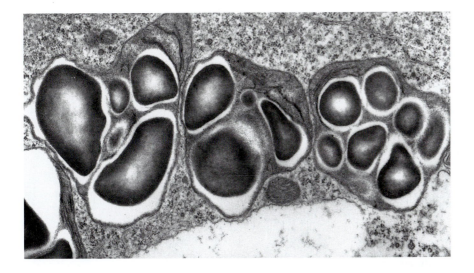

Figure 3.21 Amyloplasts, each with several prominent starch grains. (Reprinted with permission from Gunning BES, Plant Cell Biology on DVD, 2007.)

involves a transition from green to yellow, orange or red colors, which signify the conversion of chloroplasts to chromoplasts in fruits such as peppers, tomatoes, oranges, lemons, and grapefruit. The chlorophyll pigments are degraded and replaced by lycopenes, xanthophylls, and carotenes.

Amyloplasts store starch in a variety of tissues and organs

Storage tissues in seeds, tubers, bulbs, corms, stems, and roots contain numerous amyloplasts, which contain single or multiple starch grains (Figure 3.21). These are polymers of glucose, which, because they are insoluble, allow storage of high levels of sugars without imposing an unacceptable osmotic load on the cells (Chapter 7). The starch grain(s) take up most of the plastid volume, confining the stroma to a narrow layer against the envelope. Other plastids frequently contain small starch grains as a minor component of their stroma. Amyloplasts are of great importance as a food source for animals, and much research is devoted to starch chemistry in the food industry (Chapter 7).

Summary

The interior of the cell is structurally diverse, reflecting a diversity of biochemical activity. This chapter has introduced the structural complexity of plant cells. The compartmentation found in plant cells allows numerous biochemical reactions to occur simultaneously in a confined space, without mutual interference. We have seen that membranes not only separate these various compartments, but that they provide working surfaces, by virtue of their specific membrane proteins. The following chapters deal with the major biochemical reactions and pathways that support plant life. Mostly these are located inside the living cells, and often inside specific organelles. It should be emphasized here that, although extensive descriptions of plant cell structure and function are available, relatively little is known about the precise mechanisms that govern the formation and maintenance of cells.

Further Reading

Buchanan B, Gruissem R & Jones R (eds) (2000) Biochemistry and Molecular Biology of Plants. American Society of Plant Physiologists.

This book contains a chapter on plant cell structure and chapters on many aspects of plant cell function.

Gunning BES & Steer MW (1996) Plant Cell Biology: Structure and Function. Jones and Bartlett

Consult this for a collection of light and electron micrographs with accompanying text descriptions of plant cell structure.

Gunning BES (2007) Plant Cell Biology on DVD. Available from the author (brian.gunning@anu.edu.au and www.plantcellbiologyonDVD.com)

This is an invaluable DVD, with numerous very high-quality micrographs of a great range of plant cell types from light, confocal and electron microscopes and incorporating a large number of videoclips of living plant cells and computer-generated rotations of cells and cell components.

Hawes C & Satiat-Jeunemaitre B (eds) (2002) Plant Cell Biology, 2nd ed. Oxford University Press.

An excellent overview of modern methods for exploring plant cell structure and function.

Light Reactions of Photosynthesis

4

Key concepts

- Early cells evolved chlorophyll photosystems capable of capturing light energy to supply the energy needs of the cell.

- Charge separation occurs at a chlorophyll molecule in the light reaction center of the photosystems, with the reduction of an acceptor molecule, leaving the chlorophyll in an oxidized state.

- Light harvesting chlorophyll proteins extend the antenna area for collecting photons and channeling energy to the photosystems.

- Two photosystems evolved: one uses water to supply electrons that reduce the oxidized chlorophyll (with the loss of oxygen), the other uses electrons derived from reduced plastocyanin supplied from the first center to reduce $NADP^+$ to NADPH.

- Electron flow between the two centers occurs via a chain of intermediates assisted by a cytochrome protein complex.

- The splitting of water releases protons into the lumen of the photosynthetic lamellae, and more protons are pumped in during the successive reduction and oxidation steps of the intermediates, with the hydrogen atom carrier plastoquinone playing a crucial part.

- The high proton concentration gradient across the lamella membrane is used to drive a molecular turbine, ATP synthase, which generates ATP.

- Mechanisms exist that balance light energy input into the two photosystems and protect the cell from the consequences of the capture of excess light energy.

Bacteria evolved the basic photochemical pathways found in plants today

Photosynthesis has been pivotal in the development of complex life forms on this planet. This process converts light energy into chemical energy that is used to support cellular processes and to provide the basic raw materials from which cell structures are made. In addition, photosynthesis releases oxygen from water as a by-product. The oxygen has accumulated in the atmosphere, transforming the early anoxic conditions on our planet and enabling the evolution of terrestrial life forms. The metabolism of eukaryotes is invariably aerobic. This chapter will concentrate on the photochemical processes involved in photosynthesis, while Chapter 5 will deal with the subsequent chemical events involved in carbon fixation from atmospheric carbon dioxide. First, the basic features of the photochemical processes found in plants and their evolutionary origins will be reviewed. The key to the success of photosynthesis lies in the rapid separation of electrically charged products after an initial light reaction and the subsequent efficient formation of stable products that can be used to drive anabolic reactions (molecule-building reactions) in the cell. The section on plastids in Chapter 3 should be consulted before reading this chapter.

In the initial evolution of life and cellular entities, sources of chemical energy were exploited to facilitate the synthesis of more complex molecules. Initially, relatively simple chemical molecules that were available in the anoxic environment of that time served this purpose. Analogous activities are found today in organisms living in the vicinity of marine hydrothermal vents and other geothermal environments. As with all biological systems, relatively simple single chemical steps evolved, and were combined to give sequential reactions (pathways) that conferred a selective advantage in the evolutionary struggle for resources, and hence survival. It is probable that these cells evolved a high level of capability, providing the forerunners of today's organisms.

Over time (hundreds of millions of years), the chemical energy sources became depleted, providing a strong selective advantage for the evolution of organisms capable of using alternative energy sources. One pervasive energy source is sunlight. Light energy at more energetic, shorter, ultraviolet (UV) wavelengths probably promoted the synthesis of carbon and nitrogen compounds from simple precursors, such as carbon dioxide, methane, water, and ammonia. These compounds would then have been utilized by living organisms. Although the high energy quanta of UV light are favorable for generating small molecules, they are far too energetic to be used by cells for this purpose. Instead cells arose containing pigment molecules that could safely absorb and utilize light energy in the slightly longer wavelength, visible part of the spectrum; this gave them a survival advantage over their competitors.

Specifically, bacteria evolved that could use the energy trapped from light to form reduced compounds and high energy intermediates, which were then utilized in anabolic processes. The earliest such organisms used chemical sources of reducing power, for example purple photosynthetic bacteria used ferrous ions (Fe^{2+}), which were oxidized to insoluble ferric (Fe^{3+}) salts. These were deposited as iron bands in sediments. Green sulfur bacteria used hydrogen sulfide (H_2S) as a source of electrons. Elemental sulfur was discarded as a waste product leading to the creation of the large deposits of sulfur that are mined today. These organisms probably evolved early in the history of the evolution of life, about 3×10^9 years ago. They left recognizable signals in the geological record consistent with carbon fixation (reduction of atmospheric carbon dioxide to organic molecules).

These early photosynthetic systems were not sustainable on a planet-wide scale, because of limitations in the local availability of appropriate chemicals, such as H_2S. Evolution of variant forms that could utilize other electron donors was therefore favored. About 3.5×10^9 years ago cyanobacteria evolved that were able to utilize water (H_2O). Water is chemically similar to H_2S, and, as a source of electrons, it is clearly far more readily available on this planet. Chlorophyll evolved as the light-absorbing pigment, and the energy trapped was used to power ATP synthesis, while NADPH was made as the stable reduced end product. Oxygen was released as a waste product and started to accumulate in the atmosphere about 2.7×10^9 years ago.

The appearance of free oxygen, a very reactive molecule, led to dramatic chemical changes on this planet. Soluble ferrous iron salts were oxidized to insoluble ferric compounds that were laid down in vast beds of iron ore, mined today for the steel industry. Accumulation of molecular oxygen in the atmosphere shielded living organisms from the damaging effects of UVC. In addition, some of this oxygen was photochemically converted to ozone forming a layer in the upper atmosphere that provided protection against UVB radiation for the emerging life forms below. Loss of UV radiation at the surface of the planet meant that abiotic photochemical production of organic compounds was greatly reduced; therefore, the conditions that had led to the initial evolution of life were irreversibly changed.

Organisms unable to cope with the presence of oxygen, a strong oxidant, were forced to live in anaerobic environments, such as deep muds, where similar forms can be found today. Some, the aerobic organisms, evolved mechanisms to take advantage of the presence of oxygen to power oxidation processes that released all the available chemical energy in organic compounds. These were oxidized to water and carbon dioxide in a process that we call cellular respiration (Chapter 6).

How does photosynthesis fundamentally work? Trapping of photons by chlorophyll molecules raises some of their electrons to a higher energy level. These would naturally return to their ground state condition within nanoseconds (10^{-9} s), releasing the energy as heat and photons emitted as fluorescence at a longer (lower energy level) wavelength. The trick evolved by photosynthetic organisms is to move this high energy electron to an acceptor molecule on a much faster, picosecond (10^{-12} s), time scale. The acceptor molecule is thus reduced and the chlorophyll is left in a photo-oxidized state.

Two sets of reactions and components (two photosystems) evolved in cyanobacteria to take advantage of this momentary charge separation. In one photosystem, electrons had to be supplied to the oxidized chlorophyll, to replenish (reduce) the pigment molecule for the next photon excitation event, and to minimize the chance of a backflow from the acceptor molecule. This was achieved by the evolution of a water-splitting, oxygen-evolving, center that sequentially provided electrons to a pathway culminating at the chlorophyll molecule. Then the acceptor molecule had to rapidly pass on an electron, again to minimize the risk of backflow, and to enable the acceptor to receive the next electron from chlorophyll. These two sets of reactions were kept physically separated, to avoid short-circuiting the whole system, by arranging them in a linear sequence across the width of a lipid bilayer membrane.

This electron flow from chlorophyll to acceptor resulted in the formation of a stable reduced form of a copper-containing protein called plastocyanin. Plastocyanin is a relatively large molecule, whose reducing power is inaccessible for cellular metabolism. However, its utility lies in the relative ease with which the reduced form can donate electrons (much easier than water). This allows a second light-trapping chlorophyll photosystem, also located across a membrane, to pump these electrons along a carrier chain to eventually reach and reduce NADPH. NADPH is a relatively small molecule that can readily pass on its reducing power in cellular metabolism. Two systems are required because a single light reaction does not yield enough energy to extract electrons from water and move them through the entire pathway to the formation of NADPH.

Crucially, the photosynthetic membranes enclose a lumen, an internal space separated from the stroma outside. Both of these light reactions, and some of the intermediate steps, result in the accumulation of protons (H⁺) within the lumen. The concentration gradient of protons established across this membrane is then used to drive a molecular turbine, located in these membranes, which synthesizes ATP.

It is probable that the two photosystems evolved independently. The oxygen-evolving photosystem could have been derived from purple and green non-sulfur bacteria, while the NADPH-forming photosystem probably originated from green sulfur bacteria. These were combined in the cyanobacteria (presumably by lateral gene transfer) which were the first organisms to use the oxidation of water to power ATP and NADPH formation with the release of free oxygen (Figure 4.1).

Symbiotic association between these bacteria and eukaryotes led to the evolution of the eukaryotic plant cell with chloroplasts. Today the dual photochemical

Figure 4.1 Possible evolutionary pathways for the two photosystems (RCI and RCII) and pigments found in cyanobacteria (bacterial chlorophylls bch*a* and bch*g*). Lateral transfer of genetic material for both pigment synthesis and photosynthetic membrane protein complexes must have occurred at several stages. (Reprinted with permission from the *Annual Review of Plant Biology*, Volume 53, © 2003 by Annual Reviews, www.annualreviews.org.)

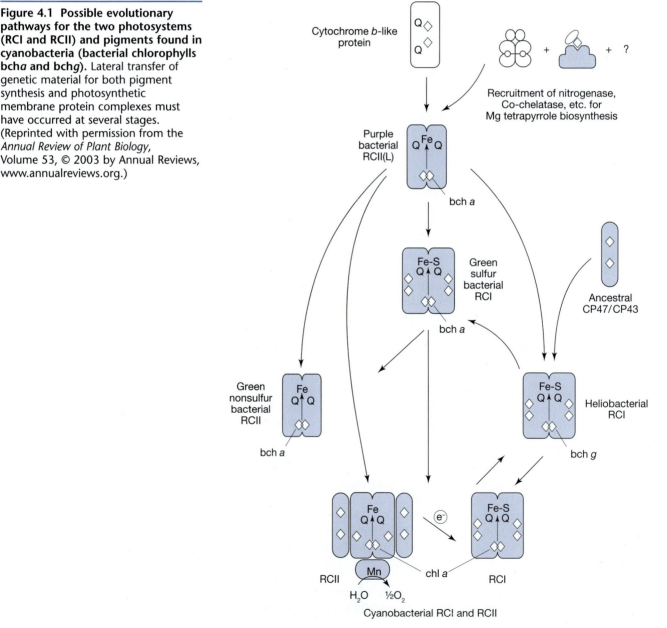

system of terrestrial, aquatic, and marine plants traps a significant proportion of the solar energy striking the planet, using it to form stable chemical products that have supported the existence and evolution of most of the life forms found on Earth today.

Remarkably, comparison of the structures of the two photosystems from higher plants with those from cyanobacteria show that over one thousand million years of evolution has resulted in very few changes. Clearly, the supposed primitive ancestors of today's plants had, in fact, evolved a truly efficient process that has stood the test of time. This account concentrates on the structure and function of higher plant photosystems as far as they are known at present. The major advance that has occurred during evolution is the development of light harvesting chlorophyll (LHC) proteins that greatly increase the area of the receiving antenna for each photosystem. This development has characterized the few changes that are found in the photosystems of modern land plants. Some polypeptides have been lost to make way for the effective docking of these new chlorophyll proteins, and others have evolved to assist the docking process. Across the range of land plants there is a remarkably small level of variation in the structure of their photosystems and the amino acid sequences of their constituent proteins. In addition to chlorophylls, other pigments, i.e. carotenoids, have been added to these proteins, which extend the range of light wavelengths that can be trapped for photosynthesis. These developments have ensured that plant leaves are able to intercept and utilize a high proportion of the visible light energy that is incident on the leaf surface.

In summary, photosynthesis starts with two linked photochemical systems that trap light energy. Water is split to release oxygen and electrons, which are ultimately used to make the reduced form of NADPH. Along the way protons (H^+) are accumulated in the grana lumen, creating a charge separation across the photosynthetic membrane that drives the ATP synthesis system.

When placed in their physical context (Figure 4.2) the two photosystems are seen to lie in the plane of the photosynthetic membrane. Both pass electrons via carrier intermediates from the lumen side to the stromal side of the membrane. The first photosystem, called Photosystem II (PSII; because it was the second to be discovered), releases protons and oxygen within the granal lumen. PSII reduces the small molecule plastoquinone (PQ) to plastoquinol (PQH_2), which is formed on the stromal side of the membrane. PQH_2 is a small lipophyllic molecule that diffuses in the plane of the membrane to a transmembrane cytochrome $b6$ complex where it enters the Q cycle. This cycle produces reduced plastocyanin on the lumen side while also pumping protons from the stroma into the lumen. Reduced plastocyanin donates

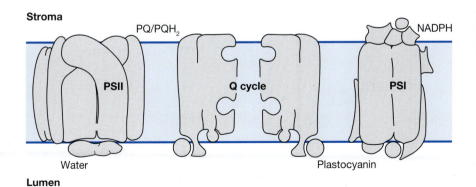

Stroma

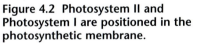

Water

Lumen

Figure 4.2 **Photosystem II and Photosystem I are positioned in the photosynthetic membrane.**

electrons to the second photosystem, Photosystem I (PSI; discovered first), which ultimately leads to the formation of reduced NADPH.

These reactions can be considered in terms of the redox potentials of the component redox couples, as each component can exist in either the reduced or the oxidized states. These drive the reactions energetically downhill from more negative potentials to more positive ones. A plot of the standard redox potentials of the components (Figure 4.3) shows that the pathway from water to reduced PSII goes from +1.1 to −0.7 electron volts (eV). The potentials of the PSII to plastocyanin pathway decline from −0.7 to +0.5 eV. PSI is boosted to −1.3 eV on light activation, and then the pathway declines to −0.4 eV with the formation of reduced NADPH. This plot is referred to as the Z scheme because of its characteristic shape, and was first proposed by Eugene Rabinowitch in 1945 and confirmed by his experiments in 1956 and 1957; and by the work of Robin Hill with Derek and Fay Bendall. Rabinowitch is also famous for his role in the development of the atomic bomb and for his campaign to the American Government for the development of peaceful uses for atomic energy. We will return to the Z scheme in more detail in a later section.

The following sections provide details of the structure and function of the two photosystems, the cytochrome complex, and the ATP-generating complex (ATP synthase). It must be emphasized that there is much that we still do not understand about photosynthesis. This is a hot research area with many papers being published each year exploring such basic issues as the molecular structure of the components, their functions, and the kinetics of the processes involved. The usual introductory textbook account of photosynthesis is a simplification that, in places, hides uncertainty and ignores fascinating, if perplexing, details.

Figure 4.3 The Z scheme summarizing the electron flow from water to NADPH plotted on the redox potential scale. The components are plotted according to the redox potentials of the component redox couples.

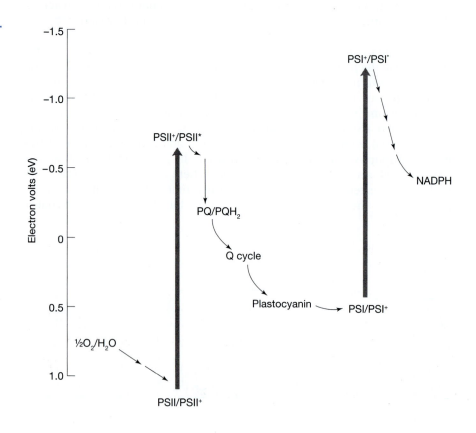

Chlorophyll captures light energy and converts it to a flow of electrons

Chlorophyll pigments have absorption maxima in the blue and red bands of the visible light spectrum and thus reflect and transmit green light. Engelmann and Sachs first discovered the dependence of photosynthesis on chlorophyll and Emerson and Arnold (in 1932) deduced that several hundred molecules of chlorophyll (a photosynthetic unit) are required for the production of one molecule of oxygen. Later, 1939, Robin Hill, working with isolated chloroplasts, demonstrated the direct connection between the light reaction steps and the release of molecular oxygen. This Hill reaction was extensively studied in both isolated chloroplasts and in whole algal cells by Otto Warburg. Eventually (1957) Emerson's work revealed that two photosystems, bearing chlorophylls of slightly different absorption properties, are involved in the photosynthetic process.

Evolution has favored the conservation of a very precise molecular configuration, so we must assume that all elements of the chlorophyll molecule are of critical importance for its functioning. Each chlorophyll molecule (Figure 4.4, see Chapter 12 for synthesis of this pigment) has a hydrophobic phytol tail attached to a hydrophilic head, which is a tetrapyrrole ring (porphyrin, more properly termed chlorin) containing a single magnesium atom (analogous heme rings contain an iron atom).

Small variations in the head pyrroles lead to formation of different chlorophyll molecules with slightly different light absorption characteristics (Figure 4.5). Analysis of chlorophyll extracts from leaves shows that chlorophyll *a* is the main form present (main absorption peak 665 nm), with chlorophyll *b* making up about 15% of the total. Chlorophyll *b* has a formyl (–CHO) group on the porphyrin ring in place of the methyl (–CH$_3$) group in this position in chlorophyll *a*. In some algal groups chlorophyll *b* is replaced by chlorophyll *c* (lacks a phytol tail) or chlorophyll *d* (has a formyl group in the ring in place of the vinyl group found in chlorophyll *a*). Chlorophyll *d* has its main absorption peak at longer wavelengths than chlorophyll *a*, beyond the visible light spectrum in the infrared (700 nm instead of 665 nm). It is found in cyanobacteria that grow as epiphytes under the fronds of red algae, and in other places, such as the undersides of didemnid ascidians of coral reefs. These are habitats where visible light wavelengths have been depleted leaving a light spectrum relatively rich in near infra-red radiation. In these cyanobacteria chlorophyll *d* replaces chlorophyll *a* in at least one of the photosystem reaction centers.

The phytol tail is a hydrophobic terpene chain 20 carbons long. The tails are identical in all forms of chlorophyll and in bacteriochlorophyll. The tails associate with a series of specific membrane proteins located in the thylakoid and stroma lamellae (frets) of chloroplasts (and with the internal membranes of cyanobacteria). Hence nearly all chlorophyll molecules are part of specific membrane protein complexes.

Two of these membrane protein complexes are the light reacting centers, Photosystems II and I (PSII and PSI) first detected by Emerson in 1957. PSI is involved in the second light reaction in the electron flow sequence. Most of the chlorophyll molecules are associated with proteins of the LHC protein complexes. These occupy most of the internal membrane surface area in the chloroplast, providing an efficient light-trapping system. See Box 4.1 for details of the structure of LHC protein complexes.

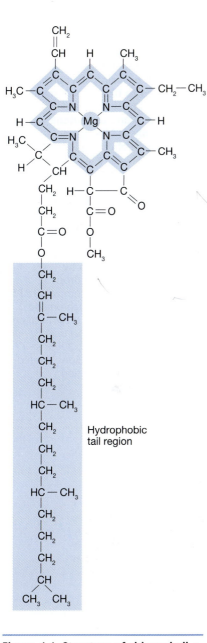

Hydrophobic tail region

Figure 4.4 Structure of chlorophyll *a*. The tetrapyrrole porphyrin head has a single magnesium atom at its center. Electrons are shared between the atoms of this ring, making it less difficult for one to be temporarily lost from the structure. The hydrophobic phytol tail associates with specific sites in the membrane proteins. (From Alberts et al., Molecular Biology of the Cell, 4th edition, New York, NY: Garland Science, 2002.)

Figure 4.5 Different pigments have different light absorption spectra, as shown here for chlorophyll *a*, chlorophyll *b* (main chlorophyll *a* peak shifted to blue green, minor peak shifted to shorter red wavelengths) and a carotene (absorbs only at blue–green end). A combination of these pigments provides a leaf with a fairly broad absorption spectrum. This is closely matched by the action spectrum showing that the energy absorbed at each wavelength is efficiently utilized in photosynthesis. Deviations from close matching of action and absorption spectra are attributable principally to some inactive absorption by carotenoids and the red drop wavelength around 700 nm, where PSII is not activated by monochromatic light at this wavelength. (From *Cell and Molecular Biology: Concepts and Experiments*, Gerald Karp, 2003. Copyright John Wiley & Sons Limited. Reproduced with permission.)

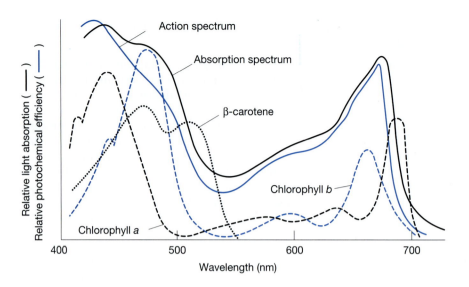

Light photons from the sun possess different energy levels, depending on the frequency or wavelength. Shorter wavelength photons (blue, 450 nm) have quanta with 1.5 times the energy level of longer wavelength photons (red, 675 nm). When a chlorophyll molecule absorbs a red wavelength photon, an electron in the porphyrin head is raised to a higher energy level (an excited singlet state, state 1). Blue light photons will raise an electron to a still higher energy level (state 2), but this state is very unstable and the electron rapidly descends to the state 1 level, with the loss of energy as heat. In free chlorophyll molecules, electrons would return from the excited state 1 to the ground state on a nanosecond (10^{-9} s) time-scale, losing energy by emitting a photon (fluorescence) at a lower energy (longer wavelength) level. When bound in a membrane protein, the excited electron is passed on to an acceptor molecule on a picosecond (10^{-12} s) time-scale. Excited state 1 singlet electrons may be temporarily unable to return to the ground state, so they are trapped at a slightly lower energy level in the triplet state. In this state the spin orientation of electrons is aligned (parallel), in contrast to state 1 electrons, which have opposite (antiparallel) spins. When these triplet electrons decay to the ground state they emit phosphorescence. This occurs later than fluorescence and is at a slightly longer wavelength than fluorescence, reflecting the lower amount of energy released. Delayed fluorescence can also take place, but in the living chloroplast fluorescence is very low compared with isolated chlorophyll in extracts, because so many excited electrons are trapped and do not return to the ground state.

The energy from excited state 1 electrons can be passed on to other pigment molecules (energy transfer), or used to reduce an acceptor molecule, converting the energy gained from light into a chemical product (photochemistry). Only a small fraction of the chlorophyll molecules is directly involved in photochemistry, the conversion of light energy to chemical energy, by reducing an acceptor molecule. As this event happens rapidly (about 1 ps) it is not a rate-limiting step. Acquisition of the appropriate photon is a much slower event, so a large number of chlorophyll molecules collaborate as an antenna, trapping photons and passing their energy on by resonance energy transfer to the core chlorophyll molecules. This ensures a high throughput of electrons to the acceptor. In fact the placement of chlorophyll molecules in the antenna molecules is so precise that there is almost 100% quantum efficiency of trapping.

Box 4.1 Structure of light-harvesting chlorophyll protein complexes and their role in photosynthesis

Most of the chlorophyll molecules in a plant are associated with light-harvesting chlorophyll protein complexes (LHC). These are divided into two main classes, those associated with Photosystem I (PSI), which are known as LHCI, and those predominantly associated with Photosystem II (PSII), termed LHCII. These proteins are very similar to each other. Both have four helices (A, B, C, and D); with A and B being closely associated transmembrane helices, C also lying perpendicular and D lying parallel to the membrane. LHCI chlorophyll pigments have absorption maxima at longer wavelengths than those of LHCII.

LHCI protein molecules contain polypeptides from a family of four 25 kDa proteins, Lhca 1–4. These bind a total of 56 chlorophylls with an a–b ratio of 3.5:1. LHC1 680 consists of homodimers of Lhca 2 and 3, while LHC1 730 consists of heterodimers of Lhca 1 and 4. Light of up to 750 nm can be used to oxidize the P700 reaction center. This high concentration of pigment molecules leads to red-shifting of their absorption spectra, enabling light harvesting to occur in dense vegetation canopies where the remaining light is richer in longer wavelengths (above 680 nm).

There have been many attempts to define the relationship between these four proteins and PSI. Some models, partly based on examination of bacterial photosystems, have four dimers (eight molecules) of Lhca associated with each PSI complex. Recent models based on higher plant material have one molecule each of all four. These form a semicircular array around one side of the PSI complex, in the sequence Lhca1 with Lhca4, and Lhca2 with Lhca3. These molecules are arranged to maximize the contact between their chlorophylls and PSI, with the D helix of one pointing towards the C helix of the next around the semi-circle from Lhca1 to Lhca3.

The main link to PSI (Box 4.4) appears to be via Lhca1 to PsaG, with the C helix interacting with two helices in PsaG. There is a weaker link from Lhca3 to PsaK at the opposite side of the semicircle (Figure 1). This arrangement facilitates changes in the precise composition of the array. There are slight differences between the four polypeptides and their chlorophylls, which affect their light absorption properties, so different combinations are optimal for different environmental conditions.

Lhca polypeptides bind 14 chlorophyll *a* or chlorophyll *b* molecules. Chlorophyll molecules also act as linkers, both between the Lhca polypeptides (two between each pair in a dimer, one between the dimers) and between these polypeptides and PSI (one on each, facing the reaction center). These have distinctive absorption spectra and serve to pass excitation energy between the polypeptides and to PSI. The absorption peaks of LHCI are at slightly shorter wavelengths (680–700 nm) than PSI (700 nm), facilitating energy flow to the reaction center.

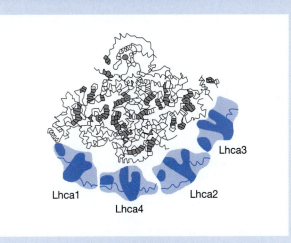

Figure 1 Four LHCI proteins form a semi-circle around one side of the PSI complex. This arrangement leaves room for LHCII trimers to bind to the other side of PSI under certain conditions. (Reprinted with permission from J. P. Dekker: *Biochimica et Biophysica Acta* 1706: 12–39, Figures 2A Page 14 and 4A, Page 18; by J.P. Dekker and E.J. Boekema, 2005.)

The LHCII complex contains the major chlorophyll and protein component of internal chloroplast membranes, accounting for more than half of all the membrane proteins and chlorophyll pigments in the plant. This major LHCII protein exists as a trimer of polypeptides, taken from the three polypeptides Lhcb1, Lhcb2, and Lhcb3. Trimers can be of homopolymers of Lhcb1, or heteropolymers of any combination of two or all three polypeptides. These major LHCII polypeptides are not present in equimolar proportions, there is twice as much Lhcb1 as of the other two combined.

Each major LHCII polypeptide consists of eight chlorophyll *a* and six chlorophyll *b* molecules, four carotenoid molecules, two complex lipids and a polypeptide chain that has three membrane-spanning α-helices. Eight chlorophylls are arranged as a layer towards the stromal side of each major polypeptide, and six towards the lumen side. Two central carotenoids are essential for the folding of LHCII polypeptides into a stable complex. A third carotenoid is associated with the chlorophyll *b* molecules. These three carotenoids serve to extend the absorption spectrum of LHCII into the blue–green part of the light spectrum, channeling the excitation energy received to the chlorophylls. The fourth carotenoid is associated with the xanthophyll cycle and the nonphotochemical quenching processes. The lipid phosphatidylglycerol is the only significant phospholipid of thylakoid membranes and is associated with LHCII. An unusual *trans*-fatty acid, 16:1Δ3, is a component of phosphatidylglycerol.

continued

Box 4.1 Structure of light-harvesting chlorophyll protein complexes and their role in photosynthesis (continued)

In the trimer the three sets of eight chlorophylls on the stromal side form two concentric rings. The inner six are chlorophyll *a*, acting to transfer energy between monomers, while the outer 18 have alternating three chlorophyll *a* and three chlorophyll *b* around the circle. These serve to collect energy from a broad spectrum and pass it on to the P680 reaction center. Chlorophylls on the lumen side pass their collected energy on to these rings on the stromal side.

There are also three minor LHCII proteins, Lhcb4, Lhcb5, and Lhcb6, which do not polymerize with each other, but form bridges between PSII and the major LHCII components. Lhcb4 and Lhcb6 lie side by side, binding to PSII on one side and to a LHCII trimer on the other (Figure 2). This trimer (M) is less strongly bound to the PSII–LHCII supercomplex than the adjacent trimer (S), which binds directly to the PSII surface beside Lhcb4. Lhcb5 is associated with this strongly bound trimer and the PSII surface. Each supercomplex consists of two PSII complexes (the PSII dimer) and four LHCII trimers, two strongly bound and two more weakly bound. Overall analysis of the total chlorophyll protein complexes shows a ratio of eight trimers per PSII dimer, so it is assumed that only half the trimers are bound specifically to the supercomplexes and the remainder are loosely associated in the surrounding membrane. As PSII is located in the appressed granal membranes, it is thought that supercomplexes in adjacent membranes lie one above the other, permitting excitation energy to flow between LHCII proteins on adjacent membranes of neighboring thylakoids.

Light harvesting by LHCII is assisted by the carotenoids, absorbing at 450–500 nm, passing energy to the chlorophylls, which absorb at 660–680 nm. This absorption

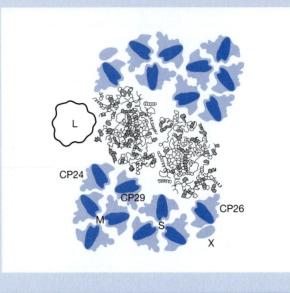

Figure 2 Major LHCII trimers (M and S) and minor proteins Lhcb4 (CP29), Lhcb5 (CP26), Lhcb6 (CP24) bound to a PSII dimer. The S trimer is strongly bound, while the M trimer shows medium binding. L indicates a possible position for a weakly bound trimer. (Reprinted with permission from J.P. Dekker: *Biochimica et Biophysica Acta* 1706: 12–39, by J.P. Dekker and E.J. Boekema, 2005.)

range is at shorter wavelengths than the PSII reaction center (680 nm), so creating a downhill energy flow from the LHC to the reaction center (shorter wavelength light is more energetic).

Carotenoids extend the spectral range of light that can be utilized in photosynthesis

Carotenoids are long chain pigments that absorb blue and green light (Figure 4.5), leaving yellow, orange, and red colors seen in such plant tissues as carrot roots (for details of carotenoid synthesis see Chapter 12). There are two types of carotenoids, carotenes, and xanthophylls. Each is a chain of 40 carbon atoms with alternating single and double bonds. Carotenes have only hydrogen atoms attached to this carbon backbone, but xanthophylls have one atom of oxygen at each end of the molecule. For example, zeaxanthin is the same as β carotene but with a hydroxyl group at each end. Carotenoids are incorporated into the chlorophyll/protein complexes where they perform two functions. They extend the range of wavelength energies (Figure 4.5) that can contribute to photosynthesis by passing on absorbed energy to neighboring chlorophylls. The xanthophyll fucoxanthin in brown seaweeds and diatoms is especially efficient in harvesting blue and green light and passing energy to chlorophylls. Also they provide protection for the reaction centers, dissipating excess energized electrons as heat and preventing the formation of damaging reactive oxygen species (see section Nonphotochemical quenching and the xanthophyll cycle).

Photosystem II splits water to form protons and oxygen, and reduces plastoquinone

PSII is located in the membranes of thylakoids making up the granal stacks. It consists of a core dimer of chlorophyll protein molecules that spans the membrane bilayer (Box 4.2). PSII uses light energy to remove electrons from water releasing protons and oxygen. The electrons are then used to reduce PQ to PQH_2. The core chlorophyll a in PSII has a maximum light absorption peak at 680 nm, and so is called P680 (pigment 680). This passes an electron on to an acceptor chlorophyll-like molecule (pheophytin, lacks magnesium), called A_0. The first photochemical reaction thus leaves the P680 molecule in an oxidized state ($P680^+$) and the acceptor in a reduced state (A_0^-). Two sets of reactions rapidly regenerate the $P680/A_0$ pair in time for the next light reaction. These are discussed on this page, and on page 78; but first we will see how electrons move along an electron transport chain.

All these electron transfer steps in photosynthesis share a common feature. They involve loss of an electron from one component, leaving it in an oxidized state indicated by a plus charge sign; and the gain of the electron by another component, leaving it in a reduced state, indicated by a minus sign. These components lie close to each other in a chain, and each has a redox potential that is slightly higher (more positive) than the previous member of the chain. This dictates the direction of flow of the charge along the chain. Typically, chain members are small molecules or atoms of metallic elements that can exist in a number of valency states. Iron would be a common example, existing in either the reduced state Fe^{2+}, or the oxidized state Fe^{3+}. We shall also meet others, such as copper and manganese. These atoms are held, either singly or in defined clusters, in an organic matrix created by polypeptides. Using iron as an example again, it is often held in place by cross-linking to sulfur atoms of cysteine residues as in Rieske proteins (2Fe–2S) and other non-heme iron proteins that carry electrons. This arrangement allows the gain or loss of an electron to be shared between the coordinately linked atoms.

In one set of reactions, $P680^+$ is reduced by the movement of an electron from an adjacent tyrosine molecule (TyrZ) in the polypeptide chain of the D1 protein of the PSII complex (Box 4.2). The oxidized tyrosine is in turn reduced by the provision of electrons from the oxygen-evolving center.

Electrons are donated from water to replenish those lost by the tyrosine molecule. Two water molecules are split simultaneously to yield one oxygen atom (O_2), four protons (H^+), and four electrons. However, tyrosine can only accept one electron at a time. So the oxygen-evolving center has evolved as a charge accumulation device, consisting of four manganese atoms held in a protein matrix with one atom each of calcium and chlorine (Box 4.2). This center sequentially provides single electrons to the tyrosine molecule, until four electrons have been donated, one from each manganese atom. Then two molecules of water are split simultaneously to replenish the four electrons donated to tyrosine. The protons are released into the lumen of the thylakoid/stroma-lamellar system. Later we will see how these electrons are utilized in ATP synthesis (see section ATP synthase utilizes the proton motive force to generate ATP).

$$2H_2O \longrightarrow 4e^- + 4H^+ + O_2$$

In the dark, the ground state of the oxygen-evolving center is always (+), so that, on illumination, only three photochemical events are needed to bring about the evolution of the first oxygen molecule. Thereafter one is formed on every fourth event. This was first observed by P. Joliot and coworkers in 1969, who illuminated a suspension of algal cells that had been kept in the dark

Box 4.2 Structure of Photosystem II

Photosystem II (PSII) is a major membrane protein complex consisting of over 20 polypeptides and about 250 chlorophyll molecules. It is located in the grana stack thylakoid membranes, where it protrudes on each side of the plane of the membrane, by about 1 nm on the stromal side and 5.5 nm on the lumen side. The boundary lipids associated with the complex provide a specific membrane environment for the hydrophobic photosystem proteins.

Fourteen of the 20 polypeptides are integral membrane proteins with a total of 36 transmembrane α-helices. The complex is bilaterally symmetrical, with a dimer of very similar proteins, D1 and D2, in the center (Figure 1). Each polypeptide has five transmembrane helices and this pair of proteins is flanked by a further pair of strongly related proteins, CP43 (adjacent to D1) and CP47 (adjacent to D2). CP43 and CP47 each consist of six transmembrane helices. CP43 has a large loop on the lumen side that is essential for completing the water splitting manganese ion cluster (Figure 2). One of the manganese ions is bound to a CP43 glutamine residue.

D1 and D2 each contain three chlorophyll molecules, while CP43 contains 14 and CP47 contains 16 chlorophyll molecules. The majority of these chlorophylls are ligated to fully conserved histidine residues in each protein. The heads of the chlorophylls lie in flat planes near each face of the membrane, but each protein has one chlorophyll located midway across the membrane, which acts as a bridge between the chlorophylls on the opposite faces. Two xanthophyll (lutein) molecules are embedded in the complex. This structure has been determined by a combination of spectroscopy and X-ray crystallography of isolated components. Functional analysis has been aided by site-directed mutagenesis, in which individual amino acids in specific proteins have been replaced and the resultant effects on photosynthetic efficiency measured.

PSII exists as a dimer surrounded by LHCII trimers to form a supercomplex. The LHCII chlorophyll proteins act as an antenna that receives photon excitation energy and passes it on to the CP43/47 chlorophylls (Box 4.1).

The P680 reaction center consists of a pair of chlorophyll molecules, one on each of the central dimer polypeptides, D1/D2, that lie very closely together (Figure 1). One of these, on D1, is the recipient of excitation energy from CP43/47 at 680 nm, forming the reduced compound P680$^+$. This in turn reduces pheophytin on D1 and then the first plastoquinone molecule, at Q_A over on D2, where it is tightly bound. It is important to note that the redox potential of the Q_A/Q_A^- is sufficiently positive relative to P680 that it decreases the probability of charge recombination (backflow). Thus part of the energy received by the reaction center is lost as stabilizing energy to decrease the probability of backflow and increase the probability of forward electron transfer down a redox potential gradient.

Q_B is the corresponding site back on D1, where plastoquinol is loosely bound, so that when plastoquinone is reduced and protonated to plastoquinol it is released. The Q_B site lies to the stromal side of the membrane. It is in the form of an elbow-shaped cavity or tunnel opening on to the stromal surface of PSII at one end and to the adjacent membrane matrix at the other. The tunnel is lipophilic, lined with the phytol tails of chlorophylls, together with some lipid groups and carotenoids. This allows the plastoquinol/plastoquinone to diffuse in and out of the photosystem, from the membrane, without encountering the aqueous phase of the stroma.

There is also an alternative route for electrons on the stromal side of PSII. They can be fed back to reduce P680$^+$ in a cyclic flow from Q_B via cytochrome b_{559}. This provides an important safety route for disposing of excess electrons that would otherwise lead to the formation of damaging intermediates, such as active oxygen species (Box 4.6).

On the inner, luminal side of D1 lies the tyrosine residue (Tyr161) that donates electrons to P680$^+$. This step is closely associated with a histidine amino acid (D1-His190) to which it donates a proton, leaving a tyrosyl residue that can oxidize the oxygen evolving center. The corresponding amino acid on D2 (D2-Tyr160, associated with D2-His189)

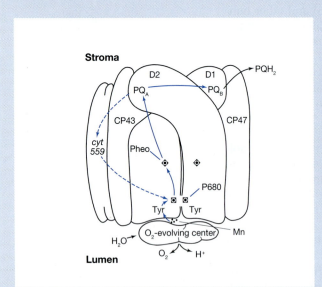

Figure 1 Photosystem II showing the main components discussed in this box. The solid blue arrows show the pathway of electrons from manganese in the oxygen-evolving center to P680, pheophytin (Pheo, A_0) and on to the site PQ$_B$ and the formation of plastoquinol (PQH$_2$). The dashed blue arrow shows a possible cyclic route for electrons from PQ$_A$ back to P680 through cytochrome 559. Black arrows indicate small molecules that diffuse towards or away from the photosystem. Black lines label components.

Box 4.2 Structure of Photosystem II (continued)

is oxidized by P680, but is not involved in water oxidation. This oxidation seems to bias electron transfer to the D1 side of the reaction center.

The oxygen evolving center consists of three extrinsic membrane proteins. These are closely bound to the projecting chains of D1 and D2 on the luminal side of PSII and serve to shield the core of the center. The core consists of four manganese atoms, a calcium atom, and a chlorine atom (Figure 2). The manganese atoms provide electrons to Tyr$_Z$, shifting progressively from state III to state IV as four electrons are lost from the cluster. The calcium and chlorine atoms provide a chemical environment that assists with the combining of oxygen atoms from water to make oxygen molecules.

The most likely arrangement of these atoms has three of the manganese atoms lying on a flat plane at the corners of the base of an imaginary triangular pyramid, with a calcium atom at its apex. In this pyramid, the calcium-manganese distances are all similar (0.34 nm), but greater than that for two of the manganese–manganese distances (0.27 nm). The third side of the manganese triangle is longer, about the same as the calcium–manganese distance (0.33 nm). The fourth manganese atom is linked by an oxygen atom to one of the manganese atoms, and lies in about the same plane. This cluster of cations is interlinked by oxygen bridges and stabilized by interactions with carboxylate groups on the adjacent polypeptide chains.

The calcium atom is linked by oxygen bridges to two of the manganese atoms and to chlorine (Figure 3). Calcium is highly electrophilic and would facilitate the withdrawal of electrons from oxygen in a water molecule and the release of protons. One of the water molecules destined for splitting is probably bound to one of the manganese atoms while the other is held within the coordination sphere of the calcium atom. Possibly this manganese atom is oxidized to state V, deprotonating the bound water molecule and leaving the oxygen atom open to forming an O=O bond with the oxygen from water bound to the calcium via a nucleophilic attack.

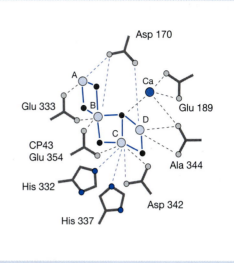

Figure 2 A proposed model of calcium and manganese (light blue) atoms in the oxygen-evolving center. Bonds from manganese to bridging oxo are in solid blue color. Bonds from manganese and calcium to amino acids are dotted. (From K. Sauer: *Science* 314: 821, Figure 4C, 11.03.2006. Reprinted with permission from AAAS (American Association for the Advancement of Science).)

This cluster of metal ions is ligated to four adjacent amino acids, three on D1 and one on CP43. Other groups provide stability via hydrogen bonds, trapped water molecules, and possibly a bicarbonate ion. All these provide an environment that can stabilize the four rounds of reduction induced by Tyr$_Z$ and then form a σ-bond between the oxygen atoms of two positioned water molecules leading to O_2 formation. The structure also has to provide an exit pathway for the H$^+$ protons from this water oxidation site to the lumen.

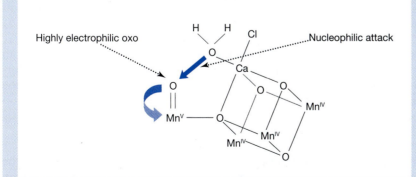

Figure 3 Model for the involvement of calcium and chlorine in holding a water molecule in position so that the oxygen atom can be released and used to form molecular oxygen. Note that the adjacent manganese atom is reduced from Mn^{IV} to Mn^V. (From *Current Opinion in Structural Biology*, "Structure of photosytem II and molecular architecture of the oxygen-evolving centre." Volume 14, Number 4, Page 7, 2004. Reprinted with permission from Elsevier Limited.)

with 20 μs flashes of light with 0.3-s dark intervals. This was interpreted by B Kok and coworkers (1970) as a cycle of successive oxidations, called the S-cycle. The ground (dark) state is S_1, and the center passes to S_2, S_3, and finally S_4 states when two water molecules are split (Box 4.2).

In the other set of reactions, A_0^- pheophytin is rapidly (picoseconds) oxidized to A_0 by passing an electron on to the first of two PQ molecules (at site Q_A, Box 4.2) and then, via an iron atom to the next PQ at site Q_B. PQ requires two electrons to become fully reduced. At Q_A, PQ is tightly bound and only undergoes single reduction events to the semiquinone state before being reoxidized by the PQ at Q_B. Each Q_B molecule undergoes two successive reduction events to become fully reduced. It then takes up two protons from the stromal side of the membrane to form PQH_2, which is released from the site and leaves the PSII complex. It is replaced by a new PQ molecule binding to the Q_B site. The PQH_2 diffuses in the lipid bilayer, acting as a mobile carrier of hydrogen atoms.

$$PQ + 2e^- + 2H^+ \longrightarrow PQH_2$$

The $P680^+/A_0^-$ pair regeneration steps are relatively slow (millisecond timescale), so that a second light reaction cannot effectively occur inside 1 ms. However, the first fairly stable end product, PQ at Q_A, is formed in less than a nanosecond, so the electron flow from the excited state of P680 is maintained.

The Q cycle uses plastoquinol to reduce plastocyanin and transport protons into the lumen

The PQH_2 formed by PSII is the substrate for the Q cycle on the integral transmembrane protein complex, cytochrome b_6f. The cycle regenerates PQ for PSII, reduces a protein electron carrier, plastocyanin, and pumps more protons from the stroma into the lumen of the grana and stroma membrane systems.

As each plastocyanin reduction requires only a single electron, the Q cycle ensures that removal of two electrons from PQH_2 is achieved safely in two steps, with one electron going to reduce plastocyanin and the other being recycled to reduce a further PQ molecule (Box 4.3).

The cytochrome b_6f complex contains two cytochromes, cytochrome b_6 and cytochrome f, linked by an iron–sulfur Rieske protein. There are two Q (quinol) binding sites on cytochrome b_6. Electrons taken from PQH_2 at the binding site Q_p (also called Q_o) on the luminal side of the cytochrome are moved via two iron heme groups to reduce PQ at the other binding site Q_n (also called Q_i) on the stromal side, to form more PQH_2.

PQH_2 is oxidized in two steps to PQ (Box 4.3, Figure 1) at the Q_p site. The first step forms plastosemiquinone, with the release of an electron that reduces plastocyanin.

In the second step, plastosemiquinone is oxidized to PQ with the release of an electron that is passed via a heme molecule to the first step of reducing a further PQ molecule at the Q_n site. Another PQH_2 molecule is oxidized in the same two steps at the Q_p site. This forms a further reduced plastocyanin molecule and completes the reduction of plastosemiquinone to PQH_2 at the Q_n site, which is fed back into the cycle.

Proton pumping is a very important feature of the Q cycle. Oxidation of the PQH_2 releases protons taken up from the stroma, during PQH_2 formation on PSII, into the thylakoid lumen. PQH_2 synthesis at Q_n also takes up protons

Box 4.3 Structure and operation of cytochrome b_6f

In the unicellular green alga *Chlamydomonas*, cytochrome b_6f exists as a dimer, each monomer being composed of four large polypeptides, cytochrome b_6, cytochrome f, an iron–sulfur Rieske protein, and subunit IV, and four small polypeptides (Figure 1). In the domains that project into the lumen side of the membrane are found a heme in cytochrome f and an Fe_2S_2 center in the Rieske protein. The cytochrome b_6f complex has two Q sites (quinol binding sites) and two heme groups within the depth of the membrane and a recently discovered heme group towards the stromal side, near the Q_n site. This heme may act as the acceptor for electrons from PSI during cyclic photophosphorylation. A chlorophyll a molecule and a carotenoid molecule are also present, whose functions are unknown. The chlorophyll tail lies near the Q_n site and the carotenoid near the Q_p site. In common with other membrane protein complexes on the photosynthetic membrane, there has been very strong conservation of the cytochrome b_6f proteins and the positions of their hemes and other cofactors from their prokaryote ancestors to single-celled eukaryote algae and on to higher plants.

The dimer arrangement is such that the Q sites face each other in a protected environment. The two opposing cytochrome f proteins form a cavity which protects the Rieske proteins (FeS in Figure 1). These Rieske proteins serve as a safety gate preventing the two electrons from plastoquinol entering the iron–sulfur center simultaneously. Each single electron donated to the heme during semiquinone formation at Q_p induces the Rieske protein to move away from cytochrome b_6 to cytochrome f, transferring the electron to the c-heme. It then returns to Q_p. This movement prevents the accumulation of the dangerous semiquinone intermediate, as reduction of plastoquinone only occurs when the b-heme of cytochrome b_6 is in the oxidized state and is therefore immediately able to oxidize the semiquinone to quinone. The Rieske protein oscillates in the cavity as electrons flow through the system. Inevitably it is not a perfect system and in time single electrons can pass to molecular oxygen, which is destructive.

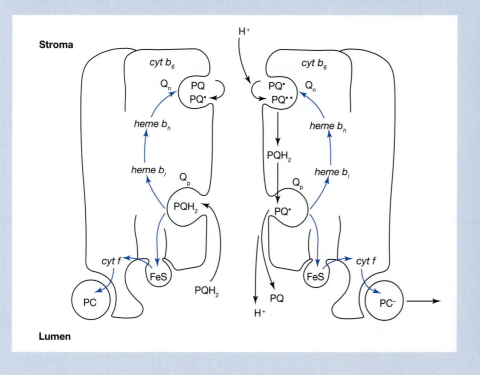

Figure 1 Cytochrome (cyt) b_6f dimer, with the Q_n and Q_p sites facing each other. Blue arrows show the pathways of electrons, black arrows show the net movements of diffusing components. In the left-hand cyt b_6f, the first step in oxidation of plastoquinol at Q_p is shown, with formation of the semiquinone (PQ•) at Q_n. The second step, in the right-hand cyt b_6f, shows formation of plastoquinone, PQ, at the Q_p site, while PQH_2 is formed at the Q_n site. Note the uptake of protons (H^+) from the stroma side during PQ reduction and their release into the lumen during oxidation. The FeS head oscillates in a cavity during electron transfer to plastocyanin (PC).

on the stromal side, while its oxidation at Q_p releases more protons on the lumen side. The Q cycle therefore acts as a proton pump, generating a trans-membrane electrochemical H^+ gradient. As we shall see later, this pump can be operated independently of PSII by taking electrons directly from PSI, a process called cyclic photophosphorylation because it leads to enhanced ATP generation without concomitant oxidation of H_2O and reduction of $NADP^+$ to NADPH.

In summary, for every two molecules of PQH_2 supplied by PSII, four molecules of reduced plastocyanin are formed, four protons are released on the lumen side, and one molecule of PQH_2 is formed with the uptake of two protons on the stroma side. Both protons will eventually end up on the lumen side as this new PQH_2 is oxidized in the same cycle.

Oxidized plastocyanin, a copper-containing protein, is reduced ($Cu^{2+} \rightarrow Cu^{+}$) by electron flow through an iron–sulfur cluster and cytochrome f, also on the lumen side. The plastocyanin diffuses in the thylakoid lumen and serves as an electron donor for PSI.

$$PQH_2 + 2Cu^{2+} \longrightarrow PQ + 2Cu^{+} + 2H^{+}$$

The Q cycle appears to present a paradox, in that the same products are cycled and remade while pumping protons into the lumen. The driving force is the downhill electrochemical gradient from PQH_2 to plastocyanin (Figure 4.3). This has driven the evolution of the cytochrome b_6f complex that takes advantage of the gradient in this step of the pathway to greatly increase the efficiency of the overall photochemical process. For each electron transported to plastocyanin, two H^{+} are transferred into the lamellar lumen. So for the four electrons released from each water oxidation step, four protons are added to the lumen at the oxygen-releasing center, and eight protons are added from the Q cycle. As these were taken up from the stroma side during PQH_2 formation, the Q cycle generates a net movement of protons into the granal lumen that are utilized in ATP formation and are crucial prerequisites for this process.

Photosystem I takes electrons from plastocyanin and reduces ferredoxin, which is used to make NADPH and other reduced compounds

PSI is located in the stroma lamellae membranes. Like PSII, it also has a dimer of membrane-spanning chlorophyll proteins at its core (Box 4.4). The reaction center consists of a dimer of chlorophyll a molecules, one on each protein molecule. These exist in a different protein environment from those in PSII, and have an absorption maximum of about 700 nm, and so the center is called P700. When energy is passed to this center from the PSI antenna chlorophyll molecules, P700 is energized and a free electron is passed on to the first acceptor molecule, A_0, another chlorophyll molecule. This in turn reduces A_1, a phylloquinone molecule, which oxidizes A_0^{-} back to A_0.

The reaction center is located close to the lumen side of the PSI complex, so that $P700^{+}$ can be reduced back to P700 by direct interaction with reduced plastocyanin diffusing from the cytochrome b_6f complex. The plastocyanin docks directly on to PSI, passing one electron from its copper atom to $P700^{+}$.

The pathway from A_1 is via a series of three further reduction–oxidation steps that moves the electron through three iron–sulfur complexes, known as F_X, F_A, and F_B, in the order that they are reduced. As P700 is located on the lumen side of PSI, the linear arrangement of the acceptor molecules and iron–sulfur complexes across the width of the membrane ensures a rapid charge separation. The result is that the reduced F_B lies close to the stroma surface. Here it reduces a bound ferredoxin protein molecule, a small water-soluble iron–sulfur protein.

Reduced ferredoxin is capable of reducing a variety of molecules via suitable reductases. Usually this will be $NADP^{+}$, which requires two electrons and a

Box 4.4 Photosystem I

Photosystem I (PSI) is an integral membrane protein complex located on the stroma lamellae lying between the grana stacks. The structure of PSI has been strongly conserved over the past 1 billion years since plants evolved from symbiotic associations between cyanobacteria and eukaryote cells. Comparison of the higher plant structure with the equivalent structure in cyanobacteria reveals minimal alterations. This is very unusual and suggests that evolution had produced a near-perfect structure long before eukaryotic plants evolved. Only three chlorophyll molecules have been lost, there are 93 in higher plants, against 96 in the prokaryote. Two of these were in small polypeptides that were deleted as part of a suite of changes that accommodated the close binding of LHC1 antenna polypeptides, which are not present in cyanobacteria.

At the core of higher plant PSI is a dimer of homologous 80 kD protein molecules, PsaA and PsaB (Figure 1). These have 11 transmembrane sequences and they bind most of the electron carriers utilized in PSI. The transmembrane sequences have strong homologies to the D1/D2 and CP43/CP47 pairs in PSII. The six transmembrane helices of CP43/CP47 are homologous to the six transmembrane helices at the N-terminal end of the PsaA/PsaB proteins, and the five helices of the D1/D2 proteins are homologous to the five helices at the C-terminal end of these proteins.

The PSI dimer core protein probably evolved by fusion of the two genes for the PSII core proteins. Depending on the species examined, there are about 13 other protein subunits associated with this PSI core pair. These provide binding sites for electron carriers on the luminal and stromal sides of the membrane complex. They also provide binding sites for light harvesting chlorophyll protein complexes (LHC1, Box 4.1).

P700 is a dimer of two chlorophylls, one on each of PsaA and PsaB at the luminal side of PSI (Plate 4.1). These chlorophylls are not identical, the PsaA one is chlorophyll a, while the one on PsaB is chlorophyll a', a C132-epimer of chlorophyll a. Their tetrapyrrole rings lie perpendicular to the plane of the membrane, with the heads and magnesium atoms stabilized by coordination with specific histidine residues and hydrogen bonds. Photoactivation yields an unpaired electron that is shared almost equally between the two molecules.

Both PsaA and PsaB carry A_0 (a chlorophyll a molecule) and A_1 (a phylloquinone molecule). Each A_0 chlorophyll is bound by axial ligands to sulfur atoms of methionine residues on the respective PsaA and PsaB proteins. Further stability is provided by the hydroxyl groups of adjacent tyrosine molecules. Both pathways appear to oxidize P700 during photosynthesis, though their properties are not identical. Both phylloquinones can reduce F_X. This is in contrast to PSII, where despite the presence of some components on both core proteins, electron flow is limited to the pathway on only one of them.

Excitation energy for P700 is derived from chlorophyll a and b molecules in the PSI complex and associated LHCI proteins. Binding sites for LHCI complexes are provided by PsaG and PsaK, which lie on opposite sides of the molecule (Box 4.1). Those for LHCII complexes are provided by PsaH (Box 4.1), which lies on the side opposite the necklace of LHCI. The core PSI complex contains 93 chlorophyll molecules, with 79 in the central dimer. Therefore, unlike PSII, these two proteins contain both antenna and reaction center chlorophylls, creating a hard-wired system on one molecule for gathering photons and for effecting charge separation.

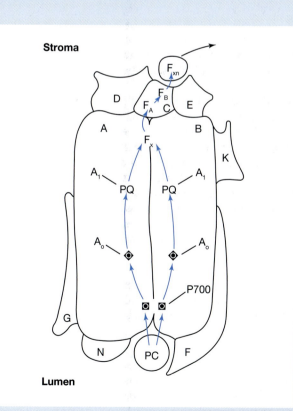

Figure 1 Photosystem I structure showing the arrangement of protein subunits (lettered A to N). The solid blue arrows show the pathway of electrons from plastocyanin (PC) to P700 and on to the formation of ferredoxin (Fxn). The black arrow indicates ferredoxin diffusing away from the photosystem. Black lines label components. The P700, A_0, A_1 pathway probably exists in parallel on both PsaA and PsaB subunits, combining at the single F_X site.

continued

Box 4.4 Photosystem I (continued)

The three F-sites are each clusters of four iron atoms linked to four sulfur atoms. The FX cluster is ligated by two cysteine residues from PsaA and PsaB. The F_A and F_B clusters (both 4Fe–4S) are bound to the PsaC protein via eight of the nine conserved cysteine residues in that protein.

The oxidized reaction center ($P700^+$) is reduced by the Cu^+-plastocyanin molecule produced by cytochrome b_6f. Plastocyanin diffuses from the granal membranes to the stroma lamellae. On the luminal side of PSI, P700 lies close to a hydrophobic surface of PsaA and PsaB, which interacts with the hydrophobic surface of plastocyanin. Docking is assisted by the PsaF protein, through the lysine-rich N-terminal region. PsaN may also be involved. The net surface charge on PSI in this location is therefore positive, and interacts with negative charges on the reduced plastocyanin. This brings the reduced copper atom into close proximity with two tryptophan residues on the PsaA/PsaB proteins, which mediate electron transfer to $P700^+$. Docking/undocking of plastocyanin, a relatively large molecule, is quite rapid. The reduced form is strongly bound, and the transfer of the electron to $P700^+$ fast (10–20 μs), while the oxidized form is rapidly released. Diffusion of this protein is a limiting factor in the functioning of P700. Plastocyanin is 3–4 nm across, and has to move in the granal lumen, which is about 4 nm in width, restricting the rate of reduction of $P700^+$. This can be shown by reducing the grana width osmotically, which

reduces the photosynthetic electron flow through P700 (see Box 4.5 for further discussion).

The Z scheme pathway for photoelectron flow, as commonly presented here and elsewhere, fails to take account of the distinction between granal and stroma membranes. A further consideration is the distinction between appressed granal membranes and granal rim membranes exposed to the stroma. PSII is located primarily in the appressed membranes, and PSI in the rim and stromal membranes. Granal rim membrane PSI is juxtaposed to PSII and the cytochromes that yield reduced plastocyanin, but stromal membrane PSI (about 70% of the total) is relatively remote from PSII electron carriers. Diffusion of plastocyanin over these long distances could significantly reduce the rate of electron flow in the system. Even with the usual stoichiometry of two PSII centers for each PSI center and four to five plastocyanin molecules for each PSI reaction center, there seems to be a bottleneck at this step in the models, leading to a 10 ms transfer time observed between cytochrome b_6f and PSI.

On the stromal side of PSI, ferredoxin is reduced. This is an iron-rich protein that docks with PSI via electrostatic charges provided by PsaD. Following electron transfer, dissociation is mediated by Arg 39 on PsaE, which provides a positive charge limiting the lifetime of the ferredoxin-PSI complex.

---

proton to yield NADPH, a process first discovered by Daniel Arnon in 1962. Arnon had previously (1951) noted that isolated chloroplasts could reduce NADP to NADPH and later (1954) found that ATP is produced at the same time. The reduction of NADP by reduced ferredoxin is mediated by ferredoxin-NADP$^+$ reductase, a flavoprotein enzyme.

$$2Fd_{reduced} + H^+ + NADP^+ \longrightarrow 2Fd_{oxidized} + NADPH$$

Note that the proton is removed from the stromal side of the membrane, enhancing the transmembrane proton gradient. Two electrons are needed for each NADPH molecule formed, so the splitting of two molecules of water at PSII provides enough electrons to form two molecules of NADPH. This stoichiometry is not exact, because other molecules can be reduced by ferredoxin, as it is not a closed system. The ferredoxin can diffuse to other reductases and reduce other molecules in the appropriate redox potential range. These include nitrite, which is used to produce amino acids (Chapter 8), and sulfate, which is reduced to sulfhydryl in cysteine (Chapter 8). It can also donate electrons to the Q cycle (see section above, The Q cycle uses plastoquinol to reduce plastocyanin and transport protons into the lumen).

Thus P700 is capable of pumping electrons from plastocyanin to achieve reduction of $NADP^+$ (Figure 4.6). The path from A_0 even involves a slight uphill segment from F_A to F_B, though this is overcome in the overall gradient (Figure 4.7).

Stroma

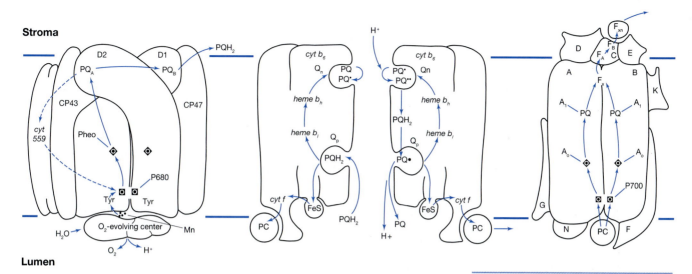

Lumen

ATP synthase utilizes the proton motive force to generate ATP

The activity of the photosystems results in a net increase in the proton content of the lumen relative to the bulk phase of the stroma. Increases within the lumen result from water splitting at PSII and PQH_2/PQ conversions in the Q cycle of cytochrome b_6f. Synthesis of NADPH leads to loss of protons in the stroma and a concentration difference of protons across the membrane, which could exist as a pH gradient or a membrane polarization gradient. This leads to a proton motive force that is the sum of the pH difference across the membrane (proton concentration difference) and the electrical potential component arising from the charge difference across the membrane. The proton motive force drives a flow of protons through a transmembrane enzyme complex generating

Figure 4.6 Summary of photochemical reactions and linking electron flow pathways. The light reactions in PSII generate PQH_2, which diffuses through the membrane to the cytochrome b_6f complex where it is oxidized back to PQ in the Q cycle. Protons are taken up from the stroma in the process and released into the lumen. Reduced plastocyanin (PC) is formed and moves to PSI, where it passes its electrons to the oxidized chlorophyll. The light reactions in PSI generate NADPH. Details of the components are presented in Boxes 4.2, 4.3, and 4.4.

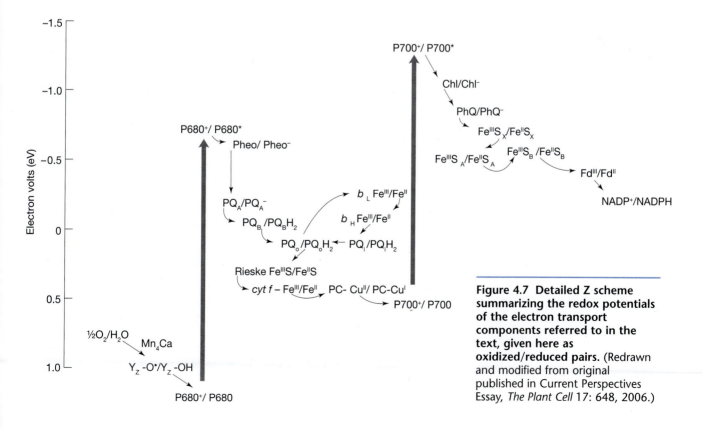

Figure 4.7 Detailed Z scheme summarizing the redox potentials of the electron transport components referred to in the text, given here as oxidized/reduced pairs. (Redrawn and modified from original published in Current Perspectives Essay, *The Plant Cell* 17: 648, 2006.)

Box 4.5 Structure of chloroplast ATP synthase

Chloroplast ATP synthase consists of an integral membrane protein complex, CF_0, attached by a stalk on the stromal side of the lamellar membrane to a large hydrophilic complex, CF_1. Protons flow across the membrane through the CF_0 complex. This induces deformations in the CF_1 complex that results in ATP synthesis.

Microscopy has been used to visualize the components using negative staining, cryomicroscopy and immunolabeling. CF_1 consists of a hexagonal ring of alternating α- and β-subunits (Figure 1). The stalk consists of one large subunit, γ, that is in contact with one of the three α/β pairs and CF_0, and ε, a smaller subunit whose placement is less clear.

CF_0 consists of a ring of subunits (called III) each composed of two transmembrane helices with a loop out to the stromal side. There is a variable number of III subunits in the ring between species and between chloroplast and mitochondrial synthases. Based on microscopy, it is reasonably certain that there are 14 in spinach CF_0. The cylindrical ring so formed is tapered, and is 5.9 nm in diameter on the lumen side and 7.4 nm diameter on the stroma side, in contact with the CF_1 complex. In addition, CF_0 has one each of subunits I, II, and IV which provide a link between the outside edge of the ring and the CF_1 hexamer. Subunit I has a single membrane spanning α

helix and a large polar domain that extends out to the CF_1 hexamer. I and II may function as a dimer and are involved in binding to the α/β pairs, III binds to ε.

The β subunits are strongly conserved across all ATP synthases. Other polypeptides are less conserved with ε the least conserved of all the polypeptides. The α subunits are chemically equivalent, but differ in their properties due to differences in their local environment induced by other subunits. There are six nucleotide-binding sites at the interfaces of the α and β subunits, which can be divided up into three distinct classes, although sites within a class can differ. These differences are due to an asymmetric interaction of the hexamer with the γ subunit. The P-loop motif GXXXGKT/S is found on the β subunits with a Glu residue. These bind di- and triphosphates. The noncatalytic sites on the α subunits do not have Glu residues.

Much of the information on ATP synthesis is derived from studies on mitochondrial synthases (MF_1MF_0, see Chapter 6). These show that three of the sites are catalytically active and three are nonactive. In CF1 there are two tightly bound ADP sites (N1 and N4), and two bind ATP or AMP-PNP tightly in presence of magnesium ions (N2 and N5). The third site (probably at N1) tightly binds Mg-ATP, but hydrolyzes it to ADP. N3 and N6 bind di- and triphosphates in a freely dissociable manner.

In the dark, ADP is bound to the sites, and on illumination it is released initially as the proton gradient becomes established, either from N1 or N4. If CF_1 activity is analogous to that of MF_1, then the active catalytic sites would be N1 and N3 and probably N6. While ATP is synthesized in the light, CF_1 can reverse the action in the dark and have ATP hydrolysis driving a proton pump that places protons in the lumen.

The three catalytic sites alternate their properties in a cycle that progresses between three states: loose binding, tight binding, and unbound or open during one cycle of activity (Figure 2). ATP is present at the tight binding site from which it is released by a conformational change in the site, to create the open state to which ADP and Pi become attached, and then loosely bound.

The only way that all three β subunits can participate in catalysis is if the properties of each change during the cycle. As interactions with γ are responsible for asymmetry around the hexamer, it follows that the CF_1 complex must rotate γ during catalysis. Obtaining convincing evidence for rotation has not been easy, and early claims for success were heavily criticized, but it is now accepted that rotation occurs. The conformation of successive β subunits is altered as they come into contact with the γ shaft on each turn of the ring. CF_0 subunits I and II act as a stator, holding the CF_1 hexamer in place relative to the CF_0 ring as the γ and ε shaft is rotated.

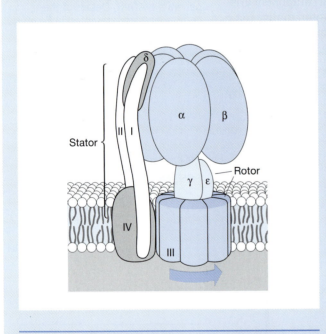

Figure 1 Chloroplast ATP synthase, showing the membrane bound CF_0 and the stromal CF_1 units. Electron flow leads to rotation of the III subunits of CF_0, which turns an off-center axle against the α and β subunits of CF_1. These are held in place by a stator linked to the CF_0 complex.

Box 4.5 Structure of chloroplast ATP synthase (continued)

Rotation is powered by the interaction of protons with the III subunit ring in the membrane. These feed through a channel one at a time, each proton ratcheting the ring around by one III subunit. Each turn of the ring completes a single turn of γ, which releases three ATP molecules from CF_1. The number of protons required to complete one turn depends crucially on the number of III subunits in the

ring. Originally this was estimated from the known stoichiometry of ATP synthesis and proton utilization as 10–12 subunits. In fact, visualization of the CF_1 from spinach using atomic force microscopy shows that there are unequivocally 14 III subunits in each ring, so 14 protons are required for a complete cycle. Therefore in spinach each ATP molecule needs 4.67 protons for its synthesis.

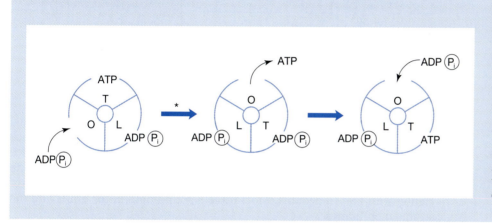

Figure 2 Diagram of CF_1 function. Three sites rotate clockwise around a central axis. As they rotate, their properties progress from tight (T) when ATP is formed and held (first diagram), to open (O), when ATP is released (first step, one-third rotation) and then ADP and P_i are bound (second step, no rotation), to loosely bound (L), when ADP and P_i are held at the site, returning to the tight condition at the end of one complete rotation.

ATP from ADP on the stromal side of the membrane. This enzyme was called coupling factor in the earlier literature, but is now more usefully labeled ATP synthase (Box 4.5). Chloroplast ATP synthase is structurally and functionally similar to mitochondrial and bacterial ATP synthases (Chapter 6).

The flow of protons is not linked directly to the synthesis of ATP from ADP and P_i. The synthase consists of a series of binding sites for ADP and P_i, and for ATP (Box 4.5). The proton gradient brings about conformational changes in the subunits binding the newly formed ATP molecule that release it from the enzyme complex. The precise number of protons required depends on the exact structure of the synthase. Using atomic force microscopy, it has been found that one molecule of ATP requires 4.67 protons for its production.

Cyclic photophosphorylation generates ATP independently of water oxidation and NADPH formation

The oxidation of two water molecules at PSII provides sufficient reductants, linked through PSI, to reduce two molecules of $NADP^+$ to NADPH and release 12 protons into the lumen. Therefore, the Z pathway alone produces slightly less than 1.5 ATP per NADPH. However, the carbon dioxide fixation cycle requires 1.5 moles of ATP for each mole of NADPH (Chapter 5). For this and other reasons additional ATP synthesis is required. The ATP deficit is made up by PSI complexes that are able to function without the involvement of PSII. Electrons from ferredoxin are diverted through the Q cycle on cytochrome b_6f to reduce plastocyanin and P700$^+$ (Figure 4.8). This is known as cyclic photophosphorylation, and it operates at the expense of NADPH synthesis. It

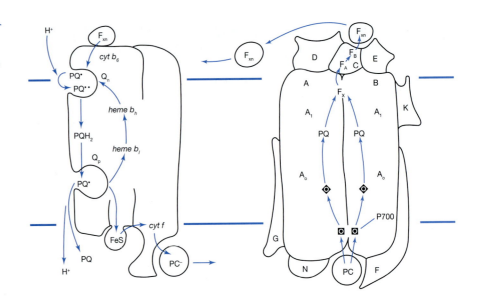

Figure 4.8 Cyclic photophosphorylation depends on diverting reduced ferredoxin (Fxn) to reduce plastoquinone (PQ) on cytochrome b_6f. This results in uptake of protons (H^+) from the stroma, and their release into the lumen as the plastoquinol (PQH_2).

results in additional PQH_2 synthesis and a net proton flow into the luminal space, which is used for ATP synthesis. This mechanism operates to balance the output of ATP and NADPH for carbon dioxide fixation (Chapter 5). Notice that cytochrome b_6f and the Q cycle are involved in two distinct pathways, the Z scheme in association with PSII, and cyclic photophosphorylation in association with PSI. Hence this cytochrome is found both in the grana and the stroma membranes.

Regulation of electron flow pathways in response to fluctuating light levels

This section explores some of the issues that affect the whole electron flow pathway from water to NADPH. Leaves in a vegetation canopy are subjected to a great range of solar radiation levels and light quality. Heavy cloud cover severely reduces the level of irradiance, so the remaining photons have to be trapped and used efficiently. In contrast, noon on a clear day will deliver very high radiation levels to the leaves at the top of the canopy. The photosynthetic system will become light-saturated. There is, however, a limit to the rate at which energy can be utilized in the production of reduced carbon compounds, mainly because of limitations on the rate of carbon dioxide diffusing into the leaf. Under these conditions the antenna chlorophylls will become saturated and deliver very high flows of excitation energy to the reaction centers that cannot be dissipated along the normal electron transport route. If these conditions are accompanied by even a temporary drought, the stomata will close, so there will be even less carbon dioxide to act as an acceptor for the reducing power generated. PSII polypeptides, especially D1, can suffer damage from oxidation by unreduced P680$^+$ and oxygen radicals. Damaged reaction centers can continue to trap light energy, which exacerbate the problem if they cannot pass on the charge to PQ. Specific proteases are on hand to excise damaged components, principally polypeptide D1. Such repairs lead to a delay in resuming full photosynthetic activity while new component synthesis is activated and replacement achieved.

In contrast, leaves in the understory will receive light that is depleted in the main absorption wavebands of chlorophyll a, and so these will have to use particular *in vivo* forms of chlorophyll and accessory pigments (carotenoids) to boost their photosynthetic output. Such leaves also develop internal anatomy and chloroplast grana organization to maximize light harvesting in their environment (shade leaves). For these leaves transient increases in light intensity, e.g. from sun flecks, can also be very damaging.

High light levels expose the cellular components to all the risks posed by photosynthesis. Significant levels of light energy end up trapped in component molecules. Excess irradiation can saturate the electron flow system leading to accumulation of molecules in the excited triplet state. This leads to formation of reactive superoxides, peroxides, and aldehydes, which have the potential to destroy the proteins and lipids in the chloroplast. Plants have evolved systems to remove these dangerous radicals and to prevent or reduce their formation in the first place. We will see that both the removal of radicals and the prevention of their formation involve many other molecules and enzymes, all working within the close confines of the thylakoid and fret membrane systems. It has therefore proved difficult to study each system in isolation and evaluate its contribution to the overall effect, so our story is incomplete at the present time. Clearly plants possess photosynthetic systems that work in a great range of habitats under a variety of physiological conditions of water and carbon dioxide supply. We might anticipate that the full range of curative and preventative measures will not be present to the same extent in all plants. Here we will describe the main processes that have been unequivocally identified and characterized in a representative number of plant species.

Oxygen molecules are normally stable and pose only a limited risk to the molecular structure of a cell. Singlet oxygen can be generated from these molecules by the input of energy, for example from light energy absorbed by chlorophyll molecules. The normal chlorophyll excited state induced by light absorption lasts for a few nanoseconds, long enough to be passed on to the reaction center. Any delay in this transfer can lead to the formation of a chlorophyll triplet state, which has a lifetime of several microseconds. These chlorophyll molecules can react with oxygen to produce singlet oxygen. This has a greatly increased oxidizing power that attacks any molecule in the vicinity. Typically, the lifetime of such oxygen radicals is about 200 ns, during which time it can diffuse as far as 10 nm. Mechanisms have evolved to rapidly remove these oxygen radicals and to curb their production at the sites of light absorption.

Singlet oxygen can be quenched directly by carotenoids, which are themselves converted to the triplet state, but this energy decays quickly being lost as heat. This can only happen if the carotenoids are close to the site of singlet production, which occurs in the LHCII antenna molecules, for example, but not in the PSII core. The carotenoids present in the PSII complex do serve to quench the P680 if charge recombination occurs to generate triplet P680 (as when the pheophytin at A_0 is not oxidized in time to receive a further reduction from P680). Singlet oxygen can also be quenched by α-tocopherol, a membrane lipid, and again the energy is dissipated as heat.

Scavenging and removal of superoxides, peroxides, and other radicals by dismutases and antioxidants

Oxidized compounds originate at PSI under high light conditions, when there is insufficient NADP to oxidize reduced ferredoxin. The electrons instead reduce molecular oxygen to superoxide, which in turn leads to peroxide formation by combination with protons. Peroxides can interact with reduced ferrous ions (Fe^{2+}) to produce hydroxyl radicals. These are extremely reactive and attack other molecules, sometimes leading to further radical formation so that a chain reaction is set up.

A number of enzymes have evolved to scavenge peroxides, including super-oxide dismutase, ascorbate peroxidase, monodehydroascorbate reductase, and glutathione reductase. Several antioxidants are available to be oxidized in these reactions, such as ascorbate, glutathione, and α-tocopherol (vitamin E). These convert the peroxide to water and oxygen. In effect this sets up a system for water breakdown (at PSII) and synthesis (at PSI), which removes excess energy trapped by the chlorophylls.

ascorbate peroxidase

$$2H_2O_2 + ascorbate \longrightarrow O_2 + 2H_2O + monodehydroascorbate$$

monodehydroascorbate reductase

$$2monodehydroascorbate + NADPH \longrightarrow 2ascorbate + NADP^+$$

Peroxide removal by ascorbate. Note that the regeneration of ascorbate leads to consumption of NADPH, removing excess reductant from the system.

Mechanisms for safely returning the levels of trapped high energy states to the ground state

Reducing the exposure of chlorophyll molecules to high light intensities can prevent or reduce singlet oxygen formation. This can be achieved by changing the leaf angle with respect to incident light, and by movement of the chloroplasts to self-shading positions along the side walls of cells. Within the chloroplast there are three basic types of mechanism for coping with high light conditions. Adjustment of the synthesis and amount of the LHCII antenna proteins, movement of the LHCII proteins from PSII to PSI (state transition) and nonphotochemical quenching when electron flow is diverted to heat generating systems.

High light conditions and/or low carbon dioxide concentrations down-regulate expression of *Lhcb* genes. The sensor mechanism is not known. One candidate is the redox potential system, e.g. the level of PQ, but this control seems to be quite loose. Other mechanisms could be the level of reactive oxygen, or the light saturation of precursors of chlorophyll. Clearly this is a slow control mechanism, depending on not replacing chlorophyll proteins lost due to turnover.

Some relief from excess energy absorption can be made within the PSII complex, by cycling electrons back from PQ via cytochrome *c*-553 to the D1 tyrosine. Another approach that has evolved is to reduce the amount of LHCII antenna proteins associated with PSII. This alters the ratio of light energy received between PSII and PSI by adjusting the extent to which each receives energy from these antenna proteins. As the two photosystems have different absorption spectra, conditions can arise where the energy flow through each is not balanced to meet the requirements of the Z scheme. This could be damaging, as high energy intermediates would accumulate in the system. The LHCII trimers serve as a simple feedback loop that adjusts the amount of antenna chlorophyll providing energy to each photosystem (state transition). If there is excess light energy flowing through PSII, then there is an excess of reduced PQ. This activates a kinase that phosphorylates some of the LHCII trimers. The extra charge causes them to dissociate from the PSII (state 2) complex and migrate towards the stroma lamellae (state 1 transition) where they bind to the PSI complex (state 1), increasing the flow through that system. Conversely, excess PSI activity leads to PQH_2 oxidation, which activates a phosphatase that removes the phosphate group and allows the trimers to return to PSII (state 2 transition).

Under high light conditions, the electron flow from PSII to PSI can be regulated by cytochrome b_6f and plastocyanin. The PQ reoxidation cycle is the slowest step in the photosynthetic electron flow pathway and the number of cytochrome b_6f complexes appears to be controlled, increasing as the flux rate increases. Plastocyanin also shows an increase with flux rate, and in leaves operating efficiently at high photosynthetic rates can reach a ratio of 4–5:1 of PSI. If the leaf's photosynthetic ability is compromised (e.g. by low or zero carbon dioxide availability), then this ratio falls back to below 1. This could limit the flow of electrons into P700. However, electron transfers involving plastocyanin (to PSI and to cytochrome b_6f) are much faster (10–100 times) than the PQ oxidation step, so the only way that reduced levels of plastocyanin could control the overall flux is if its diffusion rate to the PSI site is slow compared with the rate of electron flow.

Nonphotochemical quenching and the xanthophyll cycle

These adjustments to the electron transport system cannot entirely prevent the accumulation of excess energy in photon-trapping chlorophyll molecules. Such excited chlorophylls return to the ground state by either emitting the energy as fluorescence, or by dissipating it as heat. Molecular mechanisms for removing this trapped energy (quenching the excited state) before it is passed on down the electron transport chain are collectively termed nonphotochemical quenching. The precise details of these molecular mechanisms remain to be discovered. The following account provides a general outline of the mechanisms as far as they are understood at the present time.

In saturating light conditions, chlorophylls in the antenna proteins, which are in equilibrium with the PSII reaction center chlorophylls, are unable to pass their trapped energy on to P680. Such chlorophylls pose a risk, as they are able to activate oxygen to the singlet state. Decay of the excited chlorophylls releases energy as heat and fluorescence. This fluorescence is a temporary response to suddenly increased light levels, which declines as additional resources are developed to convert the energy to heat. A prime candidate for initiation of the controlling mechanisms is the pH of the lumen. Proton generation and pumping during the light reactions lower the lumen pH from about 7 to less than 5, providing a strong signal that can initiate a number of quenching processes based in the lumen. One of these is the activation of an epoxidase that uses electrons to reduce the xanthophyll violaxanthin to antheraxanthin and then to zeaxanthin. The zeaxanthin binds to a protein subunit of LHCII that is protonated at low pH and accepts energy transfer from excited chlorophylls. The zeaxanthin then returns to the ground state, dissipating the energy as heat. In this way excess light energy trapped by chlorophylls is diverted to heat production, away from potentially damaging photochemical events and electron transfers.

$$
\begin{array}{ccc}
\text{pH 5, de-epoxidase} \longrightarrow & & \text{pH 5, de-epoxidase} \longrightarrow \\
\text{violaxanthin} \longleftrightarrow \text{antheraxanthin} & \longleftrightarrow & \text{zeaxanthin} \\
\longleftarrow \text{pH 7, epoxidase} & & \longleftarrow \text{pH 7, epoxidase}
\end{array}
$$

LHCII** + zeaxanthin $\longrightarrow$ LHCII + zeaxanthin**

zeaxanthin** $\longrightarrow$ zeaxanthin + heat

The xanthophyll cycle and nonphotochemical quenching reactions leading to the transfer of excess excitation energy from light harvesting antenna complexes to zeaxanthin and its release as heat. This is reversed under low light conditions when the pH rises.

Antiquenching is obviously necessary when light levels fall. This is achieved by a different epoxidase, activated at higher pH levels, that converts zeaxanthin back to antheraxanthin and violaxanthin. In this way the xanthophyll cycle is thought to track the pH changes in the lumen that are directly linked to productive photosynthetic activity and regulates the level of nonphotochemical quenching. The importance of xanthophylls and other carotenoids in this protection system can be seen from the fact that, in the absence of carotenoids, photosynthetic systems can only operate in oxygen-free conditions. Also carotenoid-deficient mutants will only survive under very low light conditions; normal light levels lead to bleaching of the leaves.

There is some evidence that a very hydrophobic transmembrane protein, PSbS, is involved in nonphotochemical quenching. It is proposed that PSbS undergoes protonation at low pH, and then binds zeaxanthin into the LHCII/PSII complex facilitating quenching of the excited state molecules. Experiments in which the levels of PSbS protein are altered, or in which the protonation of PSbS at low pH is interfered with, support this hypothesis. However, the location of PSbS remains to be determined, as does the precise way in which quenching is effected.

Summary

Overall PSII and PSI operate to extract electrons from water and reduce $NADP^+$. A proton gradient is produced that is used for ATP generation, and oxygen is released as a by-product. The synthesis of these photosystems is regulated to achieve approximately equal numbers of each. Failure to channel the PSII reducing power can lead to oxygen radical production with harmful consequences for the photosynthetic machinery. This can occur in some unfavorable environmental conditions (e.g. chilling) that damage PSI, leaving the PSII reductants accumulating in the pathway (Box 4.6).

Box 4.6 Herbicides that act through the photosynthetic apparatus

As the photosynthetic electron flow pathway is the only source of energy for growth in a plant, it represents a prime target for chemicals designed to kill plants. DCMU and the herbicide atrazine bind to the Q_B site on PSII, preventing plastoquinol formation. Plastoquinone analogs (dibromothymoquinone) bind to cytochrome b_6f preventing plastoquinol oxidation at the Q_p site. Paraquat is a widely used herbicide that diverts electron flow from ferredoxin to the formation of reactive oxygen species and peroxides that damage the photosynthetic apparatus and ultimately kill the leaf. Perversely the chlorophyll pigments harvest light energy that is used to destroy the plant.

Attacking carotenoid synthesis is an alternative way of killing plants, as this will lead to loss of the nonphotochemical quenching needed to protect photosystems from the effects of relatively high light intensities. Norflurazan is a herbicide of this type and leads to bleaching of leaves that is similar to that seen when carotenoid synthesis mutants are exposed to normal light levels (see Chapter 12).

Further Reading

Evolution of the reaction centers, light harvesting photosystems, and proteins

These papers provide a good overview of the evolution of photosynthesis and would provide useful entry points for the literature both before and after 2002/2004.

De Las Rivas J, Balsera M & Barber J (2004) Evolution of oxygenic photosynthesis: genome-wide analysis of the OEC extrinsic proteins. *Trends Pl. Sci.* 9, 18–25.

Raymond J & Blankenship RE (2004) The evolutionary development of the protein complement of Photosystem II. *Biochim. Biophys. Acta* 1655, 133–139.

Xiong J & Bauer CE (2002) Complex evolution of photosynthesis. *Annu. Rev. Pl. Biol.* 53, 503–521.

Organization of the thylakoid membrane

These two papers provide an overview of photosynthetic membrane components.

Decker JP & Boekema EJ (2005) Supramolecular organization of thylakoid membrane proteins in green plants. *Biochim. Biophys. Acta* 1706, 12–39.

Merchant S & Sawaya MR (2005) The light reactions: a guide to recent acquisitions for the picture gallery. *Pl. Cell* 17, 648–663.

Photosystems

Photosystems have been the focus of research for over 50 years, so there is a very large literature about their structure and function.

Amunts A, Drory O & Nelson N (2007) The structure of a plant photosystem I supercomplex at 3.4Å resolution. *Nature* 447, 58–63.

Blankenship RE (2002) Molecular Mechanisms of Photosynthesis. Blackwell Science Ltd.

Chitnis PR (2001) Photosysytem I: function and physiology. *Annu. Rev. Pl. Physiol. Pl. Mol. Biol.* 52, 593–626.

Diner BA & Rappaport F (2002) Structure, dynamics, and energetics of the primary photochemistry of Photosystem II of oxygenic photosynthesis. *Annu. Rev. Pl. Biol.* 53, 551–580.

Minagawa J & Takahashi Y (2004) Structure, function and assembly of Photosystem II and its light-harvesting proteins. *Photosynthesis Res.* 82, 214–263.

Nelson N & Yocum CF (2006) Structure and function of Photosystems I and II. *Annu. Rev. Pl. Biol.* 57, 521–565.

Water splitting and oxygen evolution

The chemistry of water splitting in photosynthesis is fascinating, how do plants successfully break open this molecule under normal physiological conditions?

Ferreira KN, Iverson TM, Maghlaoui K, Barber J & Iwata S (2004) Architecture of the oxygen-evolving center. *Science* 303, 1831–1838.

Sauer K & Yachandra VK (2004) The water-oxidation process in photosynthesis. *Biochim. Biophys. Acta* 1655, 140–148.

Cytochromes

Processing plastoquinones and producing plastocyanins provides the vital link between PSII and PSI. New information is adding more detail to this part of the photosynthesis story.

Allen JF (2004) Cytochrome b_6f: structure for signaling and vectorial metabolism. *Trends Pl. Sci.* 9, 130–137.

Siegbahn PEM & Blomberg MRA (2004) Important roles of tyrosines in Photosystem II and cytochrome oxidase. *Biochim. Biophys. Acta* 1655, 45–50.

ATP synthase: structure and operation

An amazing molecular turbine has been uncovered that couples a proton gradient to the formation of ATP.

Kramer DM, Cruz JA & Kanazawa A (2003) Balancing the central roles of the thylakoid proton gradient. *Trends Pl. Sci.* 8, 27–32.

McCarty RE, Evron Y & Johnson EA (2000) The chloroplast ATP synthase: a rotary enzyme? *Annu. Rev. Pl. Physiol. Pl. Mol. Biol.* 51, 83–109.

Seelert H, Dencher NA & Müller DJ (2003) Fourteen protomers compose the oligomer III of the proton-rotor in spinach chloroplast ATP synthase. *J. Mol. Biol.* 333, 337–344.

World Wide Web-based information

The following is a useful site providing access to information on all aspects of photosynthesis.

http://porphy.la.asu.edu/photosyn/photoweb/

Photosynthetic Carbon Assimilation

5

Key concepts

- CO_2 entry into the leaf is accompanied by water loss through open stomata.

- All photosynthetic eukaryotes use the Calvin cycle to form carbohydrates via CO_2 assimilation.

- Radioisotope feeding experiments were crucial to the discovery of the Calvin cycle.

- There are three phases to the Calvin cycle: carboxylation, reduction, and regeneration.

- Ribulose bisphosphate carboxylase/oxygenase (Rubisco) is the enzyme responsible for CO_2 assimilation in the Calvin cycle.

- Rubisco carries out two competing reactions, the carboxylase and oxygenase reactions.

- Rubisco oxygenase activity results in the wasteful release of CO_2 in the light, in a process called photorespiration.

- C_3 plants rely entirely on the Calvin cycle for photosynthetic carbon assimilation.

- C_4 plants evolved additional reactions that result in reduced photorespiratory losses and increased water and nitrogen use efficiency.

- C_4 plants assimilate CO_2 via two spatially separated carboxylases, phospho*enol*pyruvate carboxylase and Rubisco.

- C_3–C_4 intermediate species may represent an evolutionary stage between C_3 and C_4 plants.

- Crassulacean acid metabolism is a photosynthetic pathway that enables plants to survive in arid environments.

- Crassulacean acid metabolism plants assimilate CO_2 via two temporally separated carboxylases, phospho*enol*pyruvate carboxylase, and Rubisco.

- C_3, C_4, and Crassulacean acid metabolism photosynthesis pathways each have distinct advantages and disadvantages in particular environments.

Photosynthetic carbon assimilation produces most of the biomass on Earth

About 90% of the dry weight of plants consists of carbon and oxygen obtained from the atmosphere via photosynthetic carbon assimilation. Photosynthesis provides the means for energy to enter into the global ecosystem, and carbon assimilation produces most of the global biomass. Our life depends on it.

Carbon dioxide enters the leaf through stomata but water is also lost in the process

CO_2 enters a leaf through stomatal pores on the leaf surface, formed between adjacent guard cells. Stomata open when the guard cells swell as they take in water from surrounding cells. In most plants this occurs in the light, although there are exceptions that will be described later in this chapter. The process is described in detail in Figure 5.1.

In order to gain carbon a leaf of a terrestrial plant will always face the risk of losing water. This is sometimes described as the starvation versus desiccation dilemma, or the photosynthesis/transpiration compromise. Open stomata lose water through transpiration more rapidly than they gain CO_2 because the water potential gradient is usually 100-fold greater than the inward CO_2 concentration gradient. The balance between carbon gain and water loss is represented by the term water use efficiency (WUE). This parameter can be expressed as the ratio of photosynthesis to transpiration, or sometimes as the yield (of dry matter produced) relative to water loss for a whole plant or even for an entire field of crops. Breeders are using the WUE value to select for crops that require less irrigation. As we shall see, some plants have acquired supplementary photosynthetic pathways that have resulted in improved WUE, either by concentrating CO_2 internally, or by opening their stomata at night. Before we describe these more specialized forms of photosynthesis we will first consider the essential CO_2 assimilation pathway that is common to all photosynthetic eukaryotes.

Carbon dioxide is converted to carbohydrates using energy derived from sunlight

The photosynthetic conversion of CO_2 to carbohydrates is often described by the following equation:

$$6CO_2 + 6H_2O + \text{light} \longrightarrow C_6H_{12}O_6 + 6O_2$$

This equation is misleading. It implies that glucose is produced as the end product of photosynthesis. Glucose might be produced eventually, but free glucose does not feature in the early stages of photosynthesis, nor is it a major metabolite in plants, being produced only by hydrolysis of starch or sucrose (see Chapter 6). A more useful equation to represent photosynthetic CO_2 assimilation is the van Niel equation:

$$CO_2 + 2H_2O + \text{light} \longrightarrow [CH_2O] + H_2O + O_2$$

This equation shows a generalized carbohydrate (CH_2O) as a product; as we shall see, this carbohydrate is initially a three-carbon sugar phosphate, termed a triose phosphate. Although this is still a grossly oversimplified version of the actual process, it is nevertheless fairly accurate.

Another outdated and misleading term that is still found in some textbooks is the reference to the carbon assimilation reactions as dark reactions of photosynthesis in order to distinguish them from the light reactions. Any implication that these reactions typically take place in the dark is entirely incorrect. Carbon assimilation is a light-dependent process. Not only does it require ATP and NADPH generated in the photosynthetic electron transport chain, which can only function in the light, many of the enzymes involved in carbon assimilation are themselves significantly active only in the light. We shall use the term photosynthetic carbon assimilation as a more accurate description

(a)

Guard cells turgid
Stomatal pore open

(b)

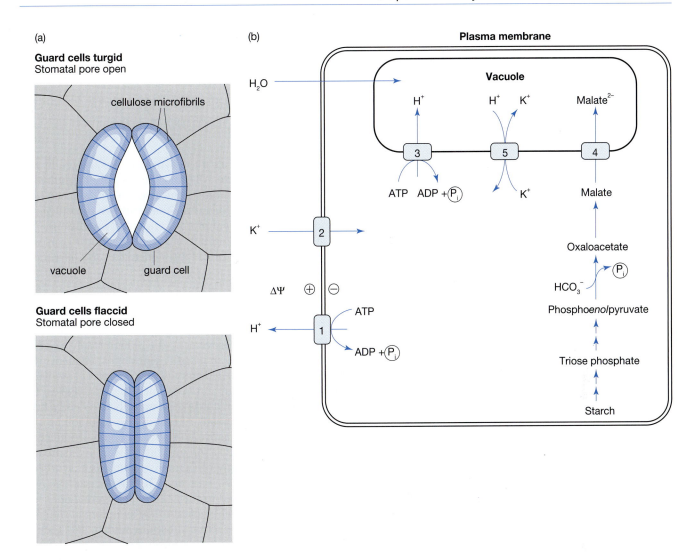

Figure 5.1 Opening and closing of stomata. (a) Guard cells are located within the leaf epidermis—either on the upper (adaxial) or lower (abaxial) surface of the leaf, or sometimes, depending on species, on both. They are always found in pairs surrounding the stomatal pores that allow gases to exchange into and out of the leaf. Cellulose microfibrils strengthen the inner guard cell walls so that swelling causes deformation into a banana shape that results in stomatal opening. When water moves out of the guard cells they become flaccid and the stomatal pore closes. (b) Stomata open in response to environmental signals such as the presence of light (especially blue light, but also red), increasing humidity and decreasing internal CO_2 concentration. This response is driven by activation of a H^+-ATPase (1) on the plasma membrane that extrudes protons out of the guard cell, producing a more negative membrane potential ($\Delta\psi$) (more negative $\ominus$ inside the cell, more positive $\oplus$ outside: changes from approximately –100 to –180 mV). This change in membrane potential triggers the opening of inward-specific K^+ channels (2) that move K^+ ions into the guard cell down its electrochemical gradient. K^+ concentrations can increase from approximately 100 mM to 800 mM. In some species, Cl^- ions may move into the guard cell to balance the positive charge of the K^+ ions. In all species, malate is formed from the breakdown of starch, and malate^{2-} ions also serve to balance the positive charge of the K^+ ions. Note that the breakdown of starch involves enzymes in both the chloroplast and cytosol, with phospho*enol*pyruvate carboxylase catalyzing the conversion of phosph*enol*pyruvate to oxaloacetate in the cytosol. A H^+-ATPase on the vacuolar membrane (3) moves protons into the vacuole, and this H^+ gradient drives malate accumulation (4) and import of K^+ ions (5) into the vacuole. Water enters the cell, moving into the vacuole in response to solute accumulation in the vacuole, causing the guard cells to swell and the stomata to open. Stomata close by the reverse process. An important signal is the hormone abscisic acid, which activates calcium channels on the plasma membrane (not shown), causing extracellular calcium to move into the cytosol. Increased cytosolic Ca^{2+} concentrations activate ion channels that extrude K^+ and anions (e.g. Cl^-) from the cell, resulting in water movement, loss of turgor and stomatal closure. Malate is converted to the osmotically inactive starch, which further reduces the osmotic potential and loss of turgor.

of the process whereby CO_2 is incorporated into carbohydrates and other photosynthetic products.

The Calvin cycle is used by all photosynthetic eukaryotes to convert carbon dioxide to carbohydrate

CO_2 is converted to carbohydrates in all photosynthetic eukaryotes by the same universal pathway, the Calvin cycle. All other photosynthetic pathways that we will discuss later, such as C_4 photosynthesis and Crassulacean acid metabolism (CAM), are really just modifications that improve the efficiency of the Calvin cycle under a particular set of environmental conditions. The Calvin cycle is essential for all photosynthetic eukaryotes; it is the only way that they can assimilate net amounts of carbon and make photosynthetic products.

Discovery of the Calvin cycle

The reactions of the Calvin cycle were discovered in the 1950s by Melvin Calvin, Andrew Benson, and James Bassham in an elegant series of experiments that used cultures of the eukaryotic, single-celled green algae *Chlorella pyrenoidosa* and *Scenedesmus*, accompanied by comparative studies on cereal leaves. The research was particularly timely. Paper chromatography had recently been developed by Martin and Synge as a means of identifying metabolites, and there was an abundant source of radioactive compounds being produced by post-war nuclear research laboratories. Calvin and his co-workers incubated the algal cells in the light, with a supply of CO_2 to support photosynthesis. Once a constant rate of photosynthesis had been reached a radioactively labeled CO_2 ($^{14}CO_2$) source was provided to the cells, which were inactivated seconds later by being dropped into boiling alcohol. By sampling at different intervals after the addition of the radioactive substrate, and by using chromatography to identify compounds that had incorporated the ^{14}C label, the progress of carbon could be tracked from CO_2 through the various reactions of the photosynthetic pathway. The first labeled intermediate was found to be a three-carbon compound, 3-phosphoglycerate (3-PGA). It is for this reason that plants using this pathway, without any modifications, are termed C_3 plants. Many crops are C_3 plants; these include wheat, barley, oats, rice, and pea.

Following the discovery of the three-carbon product, 3-PGA, Calvin and co-workers had expected to find that CO_2 reacted at first with a two-carbon substrate. However, on the basis of a careful analysis of intramolecular labeling patterns, carried out chiefly by Andrew Benson, it was shown that the CO_2 acceptor was a five-carbon compound, ribulose 1,5-bisphosphate (RuBP), which reacted with CO_2 to yield a short-lived six-carbon intermediate that split into two molecules of 3-PGA. Convincing evidence for this reaction came from experiments where the algae were transferred from light to dark. In these experiments, conducted by James Bassham, the 3-PGA pool was found to increase rapidly when the light was turned off, while, at the same time, the RuBP pool decreased. This experiment established the relationship between the two intermediates. Hence, it was concluded that 3-PGA was formed from the carboxylation of RuBP, which was itself formed as a result of a series of condensation and rearrangement reactions that originated from 3-PGA. While the formation of 3-PGA from RuBP continued for a very short period in the dark, the regeneration reactions that resulted in RuBP formation were unable to continue; hence 3-PGA levels rose while RuBP levels fell. This interpretation proved to be correct.

The remaining reactions, resulting in the formation of RuBP from 3-PGA, were identified from further radiolabeling experiments. These were carried out both under steady-state photosynthesis and during transient changes in light/dark exposure and CO_2 concentrations, which enabled changes in pool sizes and intramolecular labeling patterns to be interpreted. For example, it was during the futile search for a two-carbon CO_2 acceptor that Calvin's group discovered five-carbon and seven-carbon sugars, both mono- and diphosphates, that became radioactively labeled after a short exposure to $^{14}CO_2$. By analyzing the distribution of ^{14}C label within these molecules, Calvin and co-workers were able to deduce that the seven-carbon sugar phosphate (identified as sedoheptulose, both the mono- and bisphosphate forms) was formed from the condensation of a three-carbon compound together with a four-carbon precursor. The alternatives (i.e. six-carbon + one-carbon, or five-carbon + two-carbon) were discounted because they would not have produced the pattern of labeling of the carbon atoms that had been observed within the sedoheptulose molecule. This discovery led to the identification of the involvement of the transketolase reaction that transferred carbon residues from one intermediate to another (as described below). Although these reactions of carbon assimilation were first discovered in algae, the same reactions were confirmed in higher plants. Melvin Calvin received the Nobel Prize for Chemistry in 1961 for this work.

There are three phases to the Calvin cycle

There are 13 enzyme-catalyzed reactions in the Calvin cycle. All are located in the chloroplast and three of them, ribulose bisphosphate carboxylase/oxygenase (Rubisco), sedoheptulose 1,7-bisphosphatase, and phosphoribulokinase, are unique to the Calvin cycle. The 13 enzymes function together to assimilate CO_2 and to replenish RuBP, the CO_2 acceptor, so that the Calvin cycle can continue indefinitely in the light. Three phases of the cycle are defined: (1) carboxylation phase where CO_2 enters the pathway and is initially assimilated; (2) reduction phase, which produces triose phosphate that can in part be exported from the chloroplast to support biosynthetic reactions in the cytosol; and (3) regeneration phase, which produces more of the CO_2 acceptor, RuBP (Figure 5.2).

The carboxylation phase is catalyzed by ribulose bisphosphate carboxylase/oxygenase (Rubisco)

The first phase of the Calvin cycle is a single reaction, where 3-PGA is formed from the carboxylation of RuBP. This is the carboxylation reaction catalyzed by ribulose bisphosphate carboxylase/oxygenase (Rubisco). As its name suggests, Rubisco is capable of catalyzing two distinct reactions, acting as both a carboxylase and as an oxygenase.

It is the carboxylase reaction of Rubisco that serves as the starting point for the Calvin cycle. CO_2 itself (i.e. not bicarbonate [HCO_3^-]) is used to carboxylate RuBP to form two molecules of 3-PGA:

$$RuBP + CO_2 + H_2O \longrightarrow 3\text{-}PGA + 3\text{-}PGA + 2H^+$$

This reaction is virtually irreversible, having a large negative change in free energy for RuBP carboxylation, and proceeds through several intermediates in the process of forming 3-PGA (shown in detail in Figure 5.3).

The oxygenase reaction of Rubisco is an oddity. It uses the same substrate, RuBP, but reacts it with O_2. This reaction is catalyzed on the same active site

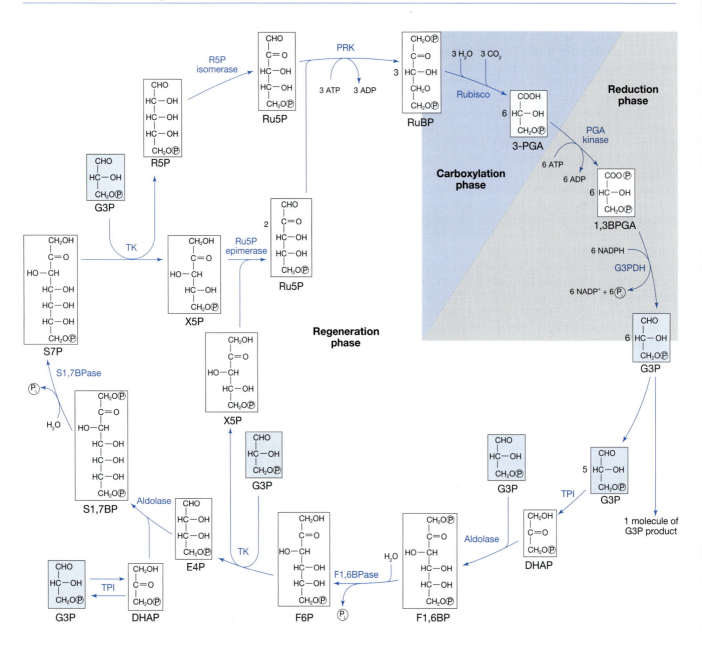

Figure 5.2. The Calvin cycle, showing reactions occurring in each of the three phases: carboxylation, reduction, and regeneration. Note that three molecules of CO_2 are shown to enter the cycle, forming six molecules of glyceraldehyde 3-phosphate following the first two phases (i.e. at the end of the reduction phase). Five of these molecules of glyceraldehyde 3-phosphate are required in the regeneration phase to replenish the three molecules of ribulose 1,5 bisphosphate and the other molecule of glyceraldehyde 3-phosphate is available for net biosynthesis (e.g. of sucrose, starch or other photosynthetic products).

For simplicity, the reactions are shown as irreversible, although all of the reactions are reversible except for those catalyzed by Rubisco, F1,6BPase, S1,7BPase and PRK.

Abbreviations: RuBP, ribulose 1,5-bisphosphate; 3-PGA, 3-phosphoglycerate; 1,3BPGA, 1,3-bisphosphoglycerate; G3P, glyceraldehyhde 3-phosphate; DHAP, dihydroxyacetone phosphate; F1,6BP, fructose 1,6-bisphosphate; F6P, fructose 6-phosphate; E4P, erythrose 4-phosphate; X5P, xylulose 5-phosphate; S1,7BP, sedoheptulose 1,7-bisphosphate; S7P, sedoheptulose 7-phosphate; Ru5P, ribulose 5-phosphate; R5P, ribose 5-phosphate; Rubisco, ribulose 1,5-bisphosphate carboxylase/oxygenase; PGA kinase, 3-phosphoglycerate kinase; G3PDH, glyceraldehyde 3-phosphate dehydrogenase; TPI, triose phosphate isomerase; F1,6BPase, fructose 1,6-bisphosphatase; TK, transketolase; S1,7BPase, sedoheptulose 1,7-bisphosphatase; Ru5P epimerase, ribulose 5-phosphate epimerase; R5P isomerase, ribose 5-phosphate isomerase; PRK, phosphoribulokinase.

(From *Biochemistry and Molecular Biology of Plants*, edited by Russell L. Jones, Bob B. Buchanan, Wilhelm Gruissem, 2000. Copyright John Wiley & Sons Limited. Reproduced with permission.)

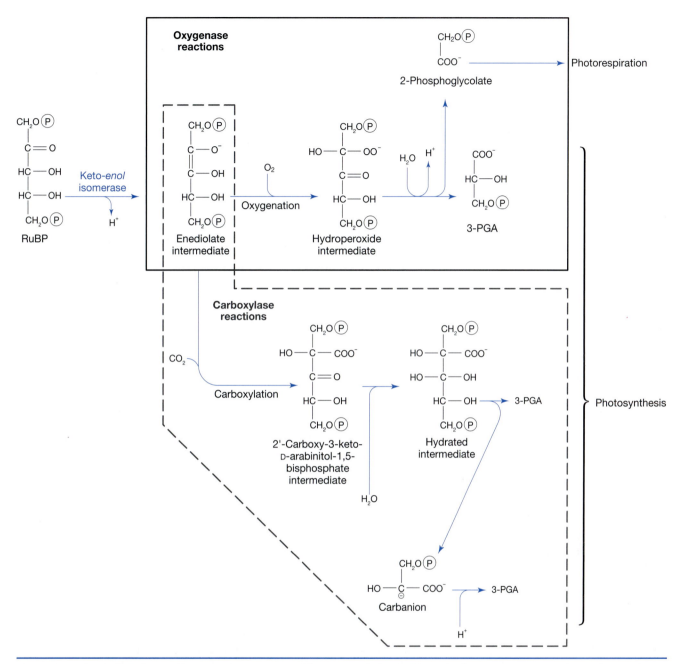

Figure 5.3 The carboxylase and oxygenase reactions of Rubisco. Common to both the carboxylase and oxygenase reactions is the formation of the 2,3-enediolate intermediate. RuBP binds to the active site of Rubisco and, provided the enzyme is active (see Figure 5.6), a proton is released from the C-3 position before RuBP undergoes a keto-enol isomerization to form the 2,3-enediolate intermediate.

In the carboxylase reaction CO_2 attacks the enediolate intermediate directly without binding to Rubisco, forming a six-carbon intermediate, 2′carboxy-3-keto-D-arabinitol-1,5-bisphosphate. This intermediate is then hydrated and cleaved between the second and third carbons to release one molecule of 3-PGA and a carbanion of 3-PGA, which is subsequently protonated to yield the second molecule of 3-PGA.

The oxygenase reaction shares the same substrate, 2,3-enediolate intermediate, with the carboxylase reaction. The precise details of the oxygenase reaction sequence still remain to be identified, but apparently the 2,3-enediolate can react

with O_2 to form a hydroperoxide intermediate. This can be hydrogenated and cleaved to produce one molecule of 3-PGA and one molecule of 2-phosphoglycolate.

The carboxylase and oxygenase reactions are catalyzed at the same active site on the enzyme, and CO_2 and O_2 are competitive substrates, hence CO_2 inhibits the oxygenase and O_2 inhibits the carboxylase reaction.

Rubisco also catalyzes a number of side reactions from intermediates of the main carboxylase and oxygenase reactions. For example, the enediolate intermediate may be converted to xylulose bisphosphate, while the hydroperoxide intermediate can generate pentodiulose bisphosphate and carboxytetritol bisphosphate. Finally, pyruvate can be formed from the carbanion intermediate, in approximately 1 of 150 of the carboxylase reactions.

Abbreviations: RuBP, ribulose 1, 5-bisphosphate; 3-PGA, 3-phosphoglycerate.

as the carboxylase reaction, hence CO_2 and O_2 are competitive substrates; CO_2 inhibits the oxygenase and O_2 inhibits the carboxylase reaction. The oxygenase reaction forms just one molecule of 3-PGA (instead of the two that would form in the carboxylase reaction) plus a two-carbon product, 2-phosphoglycolate, which is of no immediate use in the Calvin cycle:

$$RuBP + O_2 \longrightarrow 3\text{-PGA} + 2\text{-phosphoglycolate} + 2H^+$$

The reaction is shown in detail in Figure 5.3.

Each time Rubisco makes the mistake of reacting RuBP with O_2 instead of CO_2 the plant makes 50% less 3-PGA than it would have done if CO_2 had been used. This potentially eliminates the net gain in photosynthetic carbon. The 2-phosphoglycolate formed in the oxygenase reaction, however, enters into the photorespiratory pathway that eventually returns some 3-PGA to the Calvin cycle. Nevertheless, a substantial loss of carbon, as CO_2, occurs in the process, so that photosynthetic productivity is reduced. The photorespiratory pathway will be described in more detail later. For now, it is enough to appreciate that photorespiration is wasteful; it loses CO_2 that would otherwise have been used in the Calvin cycle and it functions only as an imperfect mechanism to salvage a portion of the carbon initially lost as phosphoglycolate. Nevertheless, photorespiration may serve some useful physiological functions, as discussed later.

Failing to discriminate well between CO_2 and O_2 is just one shortcoming of Rubisco. Another is its low specific activity. It is a very slow enzyme, reacting with just three molecules of CO_2 per second at each catalytic site, even when saturated with substrate. Most other enzymes achieve reaction rates in the order of 10^3 or 10^5 s^{-1}. To compensate for this sluggishness, Rubisco is produced in such massive quantity that it can contribute as much as 50% of the protein of a leaf—an enormous investment in nitrogen for one enzyme—amounting to 10 kg of Rubisco protein for every person on Earth. There is so much Rubisco protein in the chloroplast stroma that the unusual situation arises of there being a higher concentration of the enzyme than of its substrate; the stromal concentration of Rubisco catalytic sites is 4–10 mM while that of CO_2 is about 8 μM at 25°C.

Rubisco is clearly a vitally important enzyme, and yet, in a typical C_3 plant, it wastes carbon each time it reacts with O_2 and the enzyme needs to be made in large quantities to compensate for its low activity. It is chiefly for these reasons that so much research effort has been devoted to this single enzyme. Our current knowledge of Rubisco is described in the boxes in this chapter. Box 5.1 describes the structure, and Box 5.2 the synthesis of Rubisco; Box 5.3 explains the Rubisco specificity factor and how it varies between different species; Box 5.4 discusses some approaches that have been used to try and produce a Rubisco that reacts preferentially with CO_2 and not with O_2.

An improvement in the carboxylation/oxygenation activity of Rubisco (as described in Box 5.4) would result in increased efficiency of CO_2 use, saving water because the stomata would not need to open as wide, and saving nitrogen because the plant would not need to make as much Rubisco protein (see discussion of C_4 photosynthesis, later in this chapter). Whether or not these improvements also result in an increase in the yield of a crop plant will depend on the extent to which yield is limited by the rate of photosynthesis.

Box 5.1   Structure of Rubisco

There are two forms of Rubisco: form-I consists of large and small subunits while form-II consists only of large subunits.

Form-I Rubisco consists of two different types of subunit, the large subunit (LSU; 30–55 kDa) and small subunit (SSU; 12–19 kDa). Form-I is the most widespread of the two forms of Rubisco protein, being found in land plants, algae and cyanobacteria and in many photosynthetic bacteria. It is a large complex; the holoenzyme is arranged as a hexadecamer of 8 LSU and 8 SSU (L8S8; 550 kDa) showing structural symmetry, with an L2 dimer (of two large subunits) forming the functional unit. A central core of four LSU dimers is capped at the top and bottom by four SSUs (Plate 5.1). The catalytic site is located on the LSU, while the SSUs are important in holding the complex together and might also increase the specificity of Rubisco for CO_2.

Form-II Rubisco has only one type of subunit, the LSU. This form of Rubisco has been found in some prokaryotes and in eukaryotic algal dinoflagellates. Two form-II

Rubisco structures have been discovered so far. One is a homodimer (two LSUs; L2) present in some dinoflagellates and purple nonsulfur bacteria. The other, more recently discovered type of form-II Rubisco, is a decamer (10 LSU; L10) built up from a pentameric arrangement of L2 dimers, and is found in archaea such as *Thermococcus kodakaraensis*.

Form-II Rubiscos have very low specificity factors (see Box 5.3), which means that they are relatively poor at distinguishing between CO_2 and O_2, and the enzyme will be particularly sensitive to O_2 inhibition of carboxylation. Low specificity is not a problem in the absence of O_2, consequently most organisms with form-II Rubiscos are anaerobic. There are exceptions. The aerobic dinoflagellates *Symbiodinium* and *Gonyaulax* possess form-II Rubiscos, but they are thought to be able to maintain a positive carbon balance in the presence of O_2 by employing a carbon-concentrating mechanism. This mechanism would serve to increase internal CO_2 concentrations that would minimize the tendency of the form-II Rubisco to react with O_2.

Yields would only increase in response to increased photosynthesis if the plant had the capacity to allocate this additional carbon into the harvested product (e.g. the grain, tuber, or fruit). Therefore, factors such as transport, distribution, and storage of photosynthetic products are all important components, alongside photosynthetic activity, that will contribute to crop yield. Nevertheless, it is predicted that improvements in photosynthesis are likely to result in increased yield in many of our crops (see Long *et al.* 2006 in Further Reading).

The reductive phase of the Calvin cycle results in the formation of triose phosphate

In the first reaction of the reductive phase (Figure 5.2), the enzyme 3-phosphoglycerate kinase adds a P_i group on to 3-PGA to form 1,3-bisphosphoglycerate, using ATP generated in the photosynthetic electron transport chain. In the next step, 1,3-bisphosphoglycerate is reduced to glyceraldehyde 3-phosphate by NADP-dependent glyceraldehyde 3-phosphate dehydrogenase. NADPH for this reaction is also supplied by photosynthetic electron transport. Finally, triose phosphate isomerase catalyzes the isomerization of glyceraldehyde 3-phosphate to form dihydroxyacetone phosphate (DHAP). The formation of DHAP is strongly favored by the equilibrium position of the isomerase reaction. Glyceraldehyde 3-phosphate and DHAP are collectively termed triose phosphates. Both can be transported across the chloroplast envelope via the triose phosphate/phosphate translocator in exchange for P_i (Chapter 2). Export of triose phosphate to the cytosol supports the synthesis of sucrose (Chapter 7).

Box 5.2 The synthesis and assembly of Rubisco

Encoding of Rubisco large and small subunits

The Rubisco large and small subunits are encoded in the chloroplast and nuclear genomes respectively. One of the fascinating aspects of Rubisco is that two separate genomes are involved in its synthesis in land plants and green algae. The small subunit (SSU) polypeptide is encoded in the nuclear genome (gene denoted *Rbc*S) while the large subunit (LSU) polypeptide (gene denoted *rbc*L) is encoded in the chloroplast genome.

The SSU precursor polypeptide is produced on cytoplasmic ribosomes and contains an N-terminal plastid-targeting peptide that enables it to be transported into the chloroplast stroma. It was the early investigations of SSU expression that first led to the discovery of transit peptides that target proteins to organelles (see Chapter 3).

Not all photosynthetic organisms use both the nuclear and chloroplast genomes to code for Rubisco. In red and brown algae, which possess form-I Rubisco, the *rbc*L and *Rbc*S genes are both located in the chloroplast genome. This resembles the situation in prokaryotes with form-I Rubisco, where the *rbc*L and *Rbc*S genes occur adjacent to each other on the genome. All plastids share a common prokaryotic (cyanobacterial) ancestor, while separate plastid lineages arose later, leading to the green and red algal groups and their ancestors. Somewhere in this divergence the *Rbc*S gene seems to have migrated to the nucleus in the green plastid group, and remained in the plastid genome of the red group. The reason for this difference is

unclear as there is no obvious advantage, unless the nucleus provides a more effective means of regulating gene expression.

Rubisco large subunits folded within chloroplast chaperonin complexes

Rubisco LSUs are folded within chloroplast chaperonin complexes that show some similarities and differences to the GroEL/GroES complexes of prokaryotes and mitochondria. The Rubisco LSUs are synthesized on chloroplast ribosomes and then folded within chaperonin complexes. These chaperonins enable single LSU polypeptides to fold within an enclosed, hydrophilic environment in isolation from other LSUs. Without chaperonins, LSU polypeptides would associate together and the resulting aggregate would be inactive. Aggregation has been a common problem with a number of *in vitro* expression studies where Rubisco fails to fold correctly and the protein precipitates as an inactive, insoluble complex. Indeed, it has not been possible to express an active form of higher plant Rubisco in *Escherichia coli* or in any other *in vitro* translation system, although isolated chloroplasts and chloroplast extracts have been used successfully. Clearly, the biochemical machinery for the correct assembly of Rubisco is present in chloroplasts, but we still do not know enough about these components to be able to reproduce these conditions *in vitro*.

Despite the lack of a suitable *in vitro* expression system for higher plant Rubisco, experiments with isolated

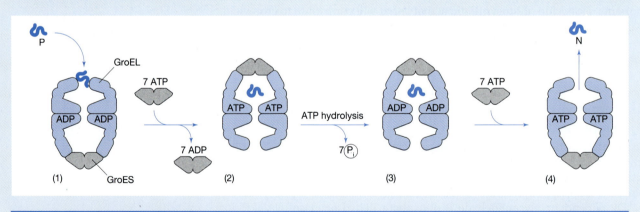

Figure 1 Model for the folding of proteins in the GroEL–GroES chaperonin cycle. Abbreviations: P, partially folded polypeptide; N, folded polypeptide. The GroES rings are shown in gray, and the GroEL ring in blue. Note that this model is based on our knowledge of prokaryotic proteins, as explained in the text. Polypeptide folding begins with the formation of a GroEL–GroES complex containing ADP, two GroEL rings and a single ring of GroES (1). The partially folded polypeptide (P) binds one of the GroEL rings (1). ATP and a second GroES ring bind and this traps the single polypeptide within a cage. The GroES ring is displaced from the opposite end of the complex and ADP is released (2). The polypeptide is now able to fold within the hydrophilic environment inside the cage, in an ATP-dependent process that involves ATPase activity within the GroEL ring (3). In the final stage (4) ATP and a further GroES ring bind and trigger the release of GroES and ADP from the opposite side of the complex. If the polypeptide is folded, it is released. If folding is incomplete, hydrophobic residues on the polypeptide will allow it to rebind to GroEL and the cycle continues until folding is complete.

Box 5.2 The synthesis and assembly of Rubisco (continued)

chloroplasts, and comparable studies on form-II Rubiscos have given us some insight into the mechanism of Rubisco subunit folding. Evidence of a chaperonin-assisted folding of LSU was obtained over 20 years ago, where it was found that the LSUs associated with a large protein complex (originally called the Rubisco-binding protein) in an ATP-dependent reaction. We now know that Rubisco assembly in the chloroplast requires homologues of the GroEL and GroES chaperonins (Figure 1) although we still do not fully understand the process in higher plants. It is likely that additional chaperonins may also be involved.

GroEL and GroES complexes

GroEL and GroES complexes are found in bacteria, mitochondria, and plastids, but the plastid forms show some distinct properties. The chloroplast GroEL homolog (originally called the Rubisco-binding protein, and also referred to as cpn60) resembles the bacterial and mitochondrial complexes in that it consists of two stacked rings, each consisting of seven 60 kDa subunits. However, in contrast to the other forms of GroEL the chloroplast complex contains two distinct subunits, α and β, with only approximately 50% homology between them. Although the structure of the bacterial GroEL complex is well characterized, we still do not fully understand how the chloroplast homolog is constructed. It appears to consist of equal numbers of α and β subunits and it has a similar, ring-like structure, to the bacterial GroEL when examined under an electron microscope.

The chloroplast GroES homologue (also called cpn10, or cpn21, or confusingly, sometimes cpn20) is also quite distinct from the bacterial and mitochondrial forms. Its subunits are more than twice the size (21–24 kDa) of other GroEL subunits (bacterial and mitochondrial, 10 kDa).

Each of the chloroplast GroEL subunits consists of two complete prokaryotic GroES-type (i.e. 10 kDa) peptides joined together head-to-tail. While the complete bacterial GroES complex consists of seven 10 kDa subunits, the chloroplast GroES homologue is thought to consist of just four of these double subunits. Nevertheless, the chloroplast GroES complex forms a similar ring structure to that of the prokaryotic GroES, and *in vitro* experiments have shown that they are able to replace one another in forming functional complexes with GroEL. Hence, despite the structural differences, chloroplast GroES would appear to function in the same way as all other GroES forms.

Although the chloroplast GroEL–GroES complex has some unique structural properties, the current view is that it functions in a similar manner to the prokaryotic and mitochondrial complexes (Figure 1).

The higher plant Rubisco holoenzyme

The higher plant Rubisco holoenzyme is assembled in the chloroplast stroma from eight LSUs and eight SSUs. Although the synthesis and assembly of higher plant Rubisco involves two separate genomes (chloroplast and nuclear) and three subcellular compartments (nucleus, chloroplast, and cytosol), the entire process is highly co-ordinated. The involvement of two separate genomes requires signals to be passed from one to the other to maintain this level of co-ordination. Both light and chloroplast development are important in regulating the expression of Rubisco subunits. Although phytochrome is involved in the processing of the light signal, we still do not understand what signals are passing between the plastid and nuclear genomes. Furthermore, we have an incomplete understanding of how the large and small subunits are assembled into the complete holoenzyme. Figure 2 summarizes these events.

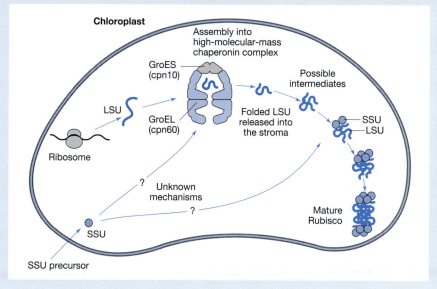

Figure 2 Assembly of the Rubisco holoenzyme in the chloroplast stroma. While the LSU is folded within the GroEL–GroES chaperonin complex, we do not know whether this is also the case with SSU. We also have very little knowledge of the mechanism that enables the subunits to form into the holoenzyme in higher plants. (From *Biochemistry and Molecular Biology of Plants*, edited by Russell L. Jones, Bob B. Buchanan, Wilhelm Gruissem, 2000. Copyright John Wiley & Sons Limited. Reproduced with permission.)

Box 5.3 Rubisco specificity

Rubisco carries out two distinct reactions; the carboxylase reaction uses CO_2 as substrate, and the oxygenase reaction uses O_2. The relative rate of these two reactions is expressed as the ratio of the carboxylation rate (v_c) to the oxygenation rate (v_o). This ratio depends on the concentration of each substrate (CO_2 and O_2) and on two particular properties of the enzyme: (1) the affinity for each of its substrates (Michaelis constant, K_c for CO_2 and K_o for O_2), and (2) the maximal velocity of each reaction (V_c for the carboxylase and V_o for the oxygenase). These two properties together define the specificity factor ($S_{c/o}$), an intrinsic property of the Rubisco enzyme:

$$S_{c/o} = (V_c \times K_o)/V_o \times K_c$$

The specificity factor is useful. It can, for example, be used to calculate the relative rate of the carboxylase and oxygenase reactions (i.e. v_c/v_o where v refers to the actual velocity of the reaction, whereas V refers to the maximum velocity) for any Rubisco operating at known concentrations of CO_2 and O_2. The following equation is used:

$$v_c/v_o = S_{c/o} \times ([CO_2]/[O_2])$$

this equation is derived from the following:

$$v_c/v_o = (V_c \times K_o)/V_o \times K_c) \times ([CO_2]/[O_2])$$

There is a wide degree of natural variation in Rubisco specificity factors, as shown in Table 1. Form-I Rubiscos (found in land plants, algae, and cyanobacteria and in many photosynthetic bacteria—see Box 5.1) tend to have a comparatively low V_c and high specificity factor, making them better able to distinguish between CO_2 and O_2. In contrast, form-II Rubiscos (found in some prokaryotes and in eukaryotic algal dinoflagellates) have a relatively high V_c and low specificity factor. This would make the carboxylase reaction of a form-II Rubisco faster than that of a form-I Rubisco at a given CO_2 concentration, providing that there was no O_2 present to inhibit it. Hence, form-II Rubisco is able to function best in anaerobic organisms. In fact, its low specificity factor means that organisms such as *Rhodospirillum rubrum* lose more carbon than they gain unless they grow under anaerobic conditions. Some prokaryotes possess both forms of Rubisco. The example here is of *Rhodopseudomonas sphaeroides*, a facultative anaerobe that benefits from using form-I Rubisco when growing aerobically, and form-II when growing anaerobically.

This natural variation in Rubisco specificity is an indication that selection pressure has resulted in an increase in discrimination between CO_2 and O_2. Rubisco is likely to have first evolved when there was no O_2 in the atmosphere, in purple photosynthetic bacteria or similar chemolithotrophs. Under these conditions there was no need for Rubisco to distinguish between CO_2 and O_2. It is only when oxygenic photosynthesis evolved, about 0.5×10^9 years later (see Chapter 4) and O_2 began to build up in the atmosphere that Rubisco would have been under any selection pressure to evolve specificity for CO_2.

Table 1 The natural variation in the specificity factor of Rubisco from a range of organisms

Rubisco source	Specificity factor ($S_{c/o}$)
Eukaryotes (all form-I Rubisco)	
Land plants	
Dicots	
Spinach	77–94
Pea	90
Soybean	82
Tobacco	77–93
Tomato	78
Monocots	
Wheat	82–90
Barley	79–87
Rice	85
Maize	78–96
Chlorophyte algae	
Scenedesmus obliquus	63
Chlamydomonas reinhardtii	61-71
Ulva pertusa	69
Euglenophyta	
Euglena gracilis	54
Chromophyta	
Cylindrotheca fusiformis	111
Olisthodiscus ruteus	101
Rhodophyta	
Porphyra yezoensis	145
Porphyridium cruentum	129
Prokaryotes	
Cyanobacteria (all form-I)	
Anacystis nidulans	44
Aphanizomenon flos-aquae	48
Coccochloris peniocystis	47
Synechococcus	41–45
Photosynthetic bacteria	
Chromatium vinosum (form-I)	44
Rhodospirillum rubrum (form-II)	12–17
Rhodopseudomonas sphaeroides form-II	9
Rhodopseudomonas sphaeroides form-I	62

Surprisingly, Rubisco has not vastly improved its specificity. There has only been, at best, a twofold increase in Rubisco specificity between cyanobacteria and land plants. Curiously, the highest Rubisco specificity factors are found in the rhodophyte and chromophyte algae (red and brown algae). Owing mainly to a high K_O value (i.e. a low affinity for O_2) these have specificity factors some 25–60% higher than those of land plants. Theoretically, if

Box 5.3 Rubisco specificity (continued)

these Rubiscos were introduced into the leaves of higher plants, it could increase photosynthesis and reduce water loss by as much as 50% (see Box 5.4). We still do not know what properties of red algal Rubisco cause this increased specificity, nor do we understand why higher plants have failed to evolve a Rubisco to match that found in these algae.

Why does Rubisco still continue to react with O_2 when, if it didn't, plants would grow faster and use water more efficiently? The answer probably lies at the heart of the Rubisco reaction. One of the intermediates, the 2,3-enediolate, is a substrate for both the oxygenase and carboxylase reactions. It reacts directly with both CO_2 and O_2 without the need for these to bind to the enzyme's active site (Figure 5.3). If any O_2 is available the 2,3-enediolate will react with it, leading to oxygenation and photorespiration. We are left with the conclusion that the oxygenase reaction is an inevitable property of the chemistry of the reaction. While the natural variation in Rubisco specificity offers promise of improving the enzyme further, this chemical inevitability suggests that it will not be possible to completely remove the oxygenase activity from Rubisco. Box 5.4 describes attempts to improve Rubisco.

Box 5.4 Attempts to improve Rubisco

A perfect Rubisco would react with CO_2 as rapidly as it diffused into its active site and would never mistake CO_2 for O_2. In other words it would have a high V_c a zero V_o and an infinite specificity factor ($S_{c/o}$)—see Box 5.3 for explanation. No such Rubisco has been found. Is it possible to produce one through genetic manipulation?

Scientists have been attempting to improve Rubisco since the 1980s. It has proved to be a challenge. Attempts to express higher plant Rubisco in *Escherichia coli* have failed because the expressed proteins do not assemble and instead form inactive aggregates. This has hampered progress with site-directed mutagenesis experiments on higher plant Rubisco, which require the protein to be expressed in its active form *in vitro*.

Studies of prokaryotic and algal Rubiscos have provided insight into the regions of the protein that affect Rubisco specificity. Mutations at the interface between the large and small subunits seem to affect specificity. So far, however, no mutations have resulted in an improved Rubisco—they have all made it worse.

One of the challenges of manipulating form-I Rubisco is the requirement for two genomes to be involved in its expression. Manipulation of the large subunit (LSU) requires the gene (*rbc*L) to be introduced into the chloroplast genome. Plastid transformation has been successful and allowed experiments to be carried out to produce mutations and deletions in the *rbc*L gene. There is still only a limited range of species for which plastid transformation has been successful, but these experiments have been informative. One approach has been to produce hybrid enzymes, with the LSU from one species and the SSU from another. Chloroplast transformation was used to replace the tobacco *rbc*L gene with the *rbc*L gene from sunflower. The result was a hybrid enzyme with sunflower LSU and tobacco SSU, having a specificity factor of 89. This is closer to that of the tobacco enzyme (85) than the sunflower enzyme (98) and indicates that the SSU can influence specificity even though the catalytic site is on the LSU. The hybrid enzyme had a substantially reduced catalytic activity, achieving only 25% of the carboxylation activity of either parent holoenzyme. Hence, an informative experiment but a poorer Rubisco.

In a further approach to generate a hybrid enzyme, the *Rbc*S gene from *Pisum sativum* has been introduced into the nucleus of arabidopsis, resulting in a hybrid Rubisco with an average of one LSU from *P. sativum* per holoenzyme. The hybrid enzyme had a correspondingly lower specificity factor indicative of the *P. sativum* enzyme. This seems to be in contrast to the previous experiment with tobacco and sunflower, where the SSU had the greater effect. Hybrid enzymes are likely to have altered interactions between the resulting LSU and SSU and, as explained earlier, this is known to have an effect on specificity.

Other hybrid Rubiscos have been produced from the cyanobacterial *Synechococcus* LSU and higher plant SSU from a range of species. All of these had specificity factors equal to or lower than that of *Synechococcus* and all had reduced carboxylation activities. It seems that changes in specificity are offset by a decrease in activity.

A consistent finding from studies of Rubiscos from a range of organisms, and from genetic engineering experiments, is that there is an inverse correlation between specificity and catalytic activity; as the capacity of the Rubiscos to discriminate between CO_2 and O_2 increases, the velocity of the carboxylase reaction decreases. This relationship between specificity and activity has frustrated attempts to improve Rubisco through genetic engineering, as each time the specificity is increased the resulting enzyme has a

continued

Box 5.4 Attempts to improve Rubisco (continued)

reduced carboxylase activity. A recent model provides a hypothesis to account for this inverse relationship between specificity and carboxylase activity (see Tcherkez *et al.*, 2006 in Further Reading). These authors propose that highly specific Rubiscos (i.e. those that favor the carboxylation reaction) form an intermediate that is so tightly bound to the enzyme that the subsequent hydrolytic and cleavage reactions are very much slower than they would be if the intermediate were more loosely bound. Hence, it may not be possible to improve both the specificity and the catalytic activity of Rubisco if these two components of the enzyme are so intrinsically linked.

The over-riding impression from these studies is that there are very few regions of the Rubisco molecule that are not important in one aspect or another of catalysis. This is in keeping with the highly interconnected nature of the Rubisco holoenzyme. In mutating a residue at one position in the protein it may be necessary to mutate residues at one or more additional sites to achieve a better Rubisco. Any change in a single amino acid causes a structural change in the protein and this affects both the interaction between subunits and the association with other proteins, such as Rubisco activase. Future approaches might be more successful if mutations in one position are balanced by mutations elsewhere so that these structural associations are maintained. Structural studies will have to accompany these experiments to determine the effect on the three-dimensional shape of the resulting enzyme.

Given that some form-I Rubiscos from non-green algae have particularly high specificity factors, another approach has been to introduce these genes into higher plants. The Rubisco of the red alga *Griffithsia monilis* has a specificity factor twice that of tobacco Rubisco. If introduced into plants it could theoretically improve CO_2 use and reduce water loss because the stomata would not need to open as wide. There would also be a saving in light requirement due to a reduction in ATP use as photosynthesis becomes more efficient. Although the *rbc*L and *Rbc*S genes from *Griffithsia monilis* have both been expressed in tobacco, only a small amount of protein was produced and there was poor incorporation into the holoenzyme.

There is still a lot that we do not know about the requirements for Rubisco expression and assembly in higher plants (see Box 5.2). Until we understand these processes better, they are likely to hamper our progress towards engineering an improved Rubisco in crop plants.

The regeneration phase of the Calvin cycle provides ribulose 1,5-bisphosphate to keep the cycle going

For the Calvin cycle to continue to turn, the CO_2 acceptor, RuBP, has to be replenished. This is accomplished by the regeneration phase of the cycle, where three-carbon sugar phosphates are converted to five-carbon compounds. If you read Box 5.5 before you continue with this section you can test whether you can design a better Calvin cycle.

The regeneration phase begins with the formation of fructose 1,6-bisphosphate from the condensation of DHAP and glyceraldehyde 3-phosphate, a reaction catalyzed by the enzyme aldolase (Figure 5.2). This enzyme can also condense DHAP with erythrose 4-phosphate to form sedoheptulose 1,7-bisphosphate, a reaction that occurs later on in the Calvin cycle. Both of these aldolase reactions are reversible, and the same enzyme plays a part in the oxidative pentose phosphate pathway (OPPP) where it catalyzes the reverse reactions (Chapter 6). The next step in the regeneration phase is the hydrolysis of fructose 1,6-bisphosphate in an irreversible reaction catalyzed by fructose 1,6-bisphosphatase, forming fructose 6-phosphate. Transketolase then removes a two-carbon unit from fructose 6-phosphate and transfers it to glyceraldehyde 3-phosphate to produce xylulose 5-phosphate and erythrose 4-phosphate. This reaction is also reversible, and operates in the reverse direction in the OPPP. Aldolase operates again, forming sedoheptulose 1,7-bisphosphate as described above. This metabolite is hydrolyzed to sedoheptulose 7-phosphate in an irreversible reaction catalyzed by sedoheptulose 1,7-bisphosphatase. Transketolase acts again, this time to remove a two-carbon unit from sedoheptulose 7-phosphate and transfer it on to glyceraldehyde 3-phosphate to produce ribose 5-phosphate and xylulose 5-phosphate.

Box 5.5 The carbon game

Your challenge is to produce a Calvin cycle that achieves the following:

• Convert five molecules of a 3-carbon compound to three molecules of a 5-carbon compound

There are some rules:

1. Use as few steps as possible
2. Keep the intermediates as small as possible
3. Never leave a carbon atom on its own—otherwise it will be lost as CO_2

Try it with building blocks:

Can you produce anything simpler than this?

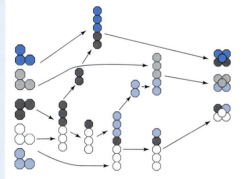

This is the **regeneration phase** of the Calvin cycle. The key reactions are:

(1) Aldolase, which joins molecules together to make larger ones, e.g.

> Two molecules of glyceraldehyde 3-phosphate (3-carbon) form one molecule of fructose 1,6-bisphosphate (6-carbon)

(2) Transketolase, which transfers 2-carbon groups from one molecule to another, e.g.

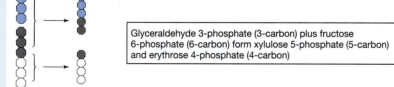

> Glyceraldehyde 3-phosphate (3-carbon) plus fructose 6-phosphate (6-carbon) form xylulose 5-phosphate (5-carbon) and erythrose 4-phosphate (4-carbon)

> These two enzyme reactions enable the molecular rearrangements to take place—rearranging the building blocks in the carbon game.

Original image courtesy of Professor Thomas D. Sharkey, Professor of Botany, University of Wisconsin-Madison, Wisconsin.

Ribulose 5-phosphate can now be formed from xylulose 5-phosphate by ribulose 5-phosphate epimerase, and from ribose 5-phosphate by ribose 5-phosphate isomerase. Finally, RuBP is formed from ribulose 5-phosphate by phosphoribulokinase in an irreversible reaction that consumes ATP.

Calvin cycle intermediates may be used to make other photosynthetic products

Although the reactions of the Calvin cycle explain how RuBP is regenerated, it is not obvious at first why so many reactions are needed and why such an array of intermediates is generated. If we take a step back and look at what is taking place, it is perhaps easier to see why. The purpose of the cycle is to convert CO_2 into triose phosphates and to regenerate the five-carbon acceptor, RuBP, so that the cycle may continue to operate in an autocatalytic manner. Five three-carbon molecules are ultimately being converted into three five-carbon molecules, without the loss of any carbon (as CO_2) at any stage. One way to understand how this is done is to play the carbon game, as described in Box 5.5. Not only do these reactions of the regeneration phase achieve the objective of forming RuBP, they also generate some useful intermediates in the process. For example, erythrose 4-phosphate can be used in the plastid-localized shikimic acid pathway to produce aromatic amino acids (see Chapters 8 and 11), while ribose 5-phosphate is a precursor for nucleotide biosynthesis.

The Calvin cycle is autocatalytic and produces more substrate than it consumes

One of the key features of the Calvin cycle is that it is autocatalytic. If nothing is drawn off from the cycle then it will produce more RuBP than it consumes. This point can be illustrated as follows. Start with five molecules of RuBP. Each can react with one molecule of CO_2, producing 10 molecules of 3-PGA, and thus retaining 30 carbon atoms in the product.

$$5 \text{ RuBP} + 5 CO_2 \longrightarrow 10 \text{ 3-PGA}$$

Without the removal of any triose phosphate for further metabolism, all 30 carbon atoms enter the regeneration phase to produce six molecules of RuBP, a gain of one RuBP from the cycle. In this way, the Calvin cycle can increase its own activity by producing more and more of its starting substrate (as an increase in substrate concentration will result in increased enzyme activity). Of course, some triose phosphate will ultimately have to be withdrawn from the cycle to achieve a net synthesis of carbohydrate; otherwise there would be no benefit to the plant. However, as long as no more than one in every six of the triose phosphates is removed, the Calvin cycle will continue to turn. These same stoichiometric constraints are applicable if and when other Calvin cycle intermediates are withdrawn for biosynthesis.

Autocatalysis can be demonstrated to occur *in vivo* and in isolated chloroplasts. If leaves have been left in the dark for some hours the Calvin cycle intermediates become depleted. When these leaves are exposed to the light there is an induction period of several minutes before photosynthesis reaches its maximum rate. This lag period is present because it takes time in the light to activate Rubisco and other Calvin cycle enzymes, to regenerate RuBP and to replenish the Calvin cycle intermediates generally.

Calvin cycle activity and regulation

The Calvin cycle is only active in the light, being regulated by thioredoxin and by stromal pH and Mg^{2+} ion concentration. Several of the enzymes that catalyze reversible reactions in the Calvin cycle also function in the OPPP, which oxidizes carbohydrates to CO_2 (see Chapter 6). Simultaneous operation of the Calvin cycle and the OPPP would result in the futile cycling of CO_2, i.e. with CO_2 being assimilated into carbohydrate by the Calvin cycle, and CO_2 being released via carbohydrate oxidation in the OPPP (Figure 5.4). Regulatory mechanisms must be in place to prevent such a futile cycle from happening. These regulatory mechanisms work to make some of the Calvin cycle enzymes active only in the light, and certain of the OPPP enzymes active only in the dark; regulation of OPPP enzymes is discussed in Chapter 6, here we will concentrate only on the regulation of Calvin cycle enzymes.

An important regulator of chloroplast enzymes is thioredoxin. Thioredoxins are small proteins (12 kDa) that consist of approximately 100 amino acids. They are found in all organisms, from Archaea to plants, and occur in most subcellular compartments (organelles). Four different forms (thioredoxins f, m, x, and y) occur in chloroplasts. Thioredoxins x and y are recent discoveries, which are thought to be involved in response to oxidative stress. Thioredoxins f and m are both involved in regulating carbon assimilation enzymes, with thioredoxin f being generally more effective with Calvin cycle enzymes. Recent proteomics studies have discovered that as many as 35 different chloroplast proteins interact with thioredoxins f and m. All thioredoxins contain the same sequence of four amino acids, cysteine–glycine–proline–cysteine, within the active site. The two cysteine groups provide a redox-sensitive mechanism that enables thioredoxin to exist in two interchangeable states. In the oxidized

Figure 5.4 The potential for futile cycling of carbon between the Calvin cycle and oxidative pentose phosphate pathway (OPPP).
The Calvin cycle and OPPP are shown in a simplified version to emphasize the potential for futile cycling between the two pathways, which would result in the simultaneous release and reassimilation of CO_2. To prevent futile cycling, key enzymes of the Calvin cycle and OPPP (shown with a bold arrow) are regulated by thioredoxin (Figure 5.5) and/or pH and Mg^{2+} ion concentration so that the Calvin cycle is active in the light, while the OPPP is active only in the dark.

Abbreviations: Rubisco, ribulose 1,5-bisphosphate carboxylase/oxygenase; G3PDH, glyceraldehyde 3-phosphate dehydrogenase; F1,6BPase, fructose 1,6-bisphosphatase; G6PDH, glucose 6-phosphate dehydrogenase; 6-PGDH, 6-phosphogluconate dehydrogenase; PRK, phosphoribulokinase.

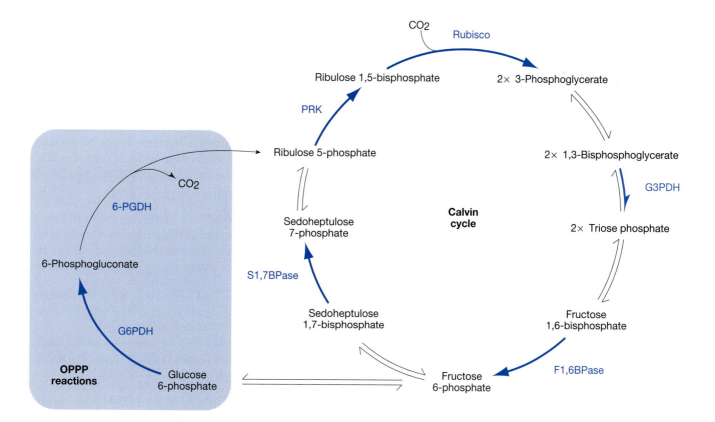

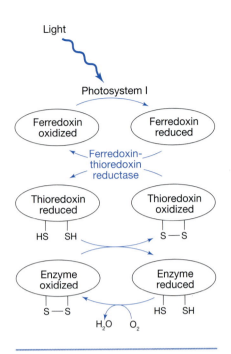

Figure 5.5 The thioredoxin/ferredoxin system for regulating enzyme activity in the light. In the light, electrons are transferred from Photosystem I to ferredoxin (NB the direction of electron transfer is indicated by the blue arrows). The reduced ferredoxin passes electrons to the enzyme, ferredoxin–thioredoxin reductase, which transfers the electrons to thioredoxin. The reduced thioredoxin, with two sulfhydryl groups (denoted as –SH) in its active site, can directly reduce disulfide groups (denoted as S–S on the oxidized enzyme) to form two sulfhydryl groups (–SH on the reduced enzyme) on target enzymes. The target enzymes and thioredoxins are oxidized by O_2 in the dark. Hence, target enzymes become reduced in the light and oxidized in the dark. Some enzymes are activated when in the reduced form (e.g. certain key enzymes of the Calvin cycle), while others are inactivated (e.g. some enzymes of the oxidative pentose phosphate pathway). This mechanism serves to regulate the activity of chloroplast enzymes in the light and dark.

form, the two cysteines are linked by a disulfide bridge (S–S). In the reduced form, the disulfide bridge is replaced with two sulfhydryl groups (SH) in the cysteine residues. Reduced thioredoxin can reduce the disulfide bonds of a large number of target proteins. It is this process of redox transfer that results in the light-dependent activation of key enzymes within the chloroplast. In the light, electrons are passed from water via Photosystems I (PSI) and II (PSII) to ferredoxin (see Chapter 4). Reduced ferredoxin passes electrons to the enzyme, ferredoxin-thioredoxin reductase, which, in turn, reduces thioredoxin. Reduced thioredoxin can then directly reduce disulfide groups on target enzymes, causing activation of some enzymes and inactivation of others (Figure 5.5). Certain of the Calvin cycle enzymes are regulated in this way, being activated by reduction in the light, and inactivated by oxidation in the dark. This type of control has been confirmed for the following enzymes: glyceraldehyde 3-phosphate dehydrogenase, fructose 1,6-bisphosphatase (which will only interact with thioredoxin *f*), phosphoribulokinase, and sedoheptulose 1,7-bisphosphatase. In contrast, some OPPP enzymes are inactivated by reduction in the light and activated by oxidation in the dark. This mechanism has been identified in the following OPPP enzymes: glucose 6-phosphate dehydrogenase (which will only interact with thioredoxin *m*), 6-phosphogluconate dehydrogenase, and transaldolase. In this way, futile cycling is prevented and the Calvin cycle is made to function only in the light (Figure 5.4). Thioredoxins also play a part in controlling some forms of the activase enzyme that activates Rubisco (see below).

A somewhat different control mechanism has arisen from the recent discovery of CP12, an 8.5 kDa 75 amino acid protein that forms a stable, 600 kDa complex with glyceraldehyde 3-phosphate dehydrogenase and phosphoribulokinase. CP12 contains two pairs of highly-conserved cysteines that can form internal disulfide bridges. When oxidized, the cysteines form a large complex with glyceraldehyde 3-phosphate dehydrogenase and phosphoribulokinase. The complex is thought to consist of two tetramers of glyceraldehyde 3-phosphate dehydrogenase, two dimers of phosphoribulokinase, and two monomers of CP12. Both glyceraldehyde 3-phosphate dehydrogenase and phosphoribulokinase are inactive when embedded in the complex. CP12 therefore provides a further mechanism for rendering the Calvin cycle inactive in the dark and active in the light. In the light, reducing conditions within the chloroplast stroma (e.g. NADPH and reduced thioredoxins) cause the CP12 complex to dissociate and the enzymes to become active. In the dark, the oxidizing conditions allow the disulfides to reassociate to form the inactive complex once again. Both phosphoribulokinase and glyceraldehyde 3-phosphate dehydrogenase are also activated by thioredoxin, as explained above. Hence, the dissociation of the CP12 complex, and the thioredoxin-dependent activation of the enzymes, both serve to turn on the Calvin cycle in the light. Formation of the CP12 complex in the dark serves as an additional mechanism to prevent futile cycling between the Calvin cycle and the OPPP. CP12 has been found in a number of higher plants, as well as in mosses, cyanobacteria, and green algae. It seems to be a particularly important means of light/dark regulation in cyanobacteria and eukaryotic algae because the Calvin cycle enzymes in these organisms are evidently redox-insensitive and are apparently not controlled directly by thioredoxin/ferredoxin.

Another means of increasing the activity of Calvin cycle enzymes in the light is provided by light-dependent changes in stromal pH and Mg^{2+} concentration. These mechanisms were first discovered in isolated chloroplasts, where the stromal pH was found to increase from approximately pH 7.2 to pH 8.0 on transfer from dark to light. Increases in stromal Mg^{2+} concentration also take place in the light. These ion movements occur as a consequence of light-driven transfer of H^+ from the stroma to the thylakoid lumen during photosynthetic electron transport (see Chapter 4). The charge balance is maintained by an influx of Mg^{2+} ions into the stroma, resulting in an increase in concentration from 1–3 mM in the dark to 3–6 mM in the light. Several Calvin cycle enzymes become activated under these conditions. Fructose 1,6-bisphosphatase, sedoheptulose 1,7-bisphosphatase, and phosphoribulokinase are all more active at alkaline pH and are stimulated by increases in Mg^{2+} concentration. This, together with their modification by reduced thioredoxin, ensures that they are significantly active only in the light.

Rubisco is a highly regulated enzyme

Regulation of Rubisco activity is a further, and important, means of controlling the Calvin cycle. The Mg^{2+} and pH changes described in the previous section also contribute to the control of Rubisco, although by a completely distinct mechanism, a process called carbamylation.

Rubisco only becomes active when CO_2 binds to form a carbamate on a particular lysine residue within the large subunit, the subunit that is directly responsible for catalysis. The CO_2 that binds here is distinct from that used as a substrate in the carboxylase reaction. The carbamate binds a Mg^{2+} ion, resulting in a conformational change that stabilizes the complex, and Rubisco becomes active (Figure 5.6). Because two protons are released when CO_2 binds, the increase in pH and Mg^{2+} concentration that occurs in the stroma in the light will favor this activation and contribute to light activation of the Calvin cycle. Rubisco, in its active, carbamylated form, will bind RuBP, and this brings about a further conformational change in the enzyme. The subsequent binding of either a CO_2 or O_2 molecule enables the Rubisco reaction to commence.

In the event that Rubisco binds to RuBP while in its decarbamylated state, a Rubisco–RuBP complex will form, which renders the enzyme inactive. Rubisco activity is restored by Rubisco activase, which releases RuBP from the complex and allows the active site to become carbamylated (Figure 5.7). In order for it to function, Rubisco activase has to associate with a specific binding site on the Rubisco large subunit polypeptide. This interaction appears to be quite specific, requiring Rubisco and Rubisco activase to recognize one another. Rubisco activase from tobacco or petunia cannot activate Rubisco from spinach or barley, for example. Similarly, Rubisco activase from barley or spinach will not activate Rubisco from petunia or tobacco. Changes in single amino acids within the large subunit can adjust the specificity. For example,

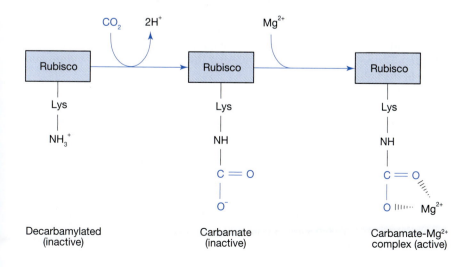

Decarbamylated (inactive)   Carbamate (inactive)   Carbamate-Mg^{2+} complex (active)

Figure 5.6 Activation of Rubisco by carbamylation. The large subunit (LSU) of higher plant Rubisco contains a lysine at position 201 of the sequence of 470 component amino acids. The ε-amino group of this lysine residue reacts with CO_2 to form a carbamate. Note that the CO_2 that binds in this reaction is entirely separate from the CO_2 used as substrate in the carboxylase reaction. Carbamylation precedes the carboxylase reaction and is an essential prerequisite for activation of all known Rubisco enzymes. The carbamate, once formed, then binds to a Mg^{2+} ion to form a stable complex that consolidates a conformational change to the LSU protein, resulting in an active Rubisco enzyme. Activation is favored by the high Mg^{2+} concentration and increased pH in the chloroplast stroma in the light.

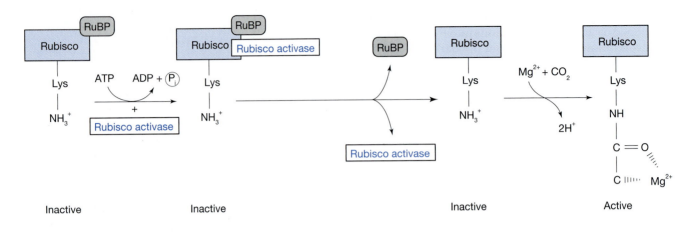

Inactive   Inactive   Inactive   Active

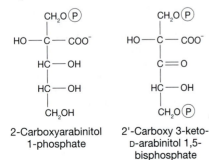

Figure 5.7 Rubisco activase removes sugar phosphates that would otherwise inactivate Rubisco. A number of sugar phosphates, RuBP in particular, are able to bind very tightly to the inactive, decarbamylated form of Rubisco. This binding blocks the active site and an amino acid loop closes over to prevent access to CO_2 and other substrates. Rubisco activase gains access to the active site by a process that requires mutual recognition between the Rubisco LSU and Rubisco activase proteins. Once bound, Rubisco activase removes RuBP in an ATP-dependent reaction, after which the activase is released to leave the active site available for activation by carbamylation: CO_2 binds to the lysine residue to form a carbamate whose structure is further consolidated by the binding of Mg^{2+} to yield active Rubisco. Rubisco activase may also remove the Rubisco inhibitor, 2-carboxyarabinitol 1-phosphate (CA1P; see Figure 5.8).

2-Carboxyarabinitol 1-phosphate

2'-Carboxy 3-keto-D-arabinitol 1,5-bisphosphate

substituting a proline for an arginine in the large subunit of *Chlamydomonas* Rubisco enables it to be activated by Rubisco activase from spinach. Hence, protein–protein recognition and interaction are important features of the Rubisco activase reaction.

The regulation of Rubisco activase provides a further mechanism for adjusting Rubisco activity in response to light. Two different processes lead to increased Rubisco activase activity, and consequently to an increased activation of Rubisco, in the light. In the first of these, the Rubisco activase reaction, which requires ATP, is subject to control by the ATP:ADP ratio. A high ATP:ADP ratio, as occurs in chloroplasts in the light, results in an increase in Rubisco activase activity. Similarly, a low ATP:ADP ratio (in the dark) results in reduced activity. A second mechanism involves a thioredoxin/ferredoxin system that results in a light-dependent activation, and a dark inactivation, of Rubisco activase. Both of these processes involve redox-sensitive cysteine residues within the Rubisco activase protein. Dark inactivation of Rubisco activase requires the formation of a disulfide between two cysteines within the C-terminus of the protein. If either of these cysteines is substituted with alanine, both the sensitivity towards the ATP:ADP ratio and the thioredoxin-dependent light/dark response of Rubisco activase are lost.

Even in its active, carbamylated form, Rubisco is subject to modification and regulation. It is strongly inhibited by the metabolite 2-carboxyarabinitol 1-phosphate (CA1P), which closely resembles 2'-carboxy 3-keto-D-arabinitol 1,5-bisphosphate, an intermediate of the carboxylation reaction (Figure 5.8). The tight binding of CA1P causes a conformational change that closes the active site of the carbamylated Rubisco, rendering it inactive. This inactivation is reversed by Rubisco activase, which releases CA1P and restores activity. CA1P has been found to accumulate in some leaves in the dark, and to be degraded in the light by a specific phosphatase. This day/night turnover of CA1P could provide a further mechanism for rendering Rubisco inactive in the dark and active in the light.

There is still considerable uncertainty about the precise *in vivo* role of Rubisco activase and of the inhibitor CA1P. Carbamylation requires mM concentrations of CO_2 for it to take place with pure Rubisco *in vitro*, i.e. with no activase

Figure 5.8 The Rubisco inhibitor 2-carboxyarabinitol 1-phosphate (CA1P) closely resembles 2'-carboxy 3 keto-D-arabinitol-1,5-bisphosphate, an intermediate of the carboxylase reaction (as shown in Figure 5.3). CA1P accumulates in the dark in some species and is removed by Rubisco activase to restore Rubisco activity in the light.

present, yet the stromal CO_2 concentration is in the μM range. Furthermore, CA1P has only been detected in a few species to date, and it remains to be seen whether it serves as a universal inhibitor of Rubisco *in vivo*.

Rubisco oxygenase: the starting point for the photorespiratory pathway

Of all of the properties of Rubisco, it is the oxygenase reaction that has in some ways been the most puzzling. Box 5.3 describes the reactions and considers why the enzyme that is so essential for CO_2 assimilation continues to make the mistake of reacting with O_2. In this section we consider the consequences of this property of Rubisco in relation to the photorespiratory pathway.

Photorespiration begins with the oxygenase reaction of Rubisco (Figure 5.9). As described above (see also Figure 5.3), this reaction forms 2-phosphoglycolate, a two-carbon compound that cannot be used in the Calvin cycle. Because it is made from RuBP, 2-phosphoglycolate can be seen as a drain that removes carbon out of the Calvin cycle. The reactions of the photorespiratory pathway eventually result in the production of 3-PGA that can be fed back into the Calvin cycle. One way to look at photorespiration, then, is as a scavenging, or salvaging pathway that mitigates the oxygenase mistake of Rubisco by diverting some of the lost carbon (2-phosphoglycolate) back into the Calvin cycle (as 3-PGA). In converting 2-phosphoglycolate back to 3-PGA there is, however, a loss of carbon as CO_2. Because the production of CO_2 only occurs in the light, the term photorespiration is used to denote this pathway. It is not at all closely related to the respiration that is described in Chapter 6 even though mitochondria play a role in both. Indeed photorespiration results in the consumption, rather than the production of ATP.

The photorespiratory pathway: enzymes in the chloroplast, peroxisome, and mitochondria

The photorespiratory pathway involves enzymes in the chloroplast, peroxisome, and mitochondria. The oxygenase reaction of Rubisco forms 2-phosphoglycolate and 3-PGA in the chloroplast stroma. While 3-PGA can be further metabolized in the Calvin cycle (Figure 5.2), 2-phosphoglycolate is converted to glycolate by phosphoglycolate phosphatase, which cleaves the phosphate group (Figure 5.10). Glycolate is exported from the chloroplast on a specific transporter, the glycolate/glycerate translocator, which transports both glycolate and glycerate. Although these two substrates may be counter-exchanged on the translocator, glycolate can also be transported separately, by co-transport with H^+ or counter-exchange with OH^-. This enables two molecules of glycolate to be exported while one molecule of glycerate returns to the chloroplast.

Glycolate is thought to enter the peroxisome by diffusion through relatively large pores, formed by transmembrane proteins called porins. Once inside the peroxisome, glycolate reacts with O_2, in a reaction catalyzed by glycolate oxidase, to produce glyoxylate and H_2O_2. These products are both highly toxic to chloroplasts, reacting with –SH groups and inactivating thioredoxin-regulated enzymes. Furthermore, H_2O_2 is a strong oxidizing agent that may damage thylakoid membranes, while glyoxylate inhibits Rubisco. This toxicity is one possible reason why this part of the photorespiratory pathway is located in peroxisomes. Another reason is the presence in peroxisomes of high concentrations of catalase, which detoxifies H_2O_2 by converting it to H_2O and O_2. Glyoxylate is converted to glycine by two different aminotransferase enzymes.

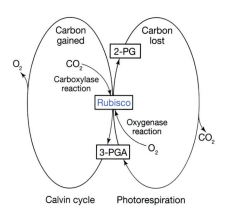

Figure 5.9 The central position of Rubisco in the reactions of the Calvin cycle and photorespiration. Rubisco reacts with CO_2 in the carboxylase reaction to produce 3-phosphoglycerate (3-PGA), a Calvin cycle intermediate used to produce photosynthetic products. Rubisco also reacts with O_2 in the oxygenase reaction to produce 2-phosphoglycolate (2-PG), an intermediate of the photorespiratory pathway. This metabolite is converted back to 3-PGA following a series of reactions (shown in detail in Figure 5.10) that release CO_2 in the process. Both photosynthesis and photorespiration occur in the light. In normal air the rate of the carboxylase reaction exceeds that of the oxygenase reaction by approximately threefold. Note that the concentrations of CO_2 and O_2 will alter the balance between the Calvin cycle and photorespiration—more CO_2 will increase Calvin cycle activity relative to photorespiration, more O_2 will have the opposite effect.

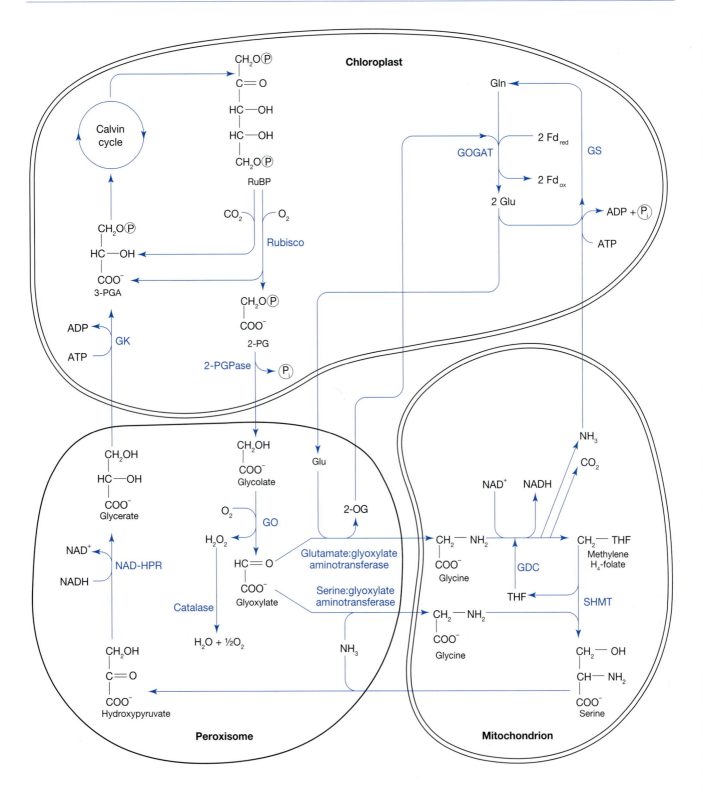

Glutamate–glyoxylate aminotransferase transfers an amino group from gluta-mate to glyoxylate to form 2-oxoglutarate and glycine, while serine–glyoxylate aminotransferase forms hydroxypyruvate and glycine from serine and glyoxy-late. Both reactions are needed, in a 1:1 ratio. Mutants lacking either enzyme have proved lethal to the plant unless it is kept under conditions that prevent the oxygenation of Rubisco and photorespiration from taking place. In the per-oxisome, hydroxypyruvate is further metabolized to glycerate; while 2-oxoglu-tarate diffuses out of the peroxisome and is transported into the chloroplast

Figure 5.10 The photorespiratory pathway. The photorespiratory pathway, or photorespiration, begins with the oxygenase reaction of Rubisco and involves reactions in the chloroplasts, peroxisomes and mitochondria. It results in the light-dependent release of CO_2 during the oxidation of glycine in the mitochondria. This reaction also releases NH_3, which is reassimilated by the combined reactions of glutamine synthetase (GS) and glutamate synthase (GOGAT) within the chloroplast. The photorespiratory pathway provides a means of returning carbon back to the Calvin cycle that would otherwise have been lost because of the mistaken reaction of Rubisco with O_2. As much as 75% of the carbon initially lost from the Calvin cycle is thus returned to it via this pathway.

Abbreviations: 3-PGA, 3-phosphoglycerate; RuBP, ribulose 1,5-bisphosphate; Methylene H_4-folate, methylene tetrahydrofolate; Rubisco, ribulose 1,5-bisphosphate carboxylase/oxygenase; GK, glycerate kinase; 2-PGPase, 2-phosphoglycolate phosphatase; 2-OG, 2-oxoglutarate; GO, glycolate oxidase; NAD-HPR, NAD-dependent hydroxypyruvate reductase; GS, glutamine synthetase; GOGAT, glutamine-oxoglutarate aminotransferase (or glutamate synthase); Fd, ferredoxin (red, reduced; ox, oxidized); GDC, glycine decarboxylase complex; SHMT, serine hydroxymethyltransferase. (From *Biochemistry and Molecular Biology of Plants*, edited by Russell L. Jones, Bob B. Buchanan, Wilhelm Gruissem, 2000. Copyright John Wiley & Sons Limited. Reproduced with permission.)

stroma to be used for glutamate formation, as described below. As well as being transaminated to glycine, glyoxylate can be reduced back to glycolate by reductase enzymes present in the chloroplast stroma and the cytosol.

Glycine leaves the peroxisomes through porins and enters the mitochondrial matrix, most probably via a glycine translocator, although this protein has yet to be characterized in detail. Glycine is oxidized within the mitochondria by a large multi-enzyme complex, glycine decarboxylase (GDC), and then converted to serine by serine hydroxymethyltransferase (SHMT).

There are four different enzymes present in the GDC complex, the H, P, T, and L proteins, and each one catalyzes a separate reaction. Collectively the action of these enzymes results in the oxidative decarboxylation and deamination of one molecule of glycine to produce one molecule of methylene tetrahydrofolate. This product is then joined to another molecule of glycine by SHMT to produce one molecule of serine. These reactions are described in detail in Figure 5.11.

The net outcome of the GDC and SHMT reactions is the production of one molecule of serine from two molecules of glycine, together with the generation of NADH and release of CO_2 and NH_3. NADH may be oxidized either via the mitochondrial electron transport chain, in which case three ATPs are produced per NADH, or by substrate-coupled reactions involving the movement of malate and oxaloacetate (OAA), sometimes called the malate/OAA shuttle. This shuttle mechanism works because of the high activity of NAD-dependent malate dehydrogenase within the mitochondrial matrix as well as in the cytosol and peroxisome. Malate dehydrogenase uses OAA to oxidize NADH to NAD^+ and to produce malate, which can be exported from the mitochondria on a translocator that exchanges malate with OAA. Malate is either oxidized in the cytosol, where the resulting NADH may be used for reducing substrates, for example, in the conversion of nitrate to nitrite by nitrate reductase (see Chapter 8). Alternatively, malate can enter the peroxisome, where its oxidation to OAA also generates NADH to be used, in this case, by hydroxypyruvate reductase to form glyoxylate. The OAA produced in either the peroxisome or cytosol moves back into the mitochondria in exchange for the export of malate. There is substantial evidence from *in vivo* and *in vitro* studies that all of these various routes for NADH oxidation can function in leaves. The extent to which each one operates at any one time will depend on the underlying metabolic demands. For example, if the mitochondria are oxidizing other substrates, then the electron transport chain may not be able to cope entirely with the demand for NADH oxidation, and substrate shuttles may start to operate. Similarly, high demands for cytosolic nitrate reduction may outcompete the requirement for NADH supply to the peroxisome. NADH can

Figure 5.11 Reactions catalyzed by the glycine decarboxylase complex (GDC) and serine hydroxymethyl transferase (SHMT). The glycine decarboxylase complex consists of four different proteins, each with separate catalytic activity. These are the P, H, T, and L proteins. The H protein is the smallest (14.1 kDa) and it interacts with each of the other proteins, in turn, via its lipoamide group (colored in blue, attached to H protein). The lipoamide group remains attached to the H protein throughout, and it becomes oxidized (in H_{ox}), attached to a methylamine group (in H_{met}) and reduced (in H_{red}) as the reaction progresses. The P protein (200 kDa homodimer) catalyzes the decarboxylation of glycine to release CO_2. The resulting methylamine residue ($CH_2NH_3^+$) is transferred to the H protein in its oxidized state (H_{ox}) where it attaches to the lipoamide arm to form a lipoamide–methylamine group on the H protein (H_{met}). Following transfer, the lipoamide–methylamine arm loops around and binds into a cleft on the H-protein surface so that it remains tightly bound to the protein. The H-protein then interacts with the T protein (45 kDa monomer). The methylene group is then transferred from the H protein to tetrahydrofolate (H_4FGlu_5) attached to the T protein, to produce CH_2-H_4FGlu_5 and NH_3. The reduced lipoamide (in blue, on the H protein, H_{red}) that results from this reaction is reoxidized by FAD bound to the L protein (homodimer of 60 kDa subunits), leading to the subsequent reduction of NAD^+ to NADH. Serine hydroxymethyl transferase (SHMT) converts another molecule of glycine to serine and regenerates H_4FGlu_5 from CH_2-H_4FGlu_5. The combined reactions lead to the following overall result:

2 glycine + NAD^+ $\longrightarrow$
serine + NADH + H^+ + CO_2 + NH_3^+

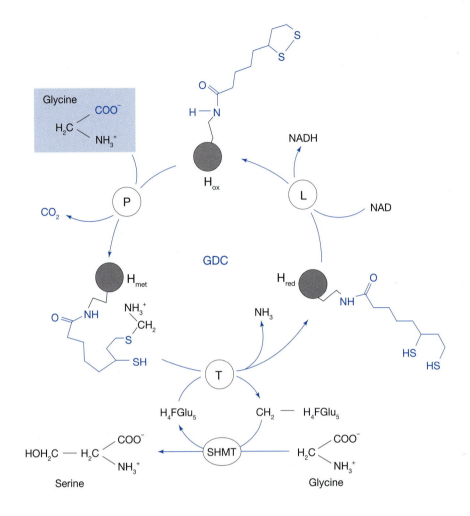

also be oxidized in association with the reduction of glyoxylate back to glycolate by glyoxylate reductase enzyme(s). These routes for NADH oxidation, along with others described in Chapter 6, provide a flexible means of supporting different metabolic pathways within the plant cell.

The GDC reaction is a key step in the photorespiratory pathway as it releases CO_2 and NH_3. The production of CO_2 is symptomatic of photorespiration, i.e. the light-dependent release of CO_2 from leaves, as discussed below. The consequences of NH_3 production would be serious if NH_3 were to simply accumulate; NH_3 production in photorespiration has been estimated to occur at 10 times the rate at which plants are able to assimilate nitrogen derived from nitrate. Fortunately, virtually all of the NH_3 produced during glycine oxidation is recaptured before it escapes from the leaf. This reassimilation takes place in the chloroplast and is catalyzed by glutamine synthetase (GS), an ATP-dependent enzyme that reacts NH_3 with glutamate to form glutamine. Glutamine, together with 2-oxoglutarate, is then converted to two molecules of glutamate by another chloroplast located enzyme, ferredoxin-dependent glutamine-oxoglutarate aminotransferase (Fd-GOGAT). One molecule of glutamate cycles back to serve as the substrate for the GS reaction. The second glutamate molecule exchanges with malate on the glutamate–malate translocator and exits the chloroplast, moving into the peroxisome to provide glutamate for the glutamate:glyoxylate aminotransferase reaction described earlier. The 2-oxoglutarate produced in this reaction enters the chloroplast, in exchange for malate export, and maintains GOGAT activity.

$$NH_3 + glutamate + ATP \longrightarrow glutamine + ADP + P_i \qquad\qquad GS$$

$$Glutamine + 2\text{-}oxoglutarate + Fd_{red} \longrightarrow 2\ glutamate + Fd_{ox} \qquad GOGAT$$

GS/GOGAT cycle:

$$NH_3 + 2\text{-}oxoglutarate + Fd_{red} + ATP \longrightarrow glutamate + ADP + P_i + Fd_{ox}$$

Serine, produced by SHMT in the mitochondria during the photorespiration process, moves into the peroxisome where it reacts with glyoxylate in the serine-glyoxylate aminotransferase reaction, forming glycine and hydroxypyruvate. Hydroxypyruvate reductase uses NADH to reduce hydroxypyruvate to D-glycerate, which diffuses out of the peroxisome and enters the chloroplast on the glycolate–glycerate translocator described above. Finally, ATP is used by glycerate kinase to convert D-glycerate to 3-PGA. This completes the scavenging function of the cycle as 3-PGA can now enter the Calvin cycle.

The isolation and analysis of mutants and the photorespiratory pathway

The isolation and analysis of mutants was a key to our understanding of the photorespiratory pathway. A screen selecting for mutants of arabidopsis that had lost the capacity to photorespire was first devised by Somerville and Ogren in the late 1970s. Mutants were produced by ethylmethane sulfonate treatment of the seeds (see Chapter 2). The mutant screen was based on the prediction that the inability to complete the full photorespiration pathway would be lethal under photorespiratory conditions. Seedlings were raised in high concentrations of CO_2 (>0.2%). This elevated concentration of CO_2 inhibited the Rubisco oxygenase reaction and prevented the operation of the photorespiration pathway. On transfer to normal air, these photorespiratory mutants showed signs of stress, such as yellowing of the leaves. Mutants were then rescued by transferring them back into high CO_2 where they would grow just as well as the wild-type plants. The mutants were then analyzed to identify which enzymes or metabolite transporters were missing. One approach was to transfer the mutants to normal air so that they would start to photorespire, and then look for any metabolite that started to accumulate in abnormal amounts. Such a metabolite would likely be a substrate for the missing enzyme. Mutants have now been found in arabidopsis, barley, pea, and tobacco, with mutations in catalase, chloroplast GS, Fd-GOGAT, GDC (P subunit), peroxisomal hydroxypyruvate reductase, phosphoglycolate phosphatase, SHMT, and serine-glyoxylate aminotransferase, and a chloroplast translocator of 2-oxoglutarate. Together these mutants have provided the definitive evidence needed to confirm the sequence of reactions that constitute the photorespiratory pathway.

Photorespiration may provide essential amino acids and protect against environmental stress

Although photorespiration is wasteful of carbon, energy, and reductant, it has not been eliminated through evolution. One of the most persuasive arguments for its retention is the advantage conferred by its scavenging role, returning carbon to the Calvin cycle that would otherwise have been lost because of the unavoidable oxygenase reaction of Rubisco. Yet, even plants that reduce photorespiratory losses by concentrating CO_2 still have the enzymic capacity to photorespire, as discussed below (C_3–C_4 and C_4

photosynthesis). We can therefore infer that photorespiration is of some benefit to the plant. The involvement of nitrogen in the cycle provides one possible function, and that is the synthesis of the two amino acids glycine and serine. However, there are alternative pathways for synthesis of glycine and serine from 3-PGA. These will operate when photorespiration is suppressed. Also, as glycine is needed for the synthesis of glutathione, which protects proteins from oxidative damage, it has been suggested that photorespiration helps plants to withstand environmental stress. One of the most damaging effects of stresses such as high light, low or high temperature, and water stress, is photoinhibition, involving photooxidative damage, which occurs when the photosynthetic electron transport chain is providing more reductant than is required for carbon assimilation (see Chapter 4). Such conditions may arise particularly when the stomata close in response to water or temperature stress, or if the light is simply too bright. Photorespiration is thought to mitigate these problems by using up excess ATP and reductant.

Photorespiration and the loss of photosynthetically fixed carbon

Photorespiration results in the loss of photosynthetically fixed carbon, and this gets worse with increasing temperature. The proportion of photosynthetically fixed carbon that is lost through photorespiration will principally depend on the relative rates of the carboxylase and oxygenase reactions of Rubisco. This ratio is determined by the ratio of the substrates CO_2 and O_2 in solution in the cell, as well as the inherent specificity factor of the Rubisco enzyme. The $[CO_2]$ to $[O_2]$ ratio in solution is influenced by temperature. The solubility of a gas decreases with increasing temperature. As the solubility of CO_2 decreases more than that of O_2 with increasing temperature, the ratio of $[CO_2]$ to $[O_2]$ decreases as the temperature rises (Table 5.1).

The relative carboxylase:oxygenase ratio will also decrease, not only because of the change in $[CO_2]$ to $[O_2]$ but also because the affinity of Rubisco for O_2 (K_o) is less sensitive to temperature than is the affinity for CO_2 (K_c). These two factors result in an increase in photorespiration relative to photosynthesis so that an increasing proportion of carbon is lost as the temperature rises.

To calculate the proportion of photosynthetically fixed carbon that is lost in photorespiration, it is necessary to determine the relative rate of the oxygenase and carboxylase reactions of Rubisco. As described in Box 5.3, this will depend on the Rubisco specificity factor ($S_{c/o}$) and on the relative concentration of CO_2 and O_2 to which it is exposed. Rubisco in the land plants has a specificity factor of approximately 80, so the rate of carboxylation (v_c) will be

Table 5.1 Temperature effect on the ratio of CO_2 and O_2 in solution			
Temperature °C	$[CO_2]$ in water (μM)	$[O_2]$ in water (μM)	$[CO_2]/[O_2]$
5	21.93	401.2	0.0515
15	15.69	319.8	0.0462
25	11.68	264.6	0.0416
35	9.11	228.2	0.0376

three times that of oxygenation (v_o) at 25°C as calculated from:

$$v_c/v_o = S_{c/o} \times ([CO_2]/[O_2]) \text{ (see Box 5.3)}$$

using a value of 80 for $S_{c/o}$ and 0.0416 for $[CO_2]/[O_2]$ (Table 5.1).

From the pathway shown in Figure 5.10 it can be seen that for every oxygenation reaction 0.5 molecules of CO_2 are released in the mitochondrial GDC reaction. So, if carboxylase is operating at three times the rate of oxygenase, for every CO_2 gained in photosynthesis (one carboxylase reaction) 0.166 molecules of CO_2 will be released in photorespiration (i.e. 0.33 oxygenase reactions for every carboxylase reaction, with 0.5 moles of CO_2 released per oxygenase: $0.33 \times 0.5 = 0.166$ molecules of CO_2), which amounts to 16.6% of the rate of photosynthesis. This is reasonably close to the range of *in vivo* measurements of photorespiration, which typically fall between 18 and 27% of net photosynthesis. Not all of this CO_2 is released from leaves, however. From gas exchange measurements we know that photorespiratory CO_2 can enter the intercellular spaces where it competes with external CO_2 for assimilation or diffusion out to the surrounding air. Reassimilation will therefore reduce the absolute photorespiratory losses of CO_2.

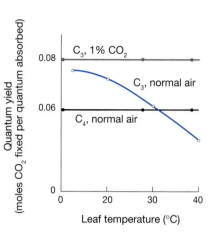

Photorespiration uses ATP and reductant

Photorespiration uses ATP and reductant and this lowers the quantum efficiency of carbon assimilation. Photorespiration uses ATP to convert glycerate to phosphoglycerate, and in the GS reaction to reassimilate NH_3 resulting from glycine oxidation. ATP is also required to convert 3-PGA, produced in the oxygenase reaction, into RuBP. Reducing equivalents, in the form of reduced ferredoxin (Fd_{red}) and NADPH, are needed for the Fd-GOGAT reaction and for the conversion of 3-PGA to RuBP. If photorespiration is operating at 25% of the net rate of photosynthesis, 5.375 moles of ATP and 3.5 moles of NADPH are used for every mole of CO_2 that is fixed. This compares with 3 moles ATP and 2 moles NADPH per mole of CO_2 if the Calvin cycle is operating in the absence of photorespiration. The additional demand for ATP and NADPH associated with photorespiration results in an increased requirement for the production of ATP and NADPH by the light reactions of photosynthesis. Photorespiration therefore reduces the efficiency with which light energy is used to fix carbon, which is measured as quantum efficiency (moles of CO_2 fixed per quantum of light absorbed). This effect is illustrated in Figure 5.12 where an increase in temperature lowers the quantum efficiency because photorespiration increases relative to photosynthesis. If plants are treated in the same way but in high CO_2 to suppress photorespiration, this effect is removed and the quantum efficiency is similar and higher at all temperatures.

Decreasing global carbon dioxide concentrations caused a rapid evolution of C₄ photosynthesis

The balance between photosynthesis and photorespiration depends on the relative rates of the carboxylase and oxygenase reactions of Rubisco, which in turn are dependent on the proportion of CO_2 and O_2 at the enzyme's active site, as discussed earlier. It follows that C_3 plants will become less and less carbon-efficient as either CO_2 concentrations fall, or O_2 concentrations rise. A fall in atmospheric CO_2 and rise in O_2 300 million years ago is thought to have led to the selection pressure that resulted in the evolution of mechanisms to concentrate CO_2 around Rubisco. One such CO_2 concentrating mechanism is provided by C_4 photosynthesis, a specialized form of photosynthesis that has evolved independently at least 45 times in 19 families of

Figure 5.12 The effect of temperature on the quantum yield of photosynthesis in C₃ and C₄ plants. C_3 plants photosynthesizing in normal air become less efficient at using light energy as the temperature rises. This response occurs because increasing temperatures cause a decrease in the CO_2 to O_2 ratio in solution (see Table 5.1), resulting in an increase in photorespiration relative to photosynthesis. Consequently, the quantum efficiency of photosynthesis is reduced, because additional light energy will be needed to assimilate a given amount of CO_2 when some of this CO_2 is being lost through photorespiration. When photorespiration is prevented, by exposing the plants to high CO_2 concentrations (>0.2% CO_2 prevents Rubisco from reacting with O_2), the quantum yield of C_3 photosynthesis remains constant, as photorespiration is no longer a factor. Note that C_4 plants do not show any reduction in quantum efficiency with increasing temperature because they do not significantly photorespire. Note also that C_4 plants have a lower quantum yield than C_3 plants at all but the highest temperatures because the C_4 pathway has a higher demand for ATP, and hence requires more light energy to assimilate a given quantity of CO_2. (From *Biochemistry and Molecular Biology of Plants*, edited by Russell L. Jones, Bob B. Buchanan, Wilhelm Gruissem, 2000. Copyright John Wiley & Sons Limited. Reproduced with permission.)

angiosperms. C_4 photosynthesis is found in just 3% of approximately 450,000 vascular plant species. Monocots make up about 75% of all C_4 species, and these occur within two families, the Poaceae (grasses) and Cyperaceae (sedges). Within the dicots, a single family, the Amaranthaceae accounts for 50% of all dicot C_4 species.

Definitive evidence of C_4 plants in the fossil record dates back to 7 million years ago (see the section on stable isotope analysis, later in this chapter), while the specialized leaf anatomy, described below, can be identified in fossils dating back 12.5 million years. It was during the Miocene period that the burgeoning of C_4 species took place. C_4 plants were emerging across the globe, in response to CO_2 concentrations that fell to as little as 180 p.p.m. between 6 and 22 million years ago. Fossil records show that C_4 plants were present in North and South America, Pakistan and Africa between 5 and 7 million years ago. C_4 was clearly a pathway beneficial to plant survival as it evolved independently on many occasions. The coincidence of the appearance of C_4 plants with periods of increased drought and decreased atmospheric CO_2 concentrations provides a clue to the benefits of C_4 photosynthesis.

C_4 photosynthesis concentrates carbon dioxide at the active site of Rubisco

C_4 photosynthesis acts to concentrate CO_2 around Rubisco. This CO_2-concentrating mechanism has two major benefits to the plant. First, it virtually eliminates photorespiration, by inhibiting the oxygenase reaction of Rubisco. Second, it provides a means of reducing the stomatal aperture, and hence of cutting water losses, while still maintaining a high rate of photosynthesis, as explained below.

The C_4 photosynthesis pathway was characterized by Hatch and Slack in the 1960s and 1970s. Their work extended earlier observations made in the 1950s by Kortschak and co-workers at the Hawaiian Sugar Planters Association Experimental Station. Kortschak reported surprising results in experiments involving feeding $^{14}CO_2$ to sugarcane leaves in order to identify the photosynthetic intermediates, as Calvin and co-workers had done some years before with other species. The early photosynthetic products observed in sugarcane were the four-carbon compounds malate and aspartate and not 3-PGA as had been expected. The reason for this result remained unclear for another 10 years until Hatch and Slack carried out a series of radiolabeling experiments with sugarcane and maize. It became apparent that CO_2 was first being incorporated into OAA and then into malate. The subsequent decarboxylation of this 4-carbon acid released CO_2 deep within the leaf where Rubisco then incorporated it into 3-PGA for further metabolism in the Calvin cycle. The pathway became known as the C_4 photosynthetic pathway. Plants that use this pathway for photosynthesis are referred to as C_4 plants and they are particularly widespread among the Gramineae (Poaceae) (e.g. sugarcane, maize, *Spartina*), Amaranthaceae (e.g. *Atriplex*), and Cyperaceae (the sedge family). The isotopic labeling data were soon followed by analysis of the enzyme complement of the leaves, leading to the understanding that photosynthesis is shared between two different types of cells in C_4 plants.

Spatial separation of the two carboxylases in C_4 leaves

Leaves of C_4 plants have a specialized anatomy that provides spatial separation of the two CO_2 fixing enzymes, phospho*enol*pyruvate carboxylase (PEP

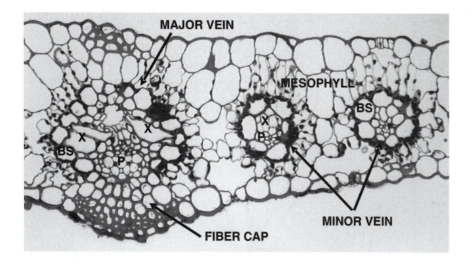

Figure 5.13. The specialized leaf anatomy that is typical of a C₄ leaf. Light micrograph of a cross-section through a C₄ leaf (Sorghum) showing the two types of photosynthetic cell; the bundle sheath cells that surround the vascular cells, and the mesophyll cells that completely enclose the bundle sheath and present a barrier through which gases must first enter before they can access the bundle sheath. Abbreviations: BS, bundle sheath; P, phloem; x, xylem. (Image produced by Dr. Vincent Franceschi. Original image courtesy of Prof. Gerald Edwards, School of Biological Sciences and Center for Integrated Biotechnology, Washington State University.)

carboxylase), and Rubisco. A cross-section through a leaf of a C₄ plant (Figure 5.13) reveals a distinct arrangement of the chloroplast-containing cells. This has been termed Kranz anatomy (German *Kranz* translates as *wreath*), a good description of the layout of the cells as the bundle sheath cells surround the veins in a wreath-like arrangement. The mesophyll cells are more generally distributed and are in contact with the stomatal cavity. Rubisco is absent from the mesophyll cells and present in the bundle sheath cells where it is protected in a CO_2-enriched and O_2-depleted environment. Any CO_2 that enters the stomata will first pass into the mesophyll cells (Figure 5.14). It is in these cells that bicarbonate is first assimilated, by PEP carboxylase. This enzyme is absent from the bundle sheath cells and present in the cytosol of the mesophyll cells. This compartmentalization is a key feature of C₄ photosynthesis, i.e. the two carboxylase enzymes PEP carboxylase and Rubisco are spatially separated in different cells altogether.

The enzyme responsible for C₄ fixation is PEP carboxylase. PEP carboxylase is a highly regulated enzyme, as we shall see in the next section. An important

Figure 5.14 Overview of the C₄ photosynthesis pathway showing the relationship between leaf anatomy and the biochemical reactions. Air surrounding the leaf enters through open stomata in the light and moves into the mesophyll cell from the intercellular air spaces. The C₄ pathway then takes place via four distinct stages. Carbonic anhydrase (not shown) within the mesophyll cells converts CO_2 to bicarbonate (HCO_3^-), the substrate for CO_2 fixation to form a C₄ acid, which is then transported to the bundle sheath cells that lie deeper within the leaf. Here the C₄ acid is decarboxylated to release CO_2 for fixation in the Calvin cycle that is present only within these bundle sheath cells. The resulting C₃ acid is transported out into the mesophyll cell to provide the substrate used to regenerate phospho*enol*pyruvate for the initial CO_2 fixation reaction. The spatial separation of the two CO_2 fixation reactions within two distinct cell types is a characteristic feature of the C₄ pathway. The release of CO_2 within the bundle sheath cells is an effective CO_2-concentrating mechanism that enables the Calvin cycle to operate at far higher concentrations of CO_2 than it would in a C₃ leaf. (From *Biochemistry and Molecular Biology of Plants*, edited by Russell L. Jones, Bob B. Buchanan, Wilhelm Gruissem, 2000. Copyright John Wiley & Sons Limited. Reproduced with permission.)

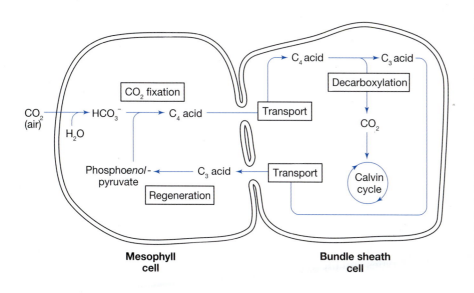

property is that, unlike Rubisco, it reacts with bicarbonate (HCO_3^-) rather than with CO_2. The advantage of this is that there is a 50-fold higher concentration of HCO_3^- than CO_2 in solution in the cytosol. This is partly because of the higher concentration of HCO_3^- in solution (e.g. at a cytosolic pH of 7.0 the ratio of [HCO_3^-] to [CO_2] in solution is approximately 5:1) and partly because of the presence in the mesophyll cell cytosol of carbonic anhydrase, which facilitates the conversion of CO_2 to HCO_3^- ($CO_2 + H_2O \rightarrow HCO_3^- + H^+$). Furthermore, the molecular structure of HCO_3^- is quite distinct from that of O_2, which more closely resembles CO_2. And, unlike Rubisco, PEP carboxylase does not react with O_2. Another important difference between PEP carboxylase and Rubisco is that, while Rubisco is regulated so that it only functions in the light, PEP carboxylase may be active in the light or in the dark, depending on the species in which it is found. As we shall see later, the form of PEP carboxylase found in C_4 plants is only active in the light. Its substrate, PEP, is generated from pyruvate (via pyruvate orthophosphate dikinase (PPDK) as discussed below), with ATP being supplied by photophosphorylation. In plants where PEP carboxylase is active in the dark (e.g. CAM plants, as explained later), PEP is provided via starch degradation, which also provides respiratory substrates for ATP synthesis.

Stages of C_4 photosynthesis and variations to the basic pathway

The four stages of the C_4 pathway are: (1) CO_2 fixation in the mesophyll cells, catalyzed by PEP carboxylase to form a 4-carbon product, OAA, which is rapidly reduced to malate and/or transaminated to aspartate; (2) transport of malate or aspartate from the mesophyll cells to the bundle sheath cells; (3) decarboxylation of malate or OAA in the bundle sheath cells, releasing CO_2 for assimilation by Rubisco into the Calvin cycle; (4) regeneration of the CO_2 acceptor, PEP, following the return of carbon from the bundle sheath (as pyruvate or alanine) to the mesophyll cells. The C_4 pathway is summarized in Figure 5.14 and shown in detail in Figure 5.15(a–c), which shows the three variations of the pathway. These pathway variations all share the same four stages described above, but they differ in the compounds that move between the mesophyll and bundle sheath, in the enzymes catalyzing the decarboxylation reactions, and in the ultrastructure of their organelles.

The NADP-malic enzyme (NADP-ME) species, which transport malate to the bundle sheath, have a particularly effective means of protecting Rubisco from O_2. The bundle sheath chloroplasts of NADP-ME species are typically agranal (i.e. they have unstacked thylakoids; Figure 5.16, page 126) and lack PSII. This prevents them from producing O_2. Although PSI is present, the absence of PSII eliminates the production of NADPH by the electron transport reactions of photosynthesis (see Chapter 4). NADP-ME therefore makes an important contribution by generating additional NADPH for the reduction of 3-PGA in the Calvin cycle. However, NADP-ME only generates about half of the NADPH needed to reduce 3-PGA within the bundle sheath chloroplasts. Consequently, some 3-PGA is transported from the bundle sheath to the mesophyll cells where it is reduced using NADPH generated via PSII. Hence, the Calvin cycle is shared between the two cell types. Further details of the pathway are shown in Figure 5.15(a). NADP-ME species include: maize (*Zea mays*) and sugarcane (*Saccharum officinale*).

In NAD-malic enzyme (NAD-ME) C_4 species the chloroplasts in both the mesophyll and bundle sheath cells contain grana and PSII activity. Hence, NADPH can be generated via the photosynthetic electron transport chain in both cell types and there is no need to shuttle reducing equivalents from one

Figure 5.15 (a)

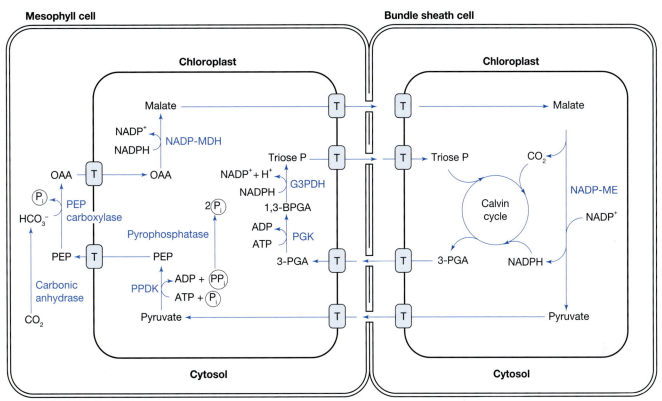

Figure 5.15 (b)

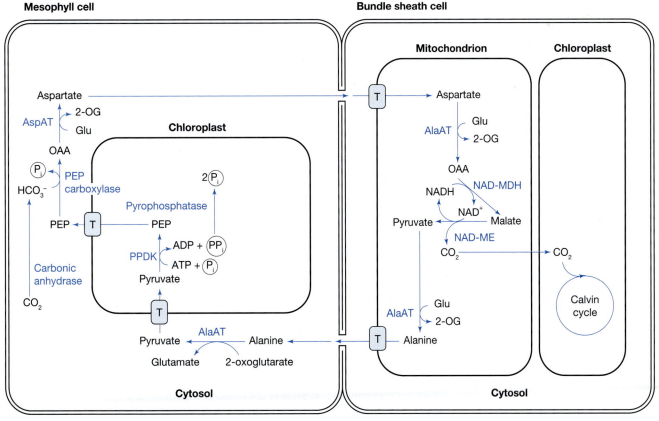

Figure 5.15 (c)

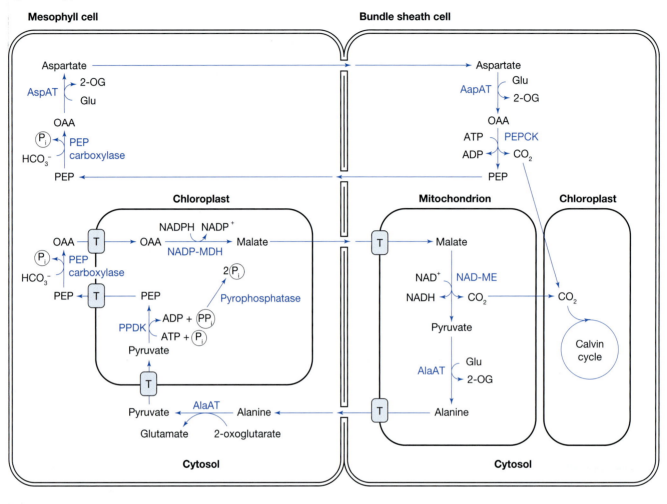

Figure 5.15 The three variants of the C$_4$ pathway. Each of the various pathways is named after the enzyme that carries out the decarboxylation reaction to produce CO$_2$ in the bundle sheath cells. The movement of metabolites between the mesophyll and bundle sheath cells is thought to occur by diffusion through the plasmodesmata. Movement of metabolites across the membranes of organelles usually involves specific translocators. These translocators are indicated by a letter T within each of the figures. In all three variants OAA is formed in the mesophyll cell cytosol as a result of the PEP carboxylase reaction, which reacts with bicarbonate (HCO$_3^-$) and PEP. (a) In NADP-ME-type plants, OAA is converted to malate via NADP-dependent MDH within the mesophyll cell chloroplasts, and malate moves to the bundle sheath chloroplasts where it is decarboxylated by NADP-ME to produce pyruvate, and CO$_2$ for Rubisco. The NADPH generated in this reaction can be used in the reductive reactions of the Calvin cycle. As explained in the text, as the NADP-ME reaction only produces 50% of the NADPH needed in the Calvin cycle, some 3-PGA is exported to the mesophyll cells where it is phosphorylated and reduced to triose phosphate within the chloroplast before moving back to the bundle sheath cells. Pyruvate, produced in the NADP-ME reaction within the bundle sheath chloroplast, moves to the mesophyll cells where it is converted to PEP by PPDK, a reaction that occurs in all three variants, within the mesophyll chloroplasts. This reaction generates PEP, which is transported out of the chloroplast via the triose phosphate/phosphate translocator for use in the PEP carboxylase reaction in the cytosol. (b) In the NAD-ME-type plants, OAA is converted to aspartate by aspAT that reacts OAA with glutamate to form aspartate and 2-oxoglutarate in the mesophyll cell cytosol. Aspartate moves from the mesophyll to the bundle sheath where it is converted back to OAA by the reverse reaction of aspAT, this time within the mitochondria. OAA is subsequently converted to malate by a mitochondrially localized NAD-dependent malate dehydrogenase, with the resulting malate being decarboxylated to pyruvate in the mitochondria by NAD-ME. This reaction releases CO$_2$ that diffuses into the chloroplast where it is assimilated into the Calvin cycle by Rubisco. Pyruvate enters into another aminotransferase reaction, this time catalyzed by alaAT, which converts pyruvate and glutamate to alanine and 2-oxoglutarate, also within the bundle sheath mitochondria. Alanine moves from the bundle sheath to the mesophyll cell where the reverse reaction of alaAT produces pyruvate for PEP production in the chloroplast (via PPDK). PEP is exported from the chloroplast to the cytosol where it serves as the substrate for carbon assimilation via PEP carboxylase. (c) The PEPCK-type C$_4$ pathway involves two C$_4$ cycles, one that cycles carbon through aspartate and PEPCK, and the other that involves malate and NAD-ME. As in NAD-ME-type plants, OAA formed from PEP in the mesophyll cell cytosol, is converted to aspartate via a cytosolic aminotransferase (aspartate aminotransferase, aspAT).

Figure 5.15 (continued)

Aspartate diffuses into the bundle sheath cells where it is converted back to OAA, by another cytosolic aspAT. PEPCK, a cytosolic enzyme, converts OAA to PEP in an ATP-dependent reaction that releases CO_2 for Rubisco. PEP moves into the mesophyll cells to be used in the cytosolic PEP carboxylase reaction, thus completing the first cycle. A second cycle also takes places, involving the movement of malate and alanine between the mesophyll and bundle sheath cells. Here, the product of the cytosolic PEP carboxylase reaction, OAA, is transported into the mesophyll cell chloroplasts and converted to malate via NADP-MDH. Malate then diffuses into the bundle sheath cells where it enters the mitochondria, to be oxidized by NAD-ME. The reaction releases CO_2 that can enter the bundle sheath chloroplast to be assimilated into the Calvin cycle, and also generates NADH that may be oxidized in the mitochondrial electron transport chain to generate ATP. This NADH oxidation is an important step, as it supplies the ATP needed in the PEPCK reaction. The second cycle is completed by the conversion of pyruvate to alanine (in the bundle sheath cytosol), which moves into the mesophyll cells to be converted back to pyruvate in the cytosol. Finally, pyruvate enters the mesophyll cell chloroplast where it is converted to PEP and exported to the cytosol to provide the substrate for PEP carboxylase.

Abbreviations: OAA, oxaloacetate; PEP, phospho*enol*pyruvate; 3-PGA, 3-phosphoglycerate; 1,3-BPGA, 1,3-bisphosphoglycerate; PPDK, pyruvate phosphate dikinase; PGK, phosphoglycerate kinase; G3PDH, glyceraldehyde 3-phosphate dehydrogenase; NAD-ME, NAD-dependent malic enzyme; NADP-ME, NADP-dependent malic enzyme; NAD-MDH, NAD-dependent malate dehydrogenase; NADP-MDH, NADP-dependent malate dehydrogenase; AspAT, aspartate aminotransferase; alaAT, alanine aminotransferase; Glu, glutamate; 2-OG, 2-oxoglutarate; P_i, inorganic phosphate; PP_i, pyrophosphate; PEPCK, PEP carboxykinase.

cell to another, as occurs in NADP-ME plants (as discussed above). Aspartate, rather than malate (which is moved between cells in NADP-ME plants) moves from the mesophyll to the bundle sheath cells, where it is converted to OAA and then to malate in the mitochondria. Malate is then decarboxylated to pyruvate by a mitochondrially located NAD-dependent malic enzyme, which releases CO_2 for Rubisco. The close association of the mitochondria and chloroplasts in these cells ensures that CO_2 diffuses from the mitochondria to the chloroplasts. Pyruvate is transaminated to alanine, which diffuses back into the mesophyll cells where it is converted back to pyruvate and transported into the mesophyll chloroplast. Pyruvate is then converted to PEP by PPDK, a reaction that occurs in the mesophyll chloroplasts of all C_4 types. Further details of the NAD-ME pathway are provided in Figure 5.15(b). NAD-ME species include: purple amaranthus (*Amaranthus cruentus*) and pearl millet (*Pennisetum glaucum*).

In plants with the PEP carboxykinase (PEPCK)-type C_4 pathway the chloroplasts contain grana and PSII activity in both the mesophyll and bundle sheath cells. The PEPCK pathway is more complex than that of the NAD-ME or NADP-ME types. Two C_4 cycles occur in these plants, one that cycles carbon through aspartate via PEPCK, and the other involving malate, via NAD-ME. The NAD-ME reaction generates NADH that can be oxidized within the mitochondria to supply ATP for the PEPCK reaction within the bundle sheath cells. Both the NAD-ME and PEPCK reactions produce CO_2 in the bundle sheath cells that is assimilated into the Calvin cycle via Rubisco. The reactions are shown in detail in Figure 5.15(c). The PEPCK pathway has evolved in two subfamilies of the Poaceae, the Panicoideae and the Chloridoideae. Examples include Texas signalgrass (*Urochloa texana*) and Common cordgrass (*Spartina anglica*).

In all three variants of the C_4 pathway, the decarboxylation reactions in the bundle sheath cells release CO_2 for fixation by Rubisco. While some of this CO_2 may leak out of the bundle sheath by diffusing into the mesophyll cells, the vast majority (about 70–90%) is assimilated by Rubisco within the bundle sheath chloroplasts. The absence of carbonic anhydrase from the bundle sheath cells prevents concentrated CO_2 from converting to bicarbonate, which is not a substrate for Rubisco. Hence, Rubisco is functioning in a high CO_2 environment. In many C_4 plants, leakage of CO_2 from the bundle sheath is effectively reduced by the presence of a gas-impermeable suberin layer (consisting of a polymer of phenolics containing waxes, see page 324) between

Figure 5.16 The ultrastructure of chloroplasts in bundle sheath and mesophyll cells of a typical NADP-malic enzyme-type C₄ plant, *Zea mays.* Note the difference in ultrastructure between the bundle sheath (upper left) and mesophyll (lower right) chloroplasts. The bundle sheath chloroplast has very few stacked thylakoids (i.e. few grana) and contains a number of starch grains. These chloroplasts contain Rubisco and lack PSII activity. In contrast, the mesophyll chloroplast contains highly stacked thylakoids and lacks starch grains. (Reprinted with permission from Gunning BES, Plant Cell Biology on DVD, 2007.)

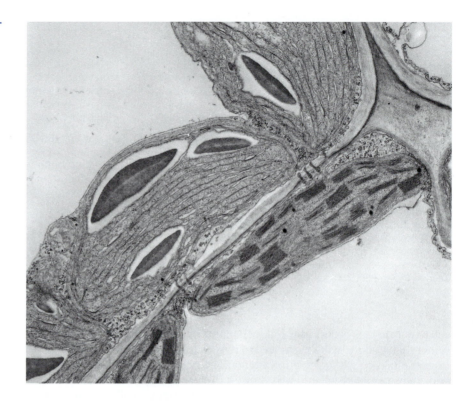

the walls of the bundle sheath and mesophyll cells. Although some gases will still diffuse through the plasmadesmata, the suberin layer minimizes the leakage by preventing CO_2 and other gases from diffusing through the cell walls.

C₃–C₄ intermediate species may represent an evolutionary stage between C₃ and C₄ plants

C_3–C_4 intermediate species have been found in 10 genera (four monocot and six dicot) and they are characterized by having features that are intermediate between those of C_3 and C_4 species. These characteristics include intermediate rates of photorespiratory CO_2 release (i.e. lower than C_3 and higher than C_4 species) as well as partial, or complete, Kranz anatomy.

The existence of C_3–C_4 intermediates has led some authors to conclude that these species represent a stage in the evolution of C_4 plants. However, while the leaves of these C_3–C_4 intermediates contain bundle sheath and mesophyll cells and, in some species, at least some of the C_4 pathway enzymes, they do not employ the C_4 pathway as a means of reducing photorespiration. Instead, many of the C_3–C_4 intermediates (species within the genera *Alternanthera, Flaveria, Mollugo, Moricandia,* and *Panicum*) achieve this reduction in photorespiration by improving the efficiency with which photorespiratory CO_2 is recaptured before it can be lost from the leaf. It is the compartmentation of GDC, together with a unique positioning of the mitochondria that results in this increased capacity for CO_2 reassimilation. GDC, the enzyme that releases CO_2 within the mitochondria during photorespiration (Figures 5.10 and 5.11), is confined to the bundle sheath mitochondria of these C_3–C_4 intermediate species. These mitochondria are therefore the site for decarboxylation of the glycine produced by photorespiratory metabolism in the whole leaf. Within the bundle sheath cells the mitochondria are arranged along the centripetal face of the cell that lies adjacent to the vascular tissue. These mitochondria are overlain by chloroplasts so that CO_2 produced in the mitochondrial GDC reaction will diffuse into the chloroplasts where it can be reassimilated into the Calvin cycle (Figure 5.17). Hence, although photorespiration occurs in these C_3–C_4 intermediates, the majority of the CO_2 that is produced is recaptured in the chloroplasts before it can escape from the leaf. This mechanism results in a low rate of photorespiratory CO_2 release in C_3–C_4 intermediate species.

Although the C_3–C_4 intermediates have been proposed to represent a step on the evolutionary path to C_4 photosynthesis, the significance of these species remains controversial. In the model that supports this proposal, increased GDC activity within the bundle sheath cells is believed to represent an early step towards a CO_2-concentrating mechanism. Subsequent steps are proposed to include the thickening of the cell walls of bundle sheath cells to

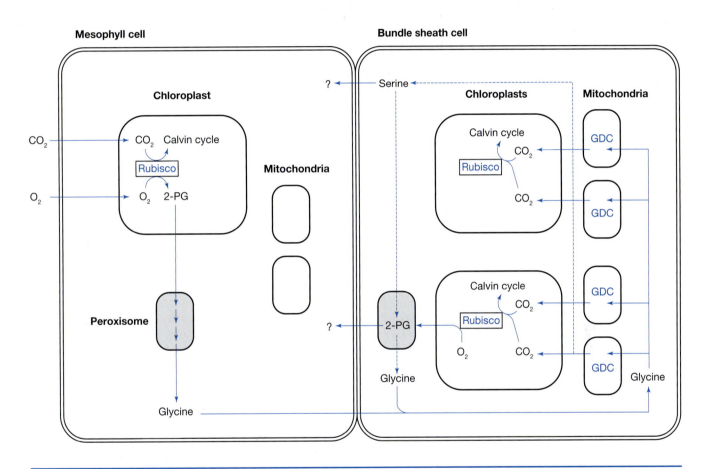

Figure 5.17 Proposed pathway for photorespiratory glycine metabolism in some C_3–C_4 intermediate species (modified from Rawsthorne (1992) *The Plant Journal 2*, 267–274). Abbreviations: 2-PG, 2-phosphoglycolate; GDC, glycine decarboxylase; ?, unknown intermediate.

The mesophyll cells of C_3–C_4 intermediates of *Moricandia* spp. contain Rubisco, which reacts with CO_2 and O_2 from the surrounding air. While CO_2 is assimilated into the Calvin cycle, the reaction of Rubisco with O_2 generates 2-phosphoglycolate that enters into the photorespiratory pathway, as in C_3 leaves. In C_3–C_4 intermediates, however, glycine decarboxylase is absent from the mesophyll cell mitochondria. Glycine therefore has to be moved from the mesophyll cells to the bundle sheath cells where it is decarboxylated within the mitochondria. The bundle sheath cells of C_3–C_4 intermediates contain a large number of chloroplasts and mitochondria. The mitochondria

are aligned along the side of the cell that faces into the leaf (i.e. towards the vascular bundles and away from the leaf surface). The chloroplasts are positioned so that they overlay the mitochondria. Hence, as glycine is decarboxylated within the bundle sheath mitochondria, the resulting CO_2 diffuses into the layer of chloroplasts where it is reassimilated by Rubisco. This close association of mitochondria and chloroplasts within the bundle sheath cells improves the efficiency of reassimilation of photorespiratory CO_2, as compared with C_3 leaves. The bundle sheath cells contain a full complement of photorespiratory enzymes, hence in the presence of O_2 these cells may form photorespiratory intermediates, such as 2-phosphoglycolate. However, photorespiratory rates are likely to be reduced as a result of the elevated CO_2 concentration within the bundle sheath chloroplast. Note that the dashed lines represent reactions that are still not fully understood.

reduce leakage of CO_2 (produced from the GDC reaction), and confinement of Rubisco to the bundle sheath, and of PEP carboxylase to the mesophyll cells. However, critics of this proposal maintain that C_3–C_4 intermediates are a failed evolutionary experiment and that these species therefore represent a stable endpoint in themselves, rather than a step towards the evolution of a C_4 pathway. There is some persuasive evidence to support this conclusion. All existing C_4 plants use a CO_2-concentrating mechanism that requires decarboxylation of a C_4 acid within the bundle sheath cells. While some C_3–C_4 intermediates may contain a partial C_4 cycle, it does not function as a CO_2-concentrating mechanism. Furthermore, it has been argued that the confinement of GDC to the bundle sheath cells of some C_3–C_4 species is not likely to be a prelude to the evolution of the C_4 pathway. This argument is based on the conclusion that the GDC model is fundamentally different from the mechanism for reducing photorespiration in C_4 plants. The latter concentrate CO_2 around Rubisco in order to reduce flux into the photorespiratory pathway. Hence, glycine production, and the demand for GDC activity, is very much reduced in C_4 plants. It therefore seems unlikely that the C_4 pathway would have evolved from a C_3–C_4 mechanism that requires the up-regulation of GDC, only to down-regulate it following the acquisition of a CO_2-concentrating mechanism.

The C_4 pathway can exist in single cells of some species

The recent discovery of a C_4 photosynthetic pathway in single chlorenchyma cells of some species challenges the classical paradigm that this pathway requires the spatial separation of Rubisco and PEP carboxylase between two different types of cell. Single-celled C_4 photosynthesis has been identified in *Bienertia cycloptera* and *Borszczowia aralocaspica*, two dicot species within the Chenopodiaceae, that grow in Central Asia. These plants are able to carry out C_4 photosynthesis within a single cell by partitioning the enzymes and organelles into two separate compartments within the cell. The precise anatomy differs between the two species. In *Borszczowia*, there is a single layer of tightly packed chlorenchyma cells within the leaf. These cells contain a dense layer of chloroplasts that lie at the proximal end of the cell, i.e. the end of the cell that faces towards the center of the leaf. The distal end of the chlorenchyma cell (i.e. facing towards the leaf surface, where CO_2 enters) contains relatively few chloroplasts. In contrast, the chlorenchyma cells of *Bienertia* (which form three cell layers within the leaf) contain a large, central cytoplasmic compartment that is densely packed with chloroplasts, as well as a few chloroplasts around the periphery of the cell. These two groups of chloroplasts are connected by channels of cytoplasm that run through the vacuole. The mitochondria show similar distribution to the chloroplasts, being concentrated at the proximal end of the cell in *Borszczowia* and within the central cytoplasmic compartment in *Bienertia*. Hence, although neither species has the typical C_4 Kranz anatomy, the distribution of the chloroplasts and mitochondria forms an analogous compartmentation within a single chlorenchyma cell.

In *Borszczowia*, CO_2 that enters the leaf will pass into the chlorenchyma cell at the distal end where it is assimilated by cytosolic PEP carboxylase. The presence of PPDK, which forms PEP, in the chloroplasts at the distal end of the cell supports this initial C_4 fixation of CO_2. Malate and/or aspartate, formed from the PEP carboxylase product, OAA, then diffuse to the proximal part of the cell where they are decarboxylated by the NAD-ME that is confined to the mitochondria in this part of the cell. The CO_2 released in this reaction is assimilated by Rubisco that is restricted to the chloroplasts that are present in the proximal part of the cell.

In *Bienertia*, CO_2 entering the leaf from the surrounding air is also assimilated by PEP carboxylase in the cytosol of the chlorenchyma cells. C_4 acids then diffuse through the cytoplasmic channels into the central cytoplasmic compartment, where they are decarboxylated by NAD-ME in the mitochondria. Rubisco, confined to the chloroplasts within this central compartment, subsequently assimilates the CO_2 released by the mitochondria.

Hence, individual chlorenchyma cells of *Bienertia* and *Borszczowia* show the same sort of spatial compartmentation that occurs between the bundle sheath and mesophyll cells of conventional C_4 species, with the same result: a C_4 pathway that provides an effective mechanism for concentrating CO_2 around Rubisco.

Some of the C_4 pathway enzymes are light-regulated

The Calvin cycle operating in C_4 plants is thought to be subject to the same light regulation as described for C_3 plants earlier in this chapter. It is therefore important that the C_4 fixation pathway is regulated in order to make sure that Rubisco receives CO_2 in the light. Three of the C_4 pathway enzymes are subject to regulation to make them active in the light. The mechanisms are different in each case.

PEP carboxylase is regulated by reversible protein phosphorylation (see also Chapter 2). A serine/threonine kinase, PEP carboxylase kinase (PEPC kinase), adds a phosphate group to a single serine residue within the PEP carboxylase protein. Phosphorylation results in an increase in PEP carboxylase activity, by increasing its affinity for PEP and by relieving it of sensitivity to inhibition by malate (Figure 5.18). Dephosphorylation is carried out by a protein phosphatase 2A. The activity of PEPC kinase is high in the light and low in the dark in C_4 plants, due to a day/night turnover (i.e. synthesis and degradation) of the PEPC kinase protein. There is also some evidence that PEPC kinase activity might be light-regulated directly by thioredoxin/ferredoxin.

PPDK is also subject to regulation by protein phosphorylation. This enzyme is present in the mesophyll chloroplasts of all C_4 types, where it produces PEP via the following reaction:

$$\text{Pyruvate} + \text{ATP} + P_i \rightleftharpoons \text{PEP} + \text{AMP} + PP_i$$

Although the PPDK reaction is freely reversible, it is pulled in the forward direction (i.e. towards PEP formation) by the accompanying hydrolysis of one of its products, pyrophosphate (PP_i), via a chloroplast-located pyrophosphatase. PPDK is regulated by reversible phosphorylation/dephosphorylation by a single bifunctional regulatory protein (RP). Phosphorylation (of a threonine residue) occurs in the dark, and results in the inactivation of PPDK. This phosphorylation reaction is unusual because it uses ADP as the phosphate donor, while most other phosphorylations use ATP. Dephosphorylation (via RP) restores PPDK activity in the light. The relatively high concentration of ADP that occurs in chloroplasts in the dark, favors the phosphorylation and inactivation of PPDK. Conversion of ADP to ATP by photophosphorylation (i.e. resulting from the photosynthetic electron transport chain; Chapter 4) causes chloroplast ADP concentrations to fall and this favors dephosphorylation, and activation, of PPDK in the light (Figure 5.19).

NADP-ME, which is located in the bundle sheath chloroplasts of NADP-ME-type C_4 plants, is regulated by the same ferredoxin/thioredoxin system that results in the light activation of Calvin cycle enzymes, as described earlier in

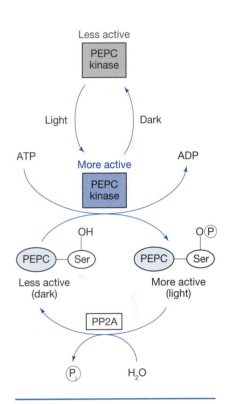

Figure 5.18. Phospho*enol*pyruvate (PEP) carboxylase regulation in C_4 plants. PEP carboxylase kinase (PEPC kinase) is a specific protein kinase that adds a phosphate group to a serine residue on PEP carboxylase. This phosphorylation modifies PEP carboxylase to make it more active. PEPC kinase is light activated, possibly via ferredoxin/thioredoxin regulation, and is also synthesized in the light and degraded in the dark. PEP carboxylase is dephosphorylated by a protein phosphatase 2A (PP2A). This reversible phosphorylation ensures that PEP carboxylase is phosphorylated and active in the light so that CO_2 assimilation coincides with the Calvin cycle in C_4 plants.

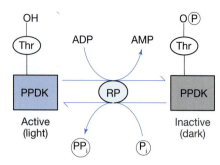

Figure 5.19 Regulation of pyruvate orthophosphate dikinase (PPDK) in C₄ plants. PPDK is regulated by reversible phosphorylation/dephosphorylation of a threonine residue. A single, bifunctional regulatory protein (RP) is responsible. It inactivates PPDK in the dark by adding a phosphate group, using ADP as the phosphate donor. PPDK is activated in the light by the same regulatory protein, which removes the phosphate group and attaches it to inorganic phosphate (P_i) to form pyrophosphate (PP_i). Photophosphorylation (via the photosynthetic electron transport chain) produces high concentrations of ATP, and low concentrations of ADP in the light, while the reverse situation in the dark, i.e. high concentrations of ADP, favor phosphorylation and inactivation of PPDK. This regulation ensures that PPDK activity is light-dependent and coincides with the reactions of the Calvin cycle in C₄ leaves.

this chapter. NADP-ME appears to associate preferentially with thioredoxin f, rather than with thioredoxin m.

Crassulacean acid metabolism as a feature of desert plants

Crassulacean acid metabolism (CAM) enables many plants to grow in extremely dry environments such as deserts. However efficient a plant might be at using water, it will always be prone to dehydration when its stomata are open on hot, dry days. Desert plants may be exposed to daytime temperatures of at least 45°C. Closing the stomata under these extreme conditions will reduce water loss but this poses another problem—How do plants continue to photosynthesize when they are unable to take up CO_2 from the surrounding air? The answer is provided by the presence of another photosynthetic pathway, CAM.

CAM was first characterized in the Crassulaceae. CAM is commonly found in desert species, such as the Cactaceae, Euphorbiaceae, and Aizoaceae; however, it is not restricted to the desert. As we shall see later, there are examples of CAM plants in tropical rainforests and in aquatic environments. Although essentially a means of surviving and growing in arid conditions, CAM also enables plants to photosynthesize at low concentrations of CO_2 because, like the C₄ pathway, it works by concentrating CO_2 for Rubisco to assimilate into the Calvin cycle.

Temporal separation of the carboxylases in CAM

CAM photosynthesis involves a temporal separation of the functioning of the two carboxylating enzymes, PEP carboxylase and Rubisco. CAM plants open their stomata at night when temperatures are usually far below the daytime maximum, so there is a reduced risk of dehydration. The Calvin cycle can only operate in the light, as explained earlier in this chapter, and Rubisco is inactive in the dark. The sequence of events is as follows: CO_2 entering the leaf at night is first assimilated (as bicarbonate) by PEP carboxylase, forming OAA from HCO_3^- and PEP as it does in C₄ plants. The regulation of PEP carboxylase is, however, quite different in CAM plants compared with C₄ plants, as the enzyme is now active in the dark and inactive in the light, as discussed later. The OAA formed in this reaction is reduced to malate, which is moved into the vacuole by a translocator mechanism and stored there overnight. This accumulation of malate, in the form of malic acid in the vacuole, can reach high concentrations so that the leaf as a whole becomes distinctly acidic.

The following morning the stomata close, the Calvin cycle enzymes are active and Rubisco is available for CO_2 assimilation. Malate moves out of the vacuole and into the cytosol where it is decarboxylated by malic enzyme. This reaction releases CO_2 and because the stomata are closed it cannot escape. Consequently, Rubisco is operating in a CO_2-enriched environment and photorespiration is kept to a minimum. Rubisco produces 3-PGA and the Calvin cycle functions exactly the same as it does in C₃ and C₄ plants. Thus, the two carboxylase reactions in CAM plants are temporally separated, with PEP carboxylase fixing carbon at night and Rubisco refixing CO_2 in the daytime (Figure 5.20).

Crassulacean acid metabolism as a flexible pathway

CAM is a flexible pathway, responding to short- and long-term changes in the environment. The night-time fixation of CO_2 through open stomata, followed

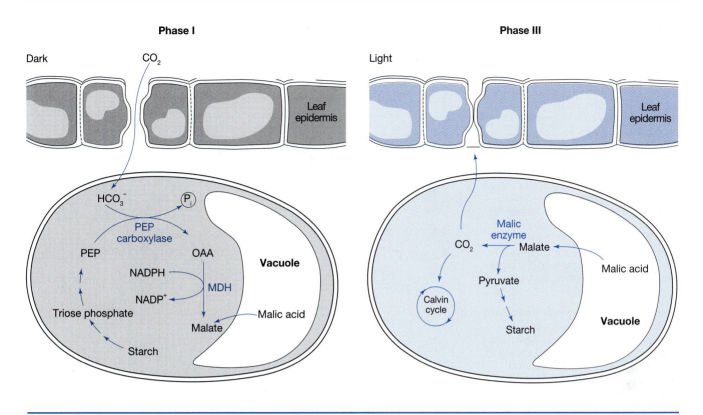

Figure 5.20 Crassulacean acid metabolism (CAM photosynthesis). CAM provides a means for the Calvin cycle to function without the need for the stomata to open in the daytime. This mechanism of photosynthesis increases water use efficiency and is a useful survival strategy when water or CO_2 are scarce. Stomata open at night (Phase I) in CAM plants when the air temperature is comparatively low and transpiration is reduced. Because the Calvin cycle is light-regulated, CO_2 entering the leaf at night cannot be fixed directly by Rubisco but is instead converted to oxaloacetate (OAA) by the cytosolic enzyme, PEP carboxylase. Its substrate, PEP, is produced from starch degradation in the dark. The OAA formed is converted to malate by NADP-malate dehydrogenase (MDH). Malate is transported to the vacuole and stored overnight in its acidic form, as malic acid. At the start of the day (Phase III), the stomata close and malic acid is released from the vacuole to be decarboxylated in the cytosol via NADP-malic enzyme. This reaction produces pyruvate and releases CO_2 for Rubisco to use in the Calvin cycle. The internal generation of CO_2 provides a concentrating mechanism that allows Rubisco to function in the light behind closed stomata. Water is thereby conserved and also photorespiratory losses are low. The reactions of PEP carboxylase, NADP-malic enzyme, and NADP-malate dehydrogenase are the same as those shown in Figure 5.15. (From *Plant Physiology*, 4th edition, edited by Lincoln Taiz & Eduardo Zeiger, Figure 8.14, 2006. Reprinted with permission from Sinauer Associates Inc.)

by the daytime refixation of internally generated CO_2 are just two (Phases I and III respectively; Figure 5.21) of four potential phases in CAM photosynthesis. In certain environmental conditions two additional phases may occur, both resulting in the opening of stomata in the light. (Figure 5.21). While the end of Phase I is marked by the closure of stomata, they may reopen briefly at the very beginning of the light period providing the plant is not experiencing drought conditions. This is defined as Phase II, where external CO_2 can be assimilated by both PEP carboxylase and Rubisco. The signal for the stomata to open under these conditions results from the combined effect of light together with low internal concentrations of CO_2. Malate decarboxylation releases CO_2 and as this builds up internally the stomata close and the plant enters Phase III. Towards the latter part of the day, as the pools of malate become depleted in the vacuole and CO_2 concentrations begin to fall again, the stomata may reopen once more in the light period. This marks the onset of Phase IV where, again, both PEP carboxylase and Rubisco are

Figure 5.21 The four phases of Crassulacean acid metabolism (CAM).
Phase I of CAM is marked by the beginning of the night period when the stomata open and CO_2 enters the leaf to be fixed by phospho*enol*pyruvate (PEP) carboxylase (PEPC), which, together with NADP-malate dehydrogenase, forms malate that accumulates throughout the night as malic acid stored in the vacuole. The progressive decrease in glucan content indicates that starch is being broken down to generate PEP as substrate for PEP carboxylase. This first phase of CAM is essential and can be used to diagnose the occurrence of CAM, simply by measuring the acidity of the cell sap resulting from malic acid accumulation at night.

Phase II does not always occur in CAM plants. The end of phase I usually coincides with the onset of dawn and the closure of stomata, as described in Figure 5.20. However, under well-watered conditions the stomata may open again at the start of the day when both PEP carboxylase and Rubisco are active. This response causes a transient increase in CO_2 uptake and fixation until the stomata close again and the cycle enters phase III.

Phase III is the other essential feature of CAM, when the stomata close in the daytime and Rubisco reacts with CO_2 generated internally via malate decarboxylation, as described in Figure 5.20. Once the stored pools of malate have been decarboxylated, the stomata may once again open in the light providing there is ample water. This change marks the onset of Phase IV.

The signal for stomatal opening in Phase IV appears to be the low internal concentration of CO_2 that occurs once malate has been depleted and Rubisco has assimilated the resulting CO_2. Both PEP carboxylase and Rubisco are active and assimilating CO_2 from the surrounding atmosphere. Under these conditions there is the possibility of photorespiration occurring because Rubisco is now exposed to atmospheric concentrations of O_2. Phase IV ends with the onset of darkness when the stomata remain open and Rubisco becomes inactive once again.

PEP carboxylase remains active in Phases II and IV despite the fact that it is phosphorylated because the malate concentration is sufficiently low that it is not completely inhibited, hence it can remain active in the light under these circumstances.

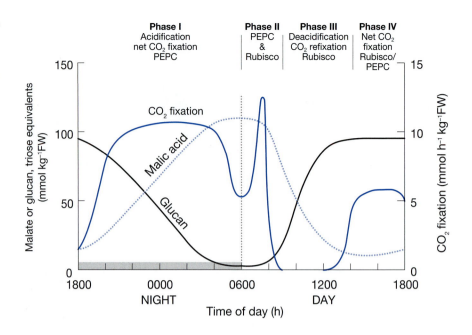

able to fix CO_2 from the surrounding air. Phase IV seems also to be a response to the environment as it will only occur when plants are provided with an adequate supply of water. The daytime opening of stomata in Phases II and IV provides an opportunity for the plant to gain additional CO_2 from the atmosphere. This is a useful mechanism in so far as it allows CAM plants to assimilate carbon and grow more rapidly when they are receiving an adequate supply of water.

A more extreme form of CAM may operate under conditions of severe drought. CAM idling takes place, where the stomata remain closed both day and night and CO_2 is recycled through respiratory release at night and photosynthetic assimilation the following day. Because no carbon is being gained, the plant is not able to grow but can nevertheless survive for prolonged periods with very little available water.

A weaker form of CAM, found in some epiphytic plants, is CAM cycling, where the stomata remain closed at night, as they do in C_3 and C_4 plants. In this case, malate is formed at night with PEP carboxylase using CO_2 released from respiration. The following day, the stomata open and Rubisco assimilates CO_2 from the surrounding air as well as the extra CO_2 released from decarboxylation of the stored malate. Some epiphytes use CAM idling when they are well-watered and revert to conventional CAM as they dry out again.

A number of species act as facultative, or inducible CAM plants. In the case of *Mesembryanthemum crystallinum*, a plant that grows in coastal regions of the Mediterranean, it changes from C_3 to CAM over the growing season as the rainfall decreases. The signal appears to be water availability, but it can be reproduced by high salt or abscisic acid treatment under laboratory conditions. *Kalanchöe blossfeldiana* (Tom Thumb variety) expresses CAM under long daylengths, coinciding with warmer temperatures and reduced water supply. This appears to be a phytochrome-regulated response, as temperature and water do not trigger a switch to CAM. Facultative CAM allows these plants to make optimum use of CO_2 and H_2O, using CAM when water is scarce and gaining most CO_2 through C_3 photosynthesis when water is adequate.

Phospho*enol*pyruvate carboxylase in crassulacean acid metabolism plants is regulated by protein phosphorylation

PEP carboxylase in CAM plants is regulated by phosphorylation to keep it active in the dark and inactive in the light The need for PEP carboxylase to be active in the dark and inactive in the light in CAM plants is quite the opposite of what occurs in C_4 plants. Despite this difference, the enzyme is regulated by phosphorylation in a similar way. PEPC kinase phosphorylates the CAM form of PEP carboxylase, making it active as it does in C_4 plants. The distinctive feature here is that PEP carboxylase is phosphorylated at night in CAM plants rather than in the daytime as occurs in C_4 plants, with ATP being supplied by respiration in the former, and by photophosphorylation in the latter. There appears to be a strong circadian rhythm to the expression of PEPC kinase in CAM plants, peaking in the middle of the night and decreasing to a minimal level of expression in the daytime. Although we still do not entirely understand the time-keeping mechanism of this circadian control, it seems to be the main process that ensures PEP carboxylase is phosphorylated and active at night. PEP carboxylase, in its daytime non-phosphorylated state, is strongly inhibited by malate, whereas the night-time phosphorylated form is relatively insensitive to malate inhibition. This control mechanism ensures that while there is malate present in the daytime, any CO_2 resulting from its decarboxylation will be assimilated by Rubisco and not by PEP carboxylase, thus enabling the Calvin cycle to gain carbon. It also explains why PEP carboxylase can be active in Phases II and IV, i.e. in the daytime. Phase II occurs right at the start of the light period, when PEP carboxylase may still be in its phosphorylated, and active state. In Phase IV, although by now phosphorylated, the malic acid concentration is sufficiently low (Figure 5.21) that PEP carboxylase is no longer inhibited.

Crassulacean acid metabolism is thought to have evolved independently on several occasions

CAM is an ancient pathway that most likely originated in the Mesozoic era, more than 200 million years ago. It occurs in about 7% of all vascular plant species, and its presence in 33 taxonomically diverse families indicates that CAM has evolved independently on a number of occasions. This hypothesis is further supported by comparison of PEP carboxylase gene sequences, which indicate a polyphyletic origin (i.e. descended from more than one ancestor) of CAM. The diverse origins are further reflected in the wide range of habitats where CAM plants are found, from arid regions to tropical rainforests and aquatic environments.

CAM is thought to have evolved in response to limitations in either water or carbon availability. While water is a scarce resource for plants growing in deserts or as epiphytes within tropical rainforests, it is the limited availability of CO_2 within the aquatic environment that is thought to have led to the evolution of CAM in aquatic plants, such as Isoetes (see later in this chapter). CO_2 may also become a limiting resource under arid conditions, where the daytime closure of stomata (in response to water deficit) will reduce CO_2 uptake. Consequently, some authors argue that CAM evolved primarily as a consequence of daytime limitations in CO_2 availability, with water availability being a secondary factor.

An early step in the evolution of CAM is thought to be the night-time fixation of respiratory CO_2 by PEP carboxylase (NB PEP carboxylase is present even in C_3 plants but at much lower levels than in C_4 or CAM plants; see Chapter 6).

Succulence may have been a pre-existing trait present in plants that were already growing in arid environments. Succulence tends to coincide with tight packing of the cells of the leaf, which also reduces the extent to which CO_2 is able to diffuse out of the leaf. Hence, succulence would enhance the capacity for the reassimilation of respiratory CO_2. A large central vacuole would also be important in the early stages of CAM evolution, and this is also thought to have coincided with the existence of succulent leaves. Any acids that leaked out of this large vacuole into the cytoplasm would have to be contained otherwise they would cause large fluctuations in the cytoplasmic pH. Hence, specific transport proteins would be required to enable acids to be returned to the vacuole, and these are also likely to have evolved at an early stage in the development of CAM.

There are far more CAM species than there are C_4 species, although only two CAM plants are of commercial value as crops; the pineapple (*Ananas comosus*), and the cactus (*Agave tequilana*), which is used to produce the alcoholic drink, tequila.

C$_3$, C$_4$, and CAM photosynthetic pathways: advantages and disadvantages

C_3, C_4, and CAM photosynthetic pathways each have distinct advantages and disadvantages in particular environments. As a general rule, C_3 plants are more abundant in temperate climates, C_4 plants in tropical climates, and CAM plants in arid regions. This distribution is related in part to the various requirements for water, nutrients, and light described below. However, it is important to appreciate that there are many exceptions to this generalization. For example, some C_4 plants grow well in high latitude climates and there are a number of cold-tolerant C_4 species. Hence, there is a current view that the present global distribution of C_4 species reflects the fact that they originated in the Tropics and they are still migrating towards the earth's poles. Although many CAM species are found in deserts, a large proportion of CAM species occur in tropical rainforests and, on the face of it, this would seem to argue against CAM being a mechanism for desiccation tolerance. Yet many of these tropical CAM plants occur as epiphytes (i.e. grow on the surface of other plants), and their water supply is subject to severe fluctuations, as their limited root development leaves them dependent on atmospheric water vapor or rain drips from surrounding vegetation. These plants are likely to experience periods of water deficit, and CAM will enable them to survive these conditions. A striking and exceptional example among CAM plants is the aquatic plant species *Isoetes* (Lycopsida division, Isoetaceae family). These fern-like plants (they are not true ferns), are more commonly known as quillworts, and all of them are aquatic or semi-aquatic. The reason for CAM is not for water conservation but because of the low daytime CO_2 concentration in the water, as a result of photosynthesis in neighboring plants. *Isoetes* benefits from nocturnal CO_2 uptake when competing species are respiring and maintaining high concentrations of dissolved CO_2 in the shallow water in which they grow.

The C$_4$ and CAM pathways require more light energy than C$_3$ photosynthesis

The Calvin cycle, if it operates at peak efficiency with no oxygenation of Rubisco, requires three moles of ATP and two moles of NADPH to fix each mole of CO_2, as described above. The Calvin cycle operates in C_3, C_4, and CAM plants, and differences in requirements are due to the additional reactions of the C_4 and CAM pathways.

The C_4 pathway requires two more moles of ATP than the C_3 pathway per mole of CO_2. This difference is due to two ATP-requiring steps:

Pyruvate orthophosphate dikinase (PPDK):

$$Pyruvate + P_i + ATP \longrightarrow PEP + AMP + PP_i$$

and adenylate kinase, which converts the resulting AMP to ADP:

$$ATP + AMP \longrightarrow 2ADP$$

Both reactions are present in all C_4 plants, irrespective of type, but PEP carboxykinase-type C_4 plants require an additional ATP in the PEP carboxykinase reaction:

$$OAA + ATP \longrightarrow PEP + CO_2 + ADP$$

The CAM pathway uses ATP for the Calvin cycle and for the two reactions already described for the C_4 pathway (i.e. PPDK and adenylate kinase) and additionally uses one mole of ATP to transport one mole of malate into the vacuole. This occurs against an increasing concentration gradient and requires a H^+-ATPase to generate a H^+ gradient that allows for uptake of the malate anion (malate^{2-}).

These additional requirements for ATP mean that both the C_4 and CAM pathways need more light energy than the C_3 pathway in order to assimilate the same quantity of CO_2. This would initially suggest a reduction in their quantum efficiency. However, if photorespiration is included in the balance sheet, which it must be, C_3 plants become progressively less quantum efficient with increasing temperature, as already discussed. Hence C_4 and CAM plants have an advantage in bright light and in warm temperatures (Table 5.2).

C_4 and crassulacean acid metabolism plants are more nitrogen and water efficient than C_3 plants

The CO_2 concentrating mechanisms of the C_4 and CAM pathways enable Rubisco to function more efficiently than in C_3 plants, and these plants compensate by producing reduced amounts of Rubisco protein compared with that of C_3 plants. Their leaf protein content is substantially lowered; hence this reduces their requirement for nitrogen. So, for a given rate of photosynthesis, both C_4 and CAM plants require less nitrogen than C_3 plants and this is expressed as a greater nitrogen use efficiency in these plants. As a result, C_4 plants tend to compete well with C_3 plants in nitrogen-poor soils, where the capacity of leaves to produce Rubisco protein is limited by nitrogen availability.

WUE is also improved in C_4 and CAM plants. The effect in C_4 plants is due to the CO_2 concentrating mechanism, once again, so that carbon assimilation occurs through a smaller stomatal aperture and water loss is consequently reduced. Although this reduction in stomatal aperture should also slow down the rate of CO_2 diffusion into the leaf, the maintenance of a very low CO_2 concentration within the mesophyll cells compensates by effectively increasing the CO_2 concentration gradient, and therefore the rate of diffusion from the air spaces into the mesophyll cells. It is the high affinity of PEP carboxylase for bicarbonate, together with carbonic anhydrase (which converts CO_2 to bicarbonate) activity that maintains the low concentration of CO_2 (approximately 5 μM) within the mesophyll cells of C_4 leaves. Under these conditions diffusion, alone, would not be sufficient to support Rubisco activity within the bundle sheath cells. Hence, the reactions that release, and concentrate, CO_2

within the bundle sheath cells are crucial to the overall process that enables C_4 leaves to photosynthesize with a relatively small stomatal aperture.

CAM plants benefit further from opening their stomata at night, so reducing transpirational water losses as the air temperature is lower. Further benefits are gained from the morphological features of many CAM plants, such as a large

Table 5.2 Comparison of some features of C_3, C_4, and CAM plants

Characteristic	C_3	C_4	CAM
Typical species/ plant groups	Wheat, barley, rice	Maize, sugarcane, millet (*Panicum* spp.), crab grass (*Digitaria sanguinalis*), Bermuda grass (*Cynodon dactylon*)	Pineapple, Cacti, succulents and many epiphytes
Leaf anatomy	Bundle sheath cells, if present, are not green (nonphotosynthetic), photosynthesis occurs in mesophyll cells	Kranz anatomy with photosynthesis occurring in mesophyll and bundle sheath cells	Photosynthesis occurs in mesophyll cells, these contain large vacuoles for malic acid storage
Carboxylating enzyme	Rubisco	Rubisco and PEP carboxylase active in the light only	Rubisco in the light, PEP carboxylase in the dark
First product of CO_2 fixation	3-PGA	OAA	OAA in the dark, 3-PGA in the light
Theoretical energy requirement (CO_2:ATP:NADPH) (with no photorespiration)	1:3:2	1:5:2 (NADP- and NAD- malic enzyme type), 1:6:2 (PEP carboxykinase type)	1:6.5:2
Water use efficiency (mg dry weight produced per g H_2O lost)	1.05–2.22	2.85–4.00	8.00–55.0
Maximum rates of net photosynthesis[a] (μmol CO_2 fixed per unit leaf area (m^2) per second)	20–40	30–60	5–12 (in the light); 6–10 (in the dark)
Photorespiration detectable?	Yes	Sometimes in bundle sheath—very low rates	Only in Phase IV
Temperature optimum for photosynthesis (°C)	15–25	30–47	35
Dry matter production (tonnes per hectare per year)	22–39	39–54	Low and very variable, (generally less than 10)
Maximum rates of productivity[b] (net assimilation rate g dry matter produced per unit leaf area (m^2) per day)	10–25	40–80	6–10
Typical range of $\delta^{13}C$ values[c]	–32 to –20‰	–17 to –9‰	–17 to –9‰ (drought) –32 to –20‰ (well-watered)

[a] Measured under conditions of saturating light, optimum temperature, adequate water, and ambient CO_2.
[b] Measured during the main growth period.   [c] See Box 5.7 for further details

volume to surface air ratio of succulent species, together with sunken stomata and waxy cuticles, which all serve to reduce water loss. However, these are not, in themselves, considered to be benefits of the CAM pathway itself, but more the consequence of independent adaptations to an arid environment.

Growth rates of CAM, C_3, and C_4 plants

CAM plants are slow-growing while C_4 plants have the highest growth rates in their natural environment. The additional ATP requirements for CAM photosynthesis in part explains why CAM plants grow more slowly than either C_3 or C_4 plants. In very arid environments, when CAM plants undergo CAM idling and close their stomata both day and night, growth stops completely. CAM is most important as a survival mechanism, and slow growth is not a fatal problem when there is little competition from other species in hostile environments such as the desert.

Under optimum conditions of light and temperature, C_4 plants photosynthesize and grow at faster rates than C_3 plants because of their greater efficiency at using CO_2, resulting primarily from the near absence of photorespiration. The high productivity of C_4 plants has attracted considerable research efforts to introduce this pathway into C_3 crops. This research is described in Box 5.6.

Box 5.6 Introducing C_4 photosynthesis into C_3 plants—is it possible?

There have been several attempts to introduce C_4 photosynthesis into C_3 crops, using both conventional breeding and transgenic approaches. Breeding was unsuccessful because of the independent inheritance of genes for morphological features, such as Kranz anatomy, and of the C_4 pathway enzymes. More recent efforts have concentrated on the production of transgenic plants, with the ultimate aim of engineering C_3 crops to have the high photosynthetic rates, improved water and nitrogen use efficiencies of C_4 plants. None of these transgenic approaches has succeeded yet.

Phospho*enol*pyruvate carboxylase overexpression

There have been several successful attempts to overexpress phospho*enol*pyruvate (PEP) carboxylase genes in C_3 plants. One complication is that PEP carboxylase is subject to regulation by protein phosphorylation, as discussed in this chapter. Although this regulation occurs in both C_3 and C_4 plants, there are some important differences. In C_4 plants, PEP carboxylase is phosphorylated in the light and dephosphorylated in the dark. In C_3 plants, nitrogen nutrition is the major signal for phosphorylation, with light exerting a lesser influence. This lack of responsiveness to light caused problems when the full-length gene from maize was introduced into rice. Although overexpression was successful, with up to a 110-fold increase in extractable activity of PEP carboxylase, the enzyme was dephosphorylated and inactive in rice leaves in the light. It was, however, phosphorylated if nitrate was fed to the plants. Hence the

host plant regulatory mechanism, rather than the C_4 one, seems to be operating in the transgenic rice plants. This C_3 regulation is an important means of increasing carbon flow into respiration (see Chapter 6) and it does not provide a means of coordinating PEP carboxylase activity with that of the Calvin cycle.

In an attempt to bypass the C_3 regulation of PEP carboxylase, genes encoding PEP carboxylase enzymes that lack phosphorylation sites have been introduced into arabidopsis and potato. Although these experiments were successful, in that the transgenic plants expressed high activities of PEP carboxylase, the plants were severely stunted, getting worse with increasing PEP carboxylase activity. This seemed to be due to depletion of PEP, which, as a substrate for aromatic amino acids and phenolics synthesis via the shikimic acid pathway (Chapters 8 and 11 respectively), resulted in amino acid deficiency and impaired growth.

Although there were a number of early reports of increased photosynthesis and reduced photorespiration in C_3 plants overexpressing PEP carboxylase genes, these have since proved not to be reproducible. The only consistent effects seem to be an increase in respiration, of up to 50%, and an alteration of stomatal response. These changes seem to be more symptomatic of an exaggerated C_3-type role for PEP carboxylase. In wild-type C_3 plants, PEP carboxylase functions to provide oxaloacetate for malate synthesis in guard cells (see Figure 5.1) and overexpression seems to interfere with this process, resulting in transient stomatal closure in the light. The increased respiration rates in overexpressing

continued

Box 5.6 Introducing C_4 photosynthesis into C_3 plants—is it possible? (continued)

lines is likely due to the enhanced flux into oxaloacetate and then into the tricarboxylic acid cycle, causing an increase in respiratory CO_2 release. This actually causes a reduction in the net rate of CO_2 assimilation, so that photosynthesis decreases with increasing PEP carboxylase expression. Hence, the introduced PEP carboxylase seems to be fulfilling its anaplerotic role, as in C_3 plants (see Chapter 6) rather than its photosynthetic function, as in C_4 plants.

It would seem, thus far, that overproduction of PEP carboxylase, by whatever means, does not improve photosynthesis in C_3 plants.

Pyruvate orthophosphate dikinase and NADP-malic enzyme overexpression

Other attempts have concentrated on the C_4 enzymes pyruvate orthophosphate dikinase (PPDK) or NADP-dependent malic enzyme. The effects of over-expressing PPDK in rice were minimal, even though activity increased by 40-fold. The PPDK reaction is freely reversible and it is possible that the reaction is consuming, rather than producing PEP when expressed in C_3 cells. Furthermore, PPDK is also regulated by phosphorylation and it may well be inactivated in the transgenic host. Overexpression of maize NADP-dependent malic enzyme in rice plants caused stunting and loss of chlorophyll. It seems that this was due to an increase in the NADPH concentration in the chloroplast that interfered

with normal functioning of the photosynthetic electron transport chain, resulting in destruction of chlorophyll through photoinhibition (see Chapter 4).

Conclusions

The results of these transgenic approaches have been uniformly disappointing. In all cases the plants either show very little improvement or grow poorly when C_4 genes are introduced. There seems little prospect of improving photosynthesis in C_3 plants by introducing a single gene for a C_4 enzyme. A current approach is to introduce groups of C_4 genes and it remains to be seen whether this strategy will be any more successful.

The introduction of the appropriate C_4 genes is not the only challenge; the resulting enzymes have to be distributed so that they provide a CO_2-enriched environment around Rubisco. While the vast majority of C_4 plants achieve this effect by expressing PEP carboxylase and Rubisco in different cells, the recent discovery of C_4 photosynthesis within single cells of *Borszczowia aralocaspica* and *Bienertia cycloptera* leaves (see page 127) indicates that Kranz anatomy is not essential for C_4 photosynthesis. On the face of it, this discovery might offer a simpler approach towards engineering C_4 photosynthesis. However, while *B. aralocaspica* and *B. cycloptera* might lack Kranz anatomy, they adopt, instead, a form of subcellular compartmentation that is likely to be just as difficult to introduce into C_3 leaves.

C_3, C_4, and CAM plants differ in their facility to discriminate between different isotopes of carbon

C_3, C_4, and CAM plants differ in the extent to which they assimilate the two naturally occurring stable isotopes of carbon (^{12}C and ^{13}C). Consequently, stable isotope analysis can be used to distinguish between C_3, C_4, and CAM plants. Although ^{12}C and ^{13}C have identical chemical properties, C_3 plants assimilate $^{12}CO_2$ more rapidly than $^{13}CO_2$. In other words, these plants discriminate against the heavier isotope, ^{13}C. The main reason for this discrimination is because Rubisco will react more rapidly with $^{12}CO_2$ than with $^{13}CO_2$. Hence, the ratio of ^{13}C to ^{12}C is lower in plants (and their products) with C_3 photosynthesis than it is in the surrounding atmosphere. In contrast, PEP carboxylase, which assimilates CO_2 from the atmosphere in both C_4 and CAM plants, discriminates very little between ^{13}C and ^{12}C. Consequently, C_4 and CAM plants assimilate $^{13}CO_2$ and $^{12}CO_2$ in a similar proportion to that of the surrounding air and the ratio of ^{13}C to ^{12}C in these plants is higher than it is for C_3 plants. While we still do not fully understand how Rubisco discriminates between these two stable isotopes of carbon, the fact that it does enables us to distinguish between C_3, C_4, and CAM plants. We can do this by analyzing the ratio of ^{13}C to ^{12}C within the plant, or plant material (see Box 5.7

Box 5.7 Stable isotope analysis and its use in studies of carbon assimilation

CO_2 in the atmosphere contains a naturally occurring mixture of carbon isotopes (isotopes are atoms with the same number of protons, but different numbers of neutrons, so that they have the same atomic number but differ in atomic mass). Two stable (i.e. non-radioactive) isotopes, ^{12}C and ^{13}C contribute 98.9% and 1.1% of the carbon atoms, respectively, while the radioactive isotope, ^{14}C contributes only one in a trillion (10^{12}) carbon atoms (i.e. 0.0000000001%). These tiny amounts of ^{14}C in the atmosphere make this isotope difficult to detect, hence the more abundant stable isotopes, ^{12}C and ^{13}C are used when analyzing plants that have grown under normal atmospheric conditions.

Given that the ^{12}C and ^{13}C isotopes are stable (i.e. they do not degrade like radioactive isotopes) the ^{13}C to ^{12}C ratio is retained over very long periods of time—even within fossil material. Indeed, fossil analysis has proved to be of immense value to investigations into the evolutionary origin of the various photosynthetic pathways, as discussed below.

Some enzymes discriminate between the different isotopes of carbon, notably Rubisco, which reacts more readily with $^{12}CO_2$ than with $^{13}CO_2$, as explained in the text. Hence, the analysis of the isotope discrimination ratio within a sample of plant or fossil material can provide evidence of the photosynthetic pathway that was operating within the living plant.

The carbon isotope composition of a sample is determined by mass spectrometry (see Chapter 2). In mass spectrometry analysis, the ratio of the two most abundant stable isotopes in the sample is expressed relative to the same ratio within an international standard. The standard sample that is used for all stable carbon isotope analyses is the Chicago Pee Dee Belemnite Marine carbonate standard, which was obtained from a marine fossil, *Belemnitella americana*, from the Pee Dee rock formation in South Carolina, USA. The carbon isotope composition of a sample is expressed as the $\delta^{13}C$ value according to the following equation:

$$\delta^{13}C = \left[\frac{^{13}C/^{12}C \text{ in sample} - {}^{13}C/^{12}C \text{ in Pee Dee standard}}{^{13}C/^{12}C \text{ in Pee Dee standard}} \right] \times 1000$$

Note that because the differences in ratios between the sample and standard are very small, they are expressed as parts per thousand, or per mil (‰)—hence the ratio is multiplied by 1000 in the above equation (rather than %, in which case the multiplication factor would be 100). Note also that the Pee Dee standard has a higher ^{13}C to ^{12}C ratio (0.0112372) than most naturally-occurring organic compounds, hence the $\delta^{13}C$ value is usually negative.

Stable isotope analysis has proved to be a valuable technique for archaeological studies and for food testing. The characteristic $\delta^{13}C$ value of plant material is largely retained in the bones of herbivores that feed on the plants, as any subsequent change in the ^{13}C to ^{12}C ratio during metabolic processing within the animal is relatively small compared with the differences between the $\delta^{13}C$ values of C_3 and C_4 plants. Hence, the original diet can be traced through a number of trophic levels through the food chain. One remarkable application of this technique has enabled archaeologists to determine the time when maize, a C_4 crop, was first cultivated in North America. Analysis of human skeletons and carbonized deposits in cooking pots provided evidence that the $\delta^{13}C$ value of the food changed from –21 to –12‰ between AD 1000 and 1200. This change, from a typical C_3 to a typical C_4 value, marks the introduction of maize into the diet and provides evidence for the transition of the human population from hunter-gatherers to the early farmers. Very recently, the same technique has been used in the UK to test whether corn-fed chickens really have been fed on maize. While corn-fed chicken commands almost three times the price of conventionally reared chicken, the only obvious difference between the two is that corn-fed chicken has a yellowy coloration to its flesh. Some farmers have been accused of adding yellow coloring to the feed of conventionally reared chicken in order to pass them off as corn-fed chickens. However, stable isotope analysis can be used to distinguish between the two, as corn-fed chicken would have a $\delta^{13}C$ value that is indicative of a largely C_4 diet (i.e. maize), while the diet of conventionally reared chickens would be C_3, as it consists of cereal grains (e.g. barley, wheat).

The analysis of fossils has also led to the identification of the period when C_4 photosynthesis is most likely to have arisen. Cerling *et al.* (see Further Reading) analyzed the $\delta^{13}C$ values of fossil soil samples and tooth enamel from living and fossilized mammals from Europe, North America, Asia, Africa, and South America. As a result, the scientists were able to determine that large mammals had a predominantly C_3 diet up until 8 million years ago, whereas mammals at low latitudes (<37°) (in Africa, North and South America, and Southern Asia) switched to a largely C_4 diet by 6 million years ago. These studies provided strong evidence for a global expansion in C_4 plants between 7 and 5 million years ago during the late Meiocine and early Pliocene epochs, when atmospheric CO_2 concentrations are thought to have fallen below a threshold that was required for C_3 photosynthesis. With current, and predicted rises in atmospheric CO_2 concentrations we may well be facing a reverse of the low CO_2 conditions that led to the evolution of C_4 plants, so that C_3 plants may begin to gain a competitive edge even within the tropical environments that normally favor the growth of C_4 plants.

for explanation of the technique). The resulting carbon isotope ratio ($\delta^{13}C$, see Box 5.7) differs between C_3, C_4, and CAM plants as follows:

- C_3 plants have $\delta^{13}C$ values that typically range between –32 and –20‰, with an average value of about –28‰.

- C_4 plants have $\delta^{13}C$ values ranging between –17 and –9‰, with an average value of about –13‰. Note that these values are higher than those for C_3 plants, as explained above.

- CAM plants can have a range of $\delta^{13}C$ values between those of C_3 and C_4 plants. In dry conditions, when CAM plants fix all of their CO_2 at night using PEP carboxylase, their $\delta^{13}C$ values are typical of C_4 photosynthesis, at –17 to –9‰. In well-watered CAM plants, the stomata may open for part of the day, in which case they assimilate CO_2 via Rubisco (i.e. in Phase IV) and have $\delta^{13}C$ values that are typical of C_3 plants, i.e. between –32 and –20‰.

Summary

The Calvin cycle is essential and is found universally in all photosynthetic eukaryotes. It is the only way that they can produce photosynthetic products from CO_2 resulting in a net gain in organic carbon. CO_2 enters leaves through open stomata, and this can also result in concomitant water losses through evapotranspiration. Rubisco, the enzyme that incorporates CO_2 into the Calvin cycle, can also catalyze the oxygenation of RuBP, and this leads to loss of carbon through photorespiration. C_4 plants have evolved to minimize photorespiratory losses, by concentrating CO_2 within the cells that contain Rubisco. This requires specialized leaf anatomy that enables the spatial separation of PEP carboxylase and Rubisco. CAM plants have evolved to minimize water loss, by opening their stomata at night when the surrounding air is cooler. CAM photosynthesis involves the temporal separation of PEP carboxylase, which functions in the dark, and of Rubisco, which assimilates CO_2 into the Calvin cycle in the light. These photosynthetic pathways confer various advantages on plants, with C_4 photosynthesis enabling plants to grow at elevated temperatures and in nitrogen-poor soils. CAM plants survive in arid environments and where daytime CO_2 supply is limited.

Further Reading

Calvin cycle discovery

Bassham JA & Calvin M (1957) The Path of Carbon in Photosynthesis. Prentice-Hall Inc.,

The book that summarizes the history, experimentation, and discovery of the Calvin cycle.

Bassham JA (2003) Mapping the carbon reduction cycle: a personal perspective. *Photosynthesis Res.* 76, 35–52.

Benson AA (2002) Following the path of carbon in photosynthesis: a personal story. *Photosynthesis Res.* 73, 29–49.

Calvin M (1989) Forty years of photosynthesis and related activities. *Photosynthesis Res.* 21, 3–16.

Personal accounts of the discovery of the Calvin cycle from each of the three key scientists

Calvin cycle regulation

Raines CA (2003) The Calvin cycle revisited. *Photosynthesis Res.* 75, 1–10.

A review of the regulation of the Calvin cycle that identifies potential target enzymes for increasing flux.

Prospects for improving C$_3$ photosynthesis

Long SP, Zhu X-G, Naidu SL & Ort DR (2006) Can improvement in photosynthesis increase crop yield? *Plant Cell Environ.* 29, 315–330.

This review considers the relationship between photosynthesis and grain yield and describes specific approaches to improving photosynthesis and productivity.

Raines CA (2006) Transgenic approaches to manipulate the environmental responses of the C$_3$ carbon fixation cycle. *Plant Cell Environ.* 29, 331–339.

A wide-ranging review that discusses approaches to improving photosynthesis, reducing photorespiratory losses and introducing C$_4$ photosynthesis into C$_3$ plants.

Rubisco reviews

Andrews TJ & Whitney SM (2003) Manipulating Ribulose bisphosphate carboxylase/oxygenase in the chloroplasts of higher plants. *Arch. Biochem. Biophys.* 414, 159–169.

This review considers how Rubisco efficiency and expression might be further improved in order to enhance crop productivity.

Parry MAJ, Andralojc PJ, Mitchell RAC, Madgwick PJ & Keys AJ (2003) Manipulation of Rubisco: the amount, activity, function and regulation. *J. Exp. Bot.* 54, 1321–1333.

This review focuses on transgenic approaches to modifying Rubisco specificity, expression, and regulation.

Spreitzer RJ & Salvucci ME (2002) Rubisco: structure, regulatory interactions, and possibilities for a better enzyme. *Annu. Rev. Pl. Biol.* 53, 449–475.

A detailed review of Rubisco properties, including its three-dimensional structure.

Tcherkez GGB, Farquhar GD & Andrews TJ (2006) Despite slow catalysis and confused substrate specificity, all ribulose bisphosphate carboxylases may be nearly perfectly optimized. *Proc. Natl Acad. Sci. USA* 103, 7246–7251.

This paper reviews the kinetic properties of a range of Rubisco proteins and asserts the hypothesis that a high specificity factor is intrinsically associated with a slow rate of carboxylation.

C$_4$ photosynthesis: discovery and manipulation

Hatch MD (1987) C$_4$ photosynthesis: a unique blend of modified biochemistry, anatomy and ultrastructure. *Biochim. Biophys. Acta* 895, 81–106.

A classic review by one of the leading scientists in C$_4$ photosynthesis.

Hatch MD (2002) C$_4$ photosynthesis: discovery and resolution. *Photosynthesis Res.* 73, 251–256.

A good account of the early experiments that led to the discovery of C$_4$ photosynthesis.

Häusler RE, Hirsch H-J, Kreuzaler F & Peterhänsel C (2002) Overexpression of C$_4$-cycle enzymes in transgenic C$_3$ plants: a biotechnological approach to improve C$_3$-photosynthesis. *J. Exp. Bot.* 53, 591–607.

A very detailed account of C$_4$ photosynthesis and its component enzymes, as well as a consideration of the approaches taken to introduce C$_4$ into C$_3$ plants.

Leegood RC (2002) C$_4$ photosynthesis: principles of CO$_2$ concentration and prospects for its introduction into C$_3$ plants. *J. Exp. Bot.* 53, 581–590.

A good account of CO$_2$ concentrating mechanisms and a consideration of the minimum requirements needed for introducing C$_4$ photosynthesis into C$_3$ plants.

Miyao M (2003) Molecular evolution and genetic engineering of C$_4$ photosynthetic enzymes. *J. Exp. Bot.* 54, 179–189.

This review considers the approaches taken to introduce C$_4$ photosynthesis into C$_3$ plants, in particular the need for high levels of expression as well as targeting to specific cellular compartments.

Suzuki S, Murai N, Kasaoka K, Hiyoshi T, Imaseki H, Burnell JN & Arai M (2006) Carbon metabolism in transgenic rice plants that express phospho*enol*pyruvate carboxylase and/or phospho*enol*pyruvate carboxykinase. *Pl. Sci.* 170, 1010–1019.

Despite achieving high expression of two C_4 enzymes in rice, the resulting plants showed no signs of carrying out C_4 photosynthesis.

C_3–C_4 photosynthesis

Rawsthorne S (1992) C_3-C_4 intermediate photosynthesis: linking physiology to gene expression. *Pl. J.* 2, 267–274.

A good review of the physiology and biochemistry of C_3–C_4 photosynthesis.

Single-cell C_4 photosynthesis

Edwards GE, Franceschi VR & Voznesenskaya EV (2004) Single-cell C_4 photosynthesis versus the dual-cell (Kranz) paradigm. *Annu. Rev. Pl. Biol.* 55, 173–196.

A recent and very thorough review of the existence of C_4 photosynthesis without the need for Kranz anatomy.

C_4 photosynthesis: evolution

Cerling TE, Wang Y, & Quade J (1993) Expansion of C_4 ecosystems as an indicator of global ecological change in the late Miocene. *Nature* 361, 344–345.

Traces the origins of C_4 photosynthesis in relation to global climate change.

Cerling TE, Harris JM, MacFadden BJ, Leakey MG, Quade J, Eisenmann V & Ehleringer JR (1997) Global vegetation change through the Miocene/Pliocene boundary. *Nature* 389, 153–158.

Explains how stable isotope analysis is used to trace the evolution of C_4 photosynthesis.

Keeley JE & Rundel PW (2003) Evolution of CAM and C_4 carbon-concentrating mechanisms. *Int. J. Pl. Sci.* 164, S55–77.

A very useful review of the evolution and ecophysiology of C_4 and CAM plants.

Sage RF (2004) The evolution of C_4 photosynthesis. *New Phytol.* 161, 341–370.

Traces the evolution of C_4 plants back to their first appearance in grasses.

Crassulacean acid metabolism

Black CC & Osmond CB (2003) Crassulacean acid metabolism photosynthesis: "working the night shift". *Photosynthesis Res.* 76, 329–341.

An entertaining and highly readable account of the science and scientists behind the discovery of CAM photosynthesis.

Cushman JC (2005) Crassulacean acid metabolism: recent advances and current opportunities. *Functional Pl. Biol.* 32, 375–380.

A useful introduction to the Special Issue of *Functional Plant Biology* devoted to CAM.

Photorespiration

Douce R & Neuburger M (1999) Biochemical dissection of photorespiration. *Curr. Opin. Pl. Biol.* 2, 214–222.

Includes a particularly good account of glycine decarboxylase.

Keys AJ (2006) The re-assimilation of ammonia produced by photorespiration and the nitrogen economy of C_3 higher plants. *Photosynthesis Res.* 87, 165–175.

A personal account of the history of photorespiration.

Keys AJ & Leegood RC (2002) Photorespiratory carbon and nitrogen cycling: evidence from studies of mutant and transgenic plants. In Photosynthetic Nitrogen Assimilation and Associated Carbon and Respiratory Metabolism (CH Foyer, G Noctor eds). Kluwer Academic Publishers.

A good source of information about the use of mutants and transgenic plants in the study of photorespiration.

Keys AJ, Bird, IF, Cornelius MJ, Lea PJ, Wallsgrove RM & Miflin BJ (1978) Photorespiratory nitrogen cycle. *Nature* 275, 741–743.

The paper that introduced the concept of ammonia release and reassimilation during photorespiration.

Somerville CR (2001) An early *Arabidopsis* demonstration. Resolving a few issues concerning photorespiration. *Pl. Physiol.* 125, 20–24.

An overview of the use of arabidopsis mutants to identify the reactions of the photorespiratory pathway.

Respiration

6

Key concepts

- Respiration involves the breakdown of organic compounds to release energy that is conserved in the formation of ATP.

- Glycolysis is the major pathway that fuels respiration, it also supplies energy and reducing power for biosynthetic reactions.

- The oxidative pentose phosphate pathway is an alternative catabolic route for glucose metabolism and, under nonphotosynthetic conditions, provides NADPH to support a number of metabolic processes.

- Pyruvate oxidation marks the link between glycolysis and the tricarboxylic acid (TCA) cycle.

- In plants, substrates for the TCA cycle are derived mainly from carbohydrates. Proteins and lipids are used during periods of carbohydrate starvation.

- The TCA cycle serves a biosynthetic function in plants.

- The mitochondrial electron transport chain oxidizes reducing equivalents and produces ATP.

- Plant mitochondria possess additional respiratory enzymes that provide a branched electron transport chain.

- ATP synthesis in plant mitochondria is coupled to the proton electrochemical gradient that forms during electron transport.

- Mitochondrial respiration interacts with photosynthesis and photorespiration in the light.

- An emerging research area that is likely to alter our understanding of plant respiration is the identification of supercomplexes and metabolons.

Overview of respiration

Aerobic respiration occurs in all multicellular eukaryotes. It is the process whereby complex organic compounds are broken down to release energy that is usually conserved in the form of ATP. As such, oxidative respiration can be seen as the reverse of photosynthesis, and is summarized in the following equation:

$$C_6H_{12}O_6 + 6O_2 + 6H_2O \longrightarrow 6CO_2 + 12H_2O$$

While it is true that carbohydrates, as shown in this equation, are the main respiratory substrates in plants, both lipids and proteins may also be respired, as we shall see later in this chapter. Hence, as with the generalized equation for photosynthesis (Chapter 5) it is important to understand that the respiratory reactions are rather more variable than might be implied by this summary equation.

The main components of plant respiration

Glycolysis, the oxidative pentose phosphate pathway (OPPP), the tricarboxylic acid (TCA) cycle, and mitochondrial electron transport chain are the main components of plant respiration. The oxidation of carbohydrates takes place via a series of reactions, starting with glycolysis. Here, glucose and fructose are oxidized to pyruvate, ATP is formed from ADP and P_i, and NAD^+ is reduced to NADH. The OPPP also results in carbohydrate oxidation but this pathway uses $NADP^+$ instead of NAD^+, and is usually considered principally as a biosynthetic, rather than a catabolic pathway, as discussed later.

Provided that oxygen is available to support aerobic respiration, pyruvate produced in glycolysis usually enters the mitochondrial TCA cycle where it is oxidized to CO_2, with the concurrent reduction of NAD^+ to NADH.

Finally, the mitochondrial electron transport chain receives electrons from NADH produced in glycolysis and the TCA cycle, and, uniquely in plants, from NADPH generated in the OPPP. Oxidation of NADH and NADPH via the electron transport chain serves to regenerate NAD^+ and $NADP^+$ respectively, allowing the oxidative reactions to continue, while at the same time releasing free energy that is used to convert ADP and P_i to ATP. Thus, a more complete respiratory equation emerges, as summarized here:

$$C_6H_{12}O_6 + 6O_2 + 6H_2O + 32ADP + 32P_i \longrightarrow 6CO_2 + 12H_2O + 32ATP$$

The production of up to 32 moles of ATP, depending on the pathway taken in mitochondrial electron transport, requires the complete oxidation of a mole of carbohydrate. This is not always the case, as respiratory intermediates are often diverted into biosynthesis rather than being fully oxidized to CO_2. Indeed, biosynthesis is an important aspect of respiration, especially in plants, as we shall see later in this chapter. Moreover, if NADH is oxidized in part via the alternative oxidase (see below) the ATP yield will be correspondingly reduced.

Plants need energy and precursors for subsequent biosynthesis

Plants synthesize every one of their carbon-containing compounds either directly from early products of photosynthesis or from stored photosynthetic products. Glycolysis and the OPPP are important metabolic pathways to generate the energy and precursors needed for subsequent biosynthesis. Glycolysis literally means lysis or breakdown of sugar. It occurs universally in all eukaryotic organisms and is generally shown as a linear sequence of 10 reactions where glucose or other monosaccharides and sugars are partially oxidized to produce the three-carbon pyruvate, ATP, and reductant in the form of NADH, which are the building blocks for anabolism. The OPPP is depicted as a cyclic process in which glucose 6-phosphate is oxidized to

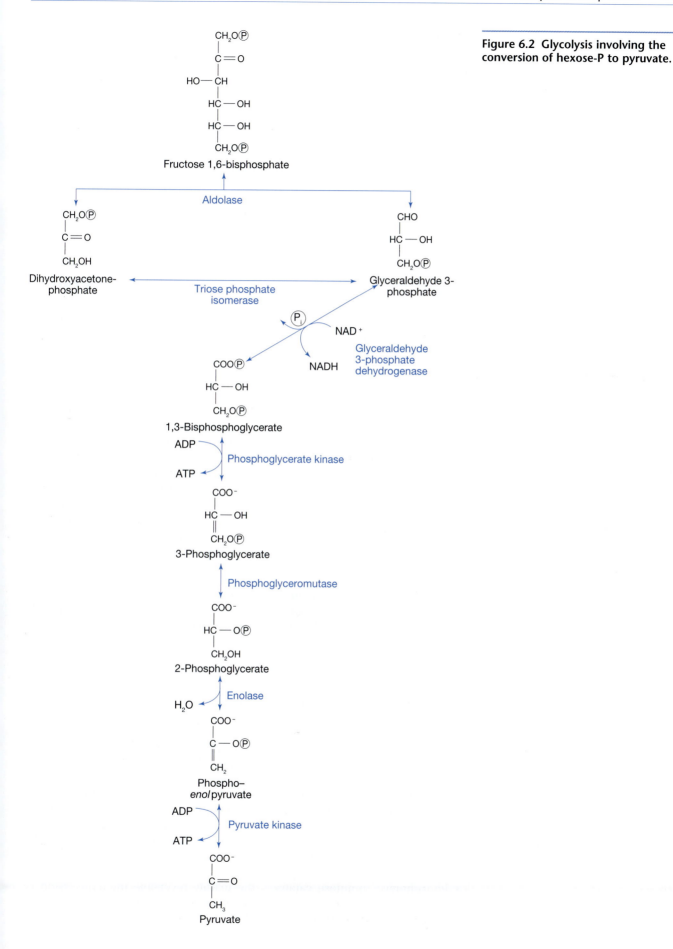

Figure 6.2 Glycolysis involving the conversion of hexose-P to pyruvate.

Hexose sugars enter into glycolysis and are converted into fructose 1,6-bisphosphate

For carbon from storage carbohydrates to enter glycolysis, glucose and fructose derived from the hydrolysis of starch, sucrose, or fructans must first be converted to fructose 1,6-bisphosphate (Figure 6.1). Energy in the form of ATP is needed for this conversion. The phosphorolytic breakdown of starch to glucose 1-phosphate offers an energetic advantage over the hydrolytic degradation in that one less ATP is required for each molecule of hexose entering via the phosphorolytic route. As glucose is a relatively stable molecule, this first stage is essentially a 'pump priming' process, with the initial phosphorylations to glucose 6-phosphate and subsequently fructose 1,6-bisphosphate being a form of activation energy. ATP is subsequently generated during the second part of glycolysis.

Fructose 1,6-bisphosphate is converted to pyruvate

Aldolase and triose phosphate isomerase participate in the reversible interconversion of fructose 1,6-bisphosphate to glyceradehyde 3-phosphate and dihydroxyacetone phosphate (Figure 6.2). The standard free energy change involving the aldol cleavage of fructose 1,6-bisphosphate to glyceraldehyde 3-phosphate and dihydroxyacetone phosphate catalyzed by aldolase indicates that the equilibrium favors fructose 1,6-bisphosphate formation. As the actual direction of the reaction is determined by the concentrations of all three compounds involved, fructose 1,6-bisphosphate only accumulates when its cleavage products achieve a threshold concentration. Triose phosphate isomerase catalyzes the readily reversible interconversion of glyceraldehyde 3-phosphate and dihydroxyacetone phosphate. In the cell the equilibrium favors dihydroxyacetone phosphate over glyceralderhyde 3-phosphate by approximately 22:1. The combined effect of both the aldolase and triose phosphate isomerase equilibria is to keep the concentration of glyceraldehyde 3-phosphate low. Because the reactions in the lower part of the glycolytic pathway consume glyceraldehyde 3-phosphate, this shifts the aldolase and triose phosphate isomerase equilibria toward glyceraldehyde 3-phosphate production and hence drives the glycolytic pathway forward.

Glyceraldehyde 3-phosphate dehydrogenase catalyzes a readily reversible phosphate-consuming reaction in which the oxidation of glyceraldehyde 3-phosphate is linked directly to the reduction of NAD^+ to NADH. The next enzyme, phosphoglycerate kinase, achieves substrate-level phosphorylation and has been studied extensively. The equilibrium position of the dehydrogenase reaction favors glyceraldehyde 3-phosphate over 1,3-bisphosphoglycerate by approximately 10:1. However, because NAD^+ and NADH bind competitively to glyceraldehyde 3-phosphate dehydrogenase, the equilibrium position is also determined by the NAD^+:NADH ratio. The concentration of NAD^+ in the cell is ordinarily maintained at approximately 10 times the concentration of NADH, thereby favoring 1,3-bisphosphoglycerate formation. The subsequent reaction catalyzed by phosphoglycerate kinase favors the formation of 3-phosphoglycerate and ATP and at prevailing ratios of ADP:ATP therefore drives the glyceradehyde 3-phosphate dehydrogenase reaction in a forward direction towards the products. Because the mixed acid anhydride bond of 1,3-bisphosphoglycerate is very unstable and has a high free energy of hydrolysis, the equilibrium strongly favors the products of the reaction, ATP and 3-phosphoglycerate.

Phosphoglyceromutase catalyzes the readily reversible interconversion of 3-phosphoglycerate and 2-phosphoglycerate. The enzyme moves the phosphate group from the 3-position to the 2-position, such that subsequent dehydration catalyzed by enolase produces a high-energy compound phospho*enol*pyruvate (PEP). This configuration is short-lived in the cell; the phosphate attached to the 2-position of pyruvate has a high negative free energy of hydrolysis, allowing transfer of the phosphate group to ADP with the resulting formation of ATP and pyruvate.

Alternative reactions provide flexibility to plant glycolysis

Under certain circumstances cytosolic plant glycolysis uses alternative enzymes to supplement or even replace those conventionally present (and described above). The existence of these alternative reactions is thought to ensure metabolic flexibility and to allow glycolysis to proceed and the plant to survive under varying and potentially stressful environmental and/or nutritional conditions. Alternative reactions involve the enzymes pyrophosphate-dependent fructose 6-phosphate 1-phosphotransferase (PFP) and non-reversible/nonphosphorylating NADP-dependent glyceraldehyde 3-phosphate dehydrogenase (NADP-G3PDH; Figure 6.3). In addition there are also potential sequences of enzymes comprising PEP carboxylase, malate dehydrogenase (MDH) and malic enzyme or PEP phosphatase capable of generating pyruvate from PEP (Figure 6.4).

PFP was identified as an alternative to ATP-dependent phosphofructokinase (PFK) in plant glycolysis. This pyrophosphate-dependent enzyme is located

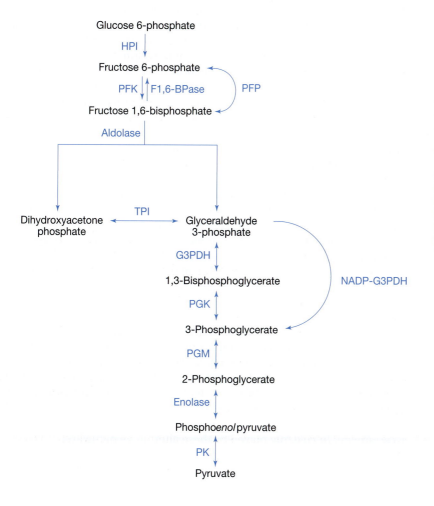

Figure 6.3 The role of pyrophosphate-dependent fructose 6-phosphate 1-phosphotransferase (PFP) and non-reversible/non-phosphorylating NADP-dependent glyceraldehyde 3-phosphate dehydrogenase (NADP-G3PDH) as alternative glycolytic reactions.

Abbreviations: F1,6-BPase, fructose 1,6-bisphosphatase; G3PDH, glyceraldehyde 3-phosphate dehydrogenase; HPI, hexose phosphate isomerase; PFK, phosphofructokinase; PGK, phosphoglycerate kinase; PGM, phosphoglyceromutase; PK, pyruvate kinase; TPI, triosephosphate isomerase.

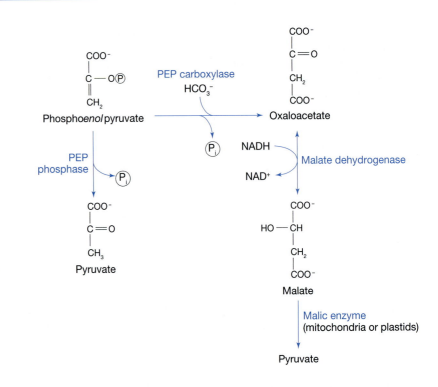

Figure 6.4 Potential role of phospho*enol*pyruvate carboxylase (PEP carboxylase), malate dehydrogenase, and malic enzyme, or phospho*enol*pyruvate (PEP) phosphatase in generating pyruvate from phospho*enol*pyruvate.

exclusively in the cytosol and catalyzes a readily reversible reaction. The activity of this enzyme is dependent on the presence of the allosteric activator fructose 2,6-bisphosphate (Chapter 7). The precise role of PFP is unclear. However, a number of studies have shown that nutritional phosphate deficiency may lead to serious reductions in the levels of cellular adenine nucleotides, including ATP, so as to prevent the operation of PFK. Under such conditions, cytosolic concentrations of pyrophosphate may be sufficient to support PFP activity in a glycolytic direction. Similarly, under phosphate-poor conditions an elevation of NADP-G3PDH takes place with a concomitant decrease in the levels of the conventional phosphate-requiring NAD-dependent glyceraldehyde 3-phosphate dehydrogenase. The cytosolic NADP-G3PDH generates 3-phosphoglyceric acid instead of 1,3-bisphosphoglycerate and can allow carbon flow through glycolysis to continue when the concentrations of ADP are too low for the synthesis of ATP to proceed (Figure 6.3).

Phosphate deficiency leading to reductions in adenine nucleotides, including ADP, could also limit pyruvate kinase activity. It has also been suggested that a PEP/malate bypass operates during conditions of phosphate deficiency (Figure 6.4). Here PEP in the cytosol may exit glycolysis via carboxylation to oxaloacetate (OAA), catalyzed by the enzyme PEP carboxylase. This OAA can then be reduced to malate via MDH. The malate can then be transported to the mitochondria and/or plastids. Alternatively PEP may enter the vacuole and be metabolized by PEP phosphatase to pyruvate.

Bypassing conventional glycolysis via PFP, NADP-G3PDH, and PEP/malate may permit the continued operation of glycolysis under conditions where low nutritional phosphate has led to deficiency in adenine nucleotides. However, poor overall concentrations of phosphate and energy status may also lead to the additional effect of depleting levels of pyridine nucleotides (i.e. NAD^+ and $NADP^+$). This would prevent the effective use of carbon substrates in biosynthetic processes. As such, more research is now needed to ascertain any *in vivo* role played by these alternative pathways.

Plant glycolysis is regulated by a bottom-up process

In contrast to animal and yeast glycolysis, where a feedforward effect is involved in activating glycolysis, with fructose 1,6-bisphosphate activating pyruvate kinase, plant pyruvate kinase is insensitive to fructose 1,6-bisphosphate, while PFK is markedly inhibited by PEP. Once PEP is converted to pyruvate via pyruvate kinase or to OAA via PEP carboxylase, the resulting fall in the concentration of PEP can feed back and relieve inhibition of PFK. This process ensures that glycolysis keeps pace with either ATP production or the provision of substrate for cellular biosynthesis. The activity of PFK similarly responds to the concentration of P_i, which is an activator.

Because the rate of glycolysis responds in this way to the level of PEP, the penultimate metabolite in the pathway, this is described as a bottom-up process of regulation.

Glycolysis supplies energy and reducing power for biosynthetic reactions

A major function of glycolysis is energy conservation, which occurs in two different ways. First, reducing-potential, in the form of NADH, is generated when glyceraldehyde 3-phosphate is converted to 1,3-bisphosphoglycerate. The NADH generated is then available to the cell as reducing potential for the synthesis of other molecules. Alternatively it may, in the presence of oxygen, be metabolized to ATP. Secondly, energy in the original hexose molecule is conserved as ATP that is generated in the reactions catalyzed by phosphoglycerate kinase and pyruvate kinase. For every molecule of hexose entering glycolysis four ATP molecules are formed via substrate-level phosphorylation. Depending on the initial storage carbohydrate and its degradation, this provides a net gain of either two or three ATP molecules.

Glycolysis takes place in both the cytosolic and plastid compartments of the plant cell. Synthesis of compounds derived from hexoses would be expected to occur in the plastids, because these organelles are responsible for photosynthesis and starch storage. In chloroplasts, photosynthesis can provide any cofactors needed. Anabolic pathways (including the production of fatty acids, amino acids, or secondary metabolites) can consume large amounts of energy, in the form of ATP, and reducing power, in the form of NAD(P)H. In nonphotosynthetic plastids, these cofactors must be generated from the activity of metabolic pathways, including plastidic glycolysis. Cytosolic glycolysis provides a means of both funneling carbon into the mitochondrion for oxidative phosphorylation and for the production of biosynthetic precursors.

The availability of oxygen determines the fate of pyruvate

Under aerobic conditions pyruvate is transported into the mitochondria where it is oxidized to CO_2 and water via the TCA cycle, and electrons are ultimately transferred to molecular oxygen via the mitochondrial electron transport chain (see later in this chapter). Under conditions of hypoxia or anoxia, e.g. in tissues/organs that are subjected to occasional anaerobic conditions (for instance in roots or waterlogged soil) there is no available oxygen to act as the terminal electron acceptor, so mitochondrial respiration shuts down and the metabolism switches to fermentation.

In fermentation, pyruvate is converted either to ethanol through the action of pyruvate decarboxylase and alcohol dehydrogenase or to lactate via lactate dehydrogenase (Figure 6.5). The main products of fermentation are usually CO_2 and ethanol, but in the early stages of anoxia lactate accumulation often precedes alcohol production. Lactate accumulation, and concomitant failure of the vacuolar H^+-translocating ATPases under conditions of ATP depletion, lowers the pH of the cytosol leading to the activation of cytosolic pyruvate decarboxylase (which has a pH optimum below 7.0) and the initiation of ethanol production. After about 90 minutes anaerobic treatment of, for example, maize seedlings, there is repression of preexisting protein synthesis and the induced synthesis of approximately 20 anaerobic proteins. The majority of the anaerobic proteins that are synthesized are involved in glycolysis and fermentation. The regulation of synthesis of these anaerobic proteins has been shown to occur as a result of transcriptional, post-transcriptional and translational controls. As oxygen limitation is the major plant stress in flooded soils, threatening the food supply of humans as well as the natural vegetation in aquatic environments, then the identification of flood tolerance genes and the mechanisms involved in coordinating the control at the level of transcription and translation is an active area of research (see Sachs *et al.* in Further Reading).

NADH produced by the oxidation of glyceraldehyde 3-phosphate in glycolysis is oxidized to NAD^+ by either fermentation reaction (i.e. alcohol dehydrogenase or lactate dehydrogenase). As such there is no net gain of reducing potential in fermentation. However, the recycling of NADH is important as the pool of NADH plus NAD^+ in the cell is fairly small and recycling NADH is necessary to maintain a supply of NAD^+ to support the continued oxidation of glyceraldehyde 3-phosphate. This ensures that glycolysis and the production of the small quantities of ATP necessary to maintain cells continues under anaerobic conditions.

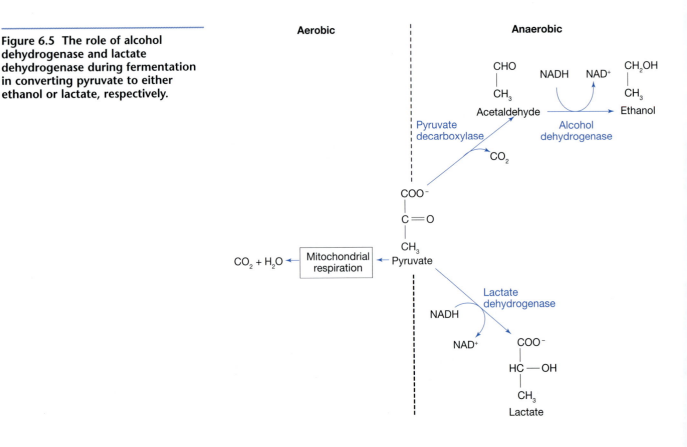

Figure 6.5 The role of alcohol dehydrogenase and lactate dehydrogenase during fermentation in converting pyruvate to either ethanol or lactate, respectively.

The oxidative pentose phosphate pathway is an alternative catabolic route for glucose metabolism

Although glycolysis is the major catabolic route for hexoses, all plant cells possess an alternative route for glucose metabolism, the oxidative pentose phosphate pathway (OPPP). Located in both the cytosol and plastids, the OPPP uses a number of shared/common intermediates with glycolysis and as such the two processes are closely integrated.

The OPPP generates NADPH, which is available as a reductant in circumstances when NADPH is not being generated by photosynthesis. As such it is particularly important in nonphotosynthetic conditions and tissues where processes such as nitrogen assimilation (see Chapter 8) and fatty acid biosynthesis (see Chapter 9) require NADPH as a reductant. NADPH is also important in maintaining an appropriate redox potential to protect against oxidative stress. Secondly, the OPPP generates ribose 5-phosphate, which is necessary for the synthesis of nucleotides and nucleic acids. Finally, the OPPP generates erythrose 4-phosphate, which is necessary for the synthesis of shikimic acid, a precursor dedicated to the synthesis of aromatic amino acids (see Chapter 8) and phenylpropanoids (Chapter 11).

The first stage of the OPPP involves the action of two enzymes whose effect is the irreversible oxidative decarboxylation of glucose 6-phosphate to ribulose 5-phosphate, resulting in the generation of two molecules of NADPH per molecule of glucose 6-phosphate. The second, reversible stage leads to the formation of fructose 6-phosphate and glyceraldehyde 3-phosphate, which in principle can then be further metabolized by glycolysis (Figure 6.6).

The irreversible oxidative decarboxylation of glucose 6-phosphate generates NADPH

The oxidation of glucose 6-phosphate to 6-phosphogluconate is the first step in the OPPP. This step is sensitive to inhibition by NADPH and is thought to determine the balance between glycolysis and the OPPP. Glucose 6-phosphate dehydrogenase catalyzes the oxidation of glucose 6-phosphate to 6-phosphoglucono-δ-lactone with the concomitant reduction of $NADP^+$ to NADPH. Reduction of the product, 6-phosphoglucono-δ-lactone, back to glucose 6-phosphate is thermodynamically feasible; however, the lactone is unstable and spontaneously hydrolyzes to form 6-phosphogluconate, making the oxidative reaction essentially irreversible. A lactonase may further increase the rate of 6-phosphogluconate formation. 6-Phosphogluconate dehydrogenase catalyzes the irreversible oxidative decarboxylation of 6-phosphogluconate to ribulose 5-phosphate and CO_2, with the concomitant reduction of $NADP^+$ to NADPH.

The second stage of the oxidative pentose phosphate pathway returns any excess pentose phosphates to glycolysis

The second stage of the OPPP leads to the formation of glyceraldehyde 3-phosphate and fructose 6-phosphate, which may be metabolized via glycolysis. The reactions catalyzed by ribulose 5-phosphate epimerase, ribose 5-phosphate isomerase, transketolase, and transaldolase are all reversible. This part of the pathway is important because it enables any excess pentose phosphate produced during NADPH generation to be returned to the glycolytic pathway. It also allows ribose 5-phosphate and erythrose 4-phosphate

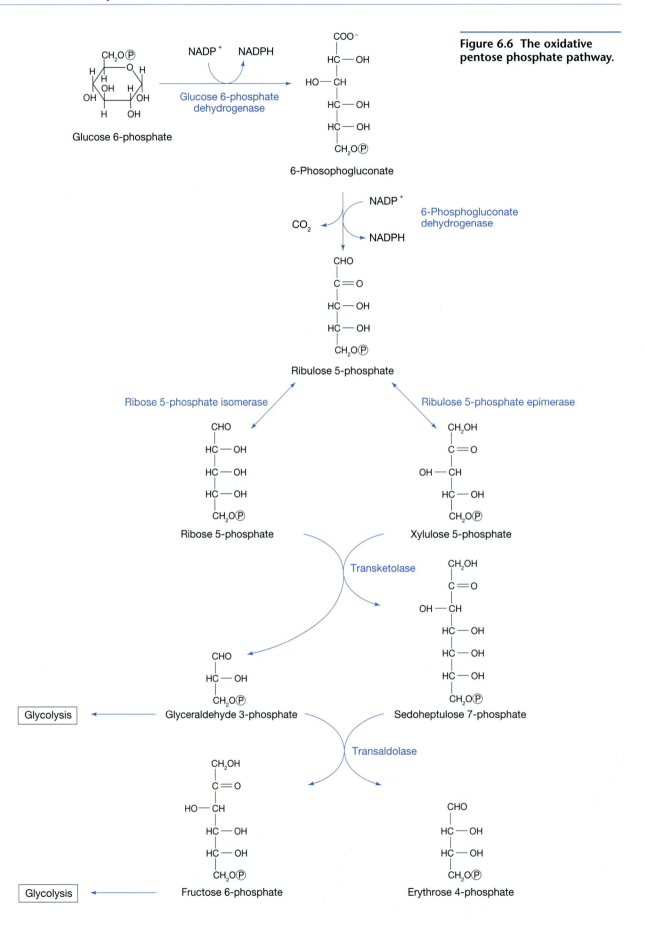

Figure 6.6 The oxidative pentose phosphate pathway.

to be formed from glyceraldehyde 3-phosphate and fructose 6-phosphate without needing to oxidize carbohydrates. Depending on the needs of the cell, differing amounts of ribose 5-phosphate and erythrose 4-phosphate will be withdrawn from the OPPP for nucleotide synthesis and for phenyl-propanoid (Chapter 11) and amino acid production (Chapter 8).

All or part of the OPPP is duplicated in the plastids and cytosol

Knowledge of the OPPP enzymes at the molecular level is not as advanced as that of other carbohydrate metabolism pathways. All the OPPP enzymes have been identified in various types of plastids, and any absences of enzymes seem to be within the cytosol. All studies have confirmed the presence of glucose 6-phosphate dehydrogenase and 6-phosphogluconate dehydrogenase in the plastids and the cytosol. In contrast the presence of the enzymes involved in the reversible reactions are less clear. In part, the distribution of these enzymes appears to depend on species, tissue type, developmental stage, and environmental conditions. The operation of a complete OPPP in the plastids and either a partial or complete pathway in the cytosol has potential implications for metabolic regulation. In addition, the need for transporters to facilitate movement of metabolites between the plastid and cytosol is essential.

The tricarboxylic acid cycle is located in the mitochondria

The tricarboxylic acid (TCA) cycle, also called the citric acid cycle, and Krebs cycle, was discovered by the scientist, Sir Hans Krebs, in 1937, using pigeon muscle homogenates as his research material (Box 6.2). The TCA cycle has now been detected in a wide range of eukaryotes where, without exception, it is localized in the mitochondria. All of the TCA cycle enzymes, except for succinate dehydrogenase (which is an integral membrane protein), are soluble enzymes located in the mitochondrial matrix, though a few may be duplicated in the cytosol. Together, these enzymes convert pyruvate to CO_2 and generate NADH, which is subsequently oxidized via the mitochondrial electron transport chain, resulting in ATP synthesis. The pathway is shown in detail in Figure 6.7, with detailed reactions in Figure 6.8.

Pyruvate oxidation marks the link between glycolysis and the tricarboxylic acid cycle

Although a cycle, by definition, should have no starting point, it is conventional to consider acetyl CoA, the product of pyruvate oxidation, as the initial substrate for the TCA cycle. Pyruvate, from glycolytic reactions in the cytosol, enters the mitochondria on the pyruvate translocator (Box 6.3).

Pyruvate is oxidized in the mitochondrial matrix by a series of reactions catalyzed by a multi-enzyme complex, the pyruvate dehydrogenase complex (PDC). The overall reaction is shown in Figure 6.8(a).

PDC, like that of other organisms, consists of three components, each of which has a separate catalytic activity: pyruvate dehydrogenase (PDH; E_1), dihydrolipoyl acetyltransferase (E_2), and dihydrolipoyl dehydrogenase (E_3). In addition, the mitochondrial complex has two associated regulatory

Box 6.2 The scientists and experiments that led to the discovery of the tricarboxylic acid (TCA) cycle

The TCA cycle is also referred to as the Krebs cycle in recognition of the contribution of Hans A Krebs, who first proposed the cycle in 1937. He called it the citric acid cycle, and a key to the elucidation of the pathway was Krebs' discovery of the reaction that formed citric acid from oxaloacetate and pyruvate. Krebs's first report on this discovery was submitted to *Nature*, who famously rejected the paper that subsequently led to the award of the 1953 Nobel Prize for Physiology or Medicine. The work was published instead in *Enzymologia* by Krebs and Johnson in 1937 (see Further Reading section in main text).

In his Nobel Lecture (1953) Krebs acknowledged the early work of several scientists whose research was fundamental to his discovery of the TCA cycle. A number of them were, like Krebs, Jewish emigrants who not only survived persecution but went on to have remarkable careers, as World-renowned researchers and, in a number of cases, as Nobel Laureates.

Thorsten Ludvig Thunberg (1873–1952), a Swedish biochemist, tested the ability of muscle tissue to oxidize a range of organic substrates. Between 1906 and 1920 he identified the rapid oxidation of succinate, fumarate, malate, and citrate, all of which are now known to be intermediates of the TCA cycle.

Frédéric Batelli and Lina Stern were working in Geneva at the time of Thunberg's experiments and were also using minced animal tissue to investigate substrate oxidation. They used the redox-sensitive dye, methylene blue, which changes color from blue to colorless upon reduction, to monitor substrate oxidation. They began to recognize the significance of these oxidative reactions and speculated that they might be directly related to the respiratory activity of the tissue itself. Lina Stern (1878–1968), a Latvian-born Jewish scientist, was the first woman to hold an academic post at the University of Geneva. She emigrated to the Soviet Union in 1925 and was the sole survivor of the Jewish Anti-Fascist Committee that was eradicated by Stalin in 1949. She was exiled to Kazakhstan, returning to Moscow after Stalin's death, to head the Department of Physiology between 1954 and 1968.

Albert Szent-Györgi (1893–1986) was a Hungarian-born Jewish scientist who won the 1937 Nobel Prize for Physiology for discovering ascorbic acid (vitamin C). During World War II, Szent-Györgi took refuge in the Swedish legation in Budapest. He escaped from a Gestapo raid, remaining in hiding for the rest of the war, before being rescued by the Russian armies and taken to Moscow. In 1947, he emigrated to the United States, where he spent the remainder of his career. In 1935, Szent-Györgi discovered that pigeon flight muscle retained its ability to oxidize substrates after being homogenized. This discovery was important for two reasons. First, the tissue was metabolically very reactive,

which made it particularly well-suited for biochemical studies. Second, homogenization resulted in soluble extracts that were more suitable for substrate measurement and inhibitor treatments than the heterogeneous minced tissue used in earlier experiments. He identified a sequence of reactions from succinate to fumarate to malate and finally to oxaloacetate. In addition to this, Szent-Györgi's particular contribution was to recognize the catalytic nature of the oxidation of organic acids. Malate and oxaloacetate caused an increase in oxygen consumption that far exceeded the amount needed to oxidize them. This catalytic effect was later interpreted by Krebs as evidence of a cycle.

Further evidence of catalysis came in 1937 from two German scientists, Carl Martius and Franz Knoop, whose contribution was recognized by Krebs as being decisive. Martius and Knoop discovered that α-ketoglutarate (2-oxoglutarate) was a product of citrate oxidation, and that *cis*-aconitate and isocitrate were intermediates. The reactions from citrate through to oxaloacetate were now largely known, but it was left to Krebs to identify and characterize the cyclic nature of the pathway.

Another key scientist in the history of the TCA cycle is Otto Warburg (1883–1970) who served in the Prussian Horse Guards during World War I and became Professor at the Kaiser Wilhelm Institute in Berlin when the war ended in 1918. His research into cancer cell physiology led to his discovery of the involvement of cytochromes and flavoproteins in respiratory oxygen consumption. He received the Nobel prize for Medicine in 1931. Because he was partly Jewish, Warburg was forced, by the Nazi regime, to decline a second Nobel Prize in 1944. He escaped further persecution because of his scientific eminence and strong personality. It was in Warburg's laboratory in Berlin that Krebs began his scientific career, as one of Warburg's research assistants. Krebs spent 4 years there, learning many of the biochemical techniques that he would later apply to his analysis of the TCA cycle. Warburg also made a number of very important contributions to the biochemistry of photosynthesis although some of his later interpretations never gained acceptance.

Hans Adolf Krebs (1900–81) was born in Germany and his first paid research post was in the laboratory of Otto Warburg, in Berlin. While working at the University of Freiburg, Germany (1931–33) Krebs discovered the ornithine cycle for urea synthesis. He was dismissed from this post during Hitler's rise to power and the removal of non-Aryans from professional occupations. Krebs moved to England, where he remained for the rest of his life. He held posts at the Universities of Cambridge, Sheffield (where he discovered the TCA cycle), and Oxford, UK.

The research carried out by Krebs's group at Sheffield was fundamental to the discovery of the TCA cycle. Three key

Box 6.2 The scientists and experiments that led to the discovery of the tricarboxylic acid (TCA) cycle (continued)

aspects were: (1) evidence that the reactions proceeded at rates that matched *in vivo* rates of respiration; (2) identification of the final step, the formation of citrate from oxaloacetate, thus completing a cycle of reactions; (3) inhibitor studies and substrate-feeding experiments that demonstrated the interdependence of the intermediates, and the cyclic nature of the pathway.

Krebs was able to demonstrate that the reaction discovered by Martius and Knoop, i.e. formation of α-ketoglutarate (2-oxoglutarate) from citrate, occurred in a range of tissues and at rates sufficient to support respiration. When malonate was added, succinate accumulated during citrate oxidation. Malonate is a competitive inhibitor of succinate dehydrogenase, hence this was evidence that succinate was formed from citrate. Malonate also led to succinate accumulation from citrate, isocitrate, *cis*-aconitate, or α-ketoglutarate as predicted. However, it was the observation that the oxidation of products of succinate oxidation, i.e. fumarate, malate, and oxaloacetate, also resulted in succinate accumulation in the presence of malonate that indicated the presence of a cycle. The breakthrough came with the discovery that citrate was not only broken down but it could also be formed, providing oxaloacetate and either pyruvate or acetate were added. Krebs concluded that oxaloacetate was condensing with pyruvate, or a derivative, to form citrate. He recognized that a cycle was occurring, rather than a simple linear pathway (Figure 1). This also explained the catalytic nature of the reactions; the addition of any of the intermediates would result in the formation of all of the others. If all of the intermediates could eventually form oxaloacetate, then the cycle was complete as oxaloacetate formed citrate, which was further oxidized through *cis* aconitate, isocitrate, α-ketoglutarate, succinic, fumaric, malic, and back to oxaloacetic acid, to regenerate the substrate for the condensation reaction once again. Hence, the catalytic nature of the reactions was explained, and the concept of the TCA cycle was formulated.

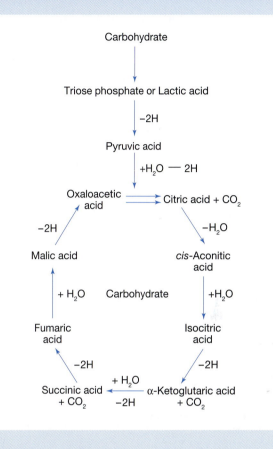

Figure 1 The original citric acid cycle. (Originally published as the *Nobel Biography of Hans Krebs,* © The Nobel Foundation, 1953.)

enzymes, pyruvate dehydrogenase kinase (PDK) and phospho-pyruvate dehydrogenase phosphatase (PDP) as explained below. The reactions are described in detail in Figure 6.9, page 163. Unique to plants is the existence of a plastidic, as well as a mitochondrial PDC. However, we will only discuss the mitochondrial PDC here, as the plastidic form is not regulated by protein phosphorylation and does not take part in respiration (being associated, instead, with fatty acid biosynthesis; see Chapter 9).

The plant mitochondrial PDC consists of a core of E_2 components that is thought to resemble the mammalian complex, with 60 copies of E_2 forming into twelve sets of five in a pentagonal dodecahedron. Such a structure has been identified from electron microscope images of purified E_2 from maize. Each E_2 component contains multiple binding domains consisting of one or

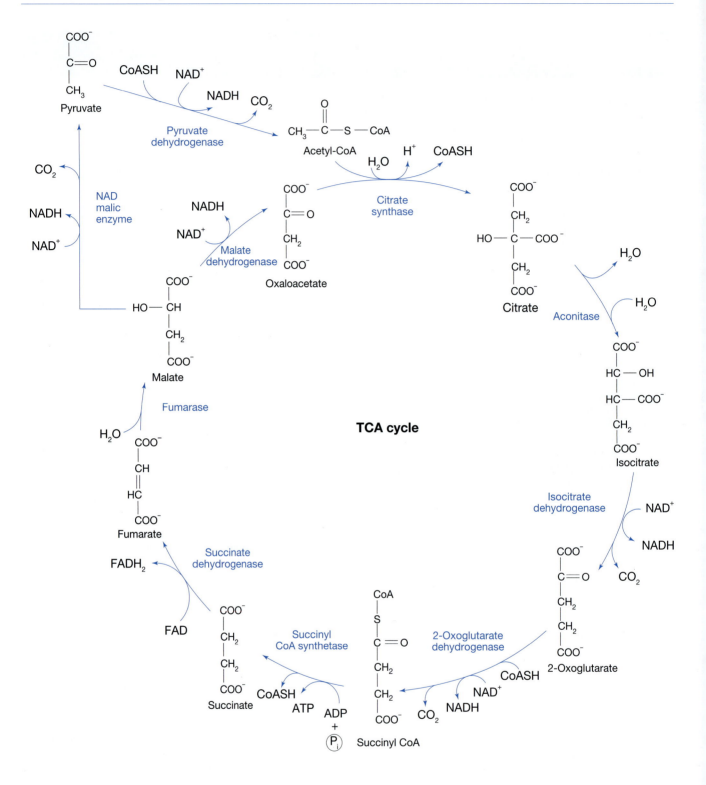

Figure 6.7 The TCA cycle. One molecule of pyruvate, if completely oxidized to three molecules of CO_2, will generate four molecules of NADH and one of $FADH_2$. Oxidation of this NADH and $FADH_2$ in the mitochondrial electron transport chain (Figure 6.16) will produce 14 molecules of ATP (12 from NADH, i.e. 4×3; two from $FADH_2$, i.e. 1×2), with an additional ATP being produced in the succinyl CoA synthetase reaction. The actual yield of ATP will depend on whether any of the nonphosphorylating branches are functioning as these will reduce the amount of ATP formed during electron transport.

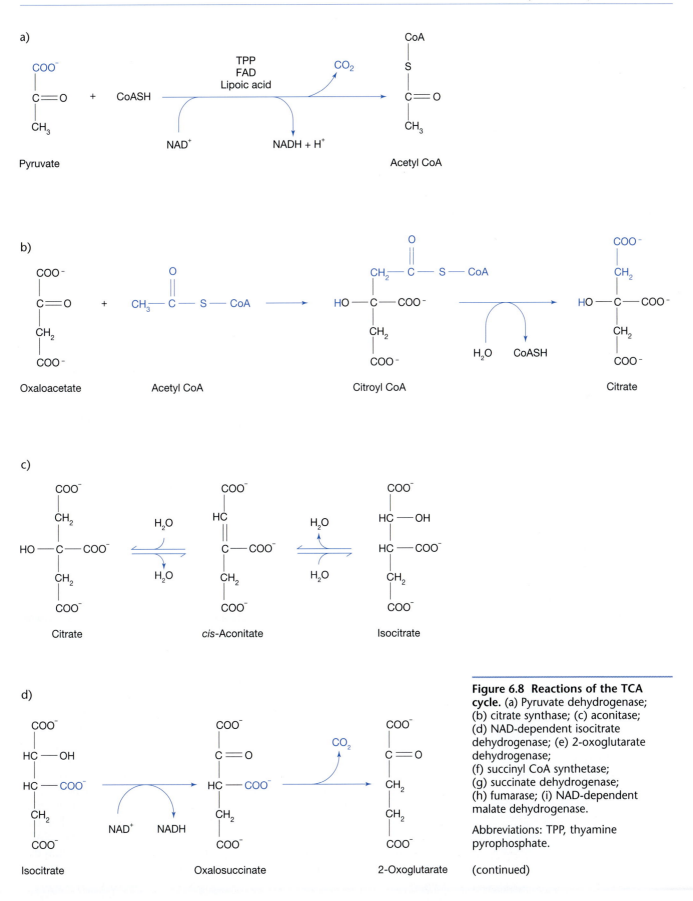

Figure 6.8 Reactions of the TCA cycle. (a) Pyruvate dehydrogenase; (b) citrate synthase; (c) aconitase; (d) NAD-dependent isocitrate dehydrogenase; (e) 2-oxoglutarate dehydrogenase; (f) succinyl CoA synthetase; (g) succinate dehydrogenase; (h) fumarase; (i) NAD-dependent malate dehydrogenase.

Abbreviations: TPP, thyamine pyrophosphate.

(continued)

e)

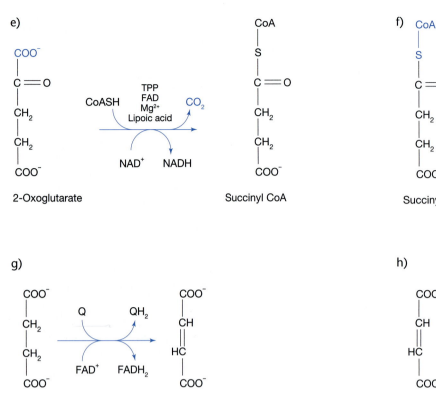

f)

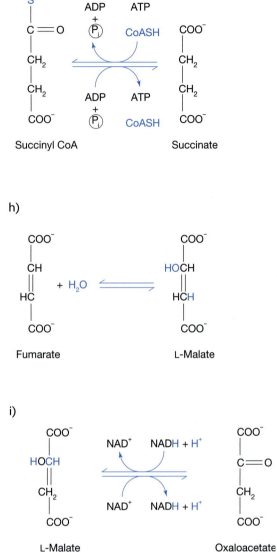

g)

Figure 6.8 Reactions of the TCA cycle (continued).

two lipoic acid-binding regions, an E_1-binding site, and a C-terminal catalytic and assembly domain. E_1 contains two types of subunit: α, which contains the serine phosphorylation sites described below, and β, which together form a heterotetramer (i.e. two of each type of subunit). Mammalian PDC contains 20–30 of these E_1 heterotetramers that are non-covalently bound to the E_2 core. The E_3 components form into dimers and there are twelve of these within the complex, also bound non-covalently to the E_2 core. The E_3 components contain FAD- and NAD-binding domains. While mammalian PDCs also contain an E_3-binding protein (E3BP, formerly called Protein X) that can recruit additional E_3 components into the complex when the NADH:NAD$^+$ ratio is high, this protein is not found in plants. Some plant PDCs contain two forms of E_2, and current thinking is that one of these forms may perform the role of E3BP instead. The capacity to recruit additional E_3 components is believed to confer added metabolic flexibility to the complex under conditions of stress.

Box 6.3 Translocation of metabolites across the inner mitochondrial membrane

Metabolites and nucleotides are transported across the inner mitochondrial membrane by specific carrier proteins, or translocators. These proteins share a number of structural features, and they are classed as a mitochondrial carrier family (MCF). Each member of the family contains six membrane-spanning α-helices (I to VI) separated by three hydrophilic regions (A, B, C) that form loops into the matrix. The domains are linked by two hydrophilic regions within the intermembrane space (a and b) as shown in Figure 1.

The arabidopsis genome contains at least 58 putative MCF members, based on sequence information. Only a few of these genes have been assigned to a physiological function. There is a need to match these *in silico* analyses with thorough biochemical studies to determine, in particular, the substrate specificity and kinetic properties of each putative translocator. Inferences of functionality, based only on gene sequences and localization studies, are insufficient for identifying function. The best approach has been to express the protein and to reconstitute it into lipid vesicles that can be used for metabolite transport studies. Several translocators have been identified in this way, including the recently discovered dicarboxylate-tricarboxylate translocator (shown in Figures 2 and 3).

Figure 1 Proposed structure of a mitochondrial carrier protein. (From *Trends in Plant Sciences*, by Picault et al., Volume 9, Issue 3, 2004, pp. 138–146, "The growing family of mitochondrial carriers in *Arabidopsis*". With permission from Elsevier.)

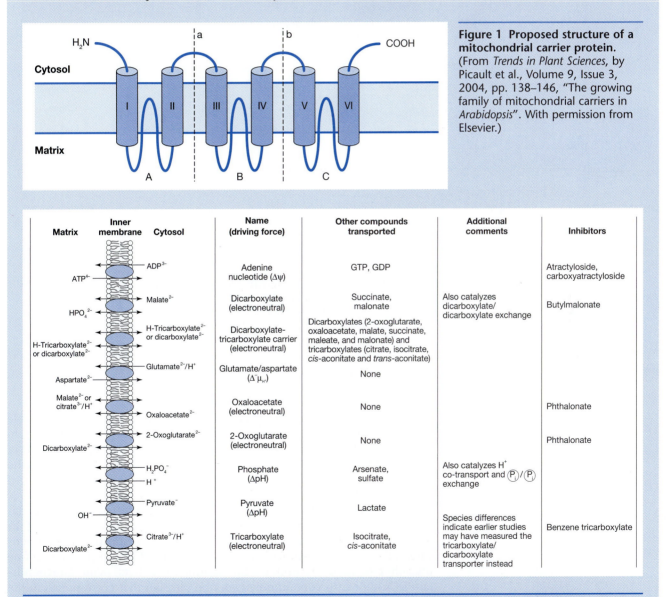

Name (driving force)	Other compounds transported	Additional comments	Inhibitors
Adenine nucleotide ($\Delta\psi$)	GTP, GDP		Atractyloside, carboxyatractyloside
Dicarboxylate (electroneutral)	Succinate, malonate	Also catalyzes dicarboxylate/ dicarboxylate exchange	Butylmalonate
Dicarboxylate-tricarboxylate carrier (electroneutral)	Dicarboxylates (2-oxoglutarate, oxaloacetate, malate, succinate, maleate, and malonate) and tricarboxylates (citrate, isocitrate, *cis*-aconitate and *trans*-aconitate)		
Glutamate/aspartate ($\Delta\bar\mu_{H^+}$)	None		
Oxaloacetate (electroneutral)	None		Phthalonate
2-Oxoglutarate (electroneutral)	None		Phthalonate
Phosphate (ΔpH)	Arsenate, sulfate	Also catalyzes H^+ co-transport and P_i/P_i exchange	
Pyruvate (ΔpH)	Lactate		
Tricarboxylate (electroneutral)	Isocitrate, *cis*-aconitate	Species differences indicate earlier studies may have measured the tricarboxylate/ dicarboxylate transporter instead	Benzene tricarboxylate

Figure 2 Metabolite translocators of the mitochondrial inner membrane. These have been identified from functional analysis, i.e. measurement of substrate transport in either reconstituted vesicles or isolated mitochondria.

Note, other potential translocators include a glycine/serine exchange, and transporters for NAD^+, coenzyme A (CoA), and thiamine pyrophosphate (TPP).

continued

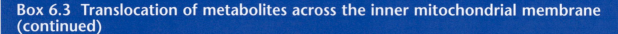

Box 6.3 Translocation of metabolites across the inner mitochondrial membrane (continued)

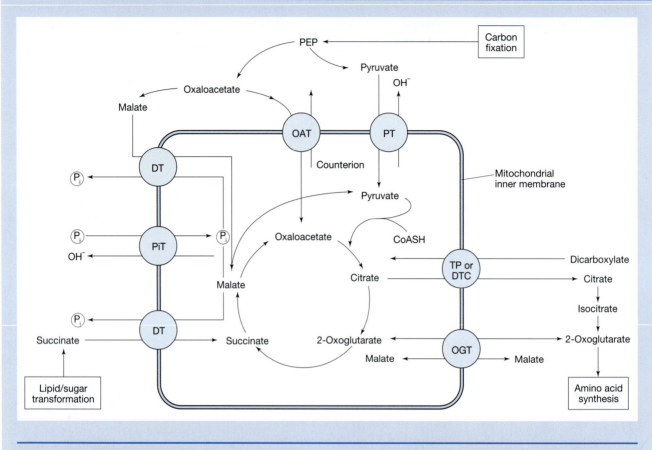

Figure 3 The involvement of mitochondrial metabolite translocators in the TCA cycle and associated reactions. DT, dicarboxylate translocator; OAT, oxaloacetate translocator; PT, pyruvate translocator; PiT, phosphate translocator; TP, tricarboxylate translocator; DTC, dicarboxylate-tricarboxylate carrier; OGT, oxoglutarate translocator; PEP, phosph*enol*pyruvate.

The forward reaction of PDH, resulting in pyruvate oxidation, is irreversible under physiological conditions. However, the reaction is very sensitive to inhibition by the products NADH and acetyl CoA, with K_i values of 20 µM in each case. This means that a high NADH:NAD$^+$ or acetyl CoA:CoA ratio in the mitochondria will inhibit pyruvate oxidation.

A further level of regulation of PDC results from the reversible phosphorylation and dephosphorylation of serine residues on the α subunit of E_1 (denoted $E_{1\alpha}$). Phosphorylation inactivates, while dephosphorylation activates the complex. Mammalian PDCs have three highly conserved serine phosphorylation sites within the $E_{1\alpha}$ subunit. In contrast, plant mitochondrial PDCs appear to contain only two: one (Ser300) that corresponds to the mammalian site 1 and a second (Ser306) that is located one residue further upstream than the mammalian site 2 position. While it is still not clear why plant PDCs lack the third phosphorylation site, the presence of more than one site is believed to confer added flexibility to the way in which pyruvate oxidation is regulated in both animal and plant cells.

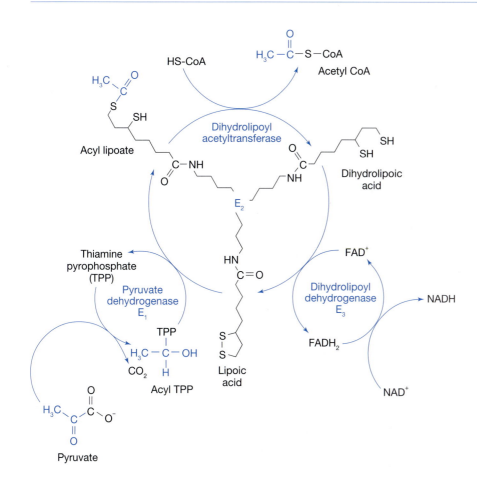

Figure 6.9 Reactions of the pyruvate dehydrogenase complex. The basic pyruvate dehydrogenase complex, present in all aerobic organisms, consists of three individual components, pyruvate dehydrogenase (E_1), dihydrolipoyl acetyltransferase (E_2), and dihydrolipoyl dehydrogenase (E_3). These combine together to oxidize pyruvate to acetyl CoA, in a coupled sequence of reactions in which the intermediates remain enzyme-bound. Pyruvate dehydrogenase (E_1) contains thiamine pyrophosphate (TPP) as its prosthetic group. Pyruvate is decarboxylated by pyruvate dehydrogenase, resulting in the release of CO_2 and the formation of hydroxyethyl TPP following the condensation of TPP with pyruvate. The hydroxyethyl group is then dehydrogenated and the resulting acetyl group is transferred to lipoic acid, the prosthetic group of the second component in the complex, dihydrolipoyl acetyltransferase (E_2). The E_2 component is linked to lipoic acid via a long chain, providing a swinging arm mechanism where the lipoyl arm channels the intermediates between each enzyme of the complex. Hence, the acetyl group, attached to lipoic acid, is transferred to the thiol group of CoA, forming acetyl CoA, while the dihydrolipoic acid remaining on the acetyltransferase is oxidized by the last enzyme in the sequence, dihydrolipoyl dehydrogenase, with NAD^+ being reduced to NADH via FAD^+.

The phosphorylation of PDC $E_{1\alpha}$ is catalyzed by PDK in an ATP-dependent reaction that results in complete inactivation of PDH. PDK binds to the inner lipoyl domain of the E_2 subunit and it is able to phosphorylate several $E_{1\alpha}$ subunits from this position. Reactivation occurs as a result of dephosphorylation, catalyzed by phospho-pyruvate dehydrogenase phosphatase (PDP; Figure 6.10). PDK activity is much higher than that of PDP and if PDK was

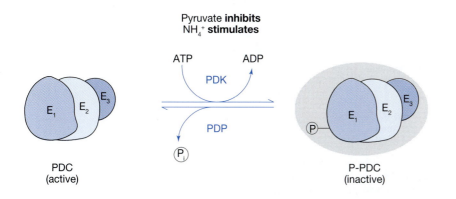

Figure 6.10 Regulation of the mitochondrial pyruvate dehydrogenase complex (PDC) by phosphorylation and dephosphorylation. In addition to the three components illustrated in Figure 6.9 mitochondrial PDC contains two associated regulatory enzymes, PDH kinase and PDH phosphatase. Mitochondrial PDC is active when in its dephosphorylated state. PDH kinase (PDK) phosphorylates PDH (i.e. the E_1 enzyme of the PDC) in an ATP-dependent reaction that is stimulated by NH_4^+ ions and inhibited by pyruvate. PDC becomes inactive when phosphorylated (P-PDC). It is dephosphorylated and reactivated by PDH phosphatase (PDP), which cleaves the P_i from E_1 and restores PDH activity. Note that the subunit composition of PDC is simplified here, and is discussed further in the text.

not regulated PDC would be phosphorylated and inactive under most physiological conditions.

PDK is inhibited by pyruvate and ADP, and stimulated by NH_4^+. This response ensures that PDC is active (i.e. PDK inhibited, PDC dephosphorylated) when pyruvate is available. Regulation by NH_4^+ provides an important control of the partitioning between respiration and photorespiration, and this is discussed further below (see Mitochondrial respiration interacts with photosynthesis and photorespiration in the light, page 187).

PDP, which dephosphorylates and reactivates PDC, does not appear to be regulated to the same extent as PDK. Hence, it seems that it is largely the activity of PDK, and its regulation, that determines the degree to which PDC is phosphorylated within plant cells.

The various mechanisms that operate to regulate pyruvate oxidation in leaves are summarized in Figure 6.11.

The product of pyruvate oxidation, acetyl CoA, enters the tricarboxylic acid cycle via the citrate synthase reaction

Acetyl CoA enters the TCA cycle by condensing with OAA to form citrate. The reaction, catalyzed by citrate synthase, is an aldol condensation between the methyl group of acetyl CoA and the carbonyl group of OAA. This produces the intermediate, citroyl CoA that is hydrolyzed to form citrate and CoA. The reaction is irreversible (Figure 6.8b). Citrate synthase is specific to the

Figure 6.11 Regulation of the mitochondrial and plastidic pyruvate dehydrogenase complexes (PDC) in photosynthetic cells in the light and dark.

Abbreviations: Arrows in blue denote high flux, while broken lines denote low or zero flux; PS, photosynthesis; PR, photorespiration; plPDC, plastid PDC; P-PDC, phosphorylated PDC; 3-PGA, 3-phosphoglycerate. The model is based on the known properties of components of PDC in plant cells. Two PDC isoforms are shown, a plastidic (plPDC) isoform, which is not regulated by phosphorylation, and a mitochondrial (PDC) isoform, which is regulated by phosphorylation as shown in Figure 6.10.

The plastidic PDC is active in the light (due to high pH and Mg^{2+} concentration in the chloroplast stroma) and forms acetyl CoA for fatty acid biosynthesis. It is inactive in the dark (lower pH and Mg^{2+} concentration).

In the light, photorespiration produces glycine that is oxidized within the mitochondria, generating ATP and releasing NH_4^+ within the mitochondria (see Chapter 5). Matrix NH_4^+ concentrations can rise to 3 mM under these conditions, sufficient to activate PDH kinase, which phosphorylates and inactivates mitochondrial PDH (shown as P-PDC) in the light. In the dark, photorespiration ceases, mitochondrial NH_4^+ concentrations fall and PDH kinase becomes inactive. PDH activity is restored by PDH phosphatase, which removes P_i from PDH (represented as the dephosphorylated complex, PDC). Because pyruvate inhibits PDH kinase, mitochondrial PDC may remain active in the light if pyruvate concentrations rise sufficiently.

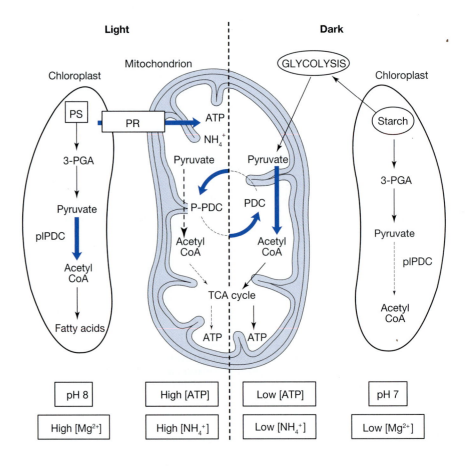

mitochondria in plants, with one exception. It is found in the glyoxysomes of germinating, oil-storing seeds, where it takes part in the glyoxylate cycle (see Chapter 9). Although citrate synthase is inhibited by ATP, this is not thought to be of physiological significance. ATP concentrations between 3 and 5 mM are needed for 50% inhibition and these are unlikely to occur in the mitochondrial matrix *in vivo.*

Citrate is converted to isocitrate by aconitase. This is a two-step isomerization reaction, first with the removal of water to form the intermediate *cis*-aconitate, which is subsequently rehydrated to form isocitrate (Figure 6.8c). The net result of the aconitase reaction is the interchange of H and OH between the two isomers, citrate and isocitrate. Aconitase is found in both the mitochondria and cytosol of plant and mammalian cells. The relative functions of these two isofoms remains uncertain, but possible roles will be discussed later in this chapter.

Isocitrate dehydrogenase (IDH) oxidizes isocitrate to 2-oxoglutarate (Figure 6.8d). This is the first of two oxidative decarboxylation reactions in the TCA cycle, the other one being catalyzed by 2-oxoglutarate dehydrogenase. In both cases, the reaction releases CO_2 and generates NADH.

Plants possess both NAD-dependent IDH (NAD-IDH) and NADP-dependent IDH (NADP-IDH) isoforms. NAD-IDH is only found in mitochondria and it is this isoform that is considered to function in the TCA cycle. The NADP-IDH occurs in the mitochondria and also in the cytosol, plastids, and peroxisomes. The precise function of the various forms of NADP-IDH is still open to question, but possible roles will be discussed in Chapter 8.

The NAD-IDH reaction is a two-step process, with isocitrate being oxidized to oxalosuccinate, which is decarboxylated while still bound to the enzyme (i.e. it is not released as a free metabolite) to produce 2-oxoglutarate.

The NAD-IDH is very sensitive to the reduction level of NAD^+ and $NADP^+$ within the mitochondrial matrix. Both NADH and NADPH inhibit the enzyme and so whenever the $NAD(P)^+$ pool becomes relatively reduced, this will inhibit isocitrate oxidation, with the potential to decrease TCA cycle flux (see page 170 for further discussion).

The second of the two oxidative decarboxylation reactions of the TCA cycle is catalyzed by 2-oxoglutarate dehydrogenase (Figure 6.8e). This is a multienzyme complex that is very similar to the PDC described above, both in terms of enzyme composition and reaction mechanism. As with PDC, the 2-oxoglutarate dehydrogenase complex uses thiamine pyrophosphate, lipoic acid, and flavin adenine dinucleotide (FAD^+), but, unlike PDC, it is not regulated by phosphorylation. Although 2-oxoglutarate dehydrogenase was successfully purified from potato some years ago, we still know very little about its regulatory properties. Early reports (from the 1970s) indicate it to be activated by AMP. In contrast, the mammalian enzyme has been studied in some detail. It is inhibited by NADH, succinyl CoA, and ATP, and activated by ADP, P_i, and Ca^{2+}, which increase its affinity for 2-oxoglutarate, while AMP has no effect.

Succinyl CoA synthetase (also called succinyl CoA ligase or succinate thiokinase) converts succinyl CoA to succinate (Figure 6.8f). While mammalian cells contain two forms, one ADP- and one GDP-dependent, plants appear only to possess the ADP-dependent form. Succinyl CoA synthetase catalyzes the only substrate-level phosphorylation in the TCA cycle; the hydrolysis of the thioester bond of succinyl CoA releases sufficient free energy to permit ATP formation from ADP and P_i. This reaction consequently contributes to the production of ATP during the operation of the TCA cycle.

All of the succinyl CoA synthetase proteins studied to date consist of two subunits, α and β. Genes encoding both subunits have now been cloned from tomato. These were the last plant genes to be characterized for TCA cycle proteins.

Succinate dehydrogenase (also called succinate:ubiquinone oxidoreducatase or Complex II) oxidizes succinate to fumarate, with the sequential reduction of enzyme-bound FAD^+ and nonheme iron (Figure 6.8g). Electrons are transferred directly to the Q pool of the mitochondrial electron transport chain, to form ubiquinol (QH_2) from Q. Succinate dehydrogenase is the only TCA cycle enzyme that is not located in the matrix. Instead, it is an integral membrane protein and a component (Complex II) of the mitochondrial electron transport chain (see page 173).

In plants, succinate dehydrogenase is activated by QH_2 and by ATP, in a similar manner to the mammalian enzyme. OAA is a strong inhibitor of all forms of the enzyme. ADP inactivates the mammalian enzyme, whereas plant succinate dehydrogenases are activated by ADP. The mechanism for ADP and ATP activation of plant succinate dehydrogenases is not fully understood but it appears to involve the release of OAA bound to the enzyme. There is evidence that the plant enzyme is regulated by the energy status of the mitochondrial inner membrane and that activation by ATP and ADP may be the indirect result of this regulation. The physiological relevance of this activation is unclear, as ATP and ADP generally show reciprocal changes in concentration within the cell, i.e. as ATP rises, ADP falls.

Fumarase catalyzes the *trans* addition of H and OH (from H_2O) to the C=C bond of fumarate. This reaction is stereospecific, as the OH is added to only one side of the double bond, so that L-malate (rather than D-malate) is always formed (Figure 6.8h). The reaction is reversible. Fumarase is unique to the mitochondria, and is often used as a mitochondrial marker in localization and organelle-isolation studies.

The TCA cycle is completed by the oxidation of malate to form OAA, catalyzed by NAD-dependent MDH (NAD-MDH). The reaction is pH sensitive and reversible; the equilibrium strongly favors the reverse reaction, i.e. malate formation from OAA (Figure 6.8i):

$$\text{Equilibrium constant } (K_{eq}) = \frac{[\text{NADH}]\,[\text{oxaloacetate}]}{[\text{malate}]\,[\text{NAD}^+]} = 2.86 \times 10^{-5} \text{ (at pH 7.0)}$$

For the reaction to function in the direction of OAA formation, the products of the reaction, OAA and NADH, have to be prevented from accumulating. Thus, efficient oxidation of NADH to NAD^+, together with the further metabolism of OAA, or its removal from the matrix, by export, ensure that the TCA cycle may continue to function.

MDH is a ubiquitous enzyme. Isoenzymes of the NAD-dependent form are found in mitochondria, peroxisomes, cytosol, and glyxoysomes of plants. An NADP-dependent MDH is found in chloroplasts, where it is involved in C_4 photosynthesis (see Chapter 5). There are also reports of low levels of NADP-dependent MDH activity in mitochondria of some plants. However, it is the mitochondrial NAD-MDH isoform that takes part in the TCA cycle. This, and all other MDH isoforms may also be involved in the indirect transfer of reducing equivalents across membranes as discussed later in this chapter.

The MDH reaction completes the TCA cycle by generating OAA for the citrate synthase reaction. Malate may also be oxidized by NAD-malic enzyme (see page 170) present in plant mitochondria.

Substrates for the tricarboxylic acid cycle are derived mainly from carbohydrates

In most plant tissues, carbohydrates are the major source of substrates for the TCA cycle. Fats and proteins usually serve only as minor or exceptional respiratory substrates, for example during germination of fatty seeds (see Chapter 9) or during leaf senescence, when proteins are remobilized from dying to growing tissue. Even under these conditions, complete respiratory breakdown of fats and proteins is unusual, as their main metabolic fate is remobilization—fats are converted to sucrose (via gluconeogenesis, glyoxylate cycle; see Chapter 9) while amino acids, formed from protein degradation, are usually moved elsewhere and reused for protein synthesis.

It is during periods of carbon starvation that proteins and lipids serve a major role as respiratory substrates. These conditions are likely to occur when photosynthetic activity is reduced, e.g. during prolonged exposure to darkness, low light, or other stresses such as drought or low temperature. Hence, as carbohydrates become depleted, fatty acids may be oxidized (via β oxidation, or in some cases α oxidation) to produce acetyl CoA, while amino acids may be either deaminated to produce glycolytic or TCA cycle intermediates, or oxidized directly within the mitochondria. The TCA cycle intermediate, 2-oxoglutarate, serves as the point of entry for the oxidation of a number of amino acids. Glutamate, for example, can be oxidized to 2-oxoglutarate via NAD-dependent glutamate dehydrogenase (Chapter 8). Proline, a non-protein amino acid that can accumulate to high concentrations during drought, is also oxidized to 2-oxoglutarate. The branched-chain amino acids, valine, isoleucine, and leucine, are also thought to be degraded within the mitochondria. The reactions remain to be fully characterized in plant mitochondria, although in mammalian cells the products can feed into the TCA cycle via acetyl CoA and succinyl CoA. The branched-chain ketoacid dehydrogenase, which catalyzes the second step in the pathway, is a multienzyme complex that closely resembles the PDH and 2-oxoglutarate dehydrogenase complexes. The amino acid lysine is also oxidized during periods of carbohydrate starvation, as well as in germinating seeds. The first step in the pathway is catalyzed by a bifunctional enzyme, lysine-ketoglutarate reductase/saccharopine dehydrogenase, which is induced during carbohydrate starvation. Lysine oxidation produces glutamate, which can enter the TCA cycle via 2-oxoglutarate, and acetyl CoA (see Further Reading).

Although most amino acid oxidation appears to occur during stressful conditions, there is one exception. The amino acid, glycine, is a major respiratory substrate in the mitochondria of photosynthetic cells of C_3 plants. However, as this reaction is a component of photorespiration, rather than respiration as such, it is described further in Chapter 5. Its impact on the TCA cycle will be considered later in this chapter.

The tricarboxylic acid cycle serves a biosynthetic function in plants

Although glycolysis, the OPPP and TCA cycle together form a chain of reactions that allows for the complete respiration of carbohydrates to CO_2 and H_2O, most plant tissues probably retain about 50% of the carbon that enters these respiratory pathways. Carbon is conserved, rather than released as CO_2, because the TCA cycle is being used to provide intermediates for biosynthesis. Carbon from pyruvate, for example, is frequently recovered in the form of amino acids, organic acids, and lipids. The TCA cycle forms a central hub from which intermediates are withdrawn to form an array of metabolic products, as summarized in Figure 6.12.

The anabolic function of the TCA cycle is extremely important to plants, as it provides a rich source of intermediates to support the biosynthesis of a huge variety of plant products. A central feature of this biosynthetic function is the flexibility with which different TCA cycle intermediates can be used. A good example of this flexibility is the way that the TCA cycle provides carbon skeletons for synthesis of the amino acid, glutamate (Figure 6.13).

Glutamate is formed from 2-oxoglutarate and glutamine, via the plastid enzyme glutamate synthase (GOGAT; see also Chapter 8). There are three potential sources of 2-oxoglutarate for this reaction; the TCA cycle NAD-IDH in the mitochondria, an NADP-IDH present in the cytosol, and a peroxisomal glutamate:glyoxylate aminotransferase (Figure 6.13). If the mitochondrial NAD-IDH is the source, 2-oxoglutarate has to be withdrawn from the TCA cycle and transported into the plastid. This movement of 2-oxoglutarate would involve the mitochondrial 2-oxoglutarate translocator and two plastidic dicarboxylate translocators, DiT1 and DiT2. The mitochondrial 2-oxoglutarate translocator exports 2-oxoglutarate in exchange with malate. The chloroplast DiT1 imports 2-oxoglutarate in exchange with malate, while DiT2 imports malate in exchange for glutamate export. The net effect of these three translocators is the import of malate into the mitochondria, together with an import of 2-oxoglutarate into, and an export of glutamate out of the chloroplast. Malate, for import into the mitochondria, may be supplied by anaplerotic reactions in the cytosol, as explained below (see page 169).

In the event that the cytosolic NADP-IDH provides 2-oxoglutarate for glutamate biosynthesis in the chloroplast, the mitochondrial TCA cycle is,

Figure 6.12 The TCA cycle forms a central hub for biosynthetic reactions. Glycolysis and the TCA cycle together generate a range of intermediates that can be used to produce amino acids (Chapter 8), fatty acids (Chapter 9), secondary metabolites, such as alkaloids (Chapter 10), and terpenoids (Chapter 12) as well as nucleic acids and porphyrins.

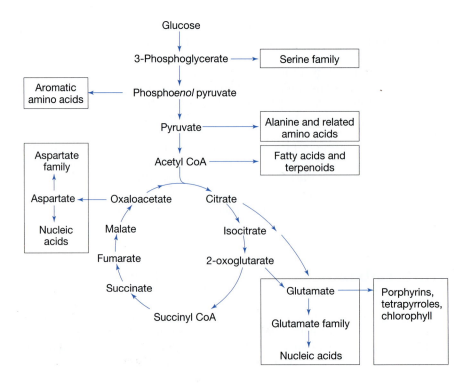

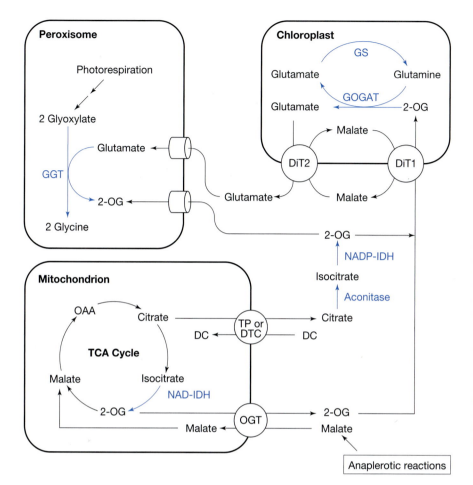

Figure 6.13 The TCA cycle can provide carbon skeletons for glutamate synthesis in the chloroplast. Either citrate or 2-oxoglutarate may be withdrawn from the TCA cycle to supply 2-oxoglutarate to the chloroplast. Anaplerotic reactions in the cytosol replace TCA cycle intermediates so that the cycle can continue to function (see Figure 6.14). While the peroxisomal glutamate: glyoxylate aminotransferase may also supply 2-oxoglutarate, this reaction consumes glutamate and does not in itself provide a net gain of carbon into amino acids. The movement of glutamate into, and 2-oxoglutarate out of the peroxisome occurs through porins, while various translocators are involved in the movement of metabolites across the chloroplast and mitochondrial membranes.

Abbreviations: GGT, glutamate: glyoxylate aminotransferase; 2-OG, 2-oxoglutarate; GS, glutamine synthetase; GOGAT, glutamine-oxoglutarate aminotransferase (also known as glutamate synthase); DiT1, chloroplast 2-oxoglutarate translocator; DiT2, chloroplast glutamate translocator; NADP-IDH, NADP-dependent isocitrate dehydrogenase; NAD-IDH, NAD-dependent isocitrate dehydrogenase; TP, mitochondrial tricarboxylate translocator; DTC, mitochondrial dicarboxylate-tricarboxylate translocator; OGT, mitochondrial 2-oxoglutarate translocator; DC, dicarboxylate; OAA, oxaloacetate.

again, responsible for supplying the substrate. The sequence of reactions begins with the withdrawal of the TCA cycle intermediate, citrate, and its export from the mitochondria via the tricarboxylate translocator or the dicarboxylate–tricarboxylate translocator. Whichever of the translocators is used, the export of citrate is accompanied by the import from the cytosol of a dicarboxylate, such as malate. Once in the cytosol, citrate may be converted to isocitrate via a cytosolic aconitase, and then to 2-oxoglutarate by cytosolic NADP-IDH. The subsequent transport of 2-oxoglutarate into the chloroplast involves DiT1 and DiT2 as explained above.

The third mechanism for supplying 2-oxoglutarate to the chloroplast involves reactions in the peroxisome. Glutamate:glyoxylate aminotransferase produces 2-oxoglutarate and glycine from glyoxylate and glutamate. Glyoxylate is produced during photorespiration (see Chapter 5) and so this set of reactions only occurs at an appreciable rate in the photosynthetic tissue of C_3 plants in the light. Furthermore, as glutamate is consumed to produce 2-oxoglutarate, this reaction does not provide any net gain of carbon into glutamate; it merely recycles existing carbon skeletons. The reactions are shown in Figure 6.13.

Anaplerotic reactions are needed to enable intermediates to be withdrawn from the tricarboxylic acid cycle

If intermediates were removed for biosynthetic purposes and not replenished, the TCA cycle would gradually become depleted of intermediates and eventually cease to function. Hence, for the TCA cycle to serve a biosynthetic role, there has to be a mechanism for restoring carbon back into the cycle to replace the intermediates that have been removed. Reactions that serve this function are termed anaplerotic (from the Greek *anaplerotikos* meaning filling up).

The main anaplerotic reaction in plant cells is that catalyzed by PEP carboxylase, which converts PEP, of glycolytic origin, to OAA:

$$PEP + HCO_3^- \longrightarrow OAA + P_i$$

The PEP carboxylase reaction takes place in the cytosol and the incorporation of carbon from bicarbonate (HCO_3^-) is an important source of additional carbon for the TCA cycle, sometimes referred to as dark CO_2 fixation (Figure 6.14a). Once formed, OAA may enter the mitochondria via a specific OAA translocator that appears to be unique to plant mitochondria. Alternatively, OAA may be converted to malate by a cytosolic MDH, with malate then entering the mitochondria via the dicarboxylate translocator (see Box 6.3).

While the PEP carboxylase reaction is a true anaplerotic reaction because it gains carbon (by reacting with bicarbonate), other reactions allow carbon to be added to the TCA cycle from stored sources. In this case, although the TCA cycle is still gaining carbon, there is no net gain to the carbon budget of the cell. For example, malate may be imported into the mitochondria from its storage site within the vacuole, via tonoplast and mitochondrial dicarboxylate translocators (Box 6.3). Once inside the matrix, malate may be oxidized to OAA by NAD-MDH. Alternatively, plant mitochondria contain an NAD-dependent malic enzyme that decarboxylates malate to form pyruvate (Figure 6.14b). The reaction is as follows:

$$Malate + NAD^+ \longrightarrow pyruvate + CO_2 + NADH$$

Pyruvate may then be converted to acetyl CoA via PDC, as described above.

While the MDH reaction enables the TCA cycle to retain all four of the carbons of malate, by converting it to OAA, the malic enzyme and PDC reactions each lose one carbon through the respective decarboxylations of malate and pyruvate (Figure 6.14a,b). Hence, the TCA cycle recovers more carbon from malate if the MDH reaction is used in preference to the combined reactions of malic enzyme and PDC. Despite this, the malic enzyme and PDC reactions together add metabolic flexibility to the TCA cycle. First, they enable plant mitochondria to oxidize both malate and citrate to CO_2 without requiring acetyl CoA derived from pyruvate produced in glycolysis. Second, along with PEP carboxylase, these enzymes enable the formation of 2-oxoglutarate that can be withdrawn for amino acid biosynthesis (Figure 6.14a–c).

The tricarboxylic acid cycle is regulated at several steps

Our knowledge of the regulation of the TCA cycle in plants is still somewhat limited, especially with respect to its operation *in vivo*. Much of our understanding has come from studies of purified enzymes and isolated mitochondria. We still do not fully understand how most of the TCA cycle enzymes are regulated in plants, although some ideas can be gained by analogy with mammalian enzymes.

The reaction catalyzed by the PDC is subject to a considerable amount of regulation, as already discussed above. In controlling the supply of acetyl CoA for citrate synthesis, PDC will exert a major influence over the balance between glycolysis and the TCA cycle. Similarly, changes in TCA cycle flux will influence PDC activity. For example, reduced TCA cycle flux will result in acetyl CoA accumulation, which inhibits PDC. Increased glycolytic flux will increase pyruvate, alleviate acetyl CoA inhibition, and stimulate flux into the TCA cycle.

The TCA cycle is particularly sensitive to the levels of the pools of NAD^+, NADH, ADP, and ATP within the mitochondrial matrix. This is true for both the plant and mammalian pathways, although there are some differences in the response of individual enzymes.

If there is a reduction in the demand for ATP in the cytosol, this will result in an accumulation of ATP and a reduction in ADP concentration. In plant cells, environmental stresses such as drought and cold can reduce the demand for ATP, resulting in an increase in the ATP:ADP ratio. In green tissue, ATP synthesis in chloroplasts in the light (i.e. photophosphorylation; Chapter 4) might also increase the cytosolic ATP:ADP ratio, although this could be balanced by an increased demand for ATP in light-dependent biosynthetic reactions. A high ATP:ADP ratio will cause the mitochondrial electron transport chain to slow down (see later in this chapter). As a result, the matrix $NADH:NAD^+$ ratio will rise and this will inhibit several TCA cycle reactions. Regulation occurs at at least three points; the PDC, NAD-IDH, and NAD-dependent MDH, all of which are sensitive to NADH inhibition. While the mammalian 2-oxoglutarate dehydrogenase is also sensitive to NADH inhibition, it remains to be seen whether this is also the case in plants. A high ATP:ADP ratio will have a direct effect on PDC, citrate synthase, and MDH, all of which are inhibited by ATP. Hence, TCA cycle flux will be inhibited at several steps in response to increases in $NADH:NAD^+$ and ATP:ADP ratios. However, one of the main differences between plant and mammalian mitochondria is the presence, in plant mitochondria, of additional, non-phosphorylating electron transport chain components (see later in this chapter). These nonphosphorylating reactions (additional NADH dehydrogenases and alternative oxidase) enable mitochondrial electron transport, and NADH oxidation, to persist free from the normal constraints imposed by ADP availability. Thus the TCA cycle in plants may well be less rigidly controlled by ATP:ADP and $NADH:NAD^+$ than in other organisms.

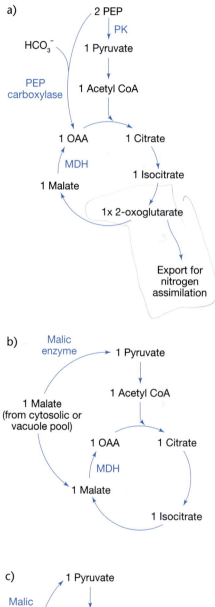

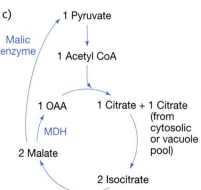

Figure 6.14 Anaplerotic reactions enable intermediates to be withdrawn for biosynthesis without depleting the TCA cycle. Malate, phospho*enol*pyruvate (PEP), and citrate may be supplied to the TCA cycle to replace intermediates that are withdrawn for biosynthesis. The reactions that replenish the TCA cycle in this way are referred to as anaplerotic reactions. Three examples are given: (a) PEP is converted to oxaloacetate (OAA) by PEP carboxylase. PEP may also be converted to pyruvate by pyruvate kinase (PK). These two reactions add OAA and pyruvate, respectively, to the TCA cycle so that 2-oxoglutarate may be withdrawn and used in biosynthetic reactions such as in nitrogen assimilation. (b) Mitochondrial NAD-dependent malic enzyme, which converts malate to pyruvate, can add carbon to the TCA cycle using malate from storage pools (e.g. vacuolar, cytosolic). As this reaction produces pyruvate, it also enables the TCA cycle to function without the need for glycolysis to supply pyruvate. (c) Citrate, from storage pools in the vacuole or cytosol, can supply the TCA cycle with additional carbon. The conversion of malate to pyruvate, by mitochondrial NAD-dependent malic enzyme, allows the TCA cycle to function without the need for pyruvate to be supplied by glycolysis.

One aspect of TCA cycle regulation that is still poorly understood is the extent to which different substrates compete for oxidation within the mitochondria. Experiments with isolated plant mitochondria, for example, indicate that OAA inhibits succinate oxidation, and that in leaf mitochondria glycine is oxidized preferentially to any of the TCA cycle intermediates. These observations have yet to be tested *in vivo*, and we await further experimentation with transgenic plants. A further, outstanding question is whether, or to what extent, the TCA cycle operates in the light in photosynthetic cells. This will be discussed later in this chapter (see Mitochondrial respiration interacts with photosynthesis and photorespiration in the light, page 187).

Recent research into a thioredoxin/NADPH redox system for regulating tricarboxylic acid cycle enzymes and other mitochondrial proteins

Recent research is beginning to uncover a thioredoxin/NADPH redox system for regulating TCA cycle enzymes and other mitochondrial proteins in plants. There is growing evidence that plant mitochondria, like those of mammals and yeasts, contain a thioredoxin system. It consists of a form of thioredoxin protein that appears to be unique to plants, and an NADPH-dependent thioredoxin reductase. An NADPH-dependent thioredoxin coupling is also present in the cytosol, whereas plastids use a ferredoxin–thioredoxin couple. In its reduced form, thioredoxin is able to reduce S–S groups on proteins, which can result in either inactivation or activation of the target enzyme (see Chapters 2 and 5). In the case of mitochondrial proteins in plants, it is too early to say how thioredoxin affects their activity, although there is recent evidence for a role in regulating the alternative oxidase (see pages 180–181). *In vitro* studies have identified that as many as 50 mitochondrial proteins in plants are able to bind to the mitochondrial thioredoxin. Of these 50 proteins, six are TCA cycle enzymes, namely aconitase, malic enzyme, succinyl CoA synthetase, and IDH, and pyruvate- and succinate dehydrogenase. Although citrate synthase was not identified as a thioredoxin-binding protein, structural models have identified a potential thioredoxin-accessible region as well as redox-sensitive cysteine residues. Thus, it would seem that mitochondrial thioredoxins have the potential to regulate TCA cycle activity in a redox-dependent manner.

The mitochondrial electron transport chain oxidizes reducing equivalents produced in respiratory substrate oxidation and produces ATP

The oxidation of carbohydrates in the glycolytic pathway generates NADH in the cytosol, forming two moles of NADH for each mole of glucose that is converted to pyruvate. Further metabolism in the mitochondria (chiefly via the TCA cycle) produces eight moles of NADH and two moles of $FADH_2$ within the mitochondrial matrix. An additional and major source of NADH in mitochondria in photosynthetic cells is glycine oxidation (see Chapter 5). For these pathways to continue to function, it is necessary to reoxidize NADH and $FADH_2$ so that the dehydrogenase reactions have a continued supply of NAD^+ and FAD^+. For the most part, this oxidation of NADH and $FADH_2$ is achieved by the mitochondrial electron transport chain.

The electron transport chain of mitochondria from all organisms serves the same fundamental function: to oxidize NADH and $FADH_2$ and to recover the energy released, in the form of ATP. Plant mitochondria are furthermore able to oxidize NADPH through the electron transport chain. Electrons are removed from NADH and $FADH_2$ (and NADPH, in the case of plants) and transferred along a series of membrane-located electron carriers to the terminal electron acceptor, oxygen, to form water. The resulting oxygen consumption can be measured to determine the rate of mitochondrial respiration (Figure 6.15).

The distribution of the electron carriers within the membrane enables protons (H^+) to be moved across the inner membrane from the matrix to the intermembrane space, thus establishing a H^+ gradient as electrons are transferred from one acceptor to another. It is this H^+ gradient that provides the mechanism for coupling electron transfer with ATP formation, as will be seen later.

Main protein complexes of the electron transport chain

There are five protein complexes of the electron transport chain that are common to mitochondria from both plants and animals. Four of these are involved in electron transfer, while the fifth complex, the ATP synthase, is responsible for ATP synthesis coupled to electron transfer through these four complexes situated in the inner mitochondrial membrane (Figure 6.16).

Complex I transfers electrons from NADH to ubiquinone

Complex I, also termed the NADH dehydrogenase complex, oxidizes NADH within the matrix to NAD^+ and transfers electrons to ubiquinone Q, reducing it to QH_2 (Figure 6.16).

Complex I is a Type I NADH dehydrogenase, found in all eukaryotes except for fermenting yeasts (e.g. *Saccharomyces cerevisiae*) and the amitochondriate (i.e. lacking mitochondria) protists (e.g. *Giardia lamblia*). Although Complex I of plant mitochondria is less well-characterized than the complex from other organisms, it appears to share common properties. It is a large complex of at least 40 different subunits. As many as nine of these may be encoded in the plant mitochondrial genome. The structure of plant Complex I is still unclear, although there appear to be two components, a membrane-spanning region that interacts with Q, and a peripheral region facing into the matrix and containing the binding site for NADH. Complex I therefore oxidizes only NADH present within the mitochondrial matrix. As electrons are passed through complex I from NADH to Q, protons ($4H^+$ per electron pair) are translocated across the membrane from the matrix to the inner membrane space.

Complex I is inhibited by rotenone (a flavonoid insecticide that can be extracted from the roots of tropical legumes such as *Derris*, *Lonchocarpus*, and *Tephrosia*), amytal (a barbiturate, sedative), and piericidin A (an antibiotic produced by *Streptomyces*).

Complex II transfers electrons from succinate to ubiquinone

Complex II contains the TCA cycle enzyme, succinate dehydrogenase. Complex II oxidizes succinate to fumarate, passing electrons from succinate to FAD^+, then through three iron–sulfur proteins to Q (Figure 6.16). Complex II consists of just four subunits, ranging in size from 13.5 to 70 kDa. It is the only one of the four respiratory complexes in plants that is

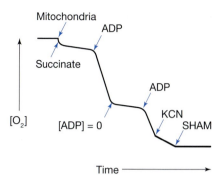

Figure 6.15 Respiration in isolated mitochondria. Mitochondrial respiration can be measured using an oxygen electrode. Isolated mitochondria are added to an isotonic buffer followed by a substrate, in this case succinate. As succinate is oxidized (via succinate dehydrogenase, Complex II) electrons begin to flow along the electron transport chain and oxygen is reduced to H_2O. The resulting oxygen uptake is seen as a downward slope in the electrode trace. Addition of ADP increases the rate of electron transport, as the electrochemical proton gradient is dissipated (due to protons moving back into the mitochondria through ATP synthase; see Box 6.5). The rapid rate of oxygen uptake in the presence of ADP is termed State 3 respiration. Once all of the added ADP has been converted to ATP, electron transport slows down again and oxygen uptake is correspondingly low. The ADP-limited rate is denoted State 4. The quality of the isolated mitochondria can be assessed by determining the ratio of the State 3 to State 4 rates (termed the respiratory control ratio, RCR). The higher the ratio, the better the quality of the mitochondria, as this is a measure of the extent to which electron transport is coupled to ATP synthesis. Note that if the nonphosphorylating reactions of the electron transport chain are operating these will result in a reduced RCR due to the reduced coupling between electron transport and ATP synthesis. The addition of potassium cyanide (KCN) inhibits electron transport through the cytochrome pathway, allowing electrons to flow only through the alternative oxidase, which can be inhibited by the addition of salicylhydroxamic acid (SHAM).

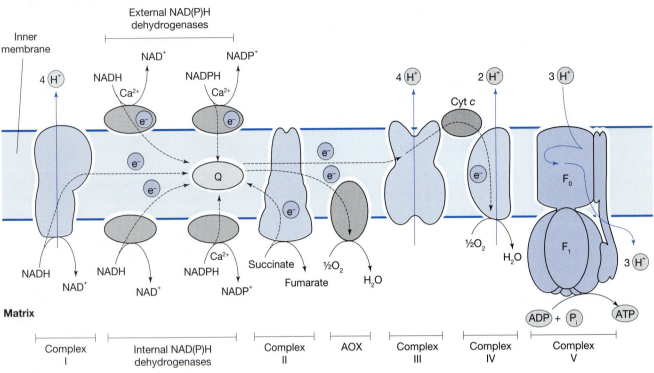

Intermembrane space

Figure 6.16 The electron transport chain in plant mitochondria. Plant mitochondria contain the same five complexes (I–V) present within virtually all mitochondria. Complexes I, III, and IV translocate protons (H^+) from the matrix into the intermembrane space (4, 4, and 2 H^+, respectively, per electron pair), hence establishing a proton electrochemical gradient (Box 6.5). Movement of protons back into the matrix through the ATP synthase (Complex V) drives ATP synthesis from ADP and P_i (Box 6.6). The ubiquinone pool (Q) is freely mobile within the inner membrane and is able to transfer electrons between complexes situated on either face of the membrane. The oxidized form, Q, accepts electrons and becomes reduced to ubiquinonol (not shown), which can transfer electrons directly to Complex III and the alternative oxidase (AOX).

In addition to Complexes I–V, plant mitochondria contain five further complexes. These do not translocate protons. Two oxidize NADH (internal and external NADH dehydrogenase) and two oxidize NADPH (internal and external NADPH dehydrogenase), while AOX accepts electrons from the ubiquinone pool and passes them directly to O_2. Further details are given in Figure 6.17.

entirely nuclear-encoded, apart from a few liverworts and algae where a small number of the subunits are mitochondrially encoded. Complex II does not translocate protons; hence succinate oxidation generates fewer moles of ATP than does NADH oxidation.

Succinate dehydrogenase is competitively inhibited by malonate and this inhibitor was used early on to help determine the sequence of reactions of the TCA cycle (See Box 6.2).

Complex III transfers electrons from ubiquinol to cytochrome *c*

Complex III consists of two *b*-type cytochromes, a cytochrome c_1, an iron–sulfur protein, and five to seven further polypeptides. It is very similar in structure to the cytochrome b_6/f complex of chloroplasts (Chapter 4).

The oxidation of NADH through Complex I, and the oxidation of succinate by Complex II, both result in the reduction of Q to QH_2. Complex III (also called the bc_1 complex or QH_2:cytochrome *c* oxidoreductase) has the function of oxidizing QH_2 back to Q, passing electrons to cytochrome *c* in the process (Figure 6.16). QH_2 reacts within the membrane phase, while cytochrome *c* is reduced at the outer surface of the inner membrane, facing into the intermembrane space.

The transfer of electrons from QH_2 to cytochrome *c* therefore results in proton translocation from the matrix across the inner membrane. The process is termed the Q cycle. For every two electrons released during oxidation of QH_2 to Q, four H^+ are released from the matrix into the intermembrane space.

Complex III is inhibited by several antibiotics, including antimycin A and myxothiazol.

Complex IV transfers electrons from cytochrome *c* to oxygen

Complex IV (also called the cytochrome *a/a₃* complex or cytochrome oxidase) removes electrons from reduced cytochrome *c*, on the outer surface of the inner membrane (facing into the inter-membrane space) and passes them on to oxygen on the matrix side, forming H_2O (Figures 6.16 and 6.17). Two molecules of H_2O are produced in the four-electron reduction of O_2. Two H^+ are pumped out of the matrix for each electron pair.

Complex IV consists of at least 10 different subunits, three of which are encoded by mitochondrial DNA. There are two catalytic subunits, I and II. The initial electron transfer reaction, from cytochrome *c*, takes place on subunit II, which contains two copper atoms, called Cu_A. From here, the remaining electron transfer reactions take place on subunit I. This subunit contains two heme centers, heme *a* (cytochrome *a*), which accepts electrons from Cu_A of subunit II, and heme a_3 (cytochrome a_3), which is part of a binuclear center formed with another copper atom (Cu_B). Electrons flow from Cu_A to cytochrome *a*, through the binuclear center and on to O_2. The binuclear a_3-Cu_B center is the site at which the cytochrome oxidase inhibitors, cyanide, azide, and carbon monoxide, inhibit the enzyme by binding and competing for electrons with oxygen. The inhibition by carbon monoxide is reversible by light, as originally shown by Warburg and Negelein.

Plant mitochondria possess additional respiratory proteins that provide a branched electron transport chain

Plant and fungal mitochondria differ from those of all other organisms in having branched electron transport pathways. As we shall see, these additional pathways, which are nonphosphorylating, enable electron transport to persist at a substantial rate in the absence of ATP synthesis. This seemingly wasteful process is thought to improve stress tolerance and, in plants, to enable mitochondria to interact with photosynthetic pathways in the chloroplast.

Figure 6.17 Additional NADH and NADPH dehydrogenases and the alternative oxidase produce routes for electron transport that bypass all of the proton-translocating complexes (Complexes I, III, and IV). One NADH dehydrogenase (NDin(NADH)), located on the inner face of the inner membrane, oxidizes matrix NADH. A second NADH dehydrogenase (NDex(NADH)) on the outer face of the inner membrane oxidizes cytosolic NADH. Two NADPH dehydrogenases are also present, although not in all plant mitochondria. One (NDin(NADPH)) on the inner face of the inner membrane oxidizes matrix NADPH (e.g. from NADP-dependent isocitrate dehydrogenase) while the other (NDex(NADPH)) on the outer face of the inner membrane, oxidizes cytosolic NADPH. All of the NAD(P)H dehydrogenases, including Complex I, pass electrons directly to the ubiquinone pool (Q) but only Complex I pumps protons across the inner membrane.

The alternative oxidase (AOX) accepts electrons directly from the Q pool and passes them directly to oxygen. It acts as an alternative terminal electron acceptor to cytochrome oxidase (Complex IV) and does not pump protons across the inner membrane. The combined action of NDin (NADH) and the AOX provides an entirely nonphosphorylating route (i.e. no ATP is synthesized) for electrons during oxidation of matrix NADH. This nonphosphorylating route is indicated by the blue arrows.

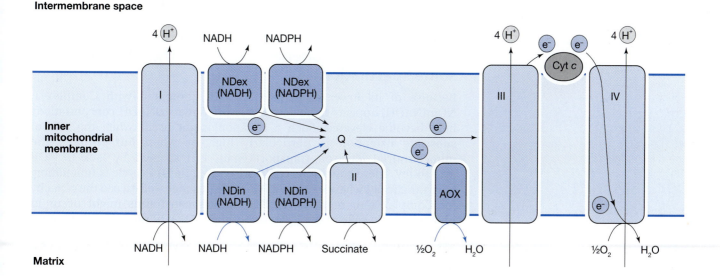

Intermembrane space

Inner mitochondrial membrane

Matrix

Plant mitochondria contain four additional NAD(P)H dehydrogenases

In addition to Complex I (NADH dehydrogenase Type I), plant and fungal mitochondria possess four further NAD(P)H dehydrogenases (Figure 6.17). These are classed as Type II NAD(P)H dehydrogenases, and are characterized by the absence of proton translocation and insensitivity to the Complex I inhibitors, rotenone and antimycin A. They are also classified according to their location within the inner membrane; NDin refers to NAD(P)H dehydrogenases that are located on the inner (i.e. matrix) face of the inner membrane, while NDex refers to dehydrogenases on the outer face of the inner membrane.

Two of these additional NAD(P)H dehydrogenases are localized on the inner face of the inner membrane. One of these inner membrane dehydrogenases, NDin(NADH), oxidizes NADH, while the other, NDin(NADPH), oxidizes NADPH present within the matrix. While NDin(NADPH) is dependent upon Ca^{2+}, the NDin(NADH) is not. NDin(NADH) has an approximately 10-fold lower affinity for NADH than Complex I.

A further two NAD(P)H dehydrogenases are located on the outer surface of the inner membrane. One of these dehydrogenases, NDex(NADH), oxidizes NADH while the other, NDex(NADPH), oxidizes NADPH. Both have access to cytosolic NAD(P)H (which readily penetrates the outer mitochondrial membrane) and they are thought to oxidize this, rather than matrix NAD(P)H. Both of these external NAD(P)H dehydrogenases are dependent on Ca^{2+}.

All four of the additional NAD(P)H dehydrogenases transfer electrons to Q but, in contrast to Complex I, none of them translocates protons (Figure 6.17). This means that there is a reduced capacity for ATP synthesis when either NADH or NADPH is oxidized through any of these alternative dehydrogenases, rather than through Complex I.

The alternative NAD(P)H dehydrogenases are encoded by a multigene family in arabidopsis. *NDA* (three genes) and *NDB* (four genes) share homology with fungal counterparts, while *NDC* (one gene) appears to have a cyanobacterial origin. It is not yet established which gene corresponds to which NAD(P)H dehydrogenase, although there is evidence that the *NDA* and *NDC* genes encode proteins targeted to the inner face of the inner membrane, while *NDB* gene products are targeted to the outer face.

The physiological function of the alternative NAD(P)H dehydrogenases remains the subject of some speculation

Because of its lower affinity for NADH, compared with Complex I, NDin(NADH) appears to be active only when matrix NADH concentrations have become elevated. Such a situation might occur when the rate of substrate oxidation is particularly high or when flux through the mitochondrial electron transport chain is restricted by lack of ADP availability. These conditions could arise when the TCA cycle is functioning biosynthetically, where increased TCA cycle flux may well exceed the capacity for Complex I to reoxidize NADH. Similar constraints might occur in photosynthetic cells in the light, when high ATP:ADP ratios, resulting from chloroplast photophosphorylation, may restrict electron transport through Complex I (see Box 6.5). Consequently, NDin(NADH) appears to serve an overflow function, by oxidizing NADH when Complex I activity is inadequate.

Experiments with isolated mitochondria have provided evidence that when malate is oxidized by malic enzyme, the resulting NADH is preferentially oxidized by NDin(NADH). Malic enzyme, with its lower affinity for NAD^+ and relative insensitivity to NADH inhibition compared with MDH (see page 166), will generate a higher matrix concentration of NADH that will enable the NDin(NADH) to function. In bypassing Complex I, malate oxidation via malic enzyme would result in a reduction in the amount of ATP generated (Table 6.1, page 184). The physiological significance is unclear. *In vivo* measurements of malate oxidation indicate that malic enzyme plays only a minor part in respiration. Furthermore, transgenic potatoes with reduced malic enzyme content were not impaired in respiration. These two studies indicate that the anaplerotic function of malic enzyme (Figure 6.14 as discussed above) is probably more significant than its respiratory function. Hence, malic enzyme is likely to be functioning within the context of a biosynthetic TCA cycle when NADH concentrations might be expected to rise, and engagement of NDin(NADH) is favored. Until definitive *in vivo* experiments are performed, however, this interaction between malic enzyme and ND(in)NADH remains speculative.

The NADH and NADPH dehydrogenases (NDex(NADH) and NDex(NADPH)) on the outer face of the inner membrane appear to be capable of oxidizing cytosolic NADH and NADPH. This makes them potentially important in regenerating NAD^+ and $NADP^+$ for cytosolic substrate oxidation, for example, in glycolysis and the OPPP. Ca^{2+}, which is a component of stress-signaling pathways in plants, stimulates the activity of both external dehydrogenases, and the internal NADPH dehydrogenase. Furthermore, the expression of genes encoding these dehydrogenases is induced by stress. It is therefore possible that these dehydrogenases might function to increase electron transport during stress, serving a similar role to the alternative oxidase in reducing free radical formation (see page 182).

Now that the genes have been identified for each of the NAD(P)H dehydrogenases, it is hoped that the creation of transgenic plants with enhanced and/or reduced expression will provide a means of investigating the *in vivo* function of each enzyme.

Plant mitochondria contain an alternative oxidase that transfers electrons from QH$_2$ to oxygen and provides a bypass of the cytochrome oxidase branch

One of the major differences between plant and mammalian mitochondria is in their response to cytochrome oxidase inhibitors, e.g. cyanide. Whereas mammalian mitochondrial respiration is inhibited completely, plant mitochondria, depending on species, show variable resistance to cyanide of between 10% and 100%. The reason for this is that all plants studied to date, along with some algae, fungi, and protozoa (including the agent of trypanosomiasis, *Trypanosoma brucei*) possess an additional terminal electron acceptor, the so-called alternative oxidase, which is insensitive to cyanide. This single enzyme has been the focus of a good deal of research interest in recent years, as an understanding of its structure, function, and regulation has begun to emerge. It is for this reason that we have chosen to describe the alternative oxidase in some detail in the sections and boxes that follow.

The alternative oxidase accepts electrons from QH_2 and transfers them directly to oxygen. Hence, the Q pool occupies a branch-point between the alternative pathway (electron flow through the alternative oxidase to oxygen) and the cytochrome pathway (electron flow through Complexes III and IV to oxygen). Unlike cytochrome oxidase (Complex IV), the alternative oxidase is insensitive to cyanide, azide, and carbon monoxide, but is blocked by other inhibitors, in particular the benzhydroxamic acids such as salicylhydroxamic acid. Another important difference is that the alternative oxidase does not translocate protons, whereas the cytochrome pathway contains two proton-translocating complexes, i.e. Complexes III and Complex IV. Hence, just as the additional NAD(P)H dehydrogenases provide nonphosphorylating branches for electrons entering into the Q pool, the alternative oxidase provides a non-phosphorylating route for electrons leaving the Q pool (Figure 6.17). Plant mitochondria therefore have the potential to oxidize reducing equivalents without generating either a proton electrochemical gradient, or ATP. Instead, the free energy is released as heat (thermogenesis) and some specialized plants have exploited this aspect of the alternative oxidase in order to volatalize chemicals that will attract pollinators (Box 6.4).

Thermogenesis remains the only clearly defined function of the alternative oxidase. A major challenge to plant biochemists has been to understand the function of the alternative oxidase in the vast majority of plants that are not thermogenic, and where respiration is, at most, 20% cyanide resistant. The most recent evidence that is emerging from the analysis of mutants and transgenic plants indicates a role in stress resistance and in regulating carbon metabolism. These functions of the alternative oxidase are discussed in more detail below.

The alternative oxidase is a dimer of two identical polypeptides with a nonheme iron center

The alternative oxidase protein has yet to be crystallized, hence our current understanding of its structure is based primarily on sequence information. The higher plant alternative oxidase is a homodimer, formed from two identical 32 kDa polypeptides. The association between the two subunits is a key feature of its regulation, as discussed below. While earlier reports suggested that the alternative oxidase was membrane-spanning, we now know this to be incorrect. It is tightly bound to the matrix face of the inner mitochondrial membrane, with its active site exposed to the matrix.

Several pieces of evidence indicate that the alternative oxidase contains two iron atoms within a nonheme center. First, the capacity to reduce oxygen is indicative of a transition metal center within the active site. Second, sequence information shows a conserved region containing an iron-binding amino acid sequence of four glutamate and two histidine residues. Finally, electron paramagnetic resonance (a spectroscopic technique for identifying unpaired electrons) analysis of arabidopsis alternative oxidase indicates that it contains a binuclear iron center.

Alternative oxidase isoforms in plants are encoded by discrete gene families

There are several alternative oxidase isoforms in plants, encoded by discrete gene families. The alternative oxidase is nuclear-encoded by two small gene subfamilies, consisting of *AOX1* and *AOX2*. While *AOX1* is induced by a range of stresses and is present in both monocots and dicots, *AOX2* genes are usually developmentally regulated, or constitutively expressed, and have yet to be found in monocots. Various subclasses, including *AOX1a–c*, *AOX2a*, and *AOX2b*, have been identified, with the number and class of genes varying according to species.

The induction of *AOX1* by a range of stresses is consistent with the proposed function of alternative oxidase in stress resistance in higher plants, as

Box 6.4 The alternative oxidase and thermogenesis

The alternative oxidase provides an alternative pathway for electrons between ubiquinone and oxygen in addition to the cytochrome pathway that occurs in all mitochondria. The alternative pathway is wasteful of energy because it bypasses two of the three proton-pumping complexes (Complexes III and IV) of the respiratory chain. Energy that would otherwise be conserved, by means of ATP synthesis, is released as heat. In some species of higher plants, high rates of respiration together with a switch from the cytochrome to the alternative pathway, results in tissue temperatures that can rise to between 10 and 25°C above ambient. This phenomenon, termed thermogenesis, was first recognized by the French scientist, Jean-Baptiste de Lamarck, in 1778 for the genus *Arum*. Thermogenesis is now known to occur in cycads and in the flowers of angiosperm species within the families Annonaceae, Araceae, Aristolochiaceae, Cyclanthaceae, Nymphaceae, and Palmae. Thermogenesis is usually associated with pollination, either as a means of attracting pollinators through volatilization of scents, or as a reward to insects by keeping them warm within the enclosed flower.

It is the Arum lilies, in particular *Arum maculatum* and *Arum italicum* (family Araceae) that have been studied the most. These have a specialized inflorescence with a unique structure, the spadix, which contains a high concentration of starch and a very large number of mitochondria. The rate of respiration within the spadix can match that of a hummingbird in flight, far exceeding that of non-thermogenic plant tissue. The spadix also contains a range of volatile chemicals, including amines and indoles that smell of rotting flesh. The rise in temperature accelerates the release of these volatiles, attracting pollinators, such as beetles and flies, which would normally feed, or lay their eggs, on carrion.

The extent to which the alternative pathway operates in the Arum spadix changes with inflorescence development.

Figure 1 The structure of the inflorescence in *Arum maculatum*. The inflorescence in *Arum maculatum* consists of a central spadix, at the base of which are found the male and female flowers. The entire structure is enclosed within a spathe, which unfolds as the inflorescence matures. As the flowers mature, respiration switches over to the alternative pathway and the spadix begins to heat up and release volatile scents. These attract beetles and flies, which crawl down into the base of the spadix where they pollinate the female flowers. The club-shaped structures ensure that the pollinator can move down, but not up the spadix, and it is only when the male flowers ripen that these structures wither and allow the insects to escape. On departure, the insects crawl over the mature male flowers, becoming coated in pollen that is then transferred to female flowers on another Arum plant. The mechanism increases the opportunity for cross-pollination. (Reprinted with permission from *The Families of Flowering Plants: descriptions, illustrations, identification and information retrieval*, L. Watson and M. J. Dallwitz, 1992 onwards. Version 21st May 2006. http://delta-intkey.com.)

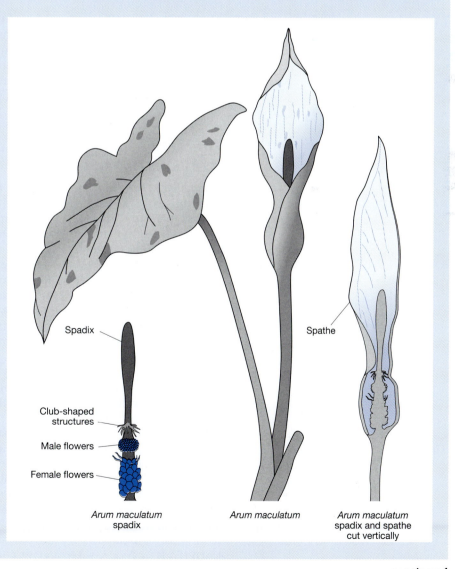

Spadix

Spathe

Club-shaped structures

Male flowers

Female flowers

Arum maculatum spadix

Arum maculatum

Arum maculatum spadix and spathe cut vertically

continued

Box 6.4 The alternative oxidase and thermogenesis (continued)

In the immature spadix, respiration is entirely dependent on the cytochrome pathway, whereas on the day of pollination (denoted D-day) it switches completely to the alternative pathway. The transition from the cytochrome to the alternative pathway appears to involve salicylic acid, which increases in concentration to reach a peak the day before D-day. The alternative pathway can be artificially induced by treating immature spadices with salicylic acid, which causes an increase in expression of alternative oxidase mRNA as well as *de novo* synthesis of the protein.

Perhaps the most spectacular species of Arum lily is the Titan Arum, *Amorphophallus titanum*, which is native to the rainforests of Western Sumatra and reputed to produce the tallest flower in the World. It is more accurate to describe the structure as an inflorescence (see Figure 1). Nevertheless, the structure is spectacular, reaching over 3 m in height. Its great height ensures that the smell of putrid flesh is carried across the dense vegetation of the rainforest, attracting beetle pollinators from a large distance. When a specimen flowered at the Royal Botanic Gardens Kew, England, in 1996 it attracted over 5000 (human!) visitors. As the inflorescence withers and dies within 24 to 48 hours, a police guard was posted to ensure that everyone had a chance to see it. The crowd was drawn partly by the promise of experiencing its horrendous smell, but, as it is night-pollinated, the smell was only strong after the visitors had gone home for the evening (see Plates 6.1, 6.2, and 6.3).

discussed below. *AOX1* has been found to be induced by pathogen infection, low and freezing temperatures, hypoxia, and nutrient deficiency (both phosphorus and nitrogen), and to be expressed during the late stages of fruit ripening immediately prior to senescence. In contrast, expression of *AOX2* genes appears to be more closely related to developmental signals, such as seed germination and early stages of fruit ripening, rather than stress.

Alternative oxidase activity is regulated by 2-oxo acids and by reduction and oxidation

Alternative oxidase activity is regulated by 2-oxo acids and by reduction and oxidation, possibly via a mitochondrial thioredoxin system. A major breakthrough in our understanding of the regulation of the alternative oxidase came with the discovery that the enzyme was modified by reduction of the disulfide bridge that linked together its two component subunits. Two highly conserved cysteine residues (Cys_I and Cys_{II}) feature in this regulation. Cys_I forms a covalent disulfide bond (–S–S–) with the corresponding Cys_I on the adjacent monomer. The alternative oxidase is inactive in this state. When reduced, the Cys_I residues hold the dimer together through noncovalent association between the two S–H groups of the monomers. It is in this noncovalent state that the dimer becomes catalytically active, and it is further activated by binding to 2-oxo acids, especially pyruvate. In this reduced, activated form, the alternative oxidase has a higher activity and a higher affinity for QH_2 (lower K_m), allowing it to compete effectively for electrons with the cytochrome pathway. The noncovalently linked dimer is the predominant form *in vivo*.

Although *in vitro* studies have added strong support for the role of reduction/oxidation in regulating alternative oxidase activity, we still do not know how the process functions *in vivo*. An emerging theory is that it involves a mitochondrial thioredoxin system. A mitochondrial NADP-dependent thioredoxin reductase has been identified in plants, as well as mitochondrial-specific forms of thioredoxin. Together, these would operate as a redox-sensing mechanism within the matrix:

$$NADPH + H^+ + Trx_{ox} \rightleftharpoons NADP^+ + Trx_{red}$$

where Trx_{ox} represents oxidized thioredoxin, and Trx_{red} reduced thioredoxin.

NADPH could be supplied by the NADP-IDH (see above), or via a NAD(P)H transhydrogenase:

$$NADH + NADP^+ \rightleftharpoons NADPH + NAD^+$$

Hence, as the reduction level of the matrix NAD(P) pool increases, the thioredoxin system would presumably operate to generate reduced thioredoxin. There is some recent evidence that one of the mitochondrial thioredoxins (Trxh2) can reduce the alternative oxidase directly, to enable it to be activated by pyruvate. The combined effect of these reactions would therefore allow the alternative oxidase to be activated as the mitochondrial $NAD(P)^+$ pool becomes reduced.

Before the effect of reduction and 2-oxo acids was known, models that sought to explain the relative contribution of the cytochrome and alternative pathways were based on the observation that the former had a higher affinity for QH_2. Thus, the alternative pathway was thought to function only when the cytochrome pathway was saturated and the Q pool was at least 50% reduced. The alternative oxidase was therefore for a long time thought to be operating as an overflow that only engaged when the capacity of the cytochrome branch was exceeded. We now know that the alternative oxidase will function when the Q pool is as little as 20% reduced, providing there is sufficient pyruvate present to activate and increase its affinity for QH_2. There is still some controversy as to whether pyruvate concentrations are sufficiently high *in vivo* to enable the alternative oxidase to be regulated in this way. Nevertheless, the current model allows for both pathways to compete for electrons from the Q pool so that the two pathways may operate simultaneously. A consequence of this model is that previous experiments that used inhibitors (e.g. salicylhydroxamic acid) to determine the *in vivo* engagement (as opposed to the maximum potential capacity) of the alternative pathway were no longer considered to be valid, as they were based on the assumption that the alternative oxidase did not compete for electrons with the cytochrome pathway. These inhibitor titration experiments were therefore superseded by the oxygen isotope fraction technique. This method is based on the observation that the alternative oxidase discriminates against ^{18}O (the heavy, stable isotope of the more abundant ^{16}O isotope) to a greater extent than does the cytochrome pathway. Samples of air, or aqueous solution in the case of isolated mitochondria, are collected and analyzed in a mass spectrometer to determine the ^{18}O to ^{16}O ratio. This value is used, together with the oxygen isotope discrimination ratio (approximately 17–20‰ for the cytochrome pathway and 23–27‰ for the alternative pathway; see Chapter 5, Box 5.7 for explanation of the technique) and the measured rate of respiration, to calculate the rate of the two branches of the respiratory chain. The extent to which each pathway functions *in vivo* will determine the extent to which ATP synthesis is coupled to electron transport, as discussed in the next section.

The alternative oxidase adds flexibility to the operation of the mitochondrial electron transport chain

While the role of the alternative oxidase in thermogenesis has been recognized for some time (see Box 6.4) it is only recently that we have begun to understand its wider physiological function in higher plants. The alternative oxidase provides an added metabolic flexibility; it enables mitochondrial respiration to function under a wider range of physiological and environmental conditions than if only the cytochrome pathway was present. The alternative oxidase appears to be important with respect to stress tolerance, and in coordinating carbohydrate metabolism in photosynthetic cells.

The alternative oxidase may prevent the formation of damaging reactive oxygen species within the mitochondria

The flow of electrons through the cytochrome pathway is subject to respiratory control, meaning that a low ADP:ATP ratio and/or a low cytosolic ADP concentration will restrict its activity. Indeed, early experiments on isolated mitochondria demonstrated that electron transport could be reversed in the presence of a sufficiently low ADP:ATP ratio. Two potential problems can arise as a result of this adenylate regulation; reactive oxygen species (ROS) may form, and NADH oxidation may become inadequate to support the regeneration of NAD^+ for the matrix NAD-dependent dehydrogenases, so that the turnover of the TCA cycle becomes restricted.

ROS (such as the hydroxyl radical, hydrogen peroxide, and superoxide) can damage proteins, lipids, and DNA. They can form within mitochondria when electrons are transferred directly to oxygen from certain reduced components of the electron transport chain. This can occur at Complexes I and III if they become highly reduced, for example, when electron transport is restricted further downstream or the rate of electron flow from NADH exceeds the capacity of either Complex. In Complex I, electrons are passed to oxygen from within the iron–sulfur clusters, while Complex III forms superoxide via autoxidation of QH_2. The alternative oxidase, in accepting electrons from QH_2, is able to maintain a flow of electrons irrespective of the ADP:ATP ratio. Hence, it is a plausible hypothesis that over-reduction of the respiratory complexes, and the ensuing ROS formation, is avoided and NADH can be oxidized at rates that are sufficient to maintain substrate oxidations (e.g. TCA cycle) within the matrix.

The most compelling evidence to support the role played by the alternative oxidase in preventing ROS formation has come from the recent creation and analysis of transgenic plants. Arabidopsis plants with reduced (antisense *AtAOX1a*) alternative oxidase expression were shown to accumulate ROS when the cytochrome pathway was inhibited with cyanide. Wild-type plants did not accumulate ROS under these conditions. The results indicate that wild-type arabidopsis contains sufficient alternative oxidase activity to prevent over-reduction of the Q pool and the associated formation of ROS when the cytochrome pathway is restricted.

Alternative oxidase appears to play a role in the response of plants to environmental stresses

A range of environmental stresses has been shown to induce expression of the alternative oxidase. These include chilling, wounding, drought, osmotic and nutrient stress, and pathogen attack, many of which can result in inhibition of mitochondrial electron transport and the subsequent risk of ROS formation. By acting as an electron drain from QH_2, the alternative oxidase might protect against its over-reduction and alleviate the risk of stress-related ROS formation. The recent production and analysis of transgenic arabidopsis plants with altered alternative oxidase expression has provided supporting evidence for this hypothesis. Transgenic plants lacking functional AOX were found to have increased amounts of ROS, while those overexpressing alternative oxidase produced reduced amounts of ROS from within the mitochondria.

Alternative oxidase and NADH oxidation

The alternative oxidase enables NADH oxidation to continue when the cellular demand for ATP may be low. We have already seen how the TCA cycle requires continuous NADH oxidation. If the mitochondrial electron transport chain is restricted, e.g. by a low ADP:ATP ratio, this could result in TCA cycle inhibition, as discussed earlier in this chapter. An increase in the NADH:NAD⁺ ratio could result in activation of both the rotenone-resistant NADH dehydrogenase and the alternative oxidase (via the proposed thioredoxin system). Reduced TCA cycle flux would also result in pyruvate accumulation, which would further activate the alternative oxidase. As a result, NADH can be oxidized and the TCA cycle may continue to operate even if the conventional electron transport chain (via the cytochrome pathway) is restricted by ADP availability.

Plant mitochondria and uncoupling proteins

Plant mitochondria also contain an uncoupling protein (UCP) that reduces the amount of ATP produced during electron transport. Mammalian mitochondria do not possess the alternative oxidase. Instead, they possess UCP that is used to manipulate the extent to which ATP synthesis is coupled to electron transport. The mitochondrial UCP works by forming a natural proton leak through the inner membrane. This short circuits the proton gradient so that energy from substrate oxidation is released as heat, rather than being conserved as ATP. The mechanism is essential for regulating the body temperature of small mammals. There is neither a mechanistic nor a structural similarity between UCP and the alternative oxidase, and yet these proteins appear to be serving analogous physiological functions. After UCPs were identified in mammals, it was somewhat surprising, therefore, to discover that they were also present in plants. Three UCP cDNA homologs have been identified in arabidopsis (*AtUCP1*, *2*, and *3*). Like their mammalian counterparts, they are stimulated by the superoxide free radical and activated by fatty acids. Transgenic potato plants overexpressing UCP1 were found to have reduced levels of ROS and to have an increased rate of pyruvate to citrate conversion. Thus, UCP in plants may serve a similar function to the alternative oxidase, in reducing ROS formation during respiration. The observed increase in pyruvate oxidation indicates that UCP might also stimulate TCA cycle flux.

There is clearly some overlap between UCP and the alternative oxidase with respect to their physiological functions. Recent localization studies indicate that, although arabidopsis contains gene families for both UCP and alternative oxidase, these are expressed in different tissues and at different stages of development. The presence of both types of energy-dissipating systems appears to provide plants with an added flexibility with which they can respond to their environment.

ATP synthesis in plant mitochondria is coupled to the proton electrochemical gradient that forms during electron transport

As explained above, the oxidation of substrates in the matrix results in the generation of reducing equivalents (NADH, NADPH, FADH₂) whose oxidation results in the release of electrons that are subsequently transferred between the respiratory complexes (I–IV) within the inner membrane. Complexes I, III, and IV all translocate protons from the matrix to the intermembrane space in the process of transporting electrons (see above), thus forming a proton gradient across the membrane, with a higher concentration of protons in the intermembrane space relative to the matrix. The concentration difference, together with the electrical potential difference (as protons carry a positive charge), both contribute to an electrochemical potential that can be used to drive ATP synthesis from ADP and P_i (Box 6.5).

The amount of ATP formed depends on the route taken by electrons within the electron transport chain. Complexes I and III both pump $4H^+$ per electron pair, while Complex IV pumps $2H^+$ (Figure 6.16). Hence, oxidation of matrix-generated NADH (e.g. resulting from malate oxidation) can result in the export of $10H^+$ from the matrix, providing electrons move through Complexes I, III, and IV. As the ATP synthase uses approximately $3H^+$ to produce one ATP (Box 6.6), and a further H^+ is needed to exchange ADP and P_i for ATP, the synthesis of one ATP requires a total of $4H^+$. Therefore the number of protons exported for each NADH oxidized ($10\,H^+$), divided by the number of protons used to synthesize one ATP ($4\,H^+$) i.e. $10 \div 4$ equals 2.5 ATPs synthesized for each NADH oxidized (i.e. per electron pair) through Complexes I, III, and IV. If NADH is oxidized through the internal NADH dehydrogenase, hence by-passing Complex I, only six protons are exported for each NADH oxidized, hence the number of protons ($6\,H^+$) exported divided by the number of protons used to synthesize one ATP ($4\,H^+$) i.e. $6 \div 4$ equals 1.5 ATPs synthesized for each NADH oxidized. By convention, the ATP to $2e^-$ ratio is usually expressed as an ADP to oxygen ratio (one oxygen atom is taken up per electron pair). Values for ADP to oxygen ratios are presented in Table 6.1.

ATP synthase uses the proton motive force to generate ATP

Synthesis of ATP is catalyzed by a proton-translocating ATP synthase (also called Complex V and F_0F_1-ATPase). If the ATP synthase (F_1 component) is isolated from the membrane it carries out the reverse reaction, i.e. it is an ATPase causing hydrolysis of ATP to ADP and P_i. It is the proton gradient that drives the ATP synthase reaction in the reverse direction that results in ATP synthesis. The link between the proton gradient and ATP synthesis was first proposed by Peter Mitchell in the chemiosmotic hypothesis, explained in more detail in Box 6.5.

Table 6.1 Theoretical values for ADP to oxygen ratios during respiration in plant mitochondria		
Substrate	**Route for electrons**	**ADP to oxygen ratio**
Malate	Complex I, III, IV	2.5
Malate	NDin, III, IV	1.5
Malate	NDin, AOX	0
Succinate	Complex II, III, IV	1.5
NADH (cytosolic)	Complex III, IV	1.5

Note that the alternative oxidase (AOX) may operate with any of these substrates, hence reducing the ADP to oxygen ratio by 1.5 as it bypasses both Complexes III and IV. Similarly, operation of the internal NADH dehydrogenase (NDin) instead of Complex I reduces the ADP to oxygen ratio by 1.0.

The mitochondrial ATP synthase is a highly conserved multienzyme complex found in all eukaryotes. A similar protein occurs in the cytoplasm of bacteria and in chloroplasts (see Chapter 4). We now have a good understanding of both the structure and mechanism of operation of the ATP synthase from a range of organisms, although there is still an incomplete knowledge of the plant mitochondrial enzyme complex.

As explained in Box 6.6 the movement of protons through the ATP synthase causes a central shaft to rotate, altering the conformation of the catalytic sites so that the reaction favors ATP synthesis rather than ATP hydrolysis. The ATP synthase complex is often described as a molecular motor because of this rotation, which can occur at rates of 300 rotations per second.

Box 6.5 The chemiosmotic hypothesis and ATP synthesis

The chemiosmotic hypothesis was formulated in 1961 by the British scientist, Peter Mitchell, while working in Edinburgh. He was awarded the Nobel Prize for Chemistry in 1978 for this discovery.

The hypothesis explains the relationship between the potential energy present within a concentration gradient of ions across a membrane, and the use of this potential energy in the formation of ATP. Central to this hypothesis is the requirement for an energy-transducing membrane that has two different types of proton pump. The primary pump generates a proton gradient across the membrane. In mitochondria, the primary pump is situated within each of the proton-translocating Complexes (I, III, and IV) and the energy to drive this pump comes from the oxidation of respiratory substrates. The secondary pump is common to all bioenergetic systems. It is the ATP synthase, or proton-translocating ATPase (also termed Complex V in mitochondria), which hydrolyzes ATP to ADP and P_i, and pumps protons—in the case of mitochondria this would be from the matrix to the intermembrane space, i.e. the same direction in which the protons move during electron transport. The chemiosmotic hypothesis predicted that the ATP synthase could be made to act in reverse, i.e. in the direction of ATP formation, if the H^+ gradient formed by the primary pump was sufficiently high (see Box 6.6 for explanation of the ATP synthase). To determine whether the H^+ gradient is, indeed, sufficiently high to permit ATP synthesis, we must first consider the thermodynamics of the relevant parts of the pathway.

As we have seen, during the operation of the mitochondrial electron transport chain, protons are translocated across the inner membrane, from the matrix to the intermembrane space (Figure 1).

The ion gradient that forms has the potential to do some work. It carries potential energy. The quantitative thermodynamic measure of this potential energy is defined as the proton electrochemical gradient $\Delta\bar{\mu}_H^+$. This is expressed in kJ mole^{-1} and is a thermodynamic measure of the extent to which the H^+ gradient is removed from equilibrium—and hence capable of doing work. It is equivalent to the change in Gibbs free energy (ΔG) associated with the formation of the gradient.

There are two components that contribute to the proton electrochemical gradient. One is due to the difference in concentration of ions (H^+) across the membrane, this is the chemical potential. The other is due to the electrical potential difference resulting from the displacement of a charged ion (in this case H^+) across the membrane, i.e. the membrane potential ($\Delta\Psi$).

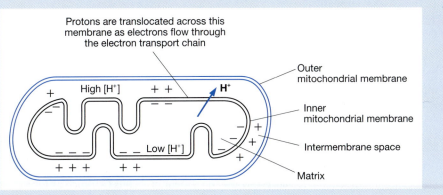

Protons are translocated across this membrane as electrons flow through the electron transport chain

High [H^+]   Low [H^+] H^+

Outer mitochondrial membrane

Inner mitochondrial membrane

Intermembrane space

Matrix

Figure 1 Movement of protons across the mitochondrial inner membrane.

continued

Box 6.5 The chemiosmotic hypothesis and ATP synthesis (continued)

We shall deal with these two components separately to begin with. The Gibbs energy change associated with the movement of a solute (in this case H^+) across a membrane is calculated from the difference in concentration of that solute across the membrane, given by:

$$\Delta G \text{ (kJ mole}^{-1}) = RT \ln \frac{[H^+]_{im}}{[H^+]_m}$$

where R = the gas constant (8.315 J mole^{-1} K^{-1}), T = 298K, $[H^+]_{im}$ is the concentration of protons in the intermembrane space, while $[H^+]_m$ is the concentration of protons in the matrix.

By converting to log base 10 this becomes:

$$\Delta G \text{ (kJ mole}^{-1}) = 2.3 \, RT \log \frac{[H^+]_{im}}{[H^+]_m}$$

This can be simplified because pH is a logarithmic function of $[H^+]$:

$$\Delta G \text{ (kJ mole}^{-1}) = 2.3 \, RT \, \Delta pH \qquad (1)$$

where ΔpH is defined as: pH in the intermembrane space – pH in the matrix.

The Gibbs free energy associated with the electrical potential component is calculated from:

$$\Delta G \text{ (kJ mole}^{-1}) = -mF \, \Delta\Psi \qquad (2)$$

where m is the charge on the ion (i.e. 1 for a H^+) and F is Faraday's constant (the charge on an electron; one mole of H^+ has a charge of 9.65×10^4 coulombs, which is equal to 9.65×10^4 J V^{-1} mole^{-1}) and $\Delta\Psi$ is the membrane potential—measured as the voltage difference across the membrane (units are mV).

Equations 1 and 2 define the chemical and electrical components, respectively, of the electrochemical gradient. In practice, both contribute to the free energy associated with the electrochemical gradient. Hence the sum of equations 1 and 2 gives the free energy change (ΔG) of the electrochemical gradient, which is equal to $\Delta\bar{\mu}_{H^+}$:

$$\Delta G \text{ (kJ mole}^{-1}) = \Delta\bar{\mu}_{H^+} = -F\Delta\Psi + 2.3 \, RT \, \Delta pH \qquad (3)$$

By convention, $\Delta\bar{\mu}_{H^+}$ is converted into units of electrical potential (mV), by dividing its value by the Faraday constant. The resulting value is defined as the proton motive force (PMF) expressed as Δp:

$$\text{PMF} = \Delta p \text{ (mV)} = \frac{-(\Delta\bar{\mu}_{H^+})}{F}$$

Hence, by dividing equation 3 through by F and substituting the values for R, T and F:

$$\Delta p \text{ (mV)} = \Delta\Psi - 59\Delta pH \qquad (4)$$

In mitochondria, the pH gradient (and hence ΔpH) across the inner membrane is actually very small, with the pH of the intermembrane space being only 0.2–0.5 pH units greater than that of the matrix. This is because the inner membrane is impermeable to anions, hence the pH gradient is limited because there is no possibility of a counter ion exchange (as occurs in chloroplasts) to balance the charge and allow H^+ to accumulate.

Formation of a pH gradient when the membrane is permeable to anions (e.g. in chloroplasts) is shown in Figure 2.

The impermeability of the mitochondrial inner membrane to anions means that a relatively small pH gradient forms, while the charge difference is greater because each proton movement carries a +ve charge that is not counter-balanced by anion (A^-) movement.

Hence, in mitochondria, the principal contributor to Δp is the electrical component, $\Delta\Psi$, which can reach values of –150 to –200 mV. Values for Δp in plant mitochondria range between –200 and –240 mV.

The value for Δp defines the proton electrochemical potential that may be used to drive ATP synthesis. It formed the central tenet of the chemiosmotic hypothesis (as explained above): Δp has to be sufficiently large to enable net ATP synthesis via the ATP synthase.

For a Δp of 200 mV this would equate to 200 kJ of energy—assuming that 10 H^+ ions are translocated per pair of

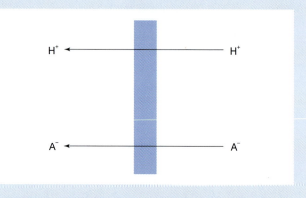

Figure 2 Anion-permeable membrane. In the above example, a relatively large difference in $[H^+]$ concentration can occur because the charge difference is counter-balanced by the movement of anions (A^-) such as Cl^-.

Box 6.5 The chemiosmotic hypothesis and ATP synthesis (continued)

electrons during electron transport $(200 \times F \times 10 = 200 \text{ kJ})$. The ΔG_p (free energy associated with ATP synthesis) per mole of ATP is approximately 60 kJ mole^{-1}. Hence, for three ATPs per pair of electrons transported, the Δp value of 200 kJ translates into the conservation of 180 kJ in the form of ATP synthesis—a highly efficient recovery. Thus, Δp is sufficiently large to drive the synthesis of ATP.

The relationship between the proton electrochemical gradient ($\Delta \bar{\mu}_H^+$) and ATP synthesis can be illustrated by the use of uncoupling reagents. Uncouplers are lipid-soluble weak acids, such as 2,4-dinitrophenol (DNP), that can exist in neutral and anionic forms (Figure 3):

DNPH (neutral form) releases a proton from the hydroxyl group to form DNP$^-$ (ionic form) in a reversible process. Both forms are soluble within lipid membranes. Consequently uncouplers such as DNP function as proton shuttles by binding protons reversibly, traveling across the inner membrane and releasing protons on the other side. Hence, as first proposed by Mitchell, uncouplers act by short-circuiting the proton electrochemical gradient (i.e. $\Delta \bar{\mu}_H^+$ becomes zero). The subsequent demonstration, in a large number of research laboratories, of the correlation between proton conduction and

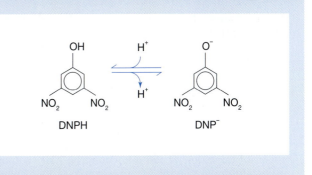

Figure 3 Neutral and anionic forms of 2,4-dinitrophenol (DNP).

uncoupling activity provided strong supporting evidence for the chemiosmotic hypothesis. In the presence of an uncoupler, the absence of a proton electrochemical gradient causes the ATP synthase to function as an ATPase so that it hydrolyzes, rather than synthesizes, ATP. The mitochondrial electron transport chain continues to function and, as it is now uncoupled from ATP synthesis, it proceeds at the same rate whether or not ADP is available.

Mitochondrial respiration interacts with photosynthesis and photorespiration in the light

The extent to which mitochondrial respiration continues in photosynthetic cells in the light has been the subject of much controversy over many decades, and there is still no clear consensus. Some studies have concluded that the TCA cycle ceases to function, while others indicate that a partial pathway operates in the light. The question of whether the mitochondrial electron transport chain operates, and the degree to which it is coupled to ATP synthesis, remains a further source of speculation. Here we shall summarize the current models that seek to explain how mitochondria interact with photosynthesis and photorespiration.

Given the sensitivity of the dehydrogenase reactions to NADH, many early reviews concluded that the TCA cycle would cease in photosynthetic tissue in the light, as a result of increased NAD(P)H:NAD(P)$^+$ ratios resulting from export of reducing equivalents from the chloroplast. The malate/OAA shuttle is the main mechanism that transfers excess reducing equivalents from the chloroplast (summarized in Figure 6.18) and current thinking is that this operates to protect the thylakoid from becoming over-reduced in the light. As already discussed, plant mitochondria may not accumulate high concentrations of NAD(P)H because of the operation of the alternative oxidase and of Type II NAD(P)H dehydrogenases (Figure 6.17). Nevertheless, recent experimental evidence indicates that only a partial TCA cycle operates in photosynthetic cells in the light. Isocitrate dehydrogenase (NAD-IDH), which is particularly sensitive to NADH, is thought to be inhibited in the light. Under these conditions,

Box 6.6 Structure and mechanism of the mitochondrial ATP synthase

ATP synthase (F_oF_1 ATP synthase) uses the energy of an electrochemical gradient to catalyze the formation of ATP from ADP and P_i in a fascinating process that has led this enzyme to be described as a rotary motor.

The F_oF_1 ATP synthase is a highly conserved enzyme, being found in the plasma membrane of prokaryotes, the inner membrane of mitochondria and the thylakoid membrane of chloroplasts (Chapter 4, Box 4.5). Here, we summarize current knowledge of the structure and mechanism of operation of the mitochondrial ATP synthase, using additional examples from bacterial proteins where relevant. Most of our understanding of the mitochondrial ATP synthase has come from studies of yeast and mammalian (mostly bovine) proteins and there is still much to learn about the plant mitochondrial ATP synthase.

All F_oF_1 ATP synthases share the same basic structure. In prokaryotes, which contain the simplest form, they consist of eight different subunits, α, β, γ, δ, ε, a, b, and c with stoichiometry of $\alpha_3:\beta_3:\gamma:\delta:\varepsilon:a:b_2:c_{10-14}$. The structure is essentially the same in chloroplasts, although different terminology is used for the subunits (see Box 4.5). Mitochondrial ATP synthases have between seven and nine additional subunits, which are thought to serve regulatory roles, and only one b-subunit. The δ-subunit of mitochondrial ATP synthase is equivalent (in form and function) to the ε-subunit of prokaryotes and chloroplasts. The mitochondrial ε-subunit is unique, having no counterpart in either prokaryotic or chloroplast ATP synthases.

The F_oF_1 ATP synthases are conventionally described as consisting of two domains, a hydrophobic, transmembrane domain (F_o consisting of $ab_{1-2}c_{10-14}$) that channels protons through the membrane, and a hydrophilic domain (F_1 consisting of $\alpha_3:\beta_3:\gamma\delta\varepsilon$) that contains the catalytic sites. In mitochondria, F_o is located in the inner membrane, while F_1 protrudes into the matrix. F_1 catalyzes both ATP formation and ATP hydrolysis, with each process being coupled to the movement of protons through F_o. ATP synthesis is coupled to an inward movement of protons (i.e. into the matrix) through F_o, while ATP hydrolysis coincides with the movement of protons out of the matrix. If F_1 is detached from the membrane it no longer forms ATP, instead it catalyzes only the reverse reaction hydrolyzing ATP to ADP and P_i. At the same time, the F_o portion that is left behind renders the membrane leaky to protons. The antibiotic, oligomycin, blocks the leak by binding to one of the F_o-subunits, the oligomycin sensitivity conferring protein. It is this effect of oligomycin that is recognized in the name, i.e. the 'o' of F_o.

The F_o, F_1 terminology is less useful when describing the way in which the ATP synthases function. The rotary motor model, that is now generally accepted, uses the terminology of a motor to describe the components of the protein.

As shown in Figure 1(a) (*Escherichia coli* ATP synthase) in prokaryotes the ring of *c*-subunits together with the central shaft, formed from the γ and ε-subunits, is the rotor. The stator (i.e. the part of a motor that remains static) consists of 2 *b* and 1 δ-subunit in prokaryotes, with the δ-subunit binding to the α-subunit at the top, while the *b*

Figure 1 Examples of prokaryotic (a; *E. coli*) and mitochondrial (b; bovine heart mitochondria) F_1F_o ATP synthases.

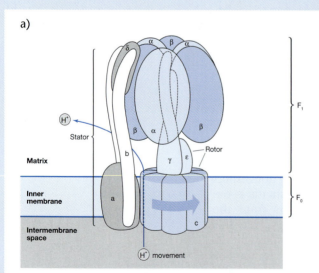

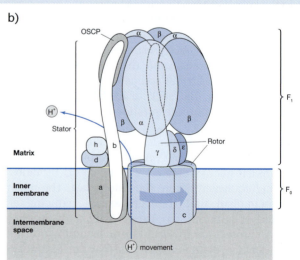

Box 6.6 Structure and mechanism of the mitochondrial ATP synthase (continued)

dimer is held in the membrane by attachment to the *a*-subunit. In mitochondria (e.g. bovine heart, Figure 1b) the α, β, γ, and δ-subunits are functionally equivalent to the α, β, γ, and ε-subunits of prokaryotes. The *c*-ring, together with the γ and δ-subunits, forms the rotor, while the mitochondrial ε-subunit is an additional component of the central stalk that has no equivalent in the bacterial protein. The stator of mitochondrial ATP synthase consists of a single *b*-subunit, while the second *b*-subunit of the prokaryotic enzyme is replaced by subunits *d* and *h* in the mitochondrial stator. As in the prokaryotic ATP synthase, the mitochondrial stator is attached to the membrane via association with the *a*-subunit, while the oligomycin sensitivity conferring protein (OSCP) (rather than the δ-subunit of prokaryotes) serves to attach the stator to the top of the $\alpha_3\beta_3$ cap. Genes encoding the mitochondrial α, β, γ, δ, and ε-subunits have been identified in arabidopsis. The remaining components have yet to be characterized and there does not appear to be a homologue of the *b*-subunit within the arabidopsis genome.

The relationship between the proton electrochemical gradient and the formation of ATP by the ATP synthase forms a central tenet of the chemiosmotic hypothesis (Box 6.5). There is widespread support for the rotary motor mechanism, where proton movement drives the rotation of components of the ATP synthase which causes a change in protein conformation that results in ATP synthesis. Paul D Boyer and John E Walker received the Nobel Prize for Chemistry in 1997 for their elucidation of the structure and the working mechanism, respectively, of the ATP synthase from bovine heart.

The rotary motor mechanism is considered to operate as follows. Protons are thought to move through a channel at the interface between the *a* and *c*-subunits of the *c*-ring rotor. This movement of protons drives the rotation of the *c*-ring and each of the *c*-subunits becomes protonated as it comes into contact with the *a*-subunit. As the *c*-ring rotates, the central shaft (γ, δ, and ε-subunits in mitochondrial ATP synthase) to which it is attached also rotates. The $\alpha_3\beta_3$ cap is prevented from rotating by its attachment to the stator that is held firm within the membrane. Consequently, the central shaft rotates within a static $\alpha_3\beta_3$ cap, causing a conformational change within each of the β-subunits as they come into contact with the shaft. In Boyer's original hypothesis, the binding change mechanism, he proposed that there were three different conformations of the catalytic sites within the β-subunits. In the open conformation, ADP and P_i bind to the catalytic site, while in the tight state, ATP is formed and then released when the conformation changes to the loose state. At the moment there is very little experimental evidence for the various states of the β-subunits. Since Boyer's original model, a number of other conformations of the β-subunit have been proposed (e.g. half-closed).

Nevertheless, all of these models remain speculative until techniques are developed that enable the detection of the fine structure of the catalytic sites within the β subunit.

It is estimated that one 360° rotation produces three ATPs, one at each catalytic site. The number of protons required for one rotation is thought to depend on the number of *c*-subunits present, as current models indicate that the passage of one proton results in axial shift of one *c*-subunit. In bovine mitochondrial ATP synthase, for example, there are 10 *c*-subunits within the ring, hence 10 H^+ are required per turn of the rotor. Therefore 10/3 = 3.33 H^+ are needed per ATP. In bacterial ATP synthase, there are nine to 12 *c*-subunits, hence the theoretical stoichiometry ranges between three and four H^+ per ATP. There is considerable debate as to whether ATP synthases that differ in the number of *c*-subunits really do differ in their H^+/ATP stoichiometry or whether adjustments occur within the rotor that result in a constant H^+/ATP stoichiometry. One such adjustment has been proposed in the elastic strain model where the energy of each *c*-subunit movement is stored as elastic energy within the flexible stator protein, until there is sufficient energy to move the central stalk through the angle required to produce one ATP molecule (i.e. through 120°). This model would allow for adjustments to be made to the stoichiometry of H^+/ATP as the stator stores and releases the energy generated within the *c* ring, irrespective of the number of *c* units within the ring.

Recent, highly innovative work has enabled measurement of the number of ATP molecules formed during a turn of the rotor. Genetically engineered, single molecules of F_1 (from a thermophilic *Bacillus* strain) were bound to the base of a tiny glass chamber via the $\alpha_3\beta_3$ cap. A magnetic bead was attached to the γ-subunit and a rotating magnetic field was applied in order to force the γ-subunit to rotate the molecule clockwise. When forced to turn in this direction, the enzyme formed ATP within the chamber. After the magnetic field was turned off, the F_1 molecule spontaneously rotated in an anticlockwise direction and hydrolyzed ATP. By measuring the number of rotations, along with the number of ATP molecules consumed, the coupling ratio, i.e. number of ATPs consumed per turn, was calculated to be 3.15. A maximum of three ATPs were synthesized for each clockwise turn of the F_1 molecule. Although this stoichiometry had been inferred from models (as discussed above), this was the first direct measurement that had ever been made. This remarkable experiment illustrates how a natural molecular motor may be manipulated artificially, opening up the field of nanotechnology. Current research is investigating the potential to use the ATP synthase molecule to move or manipulate fluids at the single molecule level, with the possibility of controlling chemical reactions and sorting and concentrating molecules on a minute scale.

Figure 6.18 Malate/oxaloacetate (OAA) shuttles transfer reducing equivalents between the chloroplast, cytosol, and mitochondrion. NADPH generated within the chloroplast may be reoxidized to NADP+ by NADP-dependent malate dehydrogenase (NADP-MDH) within the stroma, which converts OAA to malate. The malate is exported from the chloroplast, in exchange for OAA, on the dicarboxylate translocator. Once in the cytosol, malate may be transported into the mitochondria, in exchange for OAA export on the dicarboxylate translocator. Malate may then be oxidized to OAA, via NAD-dependent malate dehydrogenase (NAD-MDH), which results in the reduction of NAD+ to NADH. Thus, reducing equivalents have been transferred indirectly from the chloroplast (NADPH) into the mitochondria (NADH). The resulting NADH may be oxidized by the mitochondrial electron transport chain to produce ATP.

An alternative use of the malate released from the chloroplast into the cytosol is for it to be oxidized to OAA by NAD-MDH within the cytosol. This reaction generates cytosolic NADH, which may be used to reduce nitrate to nitrite (see Chapter 8) or oxidized by the external NADH dehydrogenase (NDex(NADH)), Figure 6.17). OAA may then enter the mitochondria, via the OAA translocator, where its reduction to malate provides a means of oxidizing NADH generated during glycine oxidation (via GDC). Hence, glycine oxidation may proceed independently of the mitochondrial electron transport chain. If, however, the NADH is oxidized by the electron transport chain, glycine oxidation will result in ATP synthesis.

Abbreviations: NADP-MDH, NADP-dependent malate dehydrogenase; NAD-MDH, NAD-dependent malate dehydrogenase; GDC, glycine decarboxylase complex.

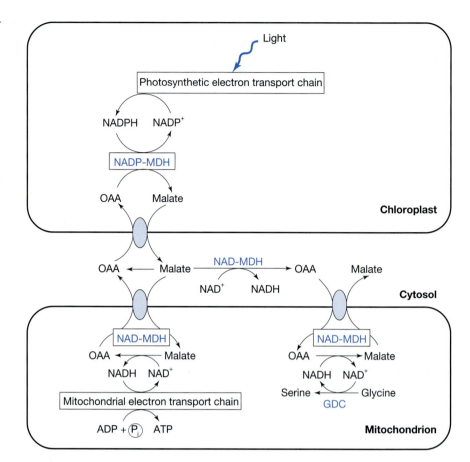

reactions proceed as far as citrate, which is presumably exported to the cytosol and used to generate carbon skeletons for nitrogen assimilation in the chloroplast (see Figure 6.13). These reactions within the chloroplast consume reduced ferredoxin (see Chapter 8) and further protect against over-reduction of the thylakoid.

The main substrate for mitochondria in photosynthetic cells in the light is the amino acid glycine, an intermediate of the photorespiratory pathway (see Chapter 5). Glycine is oxidized by the glycine decarboxylase complex (GDC) within the mitochondrial matrix. Experiments with isolated mitochondria indicate that glycine is oxidized in preference to TCA cycle substrates, further supporting the notion of limited TCA cycle flux in the light. Whether this preferential oxidation is due to structural associations between GDC and the electron transport chain (e.g. supercomplexes—see below), or whether it is due to some unidentified enzyme regulation, remains to be determined. NADH resulting from glycine oxidation may be oxidized indirectly, via substrate shuttles (e.g. OAA/malate exchange, see Figure 6.18), or directly, via the electron transport chain. In the latter case, there is some evidence that operation of the alternative oxidase and/or NDin(NADH) might enable glycine oxidation to proceed at rapid rates in the light. The recent production of transgenic potato plants with reduced GDC content provides further evidence of the interaction between glycine oxidation and respiration. The GDC-deficient plants were found to have an increased TCA cycle flux, suggesting that glycine oxidation normally inhibits TCA cycle activity in the light. This inhibition could arise from an increased NADH concentration, if glycine oxidation exceeds the capacity for NADH oxidation. Alternatively, the release of ammonium in the GDC reaction might inhibit PDC, thus reducing TCA cycle flux at the point of entry (see Figure 6.11).

Emerging research area into supercomplexes and metabolons

An emerging research area that is likely to alter our understanding of plant respiration is the identification of supercomplexes and metabolons. Proteomics analysis, together with conventional metabolic studies, have provided evidence of a structural association of some TCA cycle and glycolytic enzymes into large multi-protein complexes known as metabolons. While this information has come mostly from investigations of animals and yeast, recent evidence is also beginning to emerge from plant studies. Hexokinase was found long ago to be bound to mitochondria. A number of other glycolytic enzymes have now been found to associate with the outer membrane of arabidopsis mitochondria. It is suggested that such microcompartmentation provides a mechanism for glycolysis to channel pyruvate directly to the mitochondria while reducing competition for glycolytic intermediates with other cytosolic enzymes.

Supercomplexes between Complexes I, III, and IV have been identified in mitochondria from a range of organisms, including plants. These associations are thought to provide a means of controlling electron flow from one complex to another. There are conflicting reports of whether the alternative oxidase and the additional NAD(P)H dehydrogenases are also present in supercomplexes, and we await further developments in this emerging field.

Summary

Respiration in plant cells involves the combined operation of glycolysis, the OPPP, TCA cycle, and mitochondrial electron transport chain. Plant respiration serves both a bioenergetic (catabolic) and a biosynthetic function. It is also highly flexible, in responding to fluctuating environmental conditions as well as changes in photosynthetic and photorespiratory activity. Plant mitochondria possess additional respiratory complexes that provide an added flexibility in the coupling of respiration to ATP synthesis. This feature has been exploited by certain specialized plants as a means of generating heat, which releases volatile scents to attract pollinators. In most other plants, these additional electron transport complexes are thought to alleviate the build up of free radicals during exposure to a range of environmental stresses. In conclusion, what marks plant respiration out from that of other organisms is its flexible and versatile nature, as might be expected of autotrophic, sessile organisms that inhabit fluctuating and often unpredictable environments.

Further Reading

Glycolysis reviews

Givan CV (1999) Evolving concepts in plant glycolysis: two centuries of progress. *Biol. Rev.* 74, 277–309.

Plaxton WC (1996) The organization and regulation of plant glycolysis. *Annu. Rev. Pl. Biol.* 47, 185–214.

Both the above reviews are excellent summaries of the information available on plant glycolysis.

Sachs MM, Subbaiah CC & Saab IN (1996) Anaerobic gene expression and flooding tolerance in maize. *J. Exp. Bot.* 294, 1–15.

Useful review summarizing changes in gene expression in enzymes involved in glycolysis and related processes in response to anaerobic treatment.

TCA cycle discovery

Krebs HA & Johnson WA (1937) The role of citric acid in intermediate metabolism in animal tissues. *Enzymologia* 4, 148–156.

Krebs' first paper on the TCA cycle.

Oxidative pentose phosphate pathway

Kruger NJ & von Schaewen A (2003) The oxidative pentose phosphate pathway: structure and organization. *Curr. Opin. Pl. Biol.* 6, 236–246.

This paper considers the impact recent studies have had on the understanding of the structure and organization of OPPP. Includes useful up to date references and information on developments in this area.

Reviews of plant respiration

Fernie AR, Carrari F & Sweetlove LJ (2004) Respiratory metabolism: glycolysis, the TCA cycle and mitochondrial electron transport. *Curr. Opin. Pl. Biol.* 7, 254–261.

A brief review that focuses on specific aspects of respiration and draws information from molecular genetic studies.

Møller IM & Gardeström P (2007) Plant mitochondria- more active than ever! *Physiol. Plantarum* 129, 1–5.

An introduction to a Special Issue of this journal that contains a number of relevant reviews of plant respiration.

Mooney BP, Miernyk JA & Randall DD (2002) The complex fate of α-ketoacids. *Annu. Rev. Pl. Biol.* 53, 357–375.

A review of the α-ketoacid dehydrogenase complexes, including PDH and 2-oxoglutarate dehydrogenase.

Plaxton WC & Podestá FE (2006) The functional organisation and control of plant respiration. *Crit. Rev. Pl. Sci.* 25, 159–198.

Tovar-Méndez A, Miernyk JA & Randall DD (2003) Regulation of pyruvate dehydrogenase complex activity in plant cells. *Eur. J. Biochem.* 270, 1043–1049.

A review of the PDC and its regulation.

Interactions between respiration and photosynthesis

Nunes-Nesi A, Sweetlove LS & Fernie AR (2007) Operation and function of the tricarboxylic acid cycle in the illuminated leaf. *Physiol. Plantarum* 12, 45–56.

A thorough review of past and recent studies of the interaction between photosynthesis and respiration.

Raghavendra AS & Padmasree K (2003) Beneficial interactions of mitochondrial metabolism with photosynthetic carbon assimilation. *Trends Pl. Sci.* 8, 546–553.

This review considers the important role of mitochondrial metabolism in supporting photosynthesis.

Mitochondrial translocators

Laloi M (1999) Plant mitochondrial carriers: an overview. *Cell. Mol. Life Sci.* 56, 918–944.

A good review of the mitochondrial metabolite translocators.

ATP synthase

Berry RM (2005) ATP synthesis: the World's smallest wind-up toy. *Curr. Biol.* 15(10), R385–R387.

A brief overview of the ATPase with particular reference to the paper by Rondelez *et al.* (2005)

Capaldi RA & Aggeler R (2002) Mechanism of the F_1F_o-type ATP synthase, a biological rotary motor. *Trends Biochem. Sci.* 27, 154–160.

A review of the rotary motor hypothesis and some outstanding questions that remain about its operation.

Rondelez Y, Tresst G, Nakashima T, Kato-Yamada Y, Fujita H, Takeuchi S & Noji H (2005) Highly coupled ATP synthesis by F_1-ATPase single molecules. *Nature* 433, 773–777.

The paper that uses tiny magnets to rotate a single ATPase molecule.

Senior AE, Nadanaciva S & Weber J (2002) The molecular mechanism of ATP synthesis by F_1F_o-ATP synthase. *Biochim. Biophys. Acta* 1553, 188–211.

A very thorough review of the structure and properties of the ATP synthase together with a hypothesis of its molecular mechanism.

Alternative oxidase

Fiorani F, Umbach AL & Siedow JM (2005) The alternative oxidase of plant mitochondria is involved in the acclimation of shoot growth at low temperature. A study of Arabidopsis *AOX1a* trangenic plants. *Plant Physiol.* 139, 1795–1805.

This paper reports the effects of low temperature exposure on the physiology and biochemistry of transgenic plants with altered levels of alternative oxidase.

Gelhaye E, Rouhier N, Gérard J, Jolivet Y, Gualberto J, Navrot N, Ohlsson P-I, Wingsle G, Hirasawa M, Knaff DB, Wang H, Dizengremel P, Meyer Y & Jacquot J-P (2004) A specific form of thioredoxin *h* occurs in plant mitochondria and regulates the alternative oxidase. *Proc. Natl Acad. Sci. USA* 101, 14545–14550.

This paper provides evidence for the regulation of the alternative oxidase by a specific form of thiroedoxin.

Juszczuk IM & Rychter AM (2003) Alternative oxidase in higher plants. *Acta Biochim. Polon.* 50, 1257–1271.

A detailed account of the structure, function and regulation of the alternative oxidase.

Umbach AL, Fiorani F & Siedow JN (2005) Characterization of transformed arabidopsis with altered alternative oxidase levels and analysis of effects on reactive oxygen species in tissue. *Pl. Physiol.* 139, 1806–1820.

A further analysis of plants with altered levels of alternative oxidase that demonstrates a correlation between alternative oxidase content and the amelioration of ROS accumulation.

Other mitochondrial proteins

Balmer Y, Vensel WH, Tanaka CK, Hurkman WJ, Gelhaye E, Rouhier N, Jacquot JP, Manieri W, Schuurmann P, Droux M & Buchanan BB (2004) Thioredoxin links redox to the regulation of fundamental processes of plant mitochondria. *Proc. Natl Acad. Sci. USA* 101, 2642–2647.

Evidence for the involvement of thioredoxins in the regulation of mitochondrial metabolism.

Binder S, Knill T & Schuster J (2007) Branched-chain amino acid metabolism in higher plants. *Physiol. Plantarum* 129, 68–78.

Amino acid metabolism in mitochondria.

Eubel H, Heinemeyer J, Sunderhaus S & Braun H-P (2004) Respiratory chain supercomplexes in plant mitochondria. *Pl. Physiol. Biochem.* 42, 937–942.

Evidence for the existence of protein supercomplexes in plant mitochondria.

Millar AH, Heazlewood JL, Kristensen BK, Braun H-P & Møller I-M (2005) The plant mitochondrial proteome. *Trends Pl. Sci.* 10, 1360–1385.

A review of the current information obtained from proteomics analysis of mitochondria.

Synthesis and Mobilization of Storage and Structural Carbohydrates

7

Key concepts

- The carbohydrate source available to plants fluctuates as a result of the impact of variable light and environmental conditions on photosynthesis.

- Reduced carbon is transported from source to sink tissue in the form of sucrose.

- Breakdown of sucrose via sucrose synthase and invertase is essential to mobilize carbon resources.

- Starch, synthesized in the plastids of both source and sink tissues, is the principle storage carbohydrate in plants.

- Starch must be broken down into its component glucose units before it can be used as a source of energy and carbon skeletons.

- Carbohydrates are also stored as fructans, which are synthesized directly from sucrose and accumulate in the vacuole of photosynthetic and storage cells.

- Trehalose protects plants from stress and is synthesized in a wide range of plants.

- Plant cell walls contain large polysaccharides that are divided into two structural classes: microfibrils and matrix components.

- Cell wall polysaccharides are synthesized from ten different sugar molecules, that are derived from glucose.

- Cellulose is a straight chain polysaccharide and forms the cell wall microfibrils. It is made solely of glucose monomers and is synthesized on the outer face of the plasma membrane.

- Xyloglucans, xylans, and mannans are branched chain polysaccharides and are matrix components. They are made from a variety of sugar monomers and have a wide variety of structures.

- Pectins are acidic branched chain polysaccharides and are matrix components.

- Lignin is one of the matrix components of cell walls of vascular cells. It is hydrophobic and adds greatly to cell wall strength.

Role of carbohydrate metabolism in higher plants

Carbohydrates play a key role in living organisms (Box 7.1). As the main respiratory substrate they provide nearly all the cellular energy (see Chapter 6), and in sink tissues they are the starting point for biosynthesis of all cellular molecules. In plants, they also provide the important structural components of the cell walls leading to the form of the whole plant body. All organisms need to ensure a continuous supply of carbohydrates to their cells. These have to have a source, and the sources are usually discontinuous, with the result that mechanisms have evolved to cope with daily and seasonal variations in supply. For plants, such variations are centered on the availability of light for photosynthesis and the accompanying environmental conditions: availability of water and equable temperatures. Fluctuations in light levels occur during the day (e.g. variable cloud cover) and with the diurnal (day/night) cycle. Seasonal variations in water supply and temperature can restrict photosynthesis to particular times of the year, depending on latitude and elevation, and overlying vegetation canopies can change over the term of a growing season.

Box 7.1 The range of carbohydrates found in higher plants

Carbohydrates are polyhydroxy-aldehydes or ketones and have the basic composition $(CH_2O)_n$ and contain carbon, hydrogen, and oxygen in the ratio 1:2:1. Three carbon or triose sugars are the simplest carbohydrates and include glyceraldehyde and dihydroxyacetone.

Glyceraldehyde (Figure 1) is known as an aldo sugar with the carbonyl oxygen in glyceradehyde forming an aldehyde group (–CHO). Dihydroxyacetone (Figure 2), a structural isomer of glyceraldehyde, is known as a keto sugar with the carbonyl oxygen forming a ketone group (–C–CO–C–).

Carbohydrates with free aldehyde or ketone groups under alkaline conditions can reduce a cupric ion (Cu^{3+}) to a cuprous ion (Cu^{2+}) and are therefore also called reducing sugars. Because hydrolysis of glyceraldehyde and dihydroxyacetone yields products that are not carbohydrates, they are also described as monosaccharides. They are the units released from poly- or oligosaccharides by mild acid hydrolysis. Common monosaccharides include the four-carbon (tetrose) erythrose, the five-carbon (pentose) ribose, the six-carbon (hexose) glucose, and the seven-carbon (heptose) sedoheptulose. In higher plants most of these sugars exist in the phosphorylated state rather than as free sugars.

Triose sugars in the form of triose phosphate are important intermediates in photosynthesis (Chapter 5) and

respiration (Chapter 6). Erythrose, as erythrose-4-phosphate is an intermediate in photosynthesis (Chapter 5), some forms of respiration (Chapter 6) and is a precursor in a number of biosynthetic pathways. Pentoses are also important photosynthetic (Chapter 5) and respiratory (Chapter 6) intermediates and serve as structural elements in RNA, DNA, ATP, and nicotinamide and flavin nucleotides. Glucose, fructose, and other naturally occurring hexose sugars are represented as both the start and end product of respiration (Chapter 6) and photosynthesis (Chapter 5). They are also necessary for the synthesis of more complex carbohydrates. D-glucose, D-fructose, D-mannose, and D-galactose are the most common hexoses found in plants. Sedoheptulose 1,7-bisphosphate, a seven-carbon sugar is an intermediate in photosynthesis (Chapter 5). There is also an unusual eight-carbon sugar, octulose, found in some resurrection plants.

Individual identical or non-identical monosaccharides may link together to form oligosaccharides. The number of monosaccharides linked is identified by the type of sugar, disaccharide (2), trisaccharide (3), tetrasaccharide (4), and pentasaccharide (5).

High molecular weight polymers of monosaccharides are called polysaccharides. These usually consist of a single sugar, although polysaccharides of more than one monosaccharide species are known. Starch, fructans and cellulose are the predominant polysaccharides in higher plants.

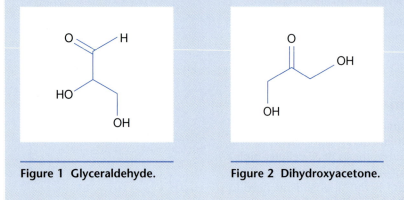

Figure 1 Glyceraldehyde.   **Figure 2 Dihydroxyacetone.**

During the day some photoassimilates are exported as sucrose from the leaf to other tissues of the plant and some temporarily accumulate in the leaf (Figure 7.1). In plants such as soybean, spinach, and tobacco, excess photoassimilates are stored mainly as starch in the chloroplast. In other plants, including the cereals wheat, barley, and oats, photoassimilates are temporarily stored as sucrose in vacuoles. In the vacuole the sucrose may be converted into fructans or hydrolyzed into glucose and fructose. Starch and sucrose synthesis are competing processes that each rely on the same pool of triose phosphates produced by photosynthesis. The role played by a variety of regulatory mechanisms via, for example, sucrose phosphate synthase, ADP-glucose pyrophosphorylase, and fructose 2,6-bisphosphate (F2,6BP), is an essential part of carbohydrate metabolism. The photoassimilates stored as starch or other carbohydrates must subsequently be broken down to their component monosaccharide units to be available to the plant to meet their respiratory needs. The focus in this chapter is on carbohydrate biosynthesis and breakdown in plants.

Sucrose is the major form of carbohydrate transported from source to sink tissue

In higher plants, disaccharides, made by joining two monosaccharides (Box 7.1), are used for short-term energy storage. Sucrose (α-D-glucopyranosyl-β-D-fructofuranoside) is the most ubiquitous and abundant disaccharide in plant tissues (Figure 7.2). It is particularly well-suited for storage and transport, being structurally very stable, highly soluble, and relatively inert in so far as it is not a reducing sugar. Glycosidic bonds join the carbonyl carbons of glucose and fructose and protect potentially reactive groups from oxidation (Figure 7.2). For example, the aldehyde group of sucrose can be spontaneously oxidized to a carboxyl group. This nonreducing configuration of sucrose is protected from nonspecific enzymatic attack making it structurally stable. Apart from possible osmotic effects sucrose does not affect most biological processes, even at high concentrations (487 g dissolve in 100 ml at 100°C, while 179 g dissolve in 100 ml at 0°C), and therefore provides a suitable form for carbon transport in plants. Although sucrose is the major form of carbohydrate translocated from source to sink tissues, it is not the only form. For example, many plants in the Rosaceae family synthesize and transport sorbitol, while Cucurbitaceae family plants transport the oligosaccharides raffinose and stachyose.

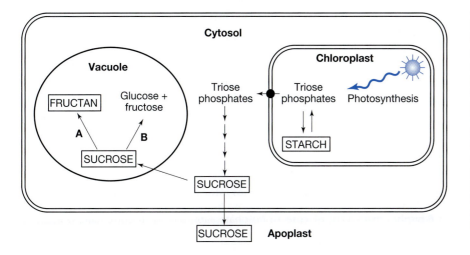

Figure 7.1 Carbohydrate metabolism in a plant cell. High photosynthetic activity is associated with high rates of carbon export, in the form of triose phosphates, from the chloroplast to the cytosol via a triose phosphate/phosphate translocator (black circle). This leads to an increase in available intermediates for sucrose synthesis. Once synthesized the sucrose is either moved to the vacuole for storage or the apoplast for export. In the vacuole the sucrose may be (A) converted into fructan by fructosyltransferases or (B) hydrolyzed to glucose and fructose by invertase.

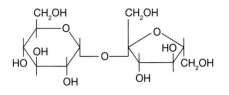

Figure 7.2 The structure of sucrose (glucose-(1α-2β)-fructose). The carbon (C-1) of glucose is joined to the carbon (C-2) of fructose by a hydrolytic bond that protects the reducing ends of both monomers.

Only plants and cyanobacteria are capable of sucrose biosynthesis. Although this suggests a possible link with the cyanobacterial ancestors of plastids and photosynthetic function, it also raises interesting questions as to why, in higher plants, sucrose biosynthesis takes place only in the cytosol. Photosynthesis in the chloroplast is linked to sucrose biosynthesis in the cytosol through the involvement of common sugar phosphates and translocators (see Chapter 2). The ability to synthesize sucrose is a widespread characteristic of higher plant cells; however, sucrose biosynthesis is at its highest in photosynthetic and gluconeogenic tissues such as post-germinating seeds and accounts for most of the carbon fixed in green leaves during photosynthesis.

Sucrose biosynthesis occurs in the cytosol utilizing triose phosphates produced in photosynthesis (Figure 7.3). Triose phosphates (for example, the isomers dihydroxyacetone phosphate, DHAP, and glyceraldehyde 3-phosphate; although DHAP predominates 22:1) are exported from the chloroplast to the cytosol via a phosphate (P_i) transporter located in the chloroplast envelope membrane (see Chapter 2). For every four molecules of DHAP exported from the chloroplast, four molecules of P_i will be imported (Figure 7.3). This gives rise to two molecules of fructose 1,6-bisphosphate (F1,6BP), which leads to the irreversible production of two molecules of fructose 6-phosphate (F6P) catalyzed by the enzyme fructose 1,6-bisphosphatase. One molecule of F6P is directly converted to sucrose 6-phosphate. The other molecule of F6P is converted to UDP-glucose, which, with F6P, is converted to a molecule of sucrose 6-phosphate. Sucrose 6-phosphate is then hydrolyzed to sucrose plus P_i. This latter step has a large negative free energy change ($G^{\circ\prime} = -18.4$ kJ mol^{-1}) compared with the small free energy changes of the other steps ($G^{\circ\prime} = -2.88$ kJ mol^{-1}; $G^{\circ\prime} = -1.46$ kJ mol^{-1}). This favors the formation of sucrose and essentially ensures that this sequence of reactions is irreversible. Recycling of P_i from sucrose biosynthesis to the chloroplast is also essential and release occurs at three steps during the overall pathway (see Figure 7.3).

Sucrose can be synthesized and degraded by both sucrose synthase and sucrose phosphate synthase (SPS). However, SPS appears to be the dominant enzyme for biosynthesis, with high activities in leaves and other sucrose synthesizing tissues. In contrast, sucrose synthase occurs primarily in nonphotosynthetic tissues and is involved in sucrose breakdown for the biosynthesis of starch, e.g. in amyloplasts of storage tissue such as potato tubers. It also plays a part in the biosynthesis of cellulose and callose where sucrose synthase is membrane bound rather than soluble (see section on cellulose synthase, page 226).

Sucrose phosphate synthase is an important control point in the sucrose biosynthetic pathway in plants

Glucose 6-phosphate (G6P) is an activator of SPS acting on an allosteric modifier site. In contrast, P_i is an inhibitor of SPS. G6P is in equilibrium with the substrate F6P via the enzyme hexose phosphate isomerase. At equilibrium the concentration of G6P:F6P is approximately at a ratio of 4:1. This means that a small change in the concentration of F6P results in a much greater change in

Figure 7.3 The pathway of sucrose biosynthesis. Abbreviations used: DHAP, dihydroxyacetone phosphate; F6P, fructose 6-phosphate; F1,6BP, fructose 1,6-bisphosphate; G1P, glucose 1-phosphate; G6P, glucose 6-phosphate; G3P, glyceraldehyde 3-phosphate; 3-PGA, 3-phosphoglycerate; Ru5P, ribulose 5-phosphate; RuBP, ribulose 1,5-bisphosphate.

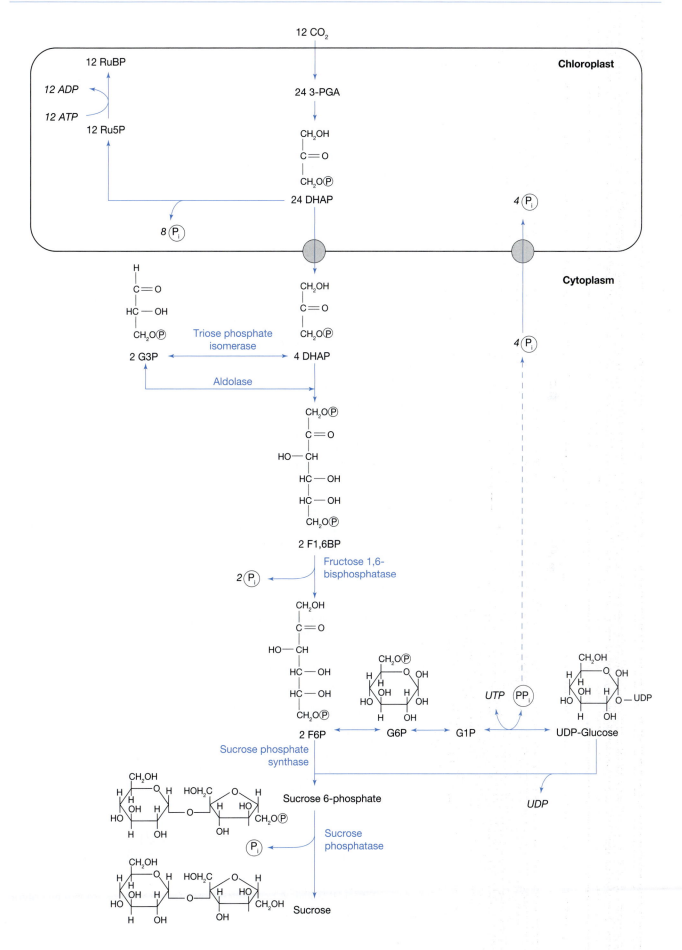

the G6P concentration. As G6P is an activator of SPS, the activity of SPS rapidly adjusts to the supply of substrates. More specifically, a decline in photosynthate supply leads to a drastic fall in G6P concentration and a reduction in SPS activity. Conversely a rise in photosynthetic activity leads to markedly increased levels of G6P and therefore SPS activity (see Fructose 2,6-bisphosphate section on page 204).

SPS activity can also be altered without a parallel change in the hexose phosphate concentration. For example, SPS mRNA and protein are controlled developmentally during leaf development and the transition from a sink (sucrose importing) to a source (sucrose exporting) leaf. A similar increase in SPS is seen during cold stress and in mature leaves in response to a range of factors including light levels and nitrogen nutrition.

Depending on the plant species, activation of SPS by protein phosphorylation may also occur in response to dark/light transitions (Box 7.2). For example, if a spinach leaf is moved from low to high light this leads to a rapid increase in net photosynthesis and the flux of carbon into sucrose. Within a few hours of transfer, the SPS protein (and V_{max} activity) remain constant, but the activation state of the enzyme increases. The biochemical mechanism involved in this increase is protein phosphorylation (Box 7.2). Over a longer period of time at higher light levels there is a gradual increase in SPS protein and mRNA. Both protein phosphorylation and the concentration of allosteric effectors contribute towards the daily variations in sucrose biosynthesis. This regulation of enzyme activity by covalent modification and by the control of SPS gene expression, functions in an integrated manner to provide short- and long-term control. Protein phosphorylation via 14-3-3 binding sites has also been implicated as being a means of regulating sucrose biosynthesis by modulating SPS and sucrose phosphatase activity (Box 7.3).

Sensing, signaling, and regulation of carbon metabolism by fructose 2,6-bisphosphate

Inorganic phosphate (P_i) is a key regulator of carbon metabolism in photosynthetic cells. P_i is required in the chloroplast stroma to support ATP synthesis, and it becomes incorporated into triose phosphates that are exported to the cytoplasm (Chapter 5). Sucrose is synthesized in the cytosol from the triose phosphates with the release of P_i (Figure 7.3), which is returned to the chloroplasts via phosphate translocators in the chloroplast envelope. Sucrose and starch synthesis are competitive processes relying on the same pool of triose phosphates produced by the tricarboxylic acid cycle. A balance is necessary to allow triose phosphates to be allocated between starch synthesis in the chloroplast and sucrose synthesis in the cytosol. To prevent inhibition of photosynthesis, which would result from either an excessive drain of metabolites from the chloroplast or from an inadequate release of P_i in the cytosol, these two processes must be closely co-ordinated. The mechanisms involved in simultaneously maintaining the amounts of stromal metabolites and P_i are complex. Many interacting factors contribute, and regulation occurs at the level of several enzymes and the metabolite F2,6BP.

Fructose 2,6-bisphosphate enables the cell to regulate the operation of multiple pathways of plant carbohydrate metabolism

F2,6BP was first discovered in liver and was found to activate the glycolytic enzyme ATP-phosphofructokinase. In contrast, plant phosphofructokinase is

Box 7.2 The role of protein phosphorylation in regulating sucrose phosphate synthase

The biochemical mechanism involved in light/dark regulation of sucrose phosphate synthase (SPS) has been demonstrated to be protein phosphorylation. This is enzyme modification by reversible covalent incorporation of phosphate and is a widespread phenomenon with important consequences for metabolic control *in vivo*. The phosphorylation reaction is catalyzed by an SPS protein kinase in an ATP-dependent manner, whereas the reverse reaction is catalyzed by SPS phosphoprotein phosphatase (Figure 1).

Protein phosphorylation results in covalent modification, where the enzyme is chemically changed so that its conformation is altered. This results in either activation or inactivation, depending on the enzyme. For SPS, the covalent phosphate incorporation was initially demonstrated experimentally by following the mechanism of radiolabel incorporation into the SPS protein subunit (Figure 2). Protein phosphorylation of SPS was further shown by experiments studying the phosphorylation of partially purified SPS protein *in vitro* using [γ^{32}P] ATP

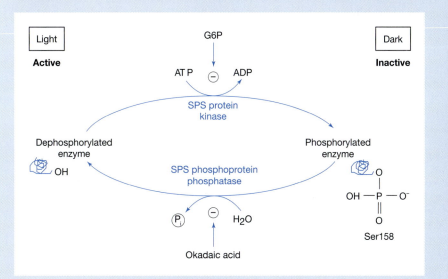

Figure 1 Protein phosphorylation as a means of regulating SPS in response to light/dark transitions. In the case of spinach SPS Ser-158 is phosphorylated by SPS protein kinase leading to a conformational change of the enzyme and its dark inactivation. The SPS protein kinase is inhibited by glucose 6-phosphate (G6P). In the light SPS is dephosphorylated by SPS phosphoprotein phosphatase leading to its activation. SPS phosphoprotein phosphatase is inhibited by okadaic acid, a type 2A protein phosphatase inhibitor.

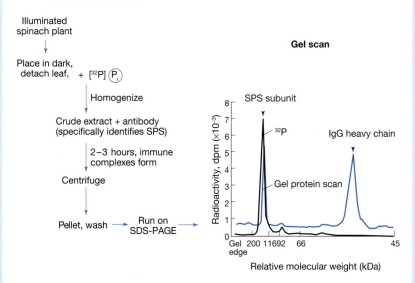

Figure 2 Demonstration of phosphorylation of SPS protein. A detached spinach leaf was placed in the dark and incubated with radioactive P$_i$ for 3 hours. The leaf was then homogenized and incubated with an antibody that specifically recognizes SPS. The antibody bound to any SPS protein to form immune complexes. The immune complexes were pelleted by centrifugation, resuspended, and then separated by sodium dodecyl sulfate–polyacrylamide gel electrophoresis. Following electrophoresis and staining, the gel was scanned for protein and radioactivity (gel scan). A protein peak (blue line) representing the SPS subunit coincided with a radioactive peak (black line). This suggested that radioactive Pi had been incorporated into the SPS subunit during the incubation of the leaf in the dark with radioactive P$_i$.

continued

Box 7.2 The role of protein phosphorylation in regulating sucrose phosphate synthase (continued)

(Figure 3). All the data obtained to date are consistent with the model of protein phosphorylation, with ATP catalyzing the conversion of the dephosphorylated active enzyme to the phosphorylated inactive enzyme, and accordingly radioactive phosphate becoming incorporated into the SPS protein.

In spinach, a particular serine residue (Ser-158) within the SPS protein is involved in light–dark regulation. The protein kinase that inactivates SPS and specifically phosphorylates Ser-158 *in vitro* has been partially purified and characterized from spinach leaves. The activity of SPS kinase is inhibited specifically by glucose 6-phosphate. The activation of SPS that occurs in a dark/light transition involves a type 2A protein phosphatase that targets the regulatory phosphorylation site. The light activation of SPS can be blocked by pre-treating leaves with a known inhibitor of type 2A protein phosphatase, okadaic acid, resulting in a decrease in the rate of sucrose biosynthesis *in situ*, thus demonstrating the importance of dephosphorylation/activation of SPS.

Protein phosphorylation of SPS is a very elegant control mechanism that has now been shown to occur in many plant species at several serine residues. It accounts not only for the activation of SPS by light, but also for other regulatory factors such as osmotic stress. The metabolic response of higher plants to osmotic stress is extremely complex. Carbon partitioning towards sucrose is increased in numerous higher plant species in response to osmotic stress. An important component of the response is thought to be the reduction in cellular water potential by osmolyte synthesis. This facilitates an increase in water uptake from the environment. If excised spinach leaves are subjected to water stress, there is a positive correlation between sucrose and SPS activity. Therefore osmotic stress leads to a higher new sucrose pool. The significance of this is that an elevated sucrose pool may act as an osmoprotector. It has been suggested that this may be due to an increased activation state of SPS. The impact of SPS activation is an increase in capacity for sucrose biosynthesis *in vivo*. ^{32}P-labeling patterns of tryptic digested ^{32}P-phosphopeptides were derived from SPS immunopurified from osmotically stressed leaves fed [^{32}P] P$_i$. The results showed there to be prominent labeling of the serine residue Ser-424 in SPS. Subsequently, a distinct kinase that phosphorylates SPS under osmotic stress has been identified. The activation of the kinase is essential and also influenced by several other factors, including an increase in cytosolic Ca^{2+} and a decrease in cytosolic pH.

It is now thought that, depending on the environmental conditions, SPS may be subject to multiple levels of regulation by phosphorylation on different serine residues. However, the true complexity of this has yet to be fully characterized.

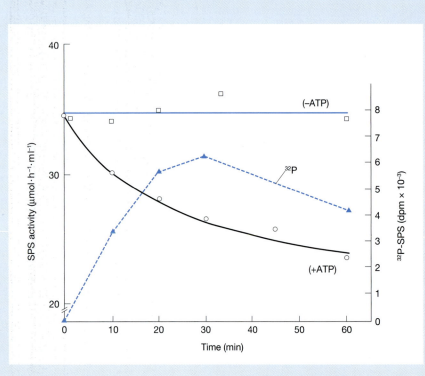

Figure 3 Purified SPS protein was incubated in the presence (+ATP, black solid line) or absence (–ATP, blue solid line) of radioactive ATP for different lengths of time. The SPS activity (solid lines) and incorporation of ^{32}P into the SPS protein (dashed line) was determined. The activity of SPS decreases with longer incubations with ATP. This inactivation of SPS is paralleled by the incorporation of radioactive ATP into the SPS protein. (Figures 2 and 3 based on figures 1 and 2 from *Archives of Biochemistry and Biophysics*, 270, 681–690, Protein phosphorylation as a mechanism for regulation of spinach leaf sucrose-phosphate synthase activity by Huber JLA, Huber SC and Neilsen TH.)

Box 7.3 Serine-229 of SPS – binding site of 14-3-3 proteins?

14-3-3 proteins are a highly conserved protein family with a central regulatory role in plant, fungal, and mammalian cells. They bind to phosphopeptide motifs in diverse target proteins and through these interactions 14-3-3 proteins have been reported to:

- modulate activities of enzymes

- regulate subcellular location of targets

- act as adaptor proteins mediating interaction between components of signal transduction pathways.

14-3-3 proteins co-purify with sucrose phosphate synthase (SPS) in a Mg^{2+}-dependent manner. There is a 14-3-3 protein-binding site (Ser-229) present in nearly all known SPSs. It has been speculated that 14-3-3 proteins may interact with SPS resulting in its inactivation. Parallel changes in expression of various sucrose phosphatase (SPP) and SPS genes have been reported, as has an association between the two enzymes (SPS and SPP) in rice. Current speculation is that SPS has an SPP-like domain that binds to SPP via 14-3-3 proteins. Such an interaction could be important for channeling metabolites between the proteins and in controlling sucrose biosynthesis.

not affected by F2,6BP. In plant tissue, F2,6BP is present at very low concentrations (1–10 μM) and can be rapidly adjusted five- to 10-fold. F2,6BP is synthesized and degraded in order to act as a regulatory signal. Where the photosynthetic rate is high, as a result of, for example, increased light intensity, the concentration of F2,6BP is lowered. This leads to a feedforward mechanism (see below), which co-ordinates sucrose biosynthesis in the cytosol with CO_2 fixation in the chloroplasts. In contrast, when the photosynthetic rate is low, the concentration of F2,6BP increases leading to a feedback mechanism (see below). With both the feedforward or feedback mechanism, regulation is through the level of F2,6BP. Critical to this regulation is consideration of how F2,6BP levels are regulated by synthesis or degradation.

Fructose 6-phosphate 2 kinase (F6P2K) catalyzes the ATP-dependent synthesis of F2,6BP from F6P (Figure 7.4). F6P2K is activated by its substrate F6P at an allosteric site, and also by P_i. F6P2K is inhibited by the three-carbon sugar phosphates, 3-phosphoglycerate and DHAP (Figure 7.5). The activity of F6P2K is also regulated by either the P_i:3-phosphoglycerate ratio or the P_i:DHAP ratio.

Fructose 2,6-bisphosphate is degraded by fructose 2,6-bisphosphatase (F2,6BPase) (Figure 7.6). F2,6BPase is inhibited by P_i and also by the product of the reaction, F6P.

Figure 7.5 summarizes the various activators and inhibitors of F2,6BP synthesis and degradation. Although a number of enzymes are either activated or inhibited by F2,6BP, the best understood effect involves controlling carbon flow from F1,6BP to F6P in a photosynthesizing leaf. This is achieved by regulating two enzymes that catalyze the interconversion of F6P to F1,6BP, fructose 1,6-bisphosphatase (F1,6BPase), and pyrophosphate F6P-1 phosphotransferase (PFP).

There are two forms of F1,6BPase, a chloroplast form involved in the tricarboxylic acid cycle (see Chapter 5) and a cytosolic form involved in sucrose synthesis (Figure 7.3). It is this latter cytosolic form that is inhibited by F2,6BP decreasing the substrate affinity 100-fold. PFP is always subject to activation by F2,6BP, though the extent of the activation varies, and is a result of a combined effect of F2,6BP on increasing the V_{max} and lowering the K_m of the enzyme.

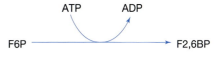

Figure 7.4 Fructose 6-phosphate 2 kinase (F6P2K) enzyme reaction.

Figure 7.5 Model summarizing the synthesis and degradation of fructose 2,6-bisphosphate (F2,6BP), the factors regulating this process and the influence of F2,6BP on fructose 1,6-bisphosphatase (F1,6BPase) and pyrophosphate F6P-1 phosphotransferase (PFP) activity. The conversion of F2,6BP from fructose 6-phosphate (F6P) is catalyzed by the enzyme fructose 6-phosphate 2 kinase (F6P2K). F6P2K is activated (+ve) by F6P and Pi and inhibited (–ve) by dihydroxyacetone phosphate (DHAP) and 3-phosphoglycerate (3-PGA). F2,6BP is degraded to F6P by the enzyme fructose 2,6-bisphosphatase (F2,6BPase). F2,6BPase is inhibited (–ve) by F6P and P_i. An increase in the concentration of F2,6BP activates (+ve) PFP and reduces the activity (–ve) of F1,6BPase. This regulation of the concentration of F2,6BP leads to a feedforward and feedback mechanism of regulation (see text for more details).

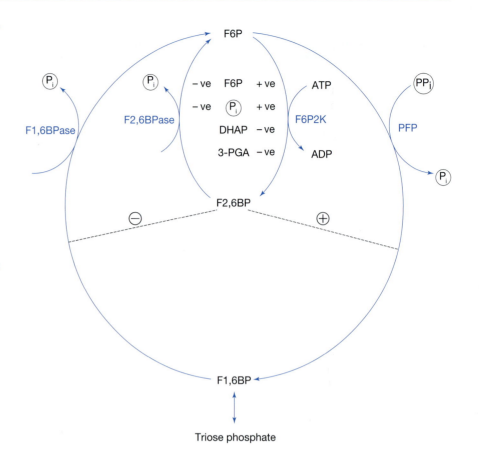

Figure 7.6 Fructose 2,6-bisphosphatase (F2,6BPase) enzyme reaction.

Fructose 2,6-bisphosphate as a regulatory link between the chloroplast and the cytosol

F2,6BP serves as a regulatory link between chloroplasts and cytosol via two mechanisms known as a feedforward and a feedback. As the rate of photosynthesis increases, in response to rising light intensity or increased CO_2 concentrations, there is an increase in the concentration of the immediate products of photosynthesis, 3-phosphoglycerate and triose phosphates, particularly DHAP. The concentration of DHAP changes with photosynthetic rate, to a greater extent than other metabolites. DHAP is exported from the chloroplast to the cytosol. An increase in the DHAP concentration inhibits F6P2K and therefore lowers the concentration of F2,6BP. The lowering of the concentration of F2,6BP relieves the inhibition of the F1,6BPase, which increases its activity. Also the lowering of the F2,6BP concentration reduces the activation of PFP. The activation of F1,6BPase results in an increased production of F6P and other hexose phosphates, including G6P. G6P in turn activates SPS. This is a feedforward mechanism, which results in an enhanced rate of sucrose biosynthesis and involves the co-ordination of the enzymes SPS and F1,6BPase. This feedforward mechanism co-ordinates sucrose biosynthesis in the cytosol with the rate of CO_2 fixation in the chloroplast (Figure 7.5).

Sucrose accumulates in leaves when photosynthesis is faster than the rate of export into the phloem. However, there comes a point where sucrose accumulation slows down or stops. At this point, surplus photosynthate is retained in the chloroplast and converted to starch. Sucrose accumulation results in the deactivation of SPS (probably by covalent modification). A decrease in SPS activity results in an accumulation of F6P, which is an activator of F6P2K and also an inhibitor of F2,6BPase. This leads to an increase in the concentration of F2,6BP, resulting in inhibition of F1,6BPase. This in turn leads to an increase in the concentration of triose phosphates in the cytosol and therefore restricts recycling of P_i to the chloroplast and leads to an inhibition of DHAP export. This feedback mechanism results in a decrease in the synthesis and accumulation of sucrose in the cytosol and an increase in starch formation in the chloroplasts (Figure 7.5).

Sucrose breakdown occurs via sucrose synthase and invertase

Sucrose can be exported from its site of formation or other source tissue to sink tissues such as areas of growth, roots, tubers, or fruits where it can then be broken down to mobilize carbon resources (Box 7.4). The cleaving of sucrose in multicellular plants is essential for the allocation of carbon resources. Two different pathways mediate sucrose breakdown *in vivo*, one involves sucrose synthase and the other is the enzyme invertase (Figure 7.7). Although sucrose synthase catalyzes a readily reversible reaction in the cytosol its primary role is believed to be degradative in producing UDP-glucose and fructose. Invertase is typically present as two isoenzymes, an acid invertase and an alkaline invertase, which function in different cellular locations to catalyze an irreversible hydrolysis of sucrose to glucose and fructose. The former has a pH optimum of 5.0 and is found in the vacuole and cell wall. The latter has a pH optimum of 7.5 and is found in the cytosol. Multiple isoforms of both sucrose synthase and invertase exist often with specific functions (Box 7.5). The expression of these enzymes is tightly controlled and has an impact on plant development (Box 7.5).

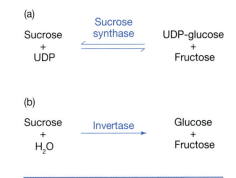

Figure 7.7 Sucrose degradation enzyme pathways catalyzed by (a) sucrose synthase and (b) invertase.

Box 7.4 The phloem tissue connects the source and sink tissues

Elongated single cell sieve elements joined by a sieve plate, consisting of diagonal cell walls perforated by pores, make a column of cells forming a sieve tube. These form the conducting system of the phloem. Sieve elements are, in the angiosperms, surrounded by companion cells connected by many plasmodesmata. Photoassimilates diffuse via the plasmodesmata to phloem parenchyma cells and are then transported symplastically via the plasmodesmata to the sieve tubes (Figure 1). Alternatively, photoassimilates are transported to an extracellular compartment, the apoplast. Because the concentration of sucrose is higher in the source cell than the apoplast, this export involves no energy expenditure.

Transport of sucrose from the apoplast to the companion cells proceeds via proton symport and is driven by a proton gradient between the apoplast and the interior of the sieve tubes (Figure 1). This is generated by a plasma membrane H^+ pumping ATPase, with mitochondrial ATP providing the necessary ATP. Where the H^+ sucrose translocator is localized in the plasma membrane of companion cells together with a large number of mitochondria, phloem loading occurs via the companion cells. In some species, for example potato leaves, a H^+ sucrose translocator has been identified in the phloem membrane of the sieve elements. The substrates for mitochondrial respiration are most

likely provided by degradation of sucrose via sucrose synthase followed by glycolytic metabolism of the hexose phosphates generated.

Depending on growth conditions and the plant, the concentration of sucrose in the phloem sap is 0.6–1.5 mol l^{-1}. Transport of sucrose can occur at rates of 30–150 cm h^{-1} and is driven by mass flow as a result of the pumping of sucrose into sieve tubes and its subsequent withdrawal at sites of consumption. Mass flow is driven by the many osmotic gradients present in a plant. Many substances present in the phloem, including macromolecules, are carried along by the pressure of the mass flow. Phloem unloading drives the direction of mass flow. Depending on needs, phloem transport may occur in an upward or downward direction towards growing shoots and flowers or the roots and storage organs, respectively. Phloem unloading at the sink is affected, at least in part, by the type of sink organ. That is, sucrose reaches the cells of vegetative tissue, for example roots or growing shoots undergoing symplastic unloading, directly from sieve elements via the plasmodesmata. While in storage tissue sucrose unloading is generally apoplastic being first transported from sieve tubes to an extracellular compartment and then taken up into the cells of the sink organ storage tissue.

continued

Box 7.4 The phloem tissue connects the source and sink tissues (continued)

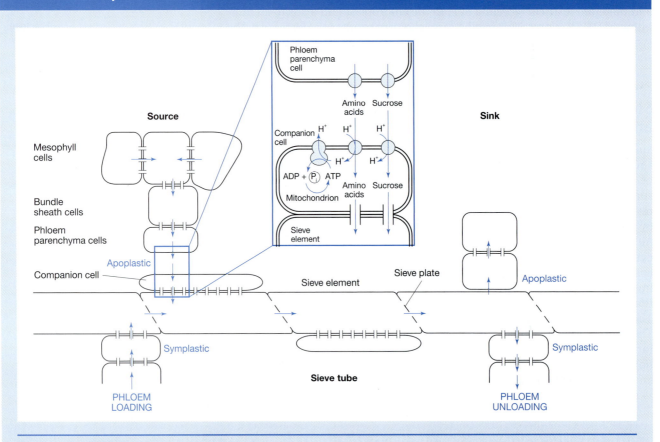

Figure 1 Apoplastic and symplastic route of sieve tube loading and unloading. Inset shows apoplast loading in more detail, with photoassimilates transferred from phloem parenchyma cells driven by a proton gradient between the apoplast and the interior of the sieve tubes.

Box 7.5 The importance of multiple forms of invertase and sucrose synthase

Biochemically distinct isoforms of invertase are found in the cytoplasm (neutral or alkaline invertase), the vacuole (vacuolar invertase), and the extracellular space (cell wall invertase; Figure 1). The corresponding genes are temporally and spatially expressed during plant development and also regulated by a number of environmental factors (Figure 2). Each isoform is thought to have a distinct metabolic role (Table 1). Vacuolar and cell wall invertases, with their acidic pH optima are also known as acid invertases, and as soluble and insoluble acid invertases, respectively. Plant acid invertases share a few highly conserved amino acid motifs and, based on comparisons of amino

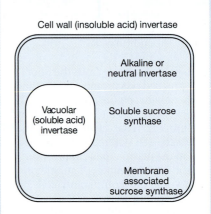

Figure 1 Invertase and sucrose synthase exist in several isoforms that have different biochemical properties and accumulate in different subcellular compartments. Invertase is found in the cytoplasm, vacuole, or extracellular space. Sucrose synthase is partially plasma membrane-associated, with the soluble form being phosphorylated on serine residues. Dephosphorylation increases the surface hydrophobicity of the polypeptide and, thereby, promotes its membrane association.

Box 7.5 The importance of multiple forms of invertase and sucrose synthase (continued)

acid sequence homo-logy, are related to yeast and bacterial invertases. In contrast, plant neutral or alkaline invertase sequences are different from each other. Sequence homologs have only been found in photosynthetic bacteria. Cleavage of sucrose by invertase and the generation of hexoses is generally correlated with an increase in respiration, growth, and cell expansion (Figure 2). The vacuolar invertase activity increases first and favors cell expansion. Later the cell wall invertase activities rise, often coinciding with increased expression of hexose transporters.

At least two isoforms of cytosolically located sucrose synthase (SuSy) have been found in most plant species (Figure 1). These isoforms are biochemically similar and have highly homologous amino acid sequences. However, the regulation of their genes is markedly different. Distinct developmental and organ-specific expression patterns of the SuSy isoforms have been found. Depending on their phosphorylation status SuSy may be soluble or tightly attached to the cellulose synthase complex at the plasma membrane or the actin cytoskeleton (Figure 1). SuSy activity becomes increasingly important as cell division and expansion are replaced by a transition towards storage and maturation. SuSy directs carbon to an ATP conserving respiratory pathway, which is advantageous in many tissues. Cleavage of sucrose by SuSy is also usually associated with anabolic processes, where UDP-glucose is the precursor of various compounds (Figure 2). However, SuSy has also been found in the sieve tube–companion cell complex of source leaves where it has been suggested to have a role in the catabolism of sucrose to generate ATP for phloem loading.

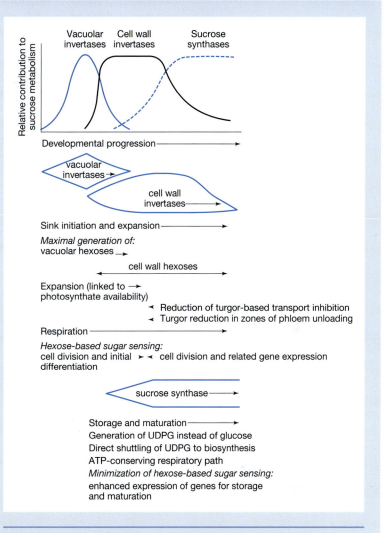

Figure 2 A generalized developmental profile showing the contributions of sucrose-cleaving enzymes invertase and sucrose synthase (SuSY) during development. Abbreviation: UDPG, UDP-glucose.

Table 1

Sucrose degradation isoform	Function
Cytosolic invertase	Channeling sucrose into metabolism (catabolism)
Vacuolar invertase	Osmoregulation and cell enlargement Control of sugar composition in fruits and storage organs Response to cold (cold sweetening)
Cell wall invertase	Sucrose partitioning between source and sink organs Response to wounding and infection Control of cell differentiation and plant development
Sucrose synthase	Channeling of sucrose into metabolism (anabolism) Sucrose partitioning between source and sink organs Response to anaerobiosis and cold

The precise contribution of invertase and sucrose synthase towards sucrose breakdown is not fully known. Both enzymes are often present in the same tissue. In some tissues, for example potato tubers, it is only sucrose synthase that is present in high enough amounts to facilitate sucrose breakdown. However, generally the capacity of both enzymes in the same tissue is sufficient to be of functional significance. The contribution of sucrose synthase and/or invertase may depend on whether the sucrose enters the cell via symplastic transport through the plasmodesmata or via an apoplastic pathway (Box 7.4 and Figure 7.8). In the former case, the sucrose is then broken down in the cytosol or sequestered in the vacuole thereby maintaining passive import down a concentration gradient. In the latter case sucrose crosses the plasma membrane, often against a concentration gradient. This is important in developing embryos, where there are no plasmodesmatal connections to the maternal plant tissues. Once transported into these cells the sucrose may remain unchanged or be converted, via an invertase located in the cell wall, to glucose and fructose. In many plants there is an inverse relationship between the acid invertase and sucrose content. This reflects the fact that sucrose stored in the vacuole is the second major source of sucrose for metabolism. Although sucrose may move out of the vacuole, its most likely route of mobilization is through hydrolysis by vacuolar acid invertase resulting in the release of hexoses into the cytosol. For example, in beetroot, acid invertase and sucrose are both essentially confined to the vacuole, a decrease in sucrose during aging is associated with an increased vacuolar invertase activity. In contrast sucrose entering the cytosol is mobilized by either alkaline invertase or sucrose synthase.

There are significant differences in energy status of the two sucrose degrading enzymes in terms of their products and in relation to the cytosolic hexose pool and the initiation of hexose-based sugar signals. Invertase produces free hexoses, which must subsequently be phosphorylated at the expense of ATP. Sucrose synthase produces UDP-glucose, which can react with PP_i to give glucose 1-phosphate and UTP. By combining with UDP-glucose pyrophosphorylase sucrose synthase provides an ATP-independent pathway for hexose

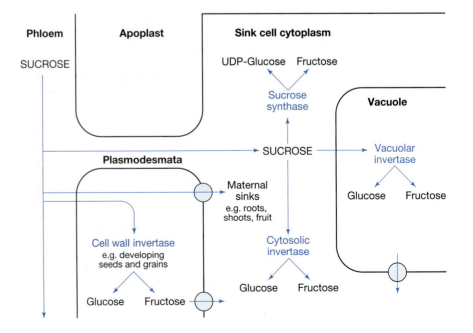

Figure 7.8 The movement of sucrose, its site of cleavage, and the enzyme involved depends on whether the sucrose enters the cell via a symplastic or apoplastic pathway. When sucrose enters the cell symplastically, via the plasmodesmata, it is either broken down by cytosolic invertase or sucrose synthase or sequestered in the vacuole where it may be broken down by vacuolar invertase. Whereas when entering via the apoplastic pathway sucrose is either broken down by cell wall invertases or directly transported into the cytoplasm via sucrose transporters. Hexoses produced are transported into the sink cells by hexose transporters and sucrose may be resynthesized for distribution by sucrose phosphate synthase.

phosphorylation. Invertase catalyzes an irreversible step and is only involved in sucrose degradation, producing glucose and fructose and therefore forming twice as many hexoses. This is important as hexose signals may regulate genes for many processes at the transcriptional and post-transcriptional levels (see Claeyssen and Rivoal (2007) in the Further Reading section).

Starch is the principal storage carbohydrate in plants

When the export of sucrose from the cell cannot keep pace with photosynthesis, so that sucrose begins to accumulate, the tricarboxylic acid cycle (Chapter 5) feeds carbon into starch synthesis. Starch builds up within chloroplasts during the day when photosynthesis exceeds the combined rates of sucrose accumulation, translocation, and respiration. Starch is easily mobilized into glucose units, providing a major metabolic substrate to support both high rates of metabolism and respiration, plus carbohydrates for export. For example, during the night, starch is rapidly degraded to form the readily transportable sucrose. In the daylight hours starch granules are synthesized and stored temporarily in the leaf, where they will grow during the day and become degraded during the following night. This is known as transitory starch because it does not persist for more than 24 hours. Alternatively reserve starch is deposited for longer periods in the amyloplasts of storage tissues such as tubers, or seeds.

Starch is a major product of photosynthesis with up to 30% of the CO_2 fixed by leaves being incorporated into starch (an annual production of 2850 million tonnes). Starch is widely distributed in plant tissues and, in most higher plants, it is the dominant plant storage polysaccharide, representing 65–80% of the dry weight in cereal grains and potato tubers. Because glucose is relatively unstable it is not suitable as a storage carbohydrate. Starch is a simple polymer of glucose molecules that are linked together in two different forms, amylose and amylopectin, providing a means for storing large amounts of carbohydrate. In some cases there is also a particularly highly branched starch present, phytoglycogen.

Amylose is essentially a linear molecule of 600–3000 1,4-α-glucosyl residues, with an occasional 1,6-α-glucosyl branchpoint occurring every 1000 residues (Figure 7.9). Amylopectin is usually much larger, 200–400 nm long and 15 nm wide, and in effect consists of 6000–60,000 1,4-α–glucosyl residues, with approximately 5% of the glucose units joined by 1,6-α–glucosyl linkages giving a branching point every 20–26 units and a single free reducing end (Figure 7.9). Amylose and amylopectin are assembled together to form a complex structure known as a semicrystalline starch granule (Figure 7.10). The ratio of amylose to amylopectin by weight is usually about 1:3. However, there is much variation in this ratio, in the molecular weight of constituent polysaccharides and in the degree and distribution of branching within the amylopectin component.

Starch synthesis occurs in plastids of both source and sink tissues

When photosynthesis is faster than the rate of export of sucrose from the cell, excess fixed carbon enters the stromal hexose phosphate pool before being converted to starch and stored in the chloroplast. In the case of the nonphotosynthetic plastids, the amyloplast, carbon is either imported from the cytosol as hexose (or hexose phosphate) molecules or, as in the case of some cereals, as ADP-glucose (Figure 7.11).

Figure 7.9 The structure of starch.
(a) The glucose molecules in starch are connected by (α-1,4) and (α-1,6) glycosidic bonds to form amylose and amylopectin. (b) Amylose is predominantly an unbranched chain of approximately 1000 glucose residues (dotted line). (c) The polyglucan chains in amylopectin have a branchpoint every 20–25 glucose residues. The glucose residue at the start of the chain (dark blue) has a reducing group. The groups at the end of the branches (light blue) act as acceptors for the addition by starch synthase of new glucose residues.

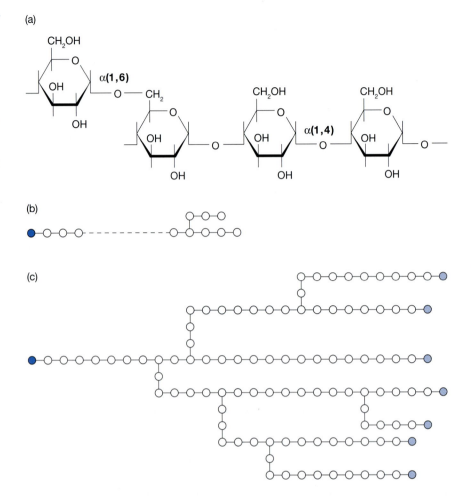

Figure 7.10 Structure of starch granule. (a) Where the amylopectin molecule (see Figure 7.9) is arranged in alternate layers representing the amorphous and semicrystalline zones. These pairs of zones make a growth ring. The amorphous zone contains amylopectin in a less-ordered structure with amylose. (b) A section through a starch granule showing the alternating amorphous and semi-crystalline layers. (c) Clusters of parallel α-1,4-linked glucan chains are packed together in a helical formation to make up the crystalline lamellae. At the base of these lamellae, α-1,6-branchpoints form the amorphous lamellae, joining the crystalline lamellae together. (From *Current Opinion in Plant Biology*, Volume 7, Number 2, "Improving starch for food and industrial applications", 2005. Reprinted with permission from Elsevier.)

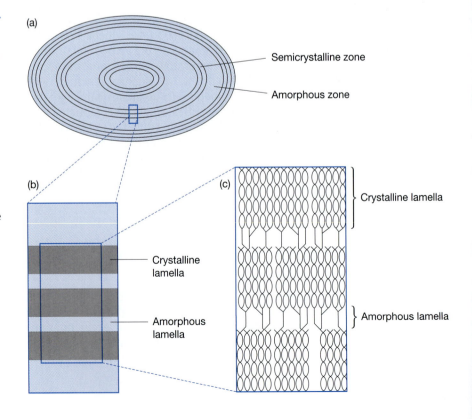

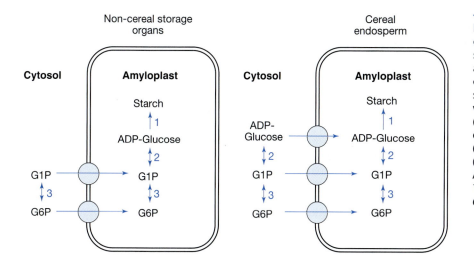

Figure 7.11 Metabolism of potential cytosolic precursors for starch synthesis in the cytosol and amyloplast stroma of non-cereal and cereal cells. The pathway of starch synthesis follows a similar pathway within either plant type. Enzymes: (1) starch synthase, starch branching enzyme, starch debranching enzyme; (2) ADP-glucose pyrophosphorylase; (3) phosphoglucomutase. Abbreviations: G1P, glucose 1-phosphate; G6P, glucose 6-phosphate.

The general pathway of starch synthesis from hexose phosphates for both chloroplasts and amyloplasts is summarized in Figure 7.12. ADP-glucose pyrophosphorylase (AGPase) catalyzes the conversion of glucose 1-phosphate and ATP to ADP-glucose (Figure 7.13). AGPase from all plant sources is heterotetrameric in structure consisting of two types of nuclear-encoded subunits (α_2, β_2) referred to as the small and large subunit. Consistent with a role in the control of flux of carbon into starch AGPase is regulated by key metabolites within the pathway of carbon assimilation. However, the nature of this regulation varies according to species (Box 7.6) and location within the plant (Box 7.7). Starch synthase subsequently transfers the glucosyl unit from ADP-glucose to the nonreducing end of a γ-glucose primer forming an additional 1,4-α-glucosidic bond with the release of ADP (Figure 7.13). Multiple isoforms of starch synthase exist, which are either exclusively bound to the starch granule, granule bound starch synthase, or both granule bound and soluble. Available evidence suggests that isoforms of starch synthase have discrete roles in the synthesis of amylose and amylopectin. Starch branching enzyme (SBE) hydrolyzes a 1,4-α-glucosyl bond and transfers the resulting short oligosaccharide to a primary hydroxyl group in a similar glucose chain to introduce the 1,6-α-glucosyl branchpoints of amylopectin (Figure 7.9) There are two classes of SBE, SBEI and SBEII, which differ in terms of the lengths of chains transferred *in vitro*. SBEII transfers shorter chains than SBEI. The temporal and spatial pattern of expression varies. In addition SBEI and II have recently been shown to be regulated differently by protein phosphorylation, though the full implications of this in terms of starch synthesis have yet to be elucidated.

Much understanding of the starch synthetic pathway has been gained from studies of available mutants (Box 7.8).

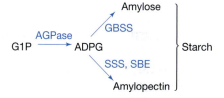

Figure 7.12 The general pathway of starch synthesis. Abbreviations used: ADPG, ADP-glucose; AGPase, ADP-glucose pyrophosphorylase; SBE, starch branching enzyme; GBSS, granule bound starch synthase; G1P, glucose 1-phosphate; SSS, soluble starch synthase.

(a)   G1P + ATP $\xrightleftharpoons{\text{AGPase}}$ ADP + PP$_i$

(b)   Glucan$_{(n)}$ + ADP $\xrightleftharpoons{\text{SS}}$ Glucan$_{(n+1)}$ + ADP

Figure 7.13 The enzyme reaction catalyzed by (a) ADP-glucose pyrophosphorylase (AGPase) and (b) starch synthase (SS). Abbreviations: ADGP, ADP-glucose; G1P, glucose 1-phosphate.

Box 7.6 The regulation of ADP-glucose pyrophosphorylase (AGPase)

AGPase from all plant sources is heterotetrameric in structure, consisting of two types of nuclear-encoded subunits (i.e. $\alpha_2\beta_2$), referred to as the small and large subunit, respectively. In an attempt to elucidate the role of each subunit, AGPase small and large subunit cDNAs from potato tubers were expressed in *Escherichia coli*. It was observed that properties of the transgenic enzyme were different from those of the enzyme extracted from potato tuber and it was proposed that this was due to the cloned small subunit lacking 10 amino acids of the N terminus when compared with the mature spinach leaf small subunit (at this time the only small subunit whose N-terminal sequence had been determined by protein sequencing). Based on the high level of conservation between the potato tuber and spinach leaf AGPase, the small subunit cDNA was extended at the N-terminus and its subsequent expression in *E. coli*, along with the large subunit cDNA, resulted in an AGPase with properties comparable with those of the potato tuber enzyme. When expressed in the absence of the large subunit, the small subunit was shown to be highly active but had a decreased sensitivity to 3-phosphoglycerate (3-PGA) activation and an increased sensitivity to P_i inhibition, compared with the recombinant heterotetrameric enzyme. The large subunit, when expressed alone, however, lacked any significant AGPase activity. These findings led to the suggestion that the small subunit's primary role is catalysis, whereas the large subunit's major function is to modify the regulatory properties of the native enzyme.

Analysis of mutants that have reduced levels of AGPase and starch indicate that this enzyme plays an important part in the control of flux of carbon into starch. For example, analysis of the *shrunken-2* mutants of maize revealed no measurable AGPase activity and starch synthesis was reduced to approximately 25% of that of a normal grain. Consistent with a role in the control of flux of carbon into starch, AGPase is regulated by key metabolites within the pathway of carbon assimilation. The nature of this regulation varies according to species and location within the plant. In photosynthetic tissues, AGPase has been shown to be highly sensitive to regulation by 3-PGA and P_i. For example, AGPase purified from wheat, spinach, and barley leaves was activated over 10-fold by 3-PGA and inhibited approximately 10-fold by P_i. Plots of activity versus 3-PGA concentration in the presence of P_i (and vice versa) resulted in sigmoidal curves (Figure 1), which are indicative of cooperation between activator- and inhibitor-binding sites. In photosynthetic tissues the physiological relevance of regulation by 3-PGA and P_i reflects the concentrations of these two effectors during the diurnal cycle. During the light period, there are relatively high stromal concentrations of 3-PGA, the first intermediate in the Calvin cycle of photosynthesis (see Chapter 5). Conversely, relatively low concentrations of P_i are present as it is used in ATP synthesis during photophosphorylation (see Chapter 5). So during light periods (high 3-PGA to P_i ratio) AGPase will be more active than during dark periods (low 3-PGA to P_i ratio). This is consistent with the role of AGPase as a starch synthetic enzyme in photosynthetic tissues.

The extent to which AGPase is subject to allosteric regulation in nonphotosynthetic tissues is less clear. AGPase purified from potato tubers and rice endosperm is activated and inhibited by 3-PGA and P_i, respectively. In contrast, AGPase from barley endosperm is insensitive to 3-PGA and P_i. This latter observation has been suggested to be indicative that in some cereal endosperms regulation of AGPase may be less important than for other organs. This is further substantiated by the discovery that high sucrose concentrations in potato tubers (but not in cereal endosperms) lead to redox activation of AGPase resulting in an increased rate of starch synthesis. This post-translational redox regulation of AGPase involves the formation of a bridge between the 2β subunits. This is activated by thioredoxin in a manner similar to the light activation of the enzymes of the Calvin cycle (see Chapter 5) and has been shown to lead to an increased sensitivity of AGPase to activation by 3-PGA.

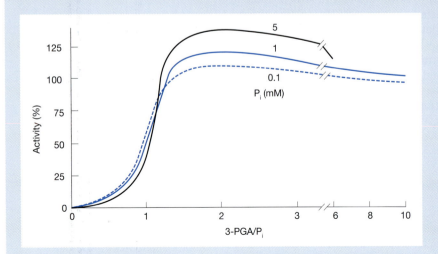

Figure 1 The regulation of barley leaf AGPase by the 3-PGA to P_i ratio. P_i concentrations were maintained at 0.1, 1 and 5 mM, whereas 3-PGA was varied from 0 to 30 mM. The activity of 100% refers to that recorded for assays with 1 mM 3-PGA in the absence of P_i.

Box 7.7 The location of ADP-glucose pyrophosphorylase

ADP-glucose pyrophosphorylase (AGPase) is known to be plastid-located in all tissues of non-cereal plants. However, analysis of the subcellular location of AGPase has revealed that cereal endosperms are unique in that they contain two distinct AGPase isoforms, one of which is located outside the plastid, in the cytosol (Figure 7.11). Evidence for this dual location comes from measurements of the recoveries of organelle-specific marker enzymes in different fractions during plastid isolation experiments (Table 1). If an enzyme is located in the plastid it will have the same pattern of recovery in the different fractions as known plastid enzymes. Obviously, the same principle can be applied for other cellular compartments such as the cytosol or mitochondria. It follows that if an enzyme has isozymes in both the cytosol and plastid then its recovery in the various fractions will be intermediate between that of a plastidic and that of a cytosolic enzyme. It has been demonstrated that in barley, maize, rice, and wheat endosperm the majority of AGPase activity is cytosolic (55–95%). These findings are consistent with the observations that ADP-glucose (ADPG) is the preferred cytosolic precursor for starch synthesis in both maize and wheat. It has since been suggested that the function of a cytosolic AGPase is to provide a more efficient partitioning of carbon from sucrose to starch. That is, where there is only a plastidial AGPase, then for sucrose to be converted to starch, hexose phosphates must be imported into the plastid and are therefore available for a range of biosynthetic pathways. In contrast, if, as in cereals, the hexose phosphates are converted to ADPG in the cytosol, and the ADPG is imported into the amyloplast, then the ADPG can only be used for starch synthesis.

Table 1 Activities of AGPase and marker enzymes in maize endosperm

Enzyme measured	Activity as a percentage of the initial extract
AGPase	3%
Cytosolic marker enzymes	
Alcohol dehydrogenase	1.4%
PFP	0.8%
Plastidial marker enzyme	
Soluble starch synthase	47%
Alkaline pyrophosphatase	24.4%

Starch formation occurs in water-insoluble starch granules in the plastids

Starch granules are composed of a mixture of amylose and amylopectin (Figures 7.9 and 7.10). A starch granule generally contains 20–30% amylose, 70–80% amylopectin, and, in some cases, up to 20% of phytoglycogen. The granule also contains small amounts of lipid and phosphate. Amylopectin synthesis occurs primarily at the surface of the granule. The granules contain the enzymes for starch biosynthesis and degradation, with granule bound starch synthase the only enzyme exclusively bound to the granule. In many cases multiple isoforms are present that may either be soluble or bound to the granules. The branchpoints within amylopectin allow the short linear chains to pack together efficiently in layers as parallel left-handed double helices. It is this packing that makes the starch granule semicrystalline in structure. The glucose residue at the start of the molecule contains a reducing group and is directed towards the inside of the structure. The glucose residues at the ends of the branches are towards the outside. The packing of neighboring branches of amylopectin forming double helices leads to the crystalline array. In contrast, amylose is present in less crystalline amorphous regions, although its precise location is still unresolved.

The particular characteristics of a starch granule may vary in terms of size and shape with different plant species (Table 7.1). This is largely determined

Box 7.8 Mutations of starch biosynthetic enzymes

A range of mutations in the starch biosynthetic pathway have been characterized in various species, such as wheat, barley, potato, pea, rice, and extensively in maize. Mutations are at loci for starch synthetic enzymes and either eliminate or reduce the activity of the specific enzyme concerned. Further understanding of starch biosynthesis may be made by careful analysis of the gene products of these mutations. Some mutations have been shown to affect only one aspect of starch structure and composition. For example, studies of mutations at the *waxy* loci of cereals have been shown to eliminate the amylose component of starch, but to have little or no effect on either the amylopectin component or on the starch granule organization. However, most mutations affect many aspects of starch structure and composition, for example, studies of the *r* locus of pea and the sugary 1

(*sul*) locus of rice show that this mutation radically affects not only the metabolic activity but also the physical appearance of the seed. The *waxy* mutant gives a waxy texture to the seed coat, while the *shrunken 1* mutant of maize gives a shrunken appearance to the seed kernels. A broad summary of the most significant mutations in the starch biosynthetic pathway is presented in Table 1 below. Despite the complex effects of many of these mutations on granule structure and composition, they often lie in genes encoding just two enzymes, starch synthase (SS) and starch branching enzyme (SBE). SS adds a glucose unit from the sugar nucleotide ADP-glucose to the non-reducing end of the glucose chain and SBE cleaves a length from the end of a chain and transfers it to the side of the same or an adjacent chain to form a branch.

Table 1 Summary of mutations affecting starch biosynthesis pathway

Mutation	Gene product affected	Phenotype
Amylose extender (ae)	Starch branching enzyme (SBEIIb isoform)	Reduction in total starch. Increased levels of amylose and decreased levels of amylopectin
Brittle2 (bt2) Shrunken1 (sh1) Shrunken2 (sh2)	AGPase (small subunit) Starch synthase (major isoform) AGPase (large subunit)	Increased soluble sugars, decreased ADP-glucose concentration and greatly reduced starch synthesis. Kernels are shrunken when dried
Shrunken x (shx)	Soluble starch synthase	Increased sugar and ADP-glucose content. Reduced starch levels. Shrunken appearance of seed kernels
Waxy (wx)	Granule bound starch synthase	Reduced levels of amylose. Waxy texture to seed
Sugary1 (su1)	Debranching enzyme	Increased amylose and reduced amylopectin levels. Presence of highly branched, soluble glucan polymer phytoglycan giving sweet taste to seed
Rugosus (rug)	Starch branching enzyme (A isoforms)	Reduction in total starch. Increased levels of amylose and decreased levels of amylopectin

by the amylose content. For example, starch granules in wheat are essentially spherical due to a lower amylose and hence a greater amylopectin content. Starch granules range in size from 1 to 100 μm and may be present as a single or multiple class. For instance, wheat has two classes of granules, one less than 10 μm in diameter and the other 10–20 μm, while potato starch granules are between 5 and 100 μm.

Table 7.1 Starch granule characteristics from important crop species				
Property	Wheat	Maize	Potato	Cassava
Type of starch	Cereal	Cereal	Tuber	Root
Granule shape	Lenticular or round	Round, polygonal	Oval, spherical	Oval, truncated
Diameter (μm)	1–45	5–30	5–100	4–35
Phosphate (% w/w)	0.06	0.02	0.08	0.01
Protein (% w/w)	0.4	0.35	0.06	0.1
Lipid (% w/w)	0.8	0.7	0.05	0.1

The composition and structure of starch affects the properties and functions of starches

Commercially more than 80% of the world market for starch is produced in the USA, primarily from maize. Europe is the major producer of wheat and potato starches, whereas cassava or tapioca starch is produced mainly in Asia. Rice and sweet potato represent only a small portion of the total starch produced.

Tuber (e.g. potato) or root (tapioca) starches have larger granules and lower protein and lipid contents than cereal (rice, wheat, maize) starches (Table 7.1). Upon processing, tuber and root starches are better for many food applications as they form clear pastes that have a bland taste. Potato starch, with a high level of phosphate groups covalently linked to the C-6 and C-3 positions of the glucose monomers, is unique among commercial starches. These phosphate groups, together with the large granule size, give potato starch a high swelling property.

Starch in its native form has only limited commercial use as a thickener or binder. When heated in water, the helices within the amylopectin of starch melt and the granule starts to swell, increasing the viscosity of the solution. With further heating and stirring the granule structure disintegrates, leading to the solubilization of the starch and a loss of viscosity. With cooling, the linear chains re-associate into aggregates, precipitate, and form a gel. The ability to control this process, primarily by chemical modification of starch in its granular form, is important for starch functionality. For instance, cross-linking glucan chains within the granule restricts swelling, while the addition of charged groups to the chains stabilizes gel formation. Starches that are modified in this way become tolerant of extremes of either hot and cold or high and low levels of pH, allowing them to be used in a range of processing conditions (Table 7.2). Furthermore, the identification and use of mutants (Box 7.8) and generation of plants by genetic modification has allowed the development of novel starches *in planta* with improved commercial uses (Box 7.9).

Table 7.2 Industrial uses of starch

Industry	Use of starch/modified starch
Adhesive	Adhesive production
Agrochemical	Mulches, pesticide delivery, seed coatings, fertilizer
Animal feed	Pellets
Building	Mineral fiber tiles, gypsum board and plaster, concrete
Cosmetics	Face and talcum powders
Detergent	Surfactants, bleaching agents and bleaching activators
Food	Viscosity modifiers, glazing agent, bread, baby food, salad dressing, mayonnaise, ketchup, soup
Beverage	Alcohol, beer, soft drinks, coffee
Confectionary	Boiled sweets, ice cream, jam, jelly gums, marshmallows
Medical	Plasma extender/replacers, transplant organ preservation, absorbent sanitary products
Oil drilling	Viscosity modifier
Energy	Fuels (bioethanol)
Paper and board	Binding, sizing, coating, corrugated board, cardboard, paper, printing paper, packaging material
Pharmaceuticals	Diluent, binder, drug delivery
Plastics	Biodegradable filler
Purification	Flocullant
Textile	Sizing, finishing, fire-resistant textiles, fabrics, baby nappy

Starch degradation is different in different plant organs

Starch accumulates in different tissues with quite distinct physiological roles. It must be broken down to its component glucose units before it can be used as a source of energy and carbon skeletons. Biochemical studies indicate that there are a range of enzymes involved in starch degradation with many isoforms capable of degrading starch and related glucans. It is only recently that researchers have started to understand the process of starch degradation reasonably clearly. Available evidence now suggests that the process of starch degradation in leaves and nonphotosynthetic organs, where starch accumulates transiently, is different to that in starch-storing organs, such as endosperms. For example, during germination of cereal grains, starch breakdown coincides with the destruction of the endosperm.

The nature and regulation of starch degradation is poorly understood

Starch degradation has been extensively studied in germinating cereal endosperm, but the nature and regulation of the process is poorly understood. Starch degradation during seed germination is extracellular and occurs via a hydrolytic route (Figure 7.14). During seed germination, glucose, produced by the action of α-amylase, β-amylase, de-branching enzyme, and α-glucosidase, is absorbed by the scutellum, converted to sucrose and then transported to the

Box 7.9 Development of novel plant starches, with improved commercial uses

The proportion of amylose:amylopectin, typically 20–30% amylose to 70–80% amylopectin, has the greatest effect on the physiochemical properties of starch. Altering the proportions of amylose and amylopectin is advantageous for certain applications. For example, in the food industry starch is often heated in water and cooled to form a gel. Amylose molecules tend to aggregate and crystallize on cooling, while amylopectin molecules generate more stable and usually desirable gels for food processing. The use of available mutant lines and now transgenic plants has led to the generation of starches with properties ideal for particular uses.

Amylose free starch

Granule bound starch synthase (GBSS) is involved in the synthesis of amylose from ADPG (Figure 7.12). The mutation of the *Waxy* locus, encoding the GBSS protein, creates an amylose-free starch. Commercially grown *Waxy* mutants of maize are used as a stabilizer and thickener in food products and as an emulsifier for salad dressings (Table 7.2). A partial *Waxy* wheat, with lower amylose content, improves the quality of pasta. Double-null *Waxy* mutants have been generated, which produce Japanese white salted noodles with desirable properties. In potato tubers, *Waxy* starches have been produced by antisense down-regulation of the *GBSS* gene. The *Waxy* potato starch has improved paste clarity and stability useful in both the food industry and in paper manufacture.

An important property of starches used in food products is improved freeze–thaw stability. Generally more stability is achieved by chemical cross-linking and stabilization. It has been argued that, from both a consumer and an environmental view, it would be best if freeze–thaw stability could be engineered without the use of these chemical treatments. Recently, a waxy starch with short chain amylopectin has been created in potato by the simultaneous antisense down-regulation of three starch synthase genes (*GBSS*, *SSII*, and *SSIII*). The starch has been shown to exhibit excellent freeze–thaw stability.

High-amylose starches

Commercially, at least 20,000 hectares of maize with amylose levels of 50%, 70%, and 90% are grown in the USA under contract each year. In the food industry, the high gelling strength of these starches makes them especially useful for producing sweets. The film-forming ability of these starches keeps the coating on fried products crispy and reduces their fat uptake upon cooking. High-amylose starches can also be processed into resistant starch, which has nutritional benefits. That is, unlike normal starch, resistant starch is not digested in the small intestine but is fermented in the large intestine by gut bacteria, producing short chain fatty acids such as butyrate that are beneficial for colon health. High-amylose starches are also used in adhesive products and in the production of corrugated board and paper (Table 7.2).

In cereals, the high-amylose phenotype is caused by the *amylose extender* (*ae*) mutation in the gene that encodes starch-branching enzyme (SBE)IIb. Down-regulation of both SBEI and SBEII expression in potato tubers using antisense techniques created starches with amylose levels of more than 60%. Recently, inhibiting gene function using single domain antibodies against SBEII produced starches that had even higher amylose levels. The starch from these lines does not swell when heated. These potatoes soften normally during cooking, indicating that the swelling pressure generated by the intracellular starch granules plays no part in this process. As these potatoes contain a lot more free water than normal potatoes do, then the texture of the resulting potato is considered more succulent. As the average chain length of this high-amylose potato is much greater than that found in cereals, then this new starch is predicted to have improved functionality.

To date, all crops that produce high-amylose starches have a much lower yield than the equivalent crop producing normal starch. Farmers must therefore be paid a premium to grow these added-value crops. Breeding programs are in place to try to minimize this yield gap for maize.

embryo. In this type of seed the mobilization of starch is controlled primarily by alterations in amounts of the relevant enzymes. During germination the activity of some enzymes increases by *de novo* synthesis, whereas for others the increase is due to activation of previously latent enzymes associated with protein bodies. These processes are generally under environmental and hormonal control and often occur in response to changes in the levels of gibberellins.

The enzymes involved in hydrolytic starch cleavage are particularly active during seed germination and expression is up-regulated by gibberellins secreted by the embryo. α-Amylase is synthesized in the aleurone layer surrounding the starchy endosperm and its *de novo* synthesis is induced by increasing the concentration of gibberellins. Inactive β-Amylase, present in

Figure 7.14 Hydrolytic starch degradation during seed germination of (a) amylose and (b) amylopectin involves the combined action of a number of hydrolytic enzymes. Amylopectin degradation is also dependent upon a debranching enzyme.

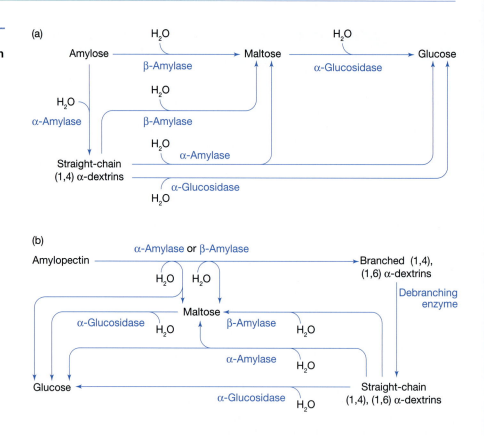

seeds prior to germination, is activated by the cleavage of a small peptide from the C-terminal of the enzyme. α-Glucosidase is present in the dried seed but, in response to increased gibberellin, large amounts are synthesized in the aleurone during germination (Figure 7.14). The seed-localized debranching enzyme and α-amylase activities are further regulated by specific disulfide proteins that act as inhibitors. Thioredoxin h reduces and inactivates these proteins early in germination.

In other seeds, where the major storage tissue remains throughout germination, starch breakdown occurs intracellularly. In some cases the amyloplast membrane is destroyed, making the starch grain accessible to cytoplasmic enzymes. The pathway of starch metabolism in these tissues is variable. For example, in pea cotyledons starch breakdown is largely phosphorylitic; while in soybean and lentil, α-amylase is the predominant activity.

Many seeds contain high β-amylase activity, but its role in starch degradation during germination is unclear. Studies on rye and soybean have shown that varieties lacking β-amylase germinate normally and mobilize starch at similar rates to varieties containing the enzyme. So although β-amylase may contribute to the degradation of oligosaccharides released by α-amylase, it is not essential for starch breakdown.

Transitory starch is remobilized initially by a starch modifying process that takes place at the granule surface during the dark period

In the above examples starch degradation is essentially irreversible and continues until all the starch is metabolized. The process is different in leaves, where starch forms a temporary reserve that either accumulates or, depending on the carbohydrate status of the cell, is mobilized. In these instances cellular integrity must be maintained during starch breakdown.

Starch accumulated in leaves during the day is degraded to act as an energy source for the night period. A starch modifying process involving phosphorylation of a proportion of glucosyl residues occurs at the granule surface. This is catalyzed by glucan-water dikinase and phosphoglucan-water dikinase, which are thought to alter the granule surface (Figure 7.15). Subsequent enzymatic attack either affects the hydrophilicity of the granule and/or the packing of the amylopectin molecule. α-Amylases and β-amylases cleave 1,4-α glucosyl bonds while the debranching class of enzymes, isoamylases and pullulanases, cleave the 1,6-α glucosyl branchpoints. The action of debranching enzymes leads to the accumulation of linear malto-oligosaccharides in the stroma. These are further metabolized, leading to maltose and glucose, both of which are exported from chloroplasts to the cytosol (Figure 7.15).

The regulation of starch degradation is unclear

Diurnal variation in starch in leaves suggests mobilization is regulated. To date, studies of the kinetic properties of the enzymes involved in starch degradation have failed to reveal any obvious regulatory features. Although control of starch degradation could be achieved solely by adjusting the rate of starch biosynthesis, with the rate of starch breakdown remaining constant, the absence of detectable starch turnover in pea and sugar beet leaves argues strongly against this possibility. The pathway of starch degradation is thought to be regulated directly. Post-translational modification, including redox modulation and protein phosphorylation, has been proposed as being an important means of control, but further evidence is needed.

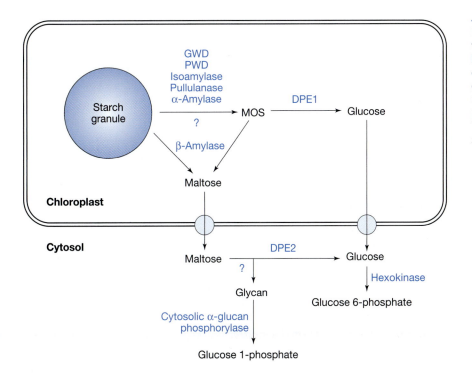

Figure 7.15 A general pathway of transitory starch degradation in leaves. Abbreviations used: DPE, disproportionating enzyme; GWD, glucan-water dikinase; MOS, linear malto-oligosaccharides; PWD, phosphoglucan-water dikinase. ? indicates that the nature of enzyme step is speculative.

Starch breakdown is normally restricted during the day, but such regulation is not achieved by light per se. This view is supported by the observation that net starch degradation can occur in leaves during continuous illumination, and by the demonstration that, under certain conditions, illuminated spinach chloroplasts catalyze the concurrent synthesis and degradation of starch. A number of studies have demonstrated that the breakdown of starch at the beginning of the night is often delayed until the concentration of sucrose in the leaf has fallen, suggesting that starch mobilization may be controlled by the requirement of the cell for a respiratory substrate.

Fructans are probably the most abundant storage carbohydrates in plants after starch and sucrose

Fructans are water-soluble polymers of fructose with different numbers of fructose molecules attached to the fructose end of sucrose. Fructans are synthesized directly from sucrose, and accumulate in the vacuole of photosynthetic and storage cells (Figures 7.1 and 7.16). They form the primary reserve carbohydrate in over 40,000 species (approximately 15%) of angiosperms, including a number of economically important species such as temperate forage grasses (*Lolium* and *Festuca*), cereals (barley, wheat, and oats), vegetables (chicory, onion, and lettuce) and ornamentals (dahlia and tulips). The presence of fructans in plant tissues may also confer drought and cold resistance.

A model has been proposed for the biosynthesis of the different fructan molecules found in plants

Fructans are synthesized from sucrose by the action of two or more fructosyltransferases, which transfer fructose from sucrose to a growing fructan chain (Figure 7.16). In contrast to bacterial fructans, which are fairly uniform in structure, plant fructans exhibit variable fructosyl chain lengths. Plant fructans have degrees of polymerization ranging from 30 to 50 fructosyl residues up to over 200. In angiosperms the fructosyl residue linkage varies and can be classified into five major and structurally distinguishable classes; levan, inulin, mixed levan, inulin neoseries, and levan neoseries (Figures 7.16 and 7.17).

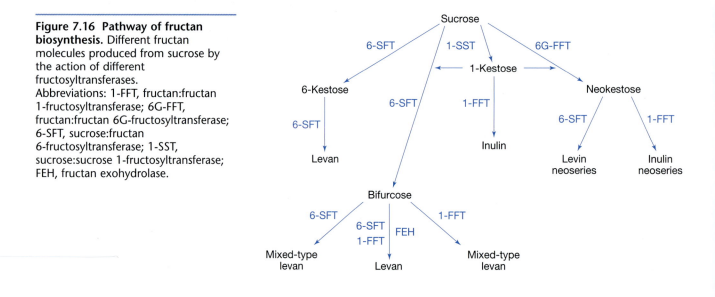

Figure 7.16 Pathway of fructan biosynthesis. Different fructan molecules produced from sucrose by the action of different fructosyltransferases.
Abbreviations: 1-FFT, fructan:fructan 1-fructosyltransferase; 6G-FFT, fructan:fructan 6G-fructosyltransferase; 6-SFT, sucrose:fructan 6-fructosyltransferase; 1-SST, sucrose:sucrose 1-fructosyltransferase; FEH, fructan exohydrolase.

(a)

(b)   (c)

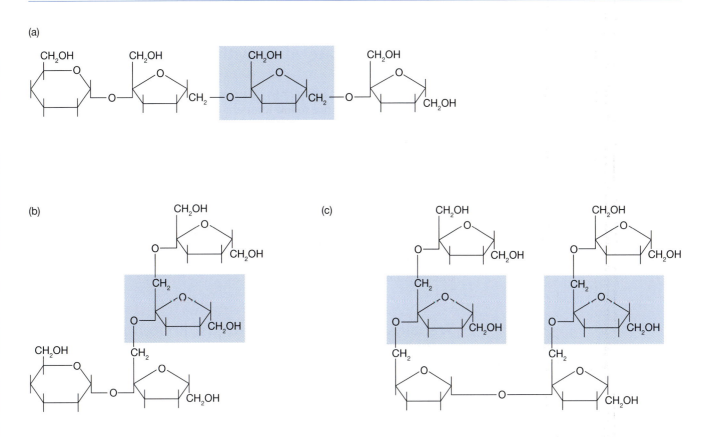

When only sucrose is available as a substrate sucrose:fructan 6-fructosyl-transferase (6-SFT) produces 6-kestose (Figure 7.16), which may be elongated by 6-SFT to produce levans. Sucrose can be converted via sucrose:sucrose 1-fructosyltransferase to 1-kestose, which is then elongated by fructan:fructan 1-fructosyltransferase (1-FFT) to inulin. Sucrose and 1-kestose can be converted by fructan:fructan 6-glucose-fructosyltransferase to neokestose (Figure 7.16), which can be elongated by 1-FFT or 6-SFT to inulin neoseries or levan neoseries respectively. Alternatively sucrose and 1-kestose can be converted by 6-SFT to bifurcose (fructose + fructose), which can be elongated by 6-SFT or 1-FFT to give mixed type levan. An alternative pathway for levan production involves the removal of the β-(2,1)-linked fructosyl residue from bifurcose either by 1-FFT or fructan exohydrolase.

In the general model for *de novo* fructan biosynthesis, sucrose is the only substrate, so there is always a terminal glucose residue. However, fructan molecules consisting of only fructose, and lacking the terminal glucose residue are found in plants. These reducing fructofuranosyl-only fructan molecules may be produced by the action of endo-inulinases or an α-glucosidase that remove the glucose moiety. Alternatively, 1-FFT is able to produce fructofuranosyl-only oligosaccharides from fructose and inulin.

The presence of fructosyltransferases and fructans in plant vacuoles has led to the suggestion that fructans are also synthesized in vacuoles. Fructan biosynthesis has been suggested to be involved in controlling sucrose concentration in the vacuole. Vacuolar fructan biosynthesis lowers the sucrose concentration in the cell, avoiding sugar-induced feedback inhibition of photosynthesis. Fructans can accumulate in the vacuole to over 70% of the plant dry weight without photosynthesis being inhibited. Fructan biosynthesis can be induced by feeding excised leaves of fructan-accumulating plants with sucrose or by exposing them to continuous illumination.

Figure 7.17 Examples of structurally different fructan molecules found in plants. Fructans are derived from ketoses and are formed by the linkage of fructose residues to a sucrose molecule. The shaded box represents fructose residues repeated n times. (a) 6-kestose consists of a (2,6)-linked β-D fructosyl unit to sucrose and is a levan-type fructans, where $n = 10$–200. (b) Trisaccharide 1-Kestose consists of a (2-1)-linked β-D fructosyl unit to sucrose and is the shortest inulin molecule, $n<50$. (c) Neokestose is the smallest inulin neoseries molecule, with a β-D fructosyl unit linked to the C-6 of the glucose moiety of sucrose, both $n<10$.

Fructan-accumulating plants are abundant in temperate climate zones with seasonal drought or frost

In fructan-accumulating plants, fructan production and photosynthesis appear to be less sensitive to low temperatures than carbohydrate production and storage systems found in other plants. As such fructan storage is advantageous in plants that are photosynthetically active during early spring or the winter. In Jerusalem artichoke (*Helianthus tuberosus*) 50% of sucrose:sucrose 1-fructosyltransferase activity is maintained at 5°C compared with the activity at the optimal temperatures of 20–25°C.

The cloning of fructosyltransferases and transformation of these cDNAs into plants that normally do not accumulate fructans have provided a clearer insight into fructan biosynthesis and its physiological role. In addition, these studies have raised the possibility that fructans may also have potential commercial value (Box 7.10).

Trehalose biosynthesis is not just limited to resurrection plants

Trehalose (α-D-glucopyranosyl-1,1-α-D-glucopyranoside) is a non-reducing disaccharide found in a wide range of organisms from bacteria to invertebrates. It plays a part as an energy source or stress protectant. Until recently trehalose was considered to have little importance in most plant species, only being identified in a limited number of desiccation-tolerant resurrection plants such as the small club mosses, or spike mosses, *Selaginella lepidophylla* and *S. sartorii* and the angiosperm *Myrothamnus flabellifolius*. However, with the complete sequencing of the arabidopsis genome in the late 1990s, it has become clear that these plants contain multiple copies of functional genes involved in trehalose biosynthesis, trehalose 6-phosphate synthase (TPS) and trehalose 6-phosphate phosphatase (TPP) (Figure 7.18). Subsequently TPS and TPP homologs have been identified in a range of higher plants, including rice, soybean, and tomato. This now raises the question as to whether these enzymes have a significant role in plant metabolism other than in desiccation-tolerant plants.

Box 7.10 Fructan-rich diet: Health-promoting effects?

Small fructans are sweet tasting, while longer fructan chains form emulsions with a neutral taste and fat-like texture. As the human body does not contain enzymes able to degrade fructans, it has been suggested by the food industry that they may be suitable as a low-calorie food ingredient. This is further supported by their efficiency as a carbon source for beneficial *Bifidobacteria* in the colon. *Bifidobacteria* are able to ferment fructans to short chain fatty acids, which have been shown to have a positive effect on systemic lipid metabolism.

Currently, chicory is the most agronomically acceptable crop for fructan production. However, the long fructan chains isolated from chicory are of limited value as they are degraded by fructan exohydralase when harvesting. Although obtaining long chain and complex branched fructans is currently difficult, they have wide applications. Fructans with high degrees of polymerization are now used in alimentary products where they can replace fat. Long chain emulsions of fructans in water have organoleptic properties similar to fat. In addition, fructans with a high degree of polymerization have potential marketing value in a range of nonfood applications.

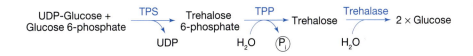

Figure 7.18 Trehalose metabolism in higher plants. Abbreviations: TPS, trehalose 6-phosphate synthase; TPP, trehalose 6-phosphate phosphatase.

Under stress conditions, and more particularly during desiccation, trehalose has a number of protective properties that are superior to those of other sugars, including sucrose, and which contribute to it being an ideal stress protectant. First, trehalose has been shown to stabilize proteins and membranes by replacing water through hydrogen bonding to polar residues. Secondly, in the dry state trehalose also forms glasses (vitrification), a process that may be required for the stabilization of dry macromolecules. Finally, trehalose remains stable at elevated temperatures and low pH.

Trehalose biosynthesis in higher plants and its role in the regulation of carbon metabolism

Early attempts to identify trehalose in plants were hampered by techniques lacking specificity or being unreliable because of possible microbial contamination. Later studies confirmed that, despite the presence of TPP and TPS genes, most higher plants did not accumulate detectable amounts of trehalose. However, it is now clear that the apparent lack of trehalose accumulation is probably due to the activity of an additional enzyme trehalase, which degrades this sugar (Figure 7.18). In studies with arabidopsis and a number of other plants, the addition of the trehalase inhibitor validamycin A to the growth medium resulted in easily detectable amounts of trehalose (approximately 15% of the sucrose content), being identified and subsequently confirmed by gas chromatography-mass spectrometry analysis, suggesting that trehalose biosynthesis is a common process in higher plants.

The degree to which trehalose formation regulates metabolism is still unknown, but trehalase activity has been proposed to be an important means of maintaining a low cellular concentration of trehalose. Where plants interact with trehalose-producing microorganisms this could be an extremely important survival process. For example, the trehalase activity is strongly induced in arabidopsis plants infected with the trehalose-producing pathogen *Plasmodiophora brassicae*. Trehalose accumulation may not be advantageous in all systems, as high trehalose concentrations have been suggested to be incompatible with the chaperone-assisted protein re-folding often necessary during stress recovery.

Current research indicates that trehalose 6-phosphate is an important signaling molecule in higher plants, regulating photosynthesis and leaf development. For example, tobacco plants that overexpress *Escherichia coli* TPS accumulate trehalose 6-phosphate (T6P) and display increased rates of photosynthesis per unit leaf area under saturating light conditions. In contrast, plants overexpressing *E. coli* TPP have decreased photosynthetic rates. This suggests that there is a correlation between photosynthetic rates and T6P levels, rather than trehalose levels. The implication is, therefore, that T6P either directly or indirectly controls sugar metabolism. Furthermore, T6P has also been shown to play a part in promoting thioredoxin-mediated redox transfer of AGPase (Box 7.6) in response to cytosolic sugar levels (see Further Reading). Although TPS appears to play an important part, this is a new area of research and much has now to be determined about the physiological role of trehalose metabolism.

Plant cell wall polysaccharides

Plant cell walls are complex physico-chemical structures that combine mechanical strength, permeability, and malleability. They are composed of large polysaccharide and protein molecules. Here we will consider the biochemistry of the polysaccharide structures and their synthesis. Of necessity this account will concentrate on general features of typical walls. In reality, as any light microscope section will reveal, cell walls are highly individual and varied in even their gross morphological characteristics. There are about 40 different cell types, some with variations between walls within a cell. Defining such walls individually at the molecular level is beyond the competence of our present technology.

Cell walls originate at cytokinesis following nuclear division. A cell plate (phragmoplast in older literature) is formed between daughter nuclei by the fusion of Golgi-derived vesicles. The membranes of these vesicles form the new plasma membrane around the disc of cell wall that expands centrifugally out to meet the walls of the original cell. New cell growth and expansion are matched by new wall growth, defined as the primary cell wall. Eventually cell expansion ceases, but wall growth may continue to form a secondary wall. Most of what follows refers to primary cell walls. The composition of secondary cell walls is dominated by cellulose microfibrils (see Chapter 3). Secondary walls have little or no capability for expansion.

Different cell wall polysaccharides are each polymers of a variety of sugar molecules joined by a variety of linkages. Specific transferases and synthases assemble each polysaccharide molecule to form a defined structure that is specifically linked (by covalent or hydrogen bonds) to other such molecules to create a massive and complex macromolecular structure, the cell wall. This structure is ultimately defined at the genome level. The mosaic of cell wall chemical types found across plant species matches the genus, Family, and Order relationships of the plant kingdom. Mutations in wall synthesis genes lead to changes in wall chemistry, often with structural consequences.

The assembled wall provides the essential aqueous and structurally robust environment (the apoplast) needed for each cell's survival. Access to the cell surface, the plasma membrane, is regulated both physically, by defining the maximum pore size that can be penetrated by inert particles (4–5 nm diameter), and chemically, by the action of ionic groups in the wall trapping incoming ions and charged molecules. Both these factors place similar restrictions over the movement of ions and molecules released to the apoplast from the cell interior.

The relative levels of solutes in the cell compared with those in the apoplast ensures that each cell has a negative water potential. The resultant osmotic uptake of water leads to cell swelling that is constrained by the wall. As most plant cells exist in tissues, osmotic forces are not, contrary to popular perceptions, an overriding issue for each individual cell. The total tissue wall pressure is resisted by the outer epidermal wall, which is much thickened and nearly impermeable to water, while internal walls experience compressive forces and are very much thinner. Inflation of these tissue balloons provides a level of structural support for aerial tissues, holding up stems, leaves and flowers. Loss of turgor, for example due to drought, results in the characteristic wilting of aerial organs. Some cells develop very much thickened secondary walls and undergo programmed cell death, losing their protoplasts and leaving a toughened skeleton behind that supports tissues and whole plants, providing the impressive architecture of trees that can rise to 100 m above the ground.

Wall polysaccharides are divided into two structural classes. The microfibrils, which lie in parallel arrays around the cell, and the matrix components that provide a gel for the microfibrils to lie in. The matrix consists of hemicelluloses and pectins. The wall therefore consists of three networks of macromolecules that are interconnected, providing the mechanical properties needed for cells to first grow to their final size and then to perform structural and transport functions for the mature plant organ. The microfibrils are made of cellulose and provide anisotropic strength in tension to a wall, resisting cell expansion parallel to their length, and allowing expansion at right angles to their preferred chain direction. Hence these are determinants of cell shape (see Chapter 3). During cell growth, there have to be regulated movements between adjacent microfibrils and a general reorientation of microfibrils as new material is added to the cell surface and the cell grows in size. The width of the cell wall is maintained and modifications are made to specific wall structures to achieve desirable functional attributes. One of these processes is the deposition of lignin prior to xylem cell death, which permanently cements the microfibrils together, producing wood and increasing structural rigidity and resistance to pathogen attack.

Synthesis of cell wall sugars and polysaccharides

About 10 different sugars are utilized by plants in the synthesis of cell wall polysaccharides, with some containing only one sugar type (homopolymers), while others contain very many different sugars. These different sugars are derived mainly from glucose by conversions involving epimerization, oxidation, and carboxylation. The epimers of D-glucose (Glc) are D-mannose (Man) and D-galactose (Gal). Oxidation of the alcohol groups to carboxylic acid at the C-6 positions of these sugars yields the corresponding uronic acids: D-glucuronic acid (GlcA) and D-galacturonic acid (GalA). The five-carbon sugar D-xylose (Xyl) is made by decarboxylation of glucuronic acid, and its epimer is D-arabinose (Ara). Alternatively these C6 alcohols can be dehydrated to methyl groups to yield the deoxy sugars L-rhamnose (Rha, 6-deoxy-L-mannose) and L-fucose (Fuc, 6-deoxy-L-galactose).

In solution, sugars predominantly exist as ring structures, either five-membered (furanose) or six-membered (pyranose) rings. These rings always contain one oxygen atom and four or five carbons. Sugars can be joined by a large number of linkages to create either homo- or heteropolymers, either straight chain or branched. Glycosidic linkages are always from a C-1 atom to another carbon (e.g. 1,3, 1,4, and 1,6 linkages). Homopolymers of pyranoses with links on opposite sides of the ring (1,4 linkages) produce linear polymers, while other links (e.g. 1,3 linkages) produce curved polymers that form helical structures. Side chains, and the size of side chains, have a large effect on the properties of the resultant molecule, not least because the side chains can limit access to the backbone links by degrading enzymes.

Cellulose

Structure and organization

Cellulose is a homopolymer of β1,4-linked glucose units forming straight chains of βD-glucan. In higher plants 30–36 chains associate to form a single cell wall microfibril. Within a microfibril all the chains have the same polarity (they all lie with their 1,4 links pointing in the same direction) and, apart from some peripheral chains, they lie in a crystal lattice stabilized by hydrogen bonds. Individual chains within the microfibril may be up to 0.1 μm long, while microfibrils can be up to 5 μm long. Microfibrils are laid down in parallel arrays within the cell wall (see Chapter 3), almost certainly in response to their interaction, direct or indirect, with cortical microtubules lying in parallel arrays beneath the plasma membrane.

Synthesis

Cellulose assembly, in higher plants, occurs at the outer face of the plasma membrane from UDP-glucose supplied by sucrose synthase at the inner surface of the membrane. Cellulose synthases are integral plasma membrane proteins that exist as very large multiprotein complexes. These are synthesized on the rough endoplasmic reticulum, pass through the Golgi complex as part of the endomembranes, and are exported to the plasma membrane in the membranes surrounding secretory vesicles.

Cellulose synthases can be visualized in freeze-etch views of the inner surface of the plasma membrane (the protoplast face) as rosettes of six large particles. Confirmation that these rosettes are cellulose synthases comes from the timing of rosette appearance during cell development, and their surface distribution in differentiating xylem tracheid cells. It has also been possible to label the particles with antibodies specific for cellulose synthase.

Larger particle arrays can be found in the plasma membranes of algae, which form larger microfibrils with a higher number of cellulose chains. In some algae freeze-etch views of the outer surface of the plasma membrane (not fractured) reveal microfibrils terminating in linear arrays of particles. All these images are interpreted to be of integral membrane proteins that span the plasma membrane from cytosol to extracellular domain.

Each of the six rosette particles consists of five or six proteins that are cellulose synthase enzymes, each synthesizing a single βD-glucan chain, so that one rosette makes 30–36 chains and a single microfibril. Chain initiation is thought to commence on the inner face of the plasma membrane (Figure 7.19) with the transfer of UDP-glucose on to a sitosterol lipid to form sitosterol-β-glucoside. This then acts as a primer for the addition of a further UDP-glucose, adding successive glucose units forming sitosterol cellodextrin. Elongation of the chain pushes the sitosterol-β-glucoside out through the membrane. At the outer face a cellulase enzyme, a hydrolase, removes the sitosterol allowing the glucan chain to spontaneously aggregate with other chains to form a microfibril. Mutants lacking this cellulase have disrupted cellulose synthesis and have defective cell walls.

Each cellulose synthase makes a single cellulose microfibril that is laid down on the plasma membrane surface. It has long been proposed that, because

Figure 7.19 Model for the possible initiation of a cellulose glycan chain. Sitosterol accepts a glucose from UDP-glucose to form sitosterol glycoside (SG). Further glucan residues are added to elongate the chain, forming S-cellodextrose (SCD). A cellulase then cuts off the terminal sitosterol and transfers the cellodextrose chain to cesA where it is elongated with UDP-glucose derived from the action of sucrose synthase (SuSy) on sucrose. (From *Science*, 295: 59, "A plant primer", 2002, Figure with caption, panel C by Steve M Read and Tony Bacic. Reprinted with permission from AAAS (American Association for the Advancement of Science).)

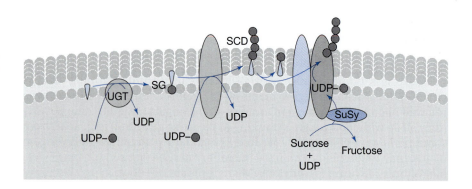

the microfibril is large compared with the synthase, the act of synthesis would provide a force that would drive the synthase complex across the cell surface, in the fluid lipid bilayer of the plasma membrane, much as the cellulose-synthesizing bacterium *Acetobacter* is driven forward on its thread. Confirmation of this proposal has been graphically demonstrated in the past few years by observation of cells in which green fluorescent protein- and yellow fluorescent protein-labeled cellulose synthases have been expressed. Individual fluorescent dots can be seen moving over the surface of living cells during cellulose synthesis. The movements are, surprisingly, not all in one direction around the cell, bidirectional movements are also seen. This work has also confirmed the close association between the cortical microtubule cytoskeleton and the cellulose synthases, which results in the orientation of the microtubules on the cytoplasmic side of the plasma membrane defining the orientation of the microfibrils in the cell wall, ultimately defining the shape of the cell. This work demonstrates the fascinating link between the biochemical process of glucose unit polymerization and the ultimate realization of complex cell structures.

Cellulose synthase activity has been difficult to demonstrate in *in vitro* preparations of plant cell extracts despite many years of research. Recently efforts to preserve the potency of cell extracts and mimic cellular conditions favorable for synthesis have proved successful and cellulose microfibrils of high structural integrity have been produced from UDP-glucoses. There are suggestions that an intact plasma membrane is required for synthesis with a membrane potential and oxidizing conditions. These preparations do, however, synthesize large amounts of callose, a β 1,3-linked glucan, that is normally laid down in the intact plant in response to wounding or pathogen attack.

The availability of cellulose-deficient mutants, genome DNA databases, and gene expression analysis have provided a molecular biological approach to the identification of cellulose synthase genes. They are all transmembrane type 2 glycosyltransferases. In arabidopsis, the *CESA* family contains at least 10 cellulose synthase genes with conserved sequences for binding UDP-glucose, for transmembrane domains and for the position of cysteine residues. In addition there are many, more than 40, *CESA-like* genes, *CSL*, which share the same conserved sequences, but have differences elsewhere. These have been grouped into a series of *CSL* families of largely unknown functions. Many are probably involved in making the backbones of noncellulosic wall components, for example, *CSLA* synthesizes mannans.

The *CESA* genes for cellulose synthesis are expressed differentially, with three being used in primary wall synthesis (*CESA1, CESA3, CESA6*) and a different set of three being used in secondary cell wall synthesis (*CESA4, CESA7, CESA8*). In each case, all three are required to be present and co-operate in the synthesis of a microfibril. For example, the three *irx* (*irregular xylem*) mutants of *Arabidopsis*, *irx1*, *irx3*, and *irx5* each cause defects in xylem formation and correspond to mutations of the *AtCESA8*, *AtCESA7*, and *AtCESA4* genes respectively. All three are expressed at the same time in cells of stems. Fluorescent labeling has shown that polypeptides from each gene initially accumulate in the cytoplasm before assembly and delivery to the cell surface. If synthesis of one of the three is inhibited, then the other two fail to associate or move to the cell surface. The *rsw1* mutation in these plants is located in the *AtCESA1* gene and the *PRC1* mutation is in the *AtCESA6* gene, both mutants have defective primary walls and decreased cell elongation. Specific inhibition of cellulose synthesis depends on binding of chemicals to specific sites on the synthase molecules. The *ixr1* and *ixr2* mutations confer resistance to this chemical binding and have been shown to be located in *AtCESA3* and *AtCESA6*, respectively.

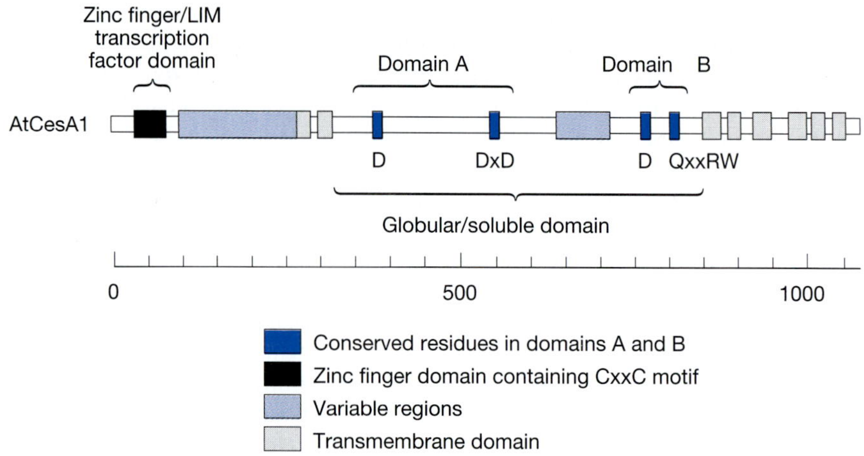

Figure 7.20 Diagram of the main features of a *CesA* gene. N-terminus to the left. There are eight transmembrane domains (gray bars), a cysteine-rich domain (black bar), and four UDP-glucose binding domains (D). (Reprinted with permission from *Genome Biology*, 1 (4): 1, 2000, Figure 3, by T Richmond.)

Zinc finger/LIM transcription factor domain

Domain A   Domain B

AtCesA1

D   DxD   D   QxxRW

Globular/soluble domain

0   500   1000

- ▦ Conserved residues in domains A and B
- ▪ Zinc finger domain containing CxxC motif
- ▨ Variable regions
- ▨ Transmembrane domain

A similar trio of genes has been located in rice and barley, corresponding to the three listed above for primary and secondary wall synthesis. Proteins from all these genes possess a RING-finger region at the N-terminal end of the sequence, consisting of four cysteine pairs, CxxC, spaced along the polypeptide chain (Figures 7.20 and 7.21). These create regions for potential dimerization assisted by zinc. In rice, OsCesA7 and –A9 proteins have identical RING regions, whereas OsCesA4 has a shorter space between the first two CxxC pairs, creating the opportunity for –A4 to dimerize with either –A7 or –A9. Mutations in this region abolish synthase activity presumably due to failed protein–protein interactions.

What is the function of each individual gene product? Different *CESA* genes seem to code for different aspects of microfibril synthesis. One seems to code for chain initiation, while another may code for chain elongation. A system for testing the activity of each *CESA* gene has been established in yeast (which does not synthesize cellulose). To date, *CESA1* from cotton has been transformed into yeast which can then elongate sitosterol-glucoside to sitosterol-cellotriose, but cannot complete chain elongation. Further work is needed to test the activity of each gene product in this system, and of different combinations of the three gene products.

Matrix components consist of branched polysaccharides

Matrix components are branched heteropolymers that form a gel medium around the cellulose microfibrils, providing strength in compression as well as flexibility. Some bind by hydrogen bonds to the microfibrils enhancing the mechanical strength of the wall. Others cross-link to each other. As there is only one reducing group available for the formation of a glycosidic bond on each sugar molecule, it is not possible for bonds to be formed between branches of adjacent macromolecules. Only the base of the branched tree will have a reducing group. Therefore most of the associations between matrix components consist of hydrogen bonds. Some of the polymers have acidic groups that can be ionically cross-linked by divalent cations, usually calcium. Boron diester bridges can also be formed between specific pairs of molecules. It is thought that every matrix polysaccharide molecule is linked into the supermolecule of the wall in some way, either ionically or by hydrogen bonds.

There is considerable variation in these polymers, both between different tissues in the same plant and between the same tissues in different plants. It

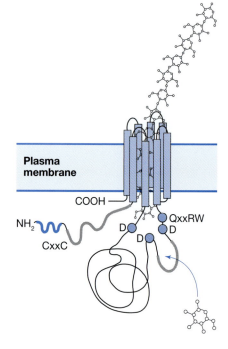

Plasma membrane

COOH

NH₂   D   D   QxxRW

CxxC   D

Figure 7.21 Model of a single CesA protein inserted into the plasma membrane and synthesizing a single glucan chain. This model does not account for the fact that three CesA proteins have to be present to achieve normal cellulose synthesis. (Reprinted with permission from *Genome Biology*, 1 (4): 1, 2000, Figure 4, by T Richmond.)

seems that the same functions can be met by a variety of different combinations of components. Previously these components were divided into two main groups, hemicelluloses and pectins. The unraveling of the chemical identity of the constituent units has allowed the use of more informative names for the different components.

Hemicelluloses are now discussed under the more specific titles of xyloglucans, of which there are two main types, heteroxylans, and heteromannans. Pectins consist of galacturonans, with three main types recognized in angiosperms.

As we shall see, some components have their origins in the earliest land plants, persisting essentially unchanged to existing plant groups, while others have evolved more recently. In general, boundaries between groups of plants classified according to their wall chemistry match taxonomic boundaries in classifications that reflect the evolutionary origins of plant groups. In particular, the evolution of the ability to synthesize xylans was pivotal in the development of secondary walls in plants able to colonize land effectively. These are absent from mosses and liverworts, but present in hornworts and all vascular plants.

Hemicelluloses

Xyloglucans are regular polymers with backbones based on glucose

The majority of xyloglucans have a basic septamer repeat unit based on a backbone of four glycosyl (G) residues (forming GGGG). Usually three bear a xylose (X) residue (forming XXXG) to give the basic heptamer unit. Note that the shorthand notation for these structures just lists the terminal residues attached to each of the four glucose residues. Galactose (L) may be attached to either or both of the xylose residues in the middle of the molecule (XXLG or XLXG or XLLG). The galactosyl residue adjacent to the terminal unbranched glycosyl residue may have a fucose (F) sugar attached to it (XXFG or XLFG). The F in this notation represents the end of a side chain, GXLF, starting at the third glucose residue. These fucogalactoxyloglucans are found in most dicotyledons and some monocotyledons (lilies and orchids for example). There are variations in the distribution of the different branch types between the major plant groups and within a single plant, with XLXG being more abundant in leaves than in stems for example.

Arabinoxyloglucans are found in some dicots (Solanales and Lamiales). These are hexamers, lacking xylose on the first glucose (GXXG) and with arabinose (A) in place of galactose on one or both of the xylose residues on the middle glucose positions (GXAG, GAXG, and GAAG).

Xylans are regular polymers with backbones based on xylose

Xylans have a rather different structure based on glucuronoarabinoxylans. Single residue side chains of glucose and arabinose are attached to a backbone of $\beta 1,4$-linked xylose units. These are found in many monocotyledonous plants, including grasses and cereals. Related glucuronoarabinoxylan polymers are also found as a minor constituent of the xyloglucan-dominated walls of dicotyledons and the other monocotyledons.

Mannans are regular polymers with backbones based on mannose or mixed mannose glucose units

Homopolymers of mannose are straight chains of $(1{\rightarrow}4)$ β-D-mannose that can hydrogen bond to form fibrils much like cellulose. Alternatively such chains can have added side branches of single galactose units at the O-6 positions on mannose to form galactomannans. A rather different molecule is formed by the inclusion of up to 50% glucose units in the main chain forming glucomannans.

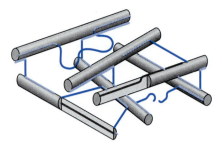

Figure 7.22 Xyloglucan chains hydrogen-bonded to cellulose microfibrils and cross-linking adjacent fibrils. Earlier models of these interactions considered that xyloglucans could coat the entire microfibril surface and even lie within the outer layers of glucose chains. The model depicted here reflects the current view, that there are no chains lying in the microfibril structure, and that a substantial portion of the microfibril surface is exposed and free of such chains.

These can still have side chains of single galactose units attached to the 6-position of the mannose, giving galactoglucomannans.

Mannans occur in small amounts in the primary walls of most angiosperms, and are significant components of secondary walls in angiosperms and gymnosperms. In some seeds, for example date seeds, extensive deposits of cell wall mannans form the main carbohydrate reserve. These render the seeds less vulnerable to insect attack than the more usual starch grain formations found in cereals.

Xyloglucans and related polymers stabilize the microfibril arrays

Xyloglucans, and related molecules, cross-link microfibrils via hydrogen bonds (i.e. noncovalent linkages) and also form such links to the pectin scaffold to form a mechanically strong but malleable cell wall (Figure 7.22). The even number of glycosyl residues in the repeating units (XXFG, etc.) leads to a regular surface structure that assists binding to cellulose. Cross-linking xyloglucans between microfibrils require enzymic modification to permit wall expansion and growth while still retaining wall strength.

Xyloglucan synthesis is poorly understood. The glucan synthase needed for the glucan backbone of the heptameric units has not been isolated. Genes for enzymes that add the xylosyl residues, xylosyltransferases, have been cloned. Labeling experiments show that these chains are assembled in the Golgi apparatus. The label appears in the wall about 20 minutes after addition to cell cultures. This is the total time for synthesis, transport between Golgi cisternae and packaging into vesicles, transport of vesicles to the plasma membrane and discharge into the wall. There is some evidence that label in xylans takes longer to appear in the wall compared with that in xyloglucans. This is in keeping with observations that β-xylosyltransferase (adds xylose to extending xylan chains) is found in *cis*- and *medial*-Golgi cisternae, while α-xylosyltransferase (adds xylose to xyloglucan chains) is found mostly in *trans*-Golgi cisternae. The time difference is due to the longer path length for xylans through the Golgi apparatus.

Fucosyltransferase activity is found throughout the Golgi apparatus, but clearly it cannot act in xyloglucan synthesis until the XLLG polymers have been assembled, so presumably it is only effective in the *trans*-Golgi cisternae. This enzyme may act in early Golgi compartments in the synthesis of other polysaccharides and *N*-glycans.

Following release into the wall, xyloglucan chains become larger, probably as a result of the activity of xyloglucan endotransglycosylases (XETs), and they become covalently bound to existing wall xyloglucans. These XETs are active during cell wall expansion, when the microfibrils need to slide within the wall matrix. The enzyme cuts the xyloglucan backbones allowing the microfibrils to move past each other, and then rejoins them to the same or different xyloglucan chains, stabilizing the new microfibril position (Figure 7.23). XET activity is highest in cells undergoing rapid cell elongation, such as in growing roots. These enzymes are found in walls that are dominated by arabinoxylans, as well as those that are predominantly xyloglucan in composition.

The XET enzyme is unusual in several respects. In contrast to xyloglucan endohydrolase, which cleaves the molecule and then adds water as the acceptor molecule to the donor end, XET shows a strong preference for transfer of the donor molecule to another xyloglucan polysaccharide as the acceptor molecule. This avoids chain cutting and hydrolysis and preserves the structural links between microfibrils. This means that the substrate and product may well be identical to each other, an unusual circumstance for a biological

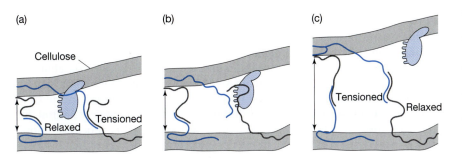

(a) (b) (c)

Cellulose

Relaxed   Tensioned

Tensioned   Relaxed

Tensioned   Relaxed

Figure 7.23 Xyloglucan endotransglycosylase (light blue) (a) cutting and (b) religating xyloglucan chain tethering microfibrils into a new position shown in (c). (Reprinted with permission from *Plant Physiology* 134: 443, 2004 by M.J. Pena, P. Ryden, M. Madson, A.C. Smith and N.C. Carpita.)

enzyme. Following cutting, the enzyme then has to find the non-reducing end of an appropriate chain or branch to which the cut end can be ligated. It has been found that the enzyme remains attached to its substrate for an unusually long period of time, perhaps as an adaptation to the time needed to find a suitable receptor.

Xyloglucans in the wall are subject to turnover and loss from the wall. Some of this may be due to trimming of segments not bound to cellulose microfibrils. Mutants lacking a galactosyl residue in the third position have been isolated. These perforce also lack the fucosyl residue that can be found in this position. This galactosyltransferase is specific for this particular xylose, and in such mutants there is an increase in the proportion of xylose groups that are galactosylated in the second position (XLXG). The effects on the strength of the cell wall range from no effect on flowering stems of *Arabidopsis* to adverse effects on hypocotyls, with bulging and increases in the diameter of cells. The reason for this difference in phenotype is that increased galactosylation does not occur in hypocotyls, demonstrating the importance of the galactose residues for wall strength. Mutants lacking the fucosyl transferase do not have the same phenotype, suggesting that the fucose residues are not important for mechanical strength.

Searches of the *Arabidopsis* genome database have located genes for nine fucosyltransferases. Mutations have located one that is essential for fucosylation of xyloglucans. This cannot be substituted for by the other genes, so their function remains unknown. Mutants lacking fucose in the xyloglucans show normal growth habit and development, so contrary to previous proposals, the fucose component of xyloglucans is not an important regulator of plant growth.

Pectins

Galacturonans are acidic polysaccharides, some have extensive complex branches

Pectins are acidic polysaccharides with a high proportion of D-galacturonic acid residues, so they are more properly termed galacturonans. The presence of –COOH carboxylic acid groups has a major effect on the cell wall charge distribution and on its interactions with cations and anions in the apoplastic water phase. These negatively charged groups perhaps provide additional strength to the wall gel by cross-links established through divalent calcium ions.

Three chemically distinct groups of galacturonans are found in plant cell walls, homogalacturonans, rhamnogalacturonan I, and rhamnogalacturonan II. These have been characterized by fractionating extracts obtained from walls by a variety of relatively mild treatments, such as with mild acid or alkaline solutions. It now seems that in the native state all these molecules may be covalently linked by a contiguous galacturonic acid backbone forming a massive single molecule.

Homogalacturonans

As their name implies, homogalacturonans are homopolymers of 1,4 αD-galacturonic acid, which form straight chains up to 200 residues long (about 100 nm). These can be modified by the addition of single xylose side chains at intervals along the backbone giving xylogalacturonans. The carboxylic acid groups are highly methylesterized (–COOCH$_3$) during synthesis in the Golgi apparatus. On insertion into the wall they are therefore relatively neutral, but are later subjected to specific methylesterase attack. This process removes methyl groups, restoring the carboxylic acid groups, which can then be cross-linked by calcium ions, strengthening the wall matrix. Over 60 probable methylesterase genes have been found in the *Arabidopsis* genome database, an indication of the importance of this process to plant growth.

Mutants that reduce the levels of galacturonic acid by 25% are dwarfed and show reduced levels of pectin synthesis and reduced levels of cell adhesion.

Rhamnogalacturonan I

Rhamnogalacturonan I (RGI) is a major component of the pectic fraction of plant cell walls. It consists of a backbone composed of alternating D-galacturonic acid and L-rhamnose residues. About half of the rhamnose residues have neutral side chains consisting of either arabinose or galactose, or both, forming arabinans, galactans, and arabinogalactans, respectively. In the latter, single arabinosyl residues are attached to the galactose side chains. In growing carrot roots, there are distinct zonations of pectin types. Antibodies against (1→4)-β-galactan are found preferentially in the root cap and in the differentiating stele and cortical cells, while those against (1→5)-α-arabinan are found in the meristematic region. Although the underlying reasons for this topology are unknown, the observations serve to emphasize that pectin distribution in cell walls is very specific.

Rhamnogalacturonan II

Rhamnogalacturonan II (RGII) is a remarkable wall polysaccharide. Although it is complex, consisting of at least 12 different glycosyls linked by more than 20 different linkages, its structure has been conserved across all vascular plants. This implies that there are strong functional constraints that severely limit variation in the molecular structure. A further unusual feature is its low content in the wall, accounting for only 1–4% of polysaccharides in the relatively pectin-rich walls of dicots and only 0.1% of the wall in pectin-poor walls of monocots. RGII forms a significant component in red wine, from which it can be extracted in gram quantities. The pectin is released from the walls during pressing and seems resistant to microbial attack during fermentation.

Among the residues found in RGII are a number of sugars rarely found elsewhere in plants: L-aceric acid, D-apiose, L-galactose, 2-keto-3-deoxy-D-lyxo-heptulosaric acid, 2-keto-3-deoxy-D-manno-octulosonic acid, 2-*O*-methyl L-fucose, and 2-*O*-methyl D-xylose.

The RGII molecule has a linear backbone repeat of eight 1,4-linked αD-galacturonic acid residues with four side chains A, B, C, and D. Chains C and D are quite simple, with two residues in each. A and B are larger and branched, with B showing some variation in the terminal members of the branch. The A side chain is unusual in having a terminal galactose residue in the L-configuration. Other galactose residues are always in the D-configuration. These two chains (A and B) are attached to the backbone via apiosyl residues (D-Apif), which are important in forming cross-links to boron (Figure 7.24).

Pairs of RGII molecules are cross-linked by boron to form dimers (Figure 7.25). The boron is in the form of a tetravalent 1:2 diester that cross links two of the

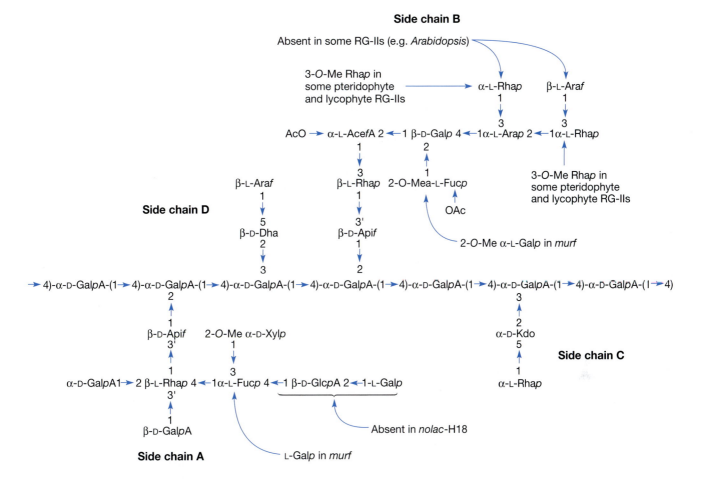

four apiosyl residues present in the dimer (one on each of the A and B chains). Current evidence favors the two A chain apiosyls, but there are claims that each dimer contains two boron atoms, so that the B chains would be cross linked as well.

The requirement for boron in the formation of RGII dimers (Figure 7.25) explains at least part, if not all of the boron-deficiency symptoms exhibited by plants, which include abnormalities in the structure and properties of primary walls. Also, monocots have both a lower RGII content in their walls and a lower boron requirement. Mutants that lack the transferase required for adding fucose to RGII cannot form these boron dimers and have similar defects in plant growth, while another mutant that cannot make borate cross-linked RGII has defects in reproductive tissues. Normal pollen cannot grow through mutant styles and pollen tube growth is inhibited when RGII synthesis is knocked out.

Functionally RGII dimers are probably covalently linked to homogalacturonans, as both have backbones of 1,4-linked galacturonic acid. These complexes also have covalent links to RGI, to make large pectic macromolecules. However, there is no hint of the reason for the high level of conservation exhibited by RGII. It is not found distributed evenly across the wall thickness,

Figure 7.24 RGII repeat structure. (Reprinted with permission from the *Annual Review of Plant Biology*, Volume 55: 109 by M.A. O'Neill, T. Ishii, P. Albersheim and G. Darvill. © 2003 by Annual Reviews, www.annualreviews.org.)

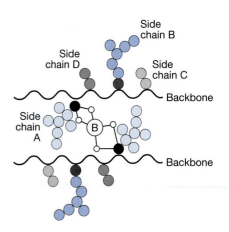

Figure 7.25 RGII dimer showing the apiosyl groups cross-linked by the boron diester. (Reprinted with permission from the *Annual Review of Plant Biology*, Volume 55: 109 by M.A. O'Neill, T. Ishii, P. Albersheim and G. Darvill. © 2003 by Annual Reviews, www.annualreviews.org.)

being concentrated near the plasma membrane surface. Mutants defective in RGII side chains fail to dimerize with boron and show reduced cell adhesion.

Expansins and extensins, proteins that play both enzymatic and structural roles in cell expansion

Expansins are proteins that promote the breakage of hydrogen bonds between microfibrils and the matrix polysaccharides xyloglucan and xylan. This allows adjacent microfibrils to slide past each other (creep), facilitating cell growth. Expansins are activated by low pH (pH 4–5), conditions that occur in the wall as a result of auxin-activation of the plasma membrane proton pump (the acid-growth hypothesis). Grasses contain another class of expansins, β-expansin, that act in the same way in grasses, but do not promote expansion of walls from other plants. There is some evidence that different members of each expansin class promote either cell extension growth or increases in cell diameter (width), or both.

Extensins are hydroxyproline-rich glycoproteins that appear to play a role in cell extension, either by interacting with microfibrils and the matrix, or by forming another cell wall meshwork. They are secreted into the cell as monomers that gradually become more cross-linked to each other. Transition from expansion growth to the final shape of the cell is accompanied by the gradual loss of solubility of the wall extensins, which may serve to lock the wall meshwork in its final form.

Lignin

Lignin is deposited in the secondary walls of xylem cells and some other vascular tissue cells, where it can make up to 15–50% of the wall mass. Lignins largely replace the xyloglucans, glucuronoarabinoxylans, and pectins in the matrix. They bind strongly to the microfibrils, preventing interfibril sliding and adding greatly to the wall strength. They are also hydrophobic, in marked contrast to the matrix components of the primary wall, and so radically alter the permeability and other physical properties of the secondary wall. This combination of orientated microfibrils and amorphous lignin in the xylem walls determines the structural strength of (woody) tree trunks, branches, and wood products.

Chemically lignin is described as a complex of phenylpropanoids, which cannot be solubilized from the walls without disrupting its structure. It is assembled from aromatic units collectively termed monolignols. These are coniferyl, ρ-coumaryl and sinapyl alcohols, products of phenylpropanoid metabolism in the cytoplasm. In lignins these are referred to as guaicyl (G), ρ-hydroxyphenyl (H), and syringyl (S) residues respectively. Secretion to the cell wall is accompanied by specific proteins which are thought to direct the polymerization processes leading to lignin formation. Although monolignols will form lignin *in vitro* under oxidative conditions, the polymer formed is more heterogeneous than that assembled *in vivo*. These oxidizing enzymes are usually peroxidases ($H_2O_2 \rightarrow 2H_2O$) but can also include laccases ($O_2 \rightarrow 2H_2O$). As polymerization takes place *in situ*, where the lignin is deposited, these enzymes must be secreted into the wall and diffuse to their site of action, initially at the middle lamella and outer regions of each wall separating a pair of cells.

There are differences between lignins of gymnosperms and angiosperms. Gymnosperm lignin is made mostly of coniferyl alcohol, with some ρ-coumaryl, while angiosperm lignin consists of equal amounts of coniferyl and sinapyl alcohols.

Summary

The key role of carbohydrates in metabolism means that it is essential that once generated they can be mobilized and broken down to useable forms within the plant as and when they are required. Plant carbohydrate metabolism is unique in that it is highly compartmentalized. The two major storage carbohydrates, sucrose and starch, are initially synthesized in the cytosol and plastids respectively. This necessitates a tightly regulated and integrated system of biochemical processes. In addition plants exhibit flexibility in carbohydrate metabolism to allow them to be adaptive to the range of environmental conditions they are subjected to. This is particularly clear with sucrose and starch degradation where multiple enzymes and isoforms may contribute to these processes. The overview of cell wall carbohydrates serves to emphasize the structural and synthetic complexity of the macromolecules underlying plant architecture. Synthesis of the sugar monomers and their precise polymerization requires a high level of control over the enzymes concerned. This control extends from the cytoplasm into the cell wall structure, with secreted enzymes playing important roles in wall assembly and wall modification during expansion growth. Our present level of understanding of the synthesis and assembly of plant cell walls is very incomplete, but we can expect that the efforts of many research laboratories around the world over the coming years will make substantial progress in resolving many of the remaining questions. We will then be able to fully appreciate the diversity and beauty of our vegetation.

Further Reading

Sucrose metabolism

Huber SC & Huber JL (1996) Role and regulation of sucrose-phosphate synthase in plants. *Annu. Rev. Pl. Physiol. Pl. Mol. Biol.* 47, 431–444.

Roitsch T & Gonzalez MC (2004) Function and regulation of plant invertases: sweet sensations. *Trends Pl. Sci.* 9, 606–613.

Sturm A & Tang G-Q (1999) The sucrose cleaving enzymes of plants are crucial for development, growth and carbon partitioning. *Trends Pl. Sci.* 4, 401–407.

Winter H & Huber SC (2000) Regulation of sucrose metabolism in higher plants: Localization and regulation of activity of key enzymes. *Crit. Rev. Biochem. Mol. Biol.* 35, 253–289.

These papers provide basic knowledge of the properties of sucrose phosphate synthase, sucrose synthase, and invertase.

Koch K (2004) Sucrose metabolism: regulatory mechanisms and pivotal roles in sugar sensing and plant development. *Curr. Opin. Pl. Biol.* 7, 235–246.

Provides a summary of recent advances in the understanding of sucrose metabolism regulation.

Starch biosynthesis, structure, and degradation

Lloyd JR, Kossmann J & Ritte G (2005) Leaf starch degradation comes out of the shadows. *Trends Pl. Sci.* 10, 130–137.

Smith AM, Zeeman SC & Smith SM (2005) Starch degradation. *Annu. Rev. Pl. Biol.* 56, 73–97.

Zeeman SC, Smith SM & Smith AM (2007) The diurnal metabolism of leaf starch. *Biochem. J.* 401, 13–28.

These reviews highlight recent advances in the understanding of starch biosynthesis, structure, and degradation.

Jobling S (2003) Improving starch for food and industrial applications. *Curr. Opin. Pl. Biol.* 7, 210–218.

Morell MK, Myers AM (2005) Towards the rational design of cereal starches. *Curr. Opin. Pl. Biol.* 8, 204–210.

These papers summarize the potential applications of starch particularly in the food and biotechnological industry.

Signals and signaling molecules in regulating carbohydrate metabolism

Cheung H-J, Sehnke PC & Ferl RJ (1999) The 14-3-3 proteins: cellular regulators of plant metabolism. *Trends Pl. Sci.* 4, 367–371.

Claeyssen E & Rivoal J (2007) Isoenzymes of plant hexokinase: occurrence, properties and functions. *Phytochemistry* 68, 709–731.

Geigenberger P, Kolbe A & Tiessen A (2005) Redox regulation of carbon storage and partitioning in response to light and sugars. *J. Exp. Bot.* 56, 1469–1479.

Huber SC (2007) Exploring the role of protein phosphorylation in plants: from signalling to metabolism. *Biochem. Soc. Trans.* 35, 28–32.

Lunn JE, Feil R & Hendriks JHM (2006) Sugar-induced increases in trehalose 6-phosphate are correlated with redox activation of ADP glucose pyrophosphorylase and higher rates of starch synthesis in *Arabidopsis thaliana. Biochem. J.* 397, 139–148.

Rolland F, Baena-Gonzalez E & Sheen J (2006) Sugar sensing and signaling in plants. Conserved and novel mechanisms. *Annu. Rev. Pl. Biol.* 57, 675–709.

Fructans

Cairns AJ (2003) Fructan biosynthesis in transgenic plants. *J. Exp. Bot.* 54, 549–567.

Ritsema T & Smeekens S (2003) Fructans: beneficial for plants and humans. *Curr. Opin. Pl. Biol.* 6, 223–230.

Vijn I, Smeekens S (1999) Fructan: more than a reserve carbohydrate? *Pl. Physiol.* 120, 351–359.

Trehalose metabolism

Eastmond PJ & Graham IA (2003) Trehalose metabolism: a regulatory role for trehalose-6-phosphate? *Curr. Opin. Pl. Biol.* 6, 231–235.

Vogel G, Fiehn O, Jean-Richard-dit-Bressel L, Boller T, Wiemken A, Aeschbacher RA & Wingler A (2001) Trehalose metabolism in Arabidopsis: occurrence of trehalose and molecular cloning and characterization of trehalose-6-phosphate synthase homologues. *J. Exp. Bot.* 52, 1817–1826.

General cell wall reference

Carpita N & McCann M (2000) The cell wall. In Biochemistry and Molecular Biology of Plants (Buchanan B, Gruissem R, Jones R eds), pp 52–108. American Society Plant Physiologists.

A good overview with excellent figures.

Cell wall sugars

Fry SC (2004) Primary cell wall metabolism. *New Phytol.* 161, 641–675.

A useful review of cell wall sugars and their arrangement in cell wall macromolecules.

Microfibril sliding in walls of expanding cells

Cosgrove D (1997) Relaxation in a high-stress environment. *Pl. Cell* 9, 1031–1041.

A key reference on this topic.

Microstructure of the cell wall

Booten TJ, Harris PJ, Melton LD & Newman RH (2004) Solid-state 13C-NMR spectroscopy shows that the primary cell walls of mung bean (Vigna radiata L.) occur in different domains. *J. Exp. Bot.* 55, 571–583.

Application of modern technology to cell walls has revealed unsuspected heterogeneity of structure within the wall.

Cellulose synthesis

Somerville C (2006) Cellulose synthesis in higher plants. *Annu. Rev. Cell Dev. Biol.* 22, 53–78.

Review of cellulose synthesis.

Richmond T & Somerville CR (2000) The cellulose synthase superfamily. *Pl. Physiol.* 124, 495–498.

Paper on cellulose synthase genes.

Tanaka K, Murata K, Yamazaki M, Onosato K, Miyao A & Hirochika H (2003) Three distinct rice cellulose synthase catalytic subunit genes required for cellulose synthesis in the secondary wall. *Pl. Physiol.* 133, 73–83.

A paper on the expression of cellulose synthase genes.

Paredez AR, Somerville CR, Ehrhardt DW (2006) Visualization of cellulose synthase demonstrates functional association with microtubules. *Science* 312, 1491–1495.

Study of *in vivo* activity of cellulose synthases revealed with a fluorescent marker tag.

Pectins

O'Neill MA, Ishii T, Albersheim P & Darvill AG (2004) Rhamnogalacturonan II: structure and function of a borate cross-linked cell wall pectic polysaccharide. *Annu. Rev. Pl. Biol.* 55, 109–139.

Scheller HV, Jensen JK, Sorensen SO, Harholt J & Geshi N (2007) Biosynthesis of pectin. *Physiol. Plantarum* 129, 283–295.

These two references provide a good introduction to the literature on pectins.

Nitrogen and Sulfur Metabolism

8

Key concepts

- Nitrogen is one of the most abundant elements in plants but is often the limiting nutrient for growth.

- Biological nitrogen fixation is performed by prokaryotes.

- In symbiotic nitrogen fixation prokaryotes supply host plants with reduced nitrogen in the form of ammonium.

- Mycorrhizae are symbiotic interactions between soil fungi and plant roots that provide the plant with nutrients, including nitrogen.

- Nitrate reductase catalyzes the reduction of nitrate to nitrite in the cytosol.

- Nitrite reductase catalyzes the reduction of nitrite to ammonium in the plastids.

- Ammonium is assimilated by the combined action of glutamine synthetase and glutamate synthase.

- Sulfur is an essential macronutrient but only represents 0.1% of plant dry matter.

- A number of sulfate transporters with different affinities for sulfate have been identified.

- Sulfate is activated by reaction with ATP to form adenosine 5'-phosphosulfate in a reaction catalyzed by ATP sulfurylase.

- Amino acids are precursors of a wide range of nitrogen-containing compounds.

- Amino acid biosynthesis requires carbon and nitrogen and must occur in a balanced manner to ensure that supply meets demand.

- Aminotransferase reactions are central to amino acid metabolism by redistributing nitrogen to a range of amino acids.

- Storage proteins are mobilized to provide essential organic building materials for new stems and leaves.

Nitrogen and sulfur must be assimilated in the plant

Plants are autotrophic organisms that are able to synthesize organic molecular components out of inorganic nutrients from the environment. This process of nutrient assimilation is essential for plant growth and development. It is also necessary for the production of a wide range of organic substances including nucleic acids, amino acids, enzyme cofactors, pigments, and lipids. Many mineral nutrients are absorbed from the soil into the roots and then incorporated directly into organic compounds. In contrast, nitrogen and sulfur must both be assimilated into organic metabolites via a complex series of high-energy requiring biochemical reactions. Nitrogen is mainly used for production of macromolecules, whereas sulfur plays a crucial part in the catalytic or electrochemical functions of biomolecules in cells. Plants need a source of inorganic nitrogen. Virtually all (99.95%) of the nitrogen on the biosphere is present in atmospheric or dissolved nitrogen, but this is only available to a few plants. Most plants get their cellular nitrogen from nitrate in cultivated or aerated neutral soils, or ammonium in acid or waterlogged soils. In some plants nitrogen is fixed via a symbiotic relationship with nitrogen-fixing bacteria, such as *Rhizobium*, or via a mychorrizal relationship between plants and actinomycete fungi, converting molecular nitrogen into ammonia. Nitrate assimilation initially converts nitrate via nitrite and ammonium into the amide nitrogen of glutamine for its subsequent incorporation into organic material. Ammonium will be used here as a collective reference to both forms found in plant cells: ammonia (NH_3) and ammonium (NH_4^+). Sulfate assimilation involves the conversion of sulfate via adenosine 5′-phosphosulfate to sulfite and sulfide before being converted into the amino acid cysteine. An indication of the large energy requirements of nitrogen and sulfur assimilation is shown by the observation that the rapid oxidation of nitrogen and sulfur compounds is the basis of many explosives. Both nitrogen and sulfur exist in multiple oxidation states and it is this property that contributes to their versatility as compounds in the cell. This chapter focuses on the primary reactions of nitrogen and sulfur assimilation. It also examines the biosynthesis of the various amino acids required by the cell. Finally, the role and types of storage proteins found in higher plants are discussed.

Apart from oxygen, carbon, and hydrogen, nitrogen is the most abundant element in plants

Nucleoside phosphates and amino acids, which form the building blocks of nucleic acids and proteins, respectively, are an example of the many nitrogen-containing biochemical compounds found in plant cells. Nitrogen is often the limiting nutrient for growth. This limitation is apparent in both agricultural and natural ecosystems, where the addition of inorganic nitrogen fertilizers often leads to marked increases in plant growth and productivity. While soil nitrogen is clearly important for plant growth, the most abundant source of global inorganic nitrogen is in the atmosphere, which contains approximately 80% by volume of nitrogen (as N_2 gas). However, much of this nitrogen is not directly accessible to living organisms. To obtain nitrogen from the atmosphere it is necessary to break an extremely stable triple covalent bond between two nitrogen atoms to produce ammonium. This reductive conversion, or nitrogen fixation, occurs through a number of alternative biological and nonbiological routes (Table 8.1). Once fixed into ammonium, nitrogen enters the biogeochemical cycle, passing through a number of organic or inorganic forms before returning to molecular nitrogen (Figure 8.1). The interconversion between different oxidation states contributes to the global nitrogen cycle (Figure 8.1). Assimilatory nitrate reduction converts nitrate (with a charge of +5) to nitrous oxide (with a charge of +1) and nitrogen (with a charge of 0). Nitrification operates in the opposite direction converting ammonium (with a charge of −3) to nitrite (with a charge of +4) and nitrate (with a charge of +5). Microorganisms have a key role in nearly all aspects of nitrogen availability and therefore support life on earth.

Table 8.1 Biological and nonbiological nitrogen fixation processes

	Total nitrogen fixed per year
Non-biological nitrogen fixation	
Atmospheric fixation via lightning and photochemical conversion of molecular nitrogen to ammonium	10 Tg
Industrial fixation of nitrogen via Haber-Bosch reaction used in nitrogenous fertilizer production	50 Tg
Fossil fuel combustion	20 Tg
Biological nitrogen fixation	
Agricultural land	90 Tg
Forest and nonagricultural land	50 Tg
Marine	35 Tg

The standard unit of measurement is the teragram, Tg, which is equivalent to 10^6 metric tons

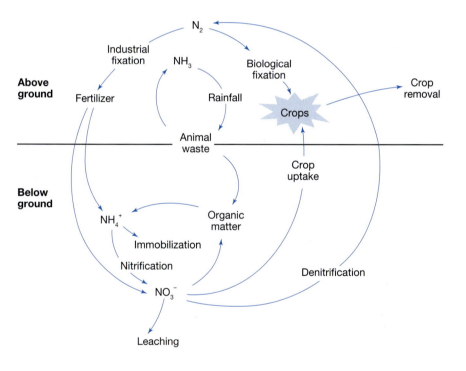

Figure 8.1 Biogeochemical nitrogen cycle.

Nitrogen fixation: some plants obtain nitrogen from the atmosphere via a symbiotic association with bacteria

During nitrogen fixation two moles of ammonia are produced for every one mole of nitrogen gas reduced. This reductive reaction uses 16 moles of ATP together with a supply of electrons and protons (H^+ ions). The overall reaction, which is catalyzed by nitrogenase is:

$$N_2 + 8e^- + 8H^+ + 16ATP \longrightarrow 2NH_3 + H_2 + 16ADP + 16P_i$$

This reaction, catalyzed by nitrogenase (Figure 8.2), is unique to prokaryotes, occurring in representatives of archaebacteria and eubacteria, including cyanobacteria. Nitrogenase occurs in symbiotic nitrogen-fixing bacteria, such as *Rhizobium* spp. (rhizobium, plural rhizobia), and is also found in a small number of species of free-living bacteria. Based on amino acid analysis, nitrogenase is highly conserved between symbiotic and free-living bacteria. Nitrogenase is made up of two easily separable proteins, dinitrogenase reductase, an iron protein also known as component II, and dinitrogenase, a molybdenum–iron protein also known as component I (Figure 8.2). The genes associated with nitrogen fixation are called *nif* and have been well characterized from studies on a variety of organisms. The highly conserved nature of these genes supports the observed similarity of nitrogenase in the different nitrogen-fixing organisms.

Dinitrogenase reductase (iron protein, component II of nitrogenase) is encoded by the *nifH* gene. It is a homodimer of 60–64 kDa with two identical subunits of 30–32 kDa forming a 4Fe–4S cluster and containing two magnesium-ATP binding sites on the iron protein. It is a one-electron carrier, which, after reduction, binds two molecules of ATP leading to a change in the protein conformation and an increase in negativity of the redox potential of the 4Fe–4S cluster from –0.25 V to –0.40 V (Figure 8.2). Once an electron has been transferred to the dinitrogenase, the ATP molecules are hydrolyzed and released from the protein. As a result the protein conformation with the lower redox potential is restored and the dinitrogenase reductase is able to take up another electron from ferredoxin or other external electron donors (Figure 8.2).

The dinitrogenase (MoFe protein, component I of nitrogenase) protein is a tetramer, $\alpha_2\beta_2$, of 220 kDa total molecular mass. The α subunit, encoded by the *nifD* gene, has a molecular mass of approximately 56 kDa. The β subunit, encoded by the *nifK* gene, has a molecular mass of about 60 kDa. Dinitrogenase has two independently acting catalytic centers, a P cluster made up of two 4Fe–4S clusters and an iron–molybdenum cofactor (Figure 8.2). The latter is a large redox center composed of an Fe_4S_3 linked to $Fe_3 MoS_3$ via three inorganic sulfide groups (Figure 8.3). Homocitrate is also present and it is linked to molybdenum via oxygen atoms within its hydroxyl and carboxyl groups. The imidazole ring of a histidine residue of the dinitrogenase protein is another ligand of molybdenum. The function of the molybdenum atom is unclear, although it has been suggested that its presence may allow a more favorable geometry and electron structure at the center of the

Figure 8.2. Nitrogenase enzyme consists of two proteins: the iron protein, dinitrogenase reductase, and the molybdenum–iron protein, dinitrogenase. The reactions occur once nitrogen is bound to the nitrogenase enzyme complex. The dinitrogenase reductase is reduced first by electrons donated by ferredoxin (Fd). Then the dinitrogenase reductase binds ATP and reduces the dinitrogenase. The dinitrogenase donates electrons to N_2, producing HN=NH. In the further cycles of this process, each requiring electrons donated by ferredoxin, HN=NH is reduced to H_2N-NH_2 and then to $2NH_3$. The reduced ferredoxin (Fd$_{red}$) supplying electrons for this process is generated by photosynthesis, respiration, or fermentation.

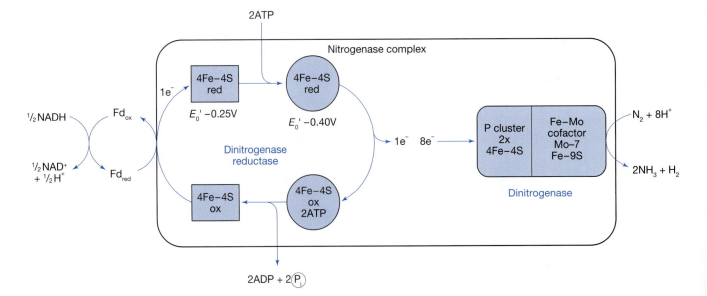

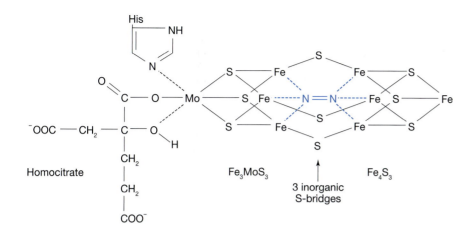

Figure 8.3 The dinitrogenase MoFe protein has three inorganic sulfide bridges linking Fe$_4$S$_3$ to Fe$_3$MoS$_3$. The molybdenum is also ligated with homocitrate and the histidine side chain of the protein. Successive uptake of electrons facilitates the binding of one N$_2$ molecule and its reduction to two molecules of ammonia.

dinitrogenase. It has also been proposed that the N$_2$ molecule could be bound in the cavity of the iron-molybdenum cofactor (FeMoCo) center (shown in blue in Figure 8.3), with the electrons needed for nitrogen fixation being transferred by the P cluster to the FeMoCo center.

Besides N$_2$, dinitrogenase is also able to reduce other substrates, such as protons, to molecular hydrogen. The unique property of the enzyme is its ability to reduce triple-bond substrates (excepting HC≡N). Under natural conditions this reduction of protons to molecular hydrogen can be substantial and can compete with N$_2$ for electrons from nitrogenase. For example, in rhizobia between 30 and 60% of the energy supplied to nitrogenase can be lost as H$_2$. Any loss of energy leads to a decreased efficiency in nitrogen fixation. To overcome these problems and improve the efficiency of nitrogen fixation, some rhizobia possess an uptake hydrogenase, which dissociates the hydrogen formed and provides electrons for nitrogen fixation.

Under limiting molybdenum some bacterial species also have an alternative dinitrogenase, containing either vanadium or iron instead of molybdenum. These alternative dinitrogenase enzymes are called vanadium nitrogenase and iron-only nitrogenase, respectively. The distribution of these alternative dinitrogenases is less uniform with, for example, *Azotobacter vinelandii* containing both, *Rhodobacter capsulatus* containing the iron-only form and *Rhizobium* spp. containing neither of these alternative forms. Both vanadium- and iron-only dinitrogenase are hexameric proteins, $\alpha_2\beta_2\delta_2$ with the δ protein having a subunit molecular mass of 10 kDa. The synthesis of the vanadium- and iron-only nitrogenases are under the control of *vnfDGK* genes. Synthesis of either the standard or alternative dinitrogenase is transcriptionally controlled. Both the vanadium- and iron-only nitrogenase have lower specific activities than the molybdenum-dependent dinitrogenase. Also in the vanadium- and iron-only nitrogenase, a greater amount of electron flow is directed towards H$_2$.

Nitrogen fixation is a unique biochemical reaction consuming energy-rich compounds but also requiring strong reductants. Nitrogenase, through the dinitrogenase-reductase component, is extremely sensitive to oxygen, being rapidly and irreversibly inactivated with a typical half decay time of 30–40 seconds. Where nitrogen fixation is occurring in anaerobic bacteria this sensitivity to oxygen is not a problem. However, unlike aerobic respiration, fermentation and anaerobic respiration do not oxidize reduced carbon very effectively, restricting the amount of available ATP to drive anaerobic nitrogen fixation. In contrast, aerobic bacteria produce high amounts of ATP from aerobic metabolism, but must overcome the disadvantage of the O$_2$ sensitivity of nitrogenase. A number of free-living nitrogen-fixing organisms have mechanical, temporal, or biochemical barriers to separate the process of nitrogen fixation from aerobic metabolism. Other organisms produce specialized structures, for example heterocysts, in which nitrogen fixation occurs. The nodule environment generated by the interaction between rhizobia and its plant hosts also serves to maintain a low oxygen environment (see page 245).

Symbiotic nitrogen fixation involves a complex interaction between host plant and microorganism

A limited number of species of free-living nitrogen-fixing prokaryotes, for example, cyanobacteria (blue–green algae), form ammonia from gaseous nitrogen (N_2). More commonly, a number of species of nitrogen-fixing bacteria have a symbiotic relationship with plants (Table 8.2). In this latter situation the bacteria fix atmospheric N_2 and provide the plant with reduced N_2 in the form of NH_4^+, while the host plant provides a specialized ecological niche and a source of energy (a carbon source) for the growth and function of the bacteria. The most common example of a host plant that enters into symbiosis with nitrogen-fixing bacteria is that of the legumes (family Leguminoseae), which house *Rhizobium* bacteria within specialized root structures, called nodules. There are approximately 12,000 species of legumes, including soybean (*Glycine max*), lentil (*Lens culinaris*), pea (*Pisum sativum*), clover (*Trifolium* spp.), lupins (*Lupinus luteus*), and lucerne/alfalfa (*Medicago sativa*). Ninety percent of these legumes form a symbiosis with species of *rhizobia* and, as seen in temperate climates, legumes used in crop rotation can lead to the fixation of 100–400 kg N_2 ha^{-1} year^{-1}. As a result rhizobia have been described as green manure. The symbiosis of *Azolla*, the water fern, with the cyanobacterium *Anabaena* provides rice fields with nitrogen, while the nitrogen-fixing actinomycetes of the genus *Frankia* form a symbiosis with woody plants, including alder (*Alnus* spp.), sweet fern (*Comptonia peregrina*), beefwood (*Casuarina* spp.), and others.

Nodule-forming bacteria (Rhizobiaceae) are composed of the three genera *Rhizobium*, *Bradyrhizobium*, and *Azorhizobium*

Different legumes associate with different species of Rhizobiaceae (Table 8.3). The Rhizobiaceae are a family of aerobic gram-negative rods that live in the soil and can grow heterotrophically, although they can only fix nitrogen when they enter into a symbiosis with a host plant.

A so-called controlled infection allows the uptake of rhizobia into the host plant (Figure 8.4). Although the molecular basis of the specificity and recognition is only partially known, the interaction begins with a specific exchange of signal

Table 8.2 Nitrogen-fixing bacteria

Free-living nitrogen-fixing bacteria		Symbiotic nitrogen-fixing bacteria with plants	
Aerobic	**Anaerobic**	**Legumes**	**Trees, woody shrubs**
Azotobacter vinelandii	Purple sulfur bacteria	Gram-negative bacteria *Rhizobium*	Gram-positive actinomycete genus (*Frankia*)
Azospirillum lipoferom	Purple nonsulfur bacteria		
Some cyanobacteria, e.g. *Anabaena, Nostoc*	Green sulfur bacteria		Cyanobacteria

Table 8.3 Association of different species of the Rhizobiaceae with legumes

Nodule-forming bacteria (Rhizobiaceae)	Legume species
Rhizobium spp.	Pea (*Pisum sativum*)
Bradyrhizobium spp.	Soybean (*Glycine max*)
Azorhizobium spp.	Tropical legumes

compounds. When soil is deficient in nitrate the host plants excrete flavonoids (Chapter 11) into the rhizosphere, which induces the rhizobia to form species-specific nodulation factors (nod factors). Nod factors are lipo-oligosaccharides that acquire a high structural specificity by acylation, acetylation, and sulfonation. *Nod* gene induction results in secretion of lipochitin oligosaccharides, which bind to specific plant receptor kinases and initiate a complex signaling pathway in the root hairs. Rhizobia have a large number of genes that are only activated upon interaction with the host. Bacterial genes for proteins involved in nitrogen fixation are *nif* and *fix* genes, and those forming root nodules are *nod* genes.

The bacteria invade the plant and cause the formation of a nodule by inducing localized and hormonally induced proliferation of the host cells. The host plant root hair curls, trapping the rhizobia, which then enter through a tubular cellulose infection thread formed by the plant (Figure 8.4). The infection thread then extends into the root cortex, forming branches before infecting the primary nodule meristem. Depending on the host plant, infection droplets form either at the tip of short intracellular infection threads or along the infection thread where the cell wall gets disrupted and rhizobial cells come into direct contact with the plasma membrane of the host cell by a process resembling endocytosis. Within the resulting symbiosome the bacteria stop dividing and begin to enlarge and to differentiate into bacteroids, which are enclosed within a plant cell membrane called the peribacteroid membrane. Depending on the legume species, one or more bacteroids may be found within the symbiosome. That is, individual symbiosomes may fuse and/or bacteroids divide within the symbiosome leading to the formation of determinate nodules. Alternatively, symbiosomes further divide together with the bacteroid leading to a symbiosome containing a single bacteroid and the formation of indeterminate nodules. Nodule-specific proteins synthesized by the host plant during

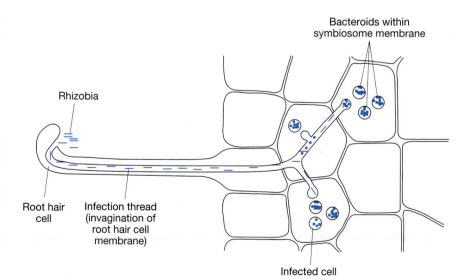

Figure 8.4 Interaction of rhizobia with a plant root hair leads to a controlled infection of the host plant cell. (Based on Plant Biochemistry and Molecular Biology, Figure 11.2, H.W. Heldt, (1997). Oxford University Press.)

nodule formation are known as nodulins. These include enzymes for carbohydrate degradation, the tricarboxylic acid cycle, the synthesis of glutamine and asparagine and, where needed, ureides.

The nodule environment is generated by interaction between legume plant host and rhizobia

In functional nodules a bacterial infected central zone is surrounded by layers of uninfected cells that occupy a region of the nodule called the nodule cortex. The nodule cortex contains phloem and xylem surrounded by a vascular endodermis. The central zones of many legume nodules contain a mixture of infected and uninfected cells. In contrast lupin (*Lupinus albus*) and peanut (*Arachis hypogaea*) can, depending on the growth conditions, have nodules that contain no uninfected cells. A network of gas-filled intracellular spaces are present throughout the entire central zone and provide a low resistance diffusion pathway for any metabolically active cells present. This network ensures an oxygen supply and facilitates the removal of N_2 and CO_2.

Rhizobia, when free-living, use a respiratory chain similar to that of the mitochondrial respiratory chain (see Chapter 6). During differentiation of some strains of rhizobia to bacteroids an additional electron transport pathway is formed. This uses, via a hydrogen oxidizing system, the hydrogen produced as a by-product of the nitrogenase-catalyzed nitrogen reduction. It is composed of an uptake hydrogenase and an electron transport chain that uses oxygen as its terminal acceptor.

Nodules have a much greater rate of respiratory oxygen consumption, approximately fourfold more than that recorded in an equivalent mass of root tissue. This oxygen consumption is needed to meet the large ATP demands within the bacteroids. However, as oxygen is also an extremely potent irreversible inhibitor of nitrogenase, then a balance is required whereby a high flux of oxygen is provided to the bacteria-infected cells but at an extremely low oxygen concentration. The entry of oxygen into the nodule is controlled in part by the nodule parenchyma, providing a variable permeability barrier while maintaining low oxygen concentrations. Leghemoglobin also plays an important part in controlling the free oxygen concentration in the nodule (see page 245). Any alteration in the external concentration of oxygen is sensed and reversed, via a currently unknown compensation mechanism.

Bacterial respiration in the nodules is much less limited by low oxygen concentrations than plant respiration. The nodule-specific bacteroid cytochrome oxidase required for nitrogen fixation has a very low K_m for oxygen of approximately 8 nM. Free-living rhizobia typically have a cytochrome oxidase with a K_m for oxygen of approximately 50 nM. In contrast, the mitochondrial cytochrome oxidase of plants has a K_m for oxygen of 100 nM. The low oxygen concentration in the nodule cells of 4–70 nM limits oxidative metabolism. When compared with the oxygen concentration of 250 µM in cells in equilibrium with the air, this suggests the outer cells of the nodule cortex could have intracellular oxygen concentrations 10,000 times higher than the infected cells. When there is a relatively rapid increase in the oxygen concentration bacterial respiration and nitrogen fixation then increase. However, the permeability of the nodule quickly decreases, returning the oxygen concentration to its original level and supporting the submaximal nitrogenase activity. It is this factor that is considered to ensure stable symbiotic nitrogen fixation. Bacterial respiration consumes oxygen and the ATP generated can be subsequently used by nitrogenase. There is a balance necessary for this entire process. If the external oxygen concentration rises too much then the nitrogenase is inhibited, leading to a reduced rate of catalysis and a decreased ATP consumption. The resulting increased ATP:ADP ratio down-regulates respiration resulting in less oxygen being consumed and therefore the concentration of oxygen increases. This leads to further inactivation of the nitrogenase and eventually leads to a metabolic collapse.

There are some key components contributing to the mechanism of regulating oxygen concentration in legume nodules. As nodule metabolism is oxygen limited, there is limited capacity to consume additional oxygen if it diffuses into the nodule. Nodules can adjust their resistance to oxygen diffusion. Although the mechanism for this latter process is unknown, the innermost region of the nodule cortex plays an important part with the tightly packed cells having few intercellular spaces. Depending on the concentration of oxygen surrounding the nodulated roots, the number of cell layers can decrease or increase. Where there are fewer smaller spaces there will be less gas-phase and more liquid-phase oxygen diffusion through the cell layer.

Leghemoglobin, an oxygen-binding protein produced by the host plant, has an active role in the regulation and delivery of oxygen into infected cells. Leghemoglobin is a part plant and part bacterium-encoded single heme moiety that binds only one oxygen molecule. It is transcribed primarily in infected nodule cells and produced in sufficient amounts to reach millimolar concentrations in the plant cytoplasm. In symbiotic root nodules the leghemoglobin, which facilitates the movement of oxygen from the surface of the cell to the symbiosome via diffusion, increases the flux of oxygen moving through the plant cytoplasm to bacteroids while controlling the free oxygen concentration. In binding oxygen, the leghemoglobin is also able to moderate changes in the oxygen concentration that result from fluctuations in the respiration rate or permeability of the diffusion barrier. The affinity of leghemoglobin for oxygen is also influenced by pH and organic acids.

The low oxygen concentration within the nodules is a key factor regulating bacteroid nitrogenase expression. In some rhizobial species an oxygen sensitive hemoprotein kinase, FixL, controls a regulating cascade that activates transcription of nitrogen-fixing genes. FixL is part of a two-component regulatory system in which FixL phosphorylates FixJ. The phosphorylated FixJ then activates the transcription of other regulatory proteins, including NifA and FixK, which control the expression of both *nif* and *fix* genes. Only when conditions are correct will bacteroids express nitrogenase and the associated proteins needed for reduction of nitrogen to ammonia.

Nitrogen fixation is energy expensive, consuming up to 20% of total photosynthates generated

Carbon and energy sources in the form of dicarboxylic acids are provided by the host cells to the bacteroids. Malate is the main substrate delivered to the bacteroids, so nodules also include a malate translocator on the symbiosome membrane. Malate is formed from sucrose and delivered to the bacteroids by the sieve tubes. Sucrose is metabolized by sucrose synthase (see Chapter 7), degraded by glycolysis to phospho*enol*pyruvate, which, after carboxylation to oxaloacetate (by phospho*enol*pyruvate carboxylase), is reduced to malate (by malate dehydrogenase). Providing there is a supply of acetyl CoA (Chapter 6), the malate taken up into the bacteroids can be oxidized by the tricarboxylic acid (TCA) cycle, with the reducing equivalents generated ultimately providing ATP for nitrogen fixation.

The product of nitrogen fixation is ammonium that enters into organic compounds by the process of ammonium assimilation. Ammonium, a product of nitrogen fixation, is delivered via a specific symbiosome membrane channel to the host cell. The ammonium is assimilated in the plant nodule cytosol and organelles via plant glutamine synthetase and NADH glutamate synthase (see page 259). The fate of the glutamine depends on the plant host and the nitrogenous transport compounds it uses (Table 8.4). Dicarboxylic acids act as the source of carbon backbones for these nitrogen-containing transport compounds. One major group of legumes (primarily temperate zone legumes) exports ammonium mainly as the amides glutamine and asparagine. Amides are readily synthesized from TCA cycle intermediates and can be metabolized in leaves by transamination and transamidation reactions. Tropical legumes mainly export nitrogen from the nodules in the form of the ureides, allantoin and allantoic acid, which have a very high nitrogen to carbon ratio and are a carbon-conserving form in which to move nitrogen around. Ureides are formed in the host cells via a complex synthetic pathway involving synthesizing purines in the infected cells and then oxidizing the purines in neighboring uninfected cells via xanthine and uric acid to allantoin and allantoic acid. Amides and ureides are both mainly exported from the nodule through the xylem. The relatively low solubility of ureides at less-than-tropical temperatures has probably militated against their use as nitrogen transporters in temperate-zone legumes.

Mycorrhizae are associations between soil fungi and plant roots that can enhance the nitrogen nutrition of the plant

Another symbiotic interaction that provides higher plants with a source of nitrogen, among other nutrients, occurs between higher plants and mycorrhizal fungi. The associations are species specific between the two partners. Some mycorrhizal fungi have a wide range of host specificity, others are found on a much more restricted range of hosts. It is probable that all plants in a natural vegetation have some sort of mycorrhizal association. Crop plants grown on soils lacking the appropriate mycorrhizal fungus may not perform as well as crops with such fungal associations. Plants benefit by increased access to soil resources, especially phosphorus and nitrogen, but also potassium and immobile metal ions such as those of copper and zinc. Orchids receive organic carbon from their fungal associations, indeed orchid seeds (which lack seed reserves) could not germinate without the correct fungal partner to provide respiratory substrates to the seed. Our knowledge of the importance of mycorrhizae to plant growth and success is far from complete. For example, it seems that in some environments mycorrhizae can be disadvantageous to some plants and advantageous to others, so they influence the composition of the vegetation that is finally established on the site.

Table 8.4 Range of forms of nitrogen exported from the nodules of different classes of legumes

Class of legume	Legume	Form of nitrogen exported from nodule
Temperate	Alfalfa (*Medicago sativa*)	Glutamine, asparagine
	Pea (*Pisum sativum*)	Glutamine, asparagine
	Trefoil (*Lotus* spp.)	Asparagine
Tropical	Soybean (*Glycine max*)	Allantoin, allantoic acid
	Cowpea (*Vigna unguiculate*)	Allantoin, allantoic acid

Two main types of mycorrhizal association are recognized, depending on the extent of the fungal association with the host root. Associations in which the fungus is kept outside the host root cells, but forms a dense sheath over the root and penetrates between the epidermal and cortical cells, are called ecto-mycorrhizae. Those in which the fungus penetrates into the cells, forming extensive intracellular hyphal growths and forming cysts (vesicles) within the cell wall are termed arbuscular mycorrhizae. Both types provide a greatly enhanced foraging capability for the plant. A much greater volume of soil can be more efficiently exploited.

Pine (*Pinus* spp.) roots that are ectomycorrhizal can take up over three times as much soil phosphorus and nearly twice as much soil nitrogen as equivalent nonmycorrhizal roots. Sitka spruce (*Picea sitchensis*) plantations are highly dependent on mycorrhizal associations for their phosphorus supply. When establishing new forests on arable, pasture, or wasteland sites that lack the appropriate soil fungi, they are often planted with a nurse tree, larch (*Larix* spp.), to assist with mycorrhizal formation.

Arbuscular mycorrhizal fungi retain their hyphal links to the network of hyphae in the soil. These provide about 80% of a plant's phosphorus requirements and 25% of its nitrogen. Both ectomycorrhizal and arbuscular mycorrhizal fungi receive carbon compounds from the host plants.

The evolution of such highly specific associations must have involved development of parallel biochemical capabilities to allow for the efficient exchange of nutrients to the benefit of both partners. This is in marked contrast to the chemical and signaling warfare between plants and pathogenic fungi. In this chapter we will only consider nitrogen compounds that are obtained or produced by the fungus and utilized by the host plant. Ectomycorrhizal fungi have evolved proteases that are secreted to release nitrogen compounds from proteins in the soil, mostly derived from dead invertebrates. *Conococcum* secretes both alkaline (94 kDa, pH optimum 8.2) and acidic (70 kDa, pH optimum 5.0) proteases, while the most active protease from *Hebeloma* is smaller, 37.8 kDa, and has a pH optimum of 2.5.

Protease secretion is induced by the presence of proteins normally encountered in the soil. Other proteins, such as casein and bovine serum albumin, both readily available commercially, are less effective in inducing secretion of proteases. Secreted proteases readily diffuse in the soil water, attacking the proteins and releasing amino acids. The host plant demand for nitrogen influences the rate of depletion of soil nitrogen by the fungus.

Amino acids released by proteases are taken up by amino acid transporters across the fungal plasma membrane. Mycorrhizal fungi also take up mineral nitrogen. Nitrate can be taken up by these fungi, but there is considerable variation in the response to added nitrates. Some mycorrhizal roots showed a significant increase in nitrate absorption compared with nonmycorrhizal roots, while others showed no effect. The nitrate is reduced to ammonium by nitrate reductase (NR) and nitrite reductase (NiR) in the fungi. In acid soils, nitrification (Figure 8.1) is inhibited and ammonium is the main form of mineral nitrogen, and this is readily taken up by the fungal partners. Ammonium is taken up by an ammonium-proton symport, K_m 150–200 μM for methyl-amine (an ammonium analog) in *Paxillus*, and 55 μM in *Laccaria*. Ammonium uptake rates in ectomycorrhizal roots are much higher than in nonectomycorrhizal roots.

Most fungi assimilate ammonium via glutamate dehydrogenase, in contrast to plant roots and shoots, which assimilate ammonium through the combined action of glutamine synthetase and glutamate synthase (see page 259). There is evidently a lack of uniformity in the transfer of fungus to host; the fungal glutamate dehydrogenase is suppressed in the *Laccaria/Eucalyptus* mycorrrhiza but active in the *Hebeloma/Eucalyptus* mycorrhiza.

The transfer from fungus to plant host in mycorrhizae involves an elaborate interface that maximizes the efficiency of transfer and minimizes losses by diffusion back to the soil. Arbuscular mycorrhizae especially have developed an intracellular interface where the wall material is reduced to a minimum and membrane-bound enzymes of both partners generate energy gradients conducive to the active transport of nutrients in both directions.

Most higher plants obtain nitrogen from the soil in the form of nitrate

In well-aerated soils, with a neutral or slightly alkaline pH, the most abundant source of inorganic nitrogen is nitrate (NO_3^-). The reason for this is that soil bacteria that carry out nitrification (Figure 8.1) are facultative aerobes that function at neutral or alkaline pH. Hence, soils that are anaerobic and/or acidic (e.g. acid bogs) tend to be high in ammonium because of the lack of nitrification. The process of nitrate assimilation can be defined as the conversion of nitrate to ammonium. It begins with the uptake of nitrate into the root cell. Nitrate usually diffuses from the soil solution into the apoplasm (cell walls and intracellular spaces) where it is taken up by specific nitrate transporters on the plasma membrane of epidermal and cortical root cells. The nitrate in the symplasm (the cytoplasm of the cells) may be reduced directly in the root tissues or it may move across the Casparian strip and into the xylem for transport into the shoots. In the case of epiphytes, or where nitrate foliar fertilizer is applied, nitrate can also be taken directly in by the leaves. Nitrate is then transported into the vacuole where it can accumulate at high concentrations of up to 20 mM. Irrespective of the initial route of nitrate uptake, the roots, stems, and leaf midribs are all possible storage organs for nitrate. Nitrate can also be transported in the xylem to the shoot. A range of physiological and environmental factors affect nitrate allocation and storage throughout the plant.

Nitrate assimilation involves two enzymes (Figure 8.5). Nitrate reductase (NR) catalyzes the first step reducing nitrate (NO_3^-) to nitrite (NO_2^-), which requires two electrons to change the charge on the nitrogen atom. NAD(P)H serves as the electron donor for this reduction, which occurs in the cytosol. The second step occurs in the chloroplast or other plastid types and is catalyzed by nitrite reductase (NiR), which reduces nitrite to ammonium (NH_4^+). This reduction requires six electrons donated by reduced ferredoxin. Ammonium is then incorporated into amino acids by the glutamine synthetase and glutamate synthase pathway (GS/GOGAT pathway, Figure 8.5). Depending on the plant and environmental conditions, nitrate assimilation will occur to different degrees

Figure 8.5 The pathway of nitrate and ammonium assimilation. Abbreviations: Gln, glutamine; Glu, glutamate.

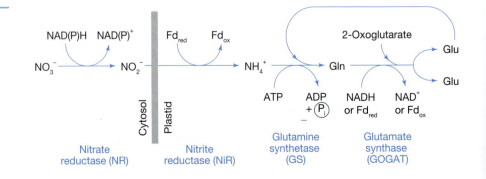

in the roots and shoots. For example, the roots of temperate legumes are the main site of nitrate assimilation at external nitrate concentrations of up to 1 mM. If the nitrate concentration is higher than 1 mM, then shoot assimilation becomes important. In contrast, the shoot is the primary site of nitrate assimilation in a number of tropical and subtropical cereals and grain legumes regardless of the external nitrate concentration. Ammonium produced by reduction of nitrate is usually assimilated in the same cells, in which it is generated; NH_4^+ is not a major form of nitrogen transported in the xylem.

In higher plants there are multiple nitrate carriers with distinct properties and regulation

If roots are immersed in nitrate solutions of 0.1–10 mM, the root epidermal and cortical cells have nitrate concentrations of 3–5 mM. This reflects active transport, i.e. nitrate is taken up against a concentration gradient. If only passive transport was occurring, then to overcome the membrane potential of –60 mV and to keep the internal concentration of nitrate at 5 mM, the external nitrate concentration would have to be greater than 50 mM. At a membrane potential of –120 mV, the external nitrate concentration would need to be in excess of 500 mM. Hence active transport is required.

Nitrate is imported and accumulated in plant cells against an electrochemical gradient. That is, upon first exposure to nitrate the plasma membrane of the cell depolarizes, with the membrane potential becoming more positive. So even though nitrate is anionic, the cytosol becomes more positively charged. The reason for this is thought to be because two protons are cotransported into the cell with every nitrate molecule, this results in a net positive import of a positive charge. As the proton motive force (PMF) across the plasma membrane enables movement of ions against a concentration gradient, i.e. moving in response to a voltage difference, then such a model facilitates active nitrate transport by either the high- or low-affinity transporters. After the onset of polarization, the plasma membrane repolarizes to its initial potential by increasing the pumping activity of the H^+-ATPase. The plasma membrane H^+-ATPase establishes the proton gradient as the main electrochemical gradient of the cell, with potentials of –100 to –250 mV.

The result of any nitrate assimilation is that hydroxyl ions will be generated. To maintain the cytoplasmic pH these hydroxyl ions must subsequently be either neutralized (e.g. by the simultaneous accumulation of a weak acid) or excreted. In root nitrate assimilation it tends to be the latter process, while in the shoots it is the former process, via the synthesis of organic acids and in particular malate.

Both high-affinity and low-affinity nitrate transporters are involved in the import of nitrate into the plant. These transporters allow the plant to cope with a broad range of external nitrate concentrations (5 µM–50 mM) without suffering from either deficiency or toxicity. The high affinity transport system (HATS) exhibits Michaelis–Menten kinetics, saturating at 0.2–0.5 mM nitrate with a K_m of between 10 and 100 µM (Figure 8.6a). There is a constitutive HATS of relatively low capacity that is expressed in the absence of nitrate and an inducible HATS, which, as its name suggests, is induced by nitrate. The low affinity transport system (LATS) has nonsaturating uptake kinetics and usually functions at nitrate concentrations of greater than 0.5 mM (Figure 8.6b). Two nitrate transporter gene families have been identified, *NRT1* and *NRT2*. *NRT1* encodes nitrate transporters with either a low affinity or both a low and high affinity for nitrate, while *NRT2* encodes high-affinity nitrate-inducible transporters. NRT1 and NRT2 proteins have some structural homology, with 12

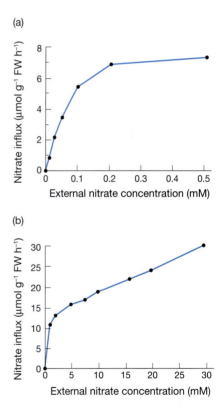

Figure 8.6 The kinetics of nitrate uptake in barley roots via (a) the high-affinity transport system (HATS), and (b) the low-affinity transport system (LATS).
Abbreviations: FW, fresh weight.

transmembrane domains distributed in two sets of six helices connected by a cytosolic loop. However, there is no primary sequence homology found between the classes of NRT1 and NRT2. It is now clear that many nitrate transport systems belonging to the two classes may be found in a single organism. This allows the plant to quickly adjust to a wide variation in the availability of nitrate and to match the demands with the requirements of the plant.

Nitrate reductase catalyzes the reduction of nitrate to nitrite in the cytosol of root and shoot cells

Within a specific organ the NR shows cell-specific expression. For example, at low nitrate concentrations NR is primarily found in root epidermal and cortical cells close to the root surface, while at higher external nitrate concentration, NR activity is more widespread. In the C_4 plant maize, NR is found in the leaf mesophyll cells and not in the bundle sheath cells. This distribution reflects a greater capacity of mesophyll chloroplasts to generate reductant via the photosynthetic electron transport chain. Most higher plant NRs use NADH, although there are some bispecific forms capable of using either NADH or NADPH. NR is described as a molybdoflavoheme-containing protein (Figure 8.7). The name reflects the requirement of NR for three cofactors that provide the redox center that facilitates the movement of electron transfer, the molybdenum (MoCo), the FAD, and the heme iron cofactor. The MoCo domain is at the N-terminal region, the heme iron domain is in the middle and the FAD domain is at the C-terminal region of the NR protein. These three functional regions are connected by two hinges (Figure 8.7). The midpoint potentials in the native enzyme for MoCo, FAD, and heme iron are −10, −160, and −272 mV, respectively.

NR forms homodimers with a binding site for nitrate and for NAD(P)H. Each NR subunit is approximately 1000 amino acids long and contains all three cofactors. Each cofactor of the NR holoenzyme is a redox center associated with a distinct functional region of the protein. Partial proteolysis of the NR protein has been used to show that discrete fragments of the enzyme have different enzyme activities. One fragment binds FAD and can use NADH to reduce the artificial electron acceptor ferricyanide. A different fragment contains the MoCo and heme iron domain and can reduce nitrate in the presence of the artificial electron donor methyl viologen. Each functional region of NR is effectively an independent unit belonging to a protein family. The MoCo region is 360–370 amino acids long and belongs to the special class of MoCo-binding proteins. These include xanthine oxidase, biotin sulfoxide reductase, dimethyl sulfoxide reductase, and sulfite oxidase. The central heme domain is 75–80 amino acid residues long and is similar to the cytochrome b_5 family of heme proteins. It also shares a number of properties with these proteins in being able to oxidize cytochrome c, and having a characteristic cytochrome b-type fold that binds the heme moiety noncovalently. The FAD binding region is 260–265 amino acids long and is similar to the ferredoxin-NADP$^+$ oxidoreductase family of flavin oxidoreductases that includes ferredoxin-NADP$^+$ oxidoreductase, cytochrome P450 reductase, nitric oxide synthase, and cytochrome b_5 reductase. Based on the crystal structure of the FAD domain it is clear that it consists of two domains, each forming a lobe that is separated by a cleft. The N-terminal lobe binds FAD, while the C-terminal lobe binds the substrate NAD(P)H. There is a cysteine in the C-terminal lobe that provides a thiol group that interacts with the NAD(P)H. It is also thought that this interaction positions the NAD(P)H so that electron transfer is facilitated to the FAD. This cysteine is not essential, as a substitution with serine leads to the NR still being functional, although less efficient, at either binding NADPH or reducing FAD. In a sense the NR enzyme can be regarded as a

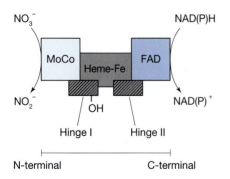

Figure 8.7 Nitrate reductase domain structure. Nitrate reductase is made up of three functional regions connected by two hinges. Abbreviations: MoCo, molybdenum cofactor.

mini-electron-transport chain that uses NAD(P)H as its source of electrons and NO_3^- as its terminal electron acceptor.

The production of nitrite is rigidly controlled by the expression, catalytic activity, and degradation of NR

In 1957 Tang and Wu demonstrated that NR is a substrate-inducible enzyme, with NR enzyme levels increasing in response to the concentration of nitrate supplied to the plant. It was subsequently shown that this regulation was at the level of transcription. When a plant is exposed to nitrate, NR mRNA accumulates within minutes and can increase in level by two orders of magnitude. The primary signal for regulation of NR is nitrate. In the absence of nitrate no other factors will regulate the level of NR. However, if nitrate is present, then other environmental signals are able to act secondarily and adjust the level of NR to meet the needs of the plant (Table 8.5). Light or photosynthetic production of sucrose induces NR transcription. Once induced, NR transcription in photosynthetic tissue shows a cyclic diurnal pattern, with maximal levels appearing just before the start of the light period. Light regulation of NR helps to integrate nitrate assimilation with photosynthesis. This integration is important because the reducing energy needed for nitrate assimilation comes from the photosynthetic apparatus. NR activity can consume up to 20% of the energy used for CO_2 fixation. Regulation of both NR transcription and activity allows a plant to fine tune the amount of nitrate reduction.

Up-regulating NR transcription potentially increases nitrite production. Under normal conditions the concentration of nitrite in a leaf is rarely above 15 nmoles/g/fresh weight and that of ammonium is rarely above 10 μmoles/g/fresh weight. Both nitrite and ammonium are toxic at higher concentrations, and a rigid control of the concentration of nitrite is achieved by a combination of the control of NR expression, NR catalytic activity, and NR protein degradation (Figure 8.8).

Since the early 1990s it has been known that the catalytic activity of the NR enzyme is rapidly and reversibly modulated when NR activity is measured in

Table 8.5 Factors which in the presence of nitrate are able to influence the transcription and activity of NR

Signal	Effect on NR transcription
Light	Up-regulates transcription and activity
Dark	Down-regulates transcription and activity
Circadian rhythm	Modulates transcription based on time of day
Sucrose	Up-regulates transcription
Cytokinin	Up-regulates transcription
High concentration CO_2	Up-regulates activity
Low concentration CO_2	Down-regulates activity
Glutamine	Down-regulates transcription
Oxygen	Down-regulates activity
Anoxia	Up-regulates activity

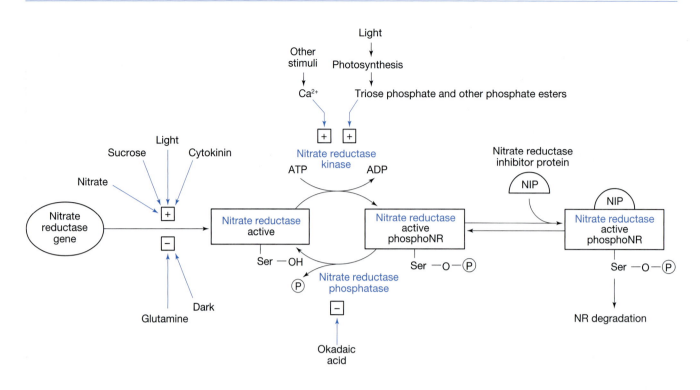

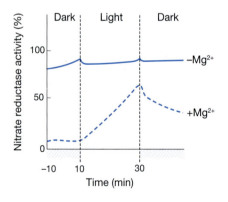

Figure 8.8 Summary of the effect of a number of factors on regulating the transcription, activity, and degradation of nitrate reductase. Activators (+) and inhibitors (–) are shown. Abbreviations: NIP, nitrate reducase inhibitor protein; NR, nitrate reductase; Ser, serine

Figure 8.9 The effect of divalent cations on nitrate reductase activity. Nitrate reductase activity was measured in spinach leaves exposed to dark/light transitions in either the presence $(+Mg^{2+})$ or absence $(-Mg^{2+})$ of the divalent magnesium cations. In the presence of Mg^{2+} there was marked activation/inactivation of nitrate reductase activity in response to light/dark transitions.

the presence of millimolar concentrations of divalent cations (Figure 8.9). The available evidence suggests that the biochemical mechanism underlying the changes in Mg^{2+} sensitivity is protein phosphorylation of a serine residue. There are two hinge regions (Figure 8.7), one connecting the MoCo and heme region (hinge I) and the other connecting the FAD and heme region (hinge II). The phosphorylated serine residue (serine 543 in spinach), is located in the hinge I region of the molecule, connecting the MoCo and heme domain (Figure 8.10). This phosphorylation of NR alone, to give a phosphoNR, has little direct effect on NR activity, and an additional inhibitor protein, or NIP (NR inhibitor protein), is required to inactivate NR. This protein has been identified as a member of a large group of regulatory 14-3-3 proteins (see Chapter 7). The binding of this 14-3-3 NIP directly to the phosphoNR in the presence of divalent cations causes a conformation change leading to the movement of the hinge and a blocking of electron flow between the Heme-Fe cytochrome and the molybdenum cofactor (Figure 8.10). This results in a completely inactive NR form that cannot transfer electrons from NAD(P)H to nitrate. This inactive phosphoNR:NIP complex, which predominates in darkened leaves and is responsible for the dark inhibition of nitrate reduction, has an absolute requirement for Mg^{2+}. In the dark the concentration of Mg^{2+} is about 0.4 mM in the leaf cytosol, while in the light it is extremely low. A low Mg^{2+} concentration promotes the dissociation of the 14-3-3 complex. Depending on the external conditions NR exists in three states: free NR (active); phosphorylated NR (phosphoNR, active); and phosphoNR:NIP complex (inactive). The ratio of these three NR forms is variable. Phosphorylation plus NIP binding does not change the substrate affinities of NR in a crude extract. The phosphoNR:NIP complex is inactivated, whereas the remaining free NR in the extract works normally. Inactivation of NR may be important for conserving reductant or preventing the build up of toxic nitrite when the ability of the plants to further reduce nitrite to ammonium is impaired. Finally, it has also been shown that NR phosphorylation and NIP binding are involved in the control of NR degradation (Figure 8.8). At present it is not clear whether the phosphoNR:NIP protects phosphoNR from degradation by shielding critical residues or whether it targets NR for degradation.

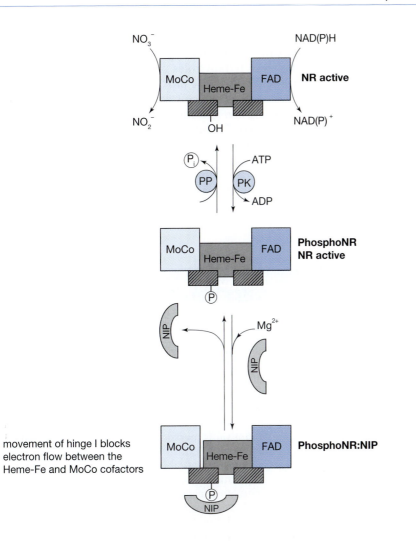

Figure 8.10 Model for post-translational modification of nitrate reductase. A serine phosphorylation site is located in hinge I of nitrate reductase. Once phosphorylated, the serine motif is recognized by a nitrate reductase inhibitor protein (NIP). The binding of NIP occurs in the presence of divalent cations and inactivates nitrate reductase by preventing electron flow from NAD(P)H to nitrate. Abbreviations: NR, nitrate reductase; PK, NR protein kinase; PP, NR protein phosphatase.

It is now known that potentially all forms of NR are capable of catalyzing, both *in vitro* and *in vivo*, the production of nitric oxide from nitrite in the presence of NAD(P)H. The role of NR in nitric oxide production, although only considered a minor one, has been implicated in stomatal regulation, germination, and pathogen responses (Box 8.1).

Nitrite reductase, localized in the plastids, catalyzes the reduction of nitrite to ammonium

This step involves six electrons that are generated via reduced ferredoxin or other nonheme iron proteins (Figure 8.5). In chloroplasts the reductant is generated photosynthetically by the noncyclic electron transport chain (Figure 8.11). In nonphotosynthetic plastids such as in the roots, the reduced ferredoxin is not generated via photosynthesis. Instead NADPH, generated via the first two steps of the oxidative pentose phosphate pathway (see Chapter 6) reduces ferredoxin via the enzyme ferredoxin NADP$^+$ oxidoreductase (Figure 8.12). This reduced ferredoxin is then available for nitrite reduction.

Nitrite reductase (NiR) is a nuclear-encoded protein translated on cytoplasmic ribosomes. It has a transit peptide to target the protein to the plastid. The mature NiR is a monomer of approximately 63 kDa (holoenzyme). NiR has two functional domains and cofactors. The N-terminal region has a ferredoxin-binding

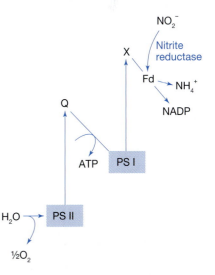

Figure 8.11 The photosynthetic electron transport chain generates reduced ferredoxin to support nitrite reduction in photosynthetic tissue. Abbreviations: Fd, ferredoxin; PS, Photosystem.

domain, while at the C-terminal region there are two redox centers composed of a 4Fe–4S center and a siroheme center (Figure 8.13), where siroheme is one of the four different classes of tetrapyrroles produced by higher plants (Box 8.2). These two redox centers are in close proximity and are linked by a sulfur ligand. There are four cysteines in two clusters, which give the bridging and sulfur ligand for the 4Fe–4S cluster. The importance of these cysteine residues in cofactor binding has been confirmed in work where mutations have been generated moving bulky side chains next to the cysteine residues. Such mutations led to a marked reduction in NiR activity. The C-terminal region of NiR shares some sequence homology with sulfite reductase (SiR, see page 269). The interaction of the functional domains are involved in the transfer of electrons from reduced ferredoxin to nitrite (Figure 8.13).

Box 8.1 Nitric oxide (NO) is one of the simplest bioactive molecules

NO was originally thought of as only a toxic air pollutant. However, in the 1990s it was discovered that NO could be produced enzymatically within cells and, furthermore, that NO acts as an essential signaling molecule. The NO free radical has a half-life of a few seconds, rapidly reacting with O_2 to form nitrogen dioxide (NO_2), which degrades to nitrite and nitrate in aqueous solutions. But the NO free radical is able to diffuse rapidly across biological membranes, allowing it to briefly be involved in cell-to-cell signaling. NO can react with the superoxide free radical O_2^-, to form the reactive peroxynitrite molecule ($ONOO^-$). NO also rapidly reacts with proteins and in particular the reactive amino acids cysteine and tyrosine, and different receptors and transcription factors.

NO is a gaseous radical, that is small, simple, and ubiquitous. There are two distinct NO-producing pathways in plants, the nitrite pathway and the arginine pathway (Figure 1). The major differences between these two pathways are that the nitrite pathway can occur in the absence of an enzyme, while the arginine pathway can only operate in the presence of oxygen. NO production from nitrite can be produced via a one-electron reduction of nitrite by either enzymatic or chemical routes. All plant nitrate reductases (NRs) can, in the presence of NAD(P)H, produce NO from nitrite. NR produces NO in the absence of photosynthetic activity or when nitrite, the substrate for NO synthesis, accumulates. However, assimilatory NR is not the only enzyme that can catalyze the nitrite pathway of NO production. Any number of enzymes with redox domains are thought to be capable of potentially reducing nitrite to NO. The chemical route of NO production only occurs in tissues under healthy conditions or in acidic compartments. However, wounding, which results in the destruction of cell compartments, may, by mixing the components nitrite (from the cytosol), acid (from the vacuole), and reductants (from the chloroplast), lead to induce NO production. The arginine pathway can be catalyzed by NO synthase (NOS). NOS activity has been identified as two distinct enzymes in plants. The enzymatic reaction has proved difficult to characterize as a number of cofactors, for example B4H (tetrahydrobiopterin), FAD, calmodulin, or Ca^{2+}, are needed. Regardless as to how it is produced, the final degradation products of NO are nitrite or nitrate. It is this complex relationship between substrates and products that has made it difficult to characterize fully the signaling pathway involving NO.

Plant metabolism is markedly affected by NO. The ability of the free radical NO to either gain or lose an electron to energetically more favorable structures, the nitrosonium cation (NO^+) and the nitroxyl radical (NO^-), facilitates both stability and reactivity. This has led to NO and its exchangeable redox-activated forms being recognized as both intra- and intercellular signaling molecules. That is NO has now been shown to participate in a wide range of physiological phenomena, including germination, induction of cell death, pathogen response, stomatal regulation, and photosynthesis regulation.

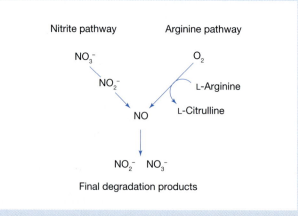

Figure 1 There are two distinct nitric oxide-producing pathways in plants, the nitrite pathway and the arginine pathway.

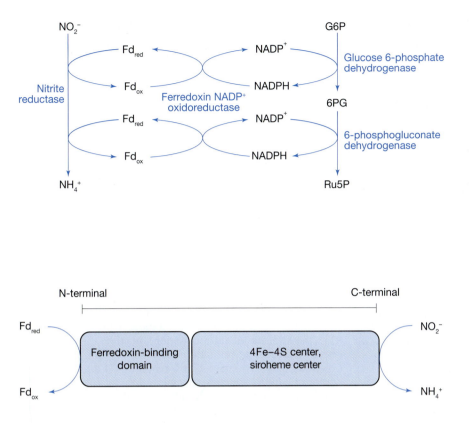

Figure 8.12 The oxidative pentose phosphate pathway generates NADPH, which is used as the initial reductant to support nitrite reduction in nonphotosynthetic tissue. Electrons are transferred from NADPH to ferredoxin via the enzyme ferredoxin NADP⁺ oxidoreductase. The reduced ferredoxin can then be used by nitrite reductase. Abbreviations: Fd, ferredoxin; G6P, glucose 6-phosphate; 6PG, 6-phosphogluconate; Ru5P, ribulose 5-phosphate.

Figure 8.13 Nitrite reductase domain structure.

NiR is regulated transcriptionally by nitrate. Its induction is usually coordinated with that of NR so that any nitrite produced by NR will be immediately reduced by NiR. This response is important as nitrite is relatively toxic. In addition to nitrate, light also affects the levels of NiR (gene expression and protein).

Plant cells have the capacity for the transport of ammonium ions

In acidic soils, when the rate of nitrification and the availability of nitrate are low, the plant will often encounter high concentrations of ammonium ions. This is an attractive form of nitrogen for uptake into roots as ammonium is permanently available as a reduced state of nitrogen. Where ammonium is at nontoxic concentrations, then the root can act as the main site for ammonium uptake from the soil. Different plant species have been examined and a number of gene families encoding putative ammonium transporters have been identified. Ammonium is taken up by plant cells via ammonium transporters in the plasma membrane and then distributed to the chloroplasts, mitochondria, and vacuoles most likely via a different set of transporters. Broadly speaking, ammonium transporters can be divided into at least two groups of high- and low-affinity transport systems. The transport system operating at low ammonium concentration in roots shows Michaelis–Menten kinetics. That operating at high ammonium concentrations exhibits linear nonsaturating uptake up to 40–50 mM. The transcript levels of the ammonium transporters strongly respond to nitrogen levels, sugars, and the time of the day.

The assimilation of ammonium leads to the generation of protons, which, to maintain the pH of the cytoplasm, must be excreted or neutralized. When

Box 8.2 Glutamate is the precursor for the biosynthesis of tetrapyrroles

Higher plants produce four different classes of tetrapyrroles; the chlorophyll tetrapyrrole, heme, phytochromobilin, and siroheme. All of these compounds are involved in essential physiological processes. Chlorophyll, of which there are two forms in higher plants, chlorophylls *a* and *b*, is the main photosynthetic pigment (Chapter 4). It consists of a magnesium-containing tetrapyrrole ring attached to a phytol (diterpenoid; see Chapter 12) side chain. Heme is a component of cytochromes (found in both chloroplasts and mitochondria) and is therefore essential for both photosynthesis and respiration. Phytochromobilin forms the chromophore (i.e. the light-absorbing structure) of phytochrome. Finally, siroheme is a prosthetic group of nitrite and sulfite reductases, as explained in this chapter. Tetrapyrrole biosynthesis is consequently of wide-ranging importance to higher plants and it is not surprising to find that this pathway has attracted considerable research interest. Most of the enzymes and genes involved in this pathway have been characterized. Pyrrole and a representative tetrapyrrole structure are shown in Figure 1.

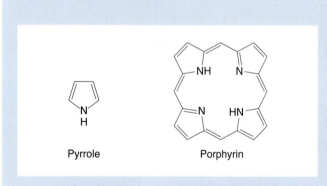

Figure 1 The structure of pyrrole and of a tetrapyrrole, porphyrin.

Glutamate is the precursor for the synthesis of tetrapyrroles in higher plants. The tetrapyrrole pathway is highly branched but the six reactions that convert glutamate to uroporphyrinogen III are common to all branches of the pathway (Figure 2).

Glutamate has to be activated before it can be modified, and this occurs in an unusual way (reaction 1), in that glutamate forms a covalent linkage to a specific tRNA (tRNA^{Glu}). The carboxyl group that is attached to the tRNA is subsequently reduced, by glutamyl-tRNA reductase (reaction 2), forming glutamate 1-semi-aldehyde (GSA) and releasing tRNA. In the next step (3), GSA is rearranged in a rare intramolecular aminotransferase reaction, catalyzed by GSA aminotransferase, where the amino group donor and acceptor are both present on the same molecule. As a result, the α amino group migrates to react with the terminal aldehyde, to form the product δ-aminolevulinic acid (ALA).

Two molecules of ALA are condensed to form a pyrrole molecule, porphobilinogen (PBG). Four molecules of PBG polymerize to form a linear tetrapyrrole, hydroxymethylbilane, which is rearranged and cyclized to form the first cyclic structure, uroporphyrinogen III (the precursor for siroheme synthesis, see below). There then follow two consecutive decarboxylation reactions (reactions 7 and 8) and a final oxidation reaction to form protoporphyrin IX. At this point the tetrapyrrole pathway branches, as protoporphyrin IX is the precursor for both chlorophyll and heme biosynthesis.

Chlorophyll branch

In the first step of this branchpoint Mg^{2+} is inserted into the center of protoporphyrin IX, in an ATP-dependent reaction catalyzed by magnesium chelatase (reaction 10). Chlorophyll synthesis involves further modifications of the ring, such as the addition of a methyl group (reaction 11), the addition of atomic oxygen (reaction 12) followed by two successive reductions catalyzed by protochlorophyllide oxidoreductase (POR, reaction 13) and divinyl-chlorophyllide reductase (reaction 14). In angiosperms the POR reaction is strictly light-dependent as it requires protochlorophyllide to be activated by the absorption of light. In gymnosperms, algae, and photosynthetic bacteria, POR is light-independent and these organisms can synthesize chlorophyll in the dark.

In the final step in the synthesis of chlorophyll *a*, the diterpenoid phytol side chain is attached to monovinyl chlorophyllide *a* in a typical terpene synthase reaction (see Chapter 12) catalyzed by chlorophyll synthase (reaction 15). Chlorophyll *a* can be converted to chlorophyll *b* and vice versa (reactions not shown).

Heme branch

In a reaction that resembles the first step in the chlorophyll branch, protoporphyrin IX ferrochelatase inserts Fe^{2+} (rather than Mg^{2+} as in chlorophyll biosynthesis) into the center of protoporphyrin IX to form protoheme (heme b; reaction 16). Other hemes are synthesized in plants (e.g. as components of cytochromes) but the reactions have yet to be identified. Protoheme is then oxidized and the ring is opened to form biliverdin IXα (reaction 17). Finally, phytochromobilin, the chromophore of phytochrome, is formed by phytochromobilin synthase (reaction 18).

Siroheme branch

Uroporphyrinogen III serves as the precursor for the biosynthesis of siroheme. This branch of the tetrapyrrole pathway is not as well characterized. The first step (19, product not shown) adds a methyl group to form dihydrosirohydrochlorin. An oxidation step follows, although the oxidase has yet to be identified. Finally, Fe^{2+} is inserted into the center of the ring to form siroheme. This reaction is catalyzed by sirohydrochlorin ferrochelatase.

Box 8.2 Glutamate is the precursor for the biosynthesis of tetrapyrroles (continued)

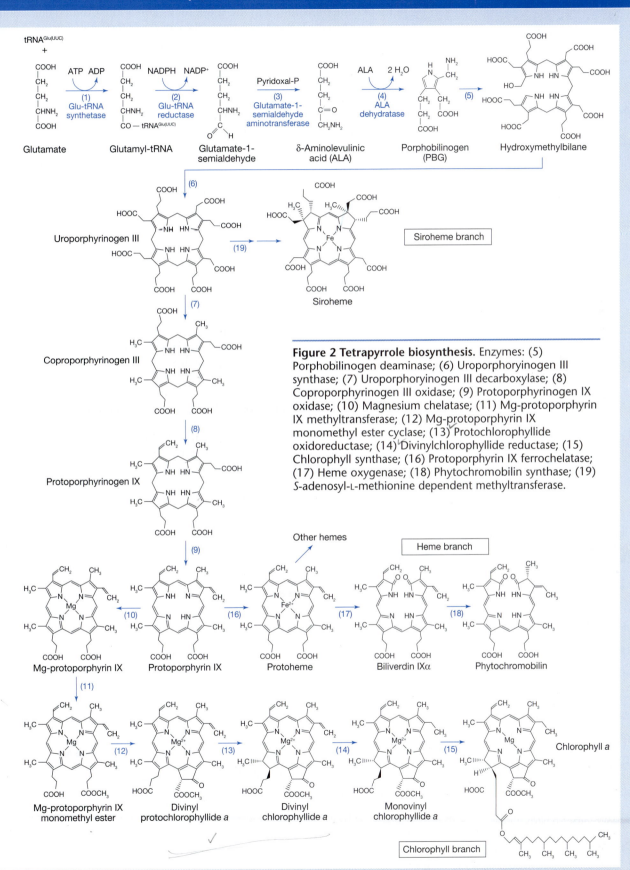

Figure 2 Tetrapyrrole biosynthesis. Enzymes: (5) Porphobilinogen deaminase; (6) Uroporphoryinogen III synthase; (7) Uroporphoryinogen III decarboxylase; (8) Coproporphyrinogen III oxidase; (9) Protoporphyrinogen IX oxidase; (10) Magnesium chelatase; (11) Mg-protoporphyrin IX methyltransferase; (12) Mg-protoporphyrin IX monomethyl ester cyclase; (13) Protochlorophyllide oxidoreductase; (14) Divinylchlorophyllide reductase; (15) Chlorophyll synthase; (16) Protoporphyrin IX ferrochelatase; (17) Heme oxygenase; (18) Phytochromobilin synthase; (19) S-adenosyl-L-methionine dependent methyltransferase.

ammonium assimilation occurs in the roots, the protons are generally extruded into the soil. Where ammonium concentrations are high, ammonium may be transported to the shoot for assimilation, proton generation by ammonium in the leaves has potentially a marked effect, causing leaf damage.

Ammonium is assimilated into amino acids

Ammonium is assimilated into amino acids, first into glutamine and glutamate, and then, through aminotransferase reactions, into other amino acids such as aspartate and asparagine (see later in this chapter). The pathway for ammonium assimilation involves just two enzymes, glutamine synthetase (GS) and glutamate synthase (GOGAT). The reactions can occur in all plant organs, but the existence of several isoenzymes, multiple gene families, and quite distinct cellular and subcellular locations adds considerable complexity to this pathway. We are now beginning to understand how this complexity reflects the differing demands for nitrogen and carbon allocation within higher plant tissues.

Ammonium originates from both primary and secondary sources

One reason for the complex distribution of ammonium assimilation enzymes is that there are a number of different sources of ammonium. Primary sources are defined as those that originate from the primary assimilation of inorganic nitrogen, for example nitrite reduction and nitrogen fixation and also direct uptake of ammonium from the soil (as described above). Assimilation of ammonium from these sources is therefore termed primary ammonium assimilation. Other sources of ammonium are secondary as the nitrogen has been derived from organic sources within the plant. These secondary sources include: (1) oxidation of glycine to serine in photorespiration (only in photosynthetic tissue); (2) degradation of transport compounds such as arginine, ureides, and asparagine; (3) reactions of amino acid metabolism, e.g. conversion of cystathionine to homocysteine in methionine biosynthesis, and threonine to 2-oxobutyrate in isoleucine biosynthesis; (4) conversion of phenylalanine to *trans* cinnamic acid by phenylalanine ammonia lyase in phenylpropanoid biosynthesis (Chapter 11); and (5) protein degradation and subsequent deamination of amino acids during senescence and seed germination. These sources of ammonium are summarized in Figure 8.14.

Figure 8.14 Examples of primary and secondary sources of ammonium in plant cells.

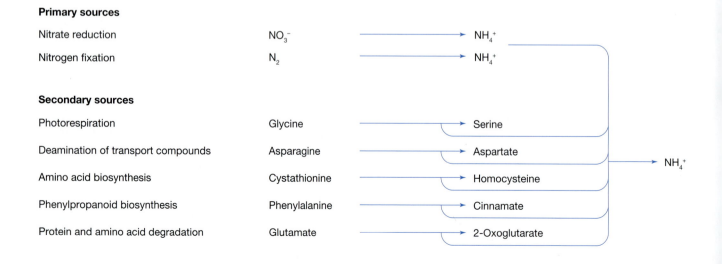

Primary sources

Nitrate reduction	NO_3^-	NH_4^+
Nitrogen fixation	N_2	NH_4^+

Secondary sources

Photorespiration	Glycine	Serine
Deamination of transport compounds	Asparagine	Aspartate
Amino acid biosynthesis	Cystathionine	Homocysteine
Phenylpropanoid biosynthesis	Phenylalanine	Cinnamate
Protein and amino acid degradation	Glutamate	2-Oxoglutarate

NH_4^+

Ammonium is assimilated by glutamine synthetase and glutamate synthase, which combine together in the glutamine synthetase/glutamate synthase cycle

In higher plants, ammonium assimilation is carried out by the combined action of two enzymes, glutamine synthetase (GS) and glutamine–oxoglutarate aminotransferase (also called glutamate synthase, and GOGAT).

GS adds NH_2 from ammonia to the δ carboxyl group of glutamate to form the amide, glutamine (Figure 8.15). The reaction is irreversible, being driven by ATP hydrolysis.

Glutamine synthetase (GS) reaction:

NH_3 + glutamate + ATP $\longrightarrow$ glutamine + ADP + P_i

GOGAT transfers the amide group of glutamine to 2-oxoglutarate to form two molecules of glutamate.

Glutamate synthase (GOGAT) reaction:

Glutamine + 2-oxoglutarate + Fd_{red} (or NADH) $\longrightarrow$ 2 glutamate + Fd_{ox} (or NAD^+)

One of the two molecules of glutamate formed in the GOGAT reaction serves as a substrate in the GS reaction. The combined action of GS and GOGAT is termed the GS/GOGAT cycle (summarized in Figure 8.15). Both forms of GOGAT, Fd-GOGAT and NADH-GOGAT, are found only in plastids, whereas GS is present in both plastids and cytosol in the form of separate isoenzymes.

GS is an octameric protein with two isoforms, localized in the cytosol and plastid

GS is an octameric protein, usually consisting of eight identical (approximately 40 kDa) subunits, with a molecular mass ranging between 320 and 380 kDa. Some heteromeric forms have been reported in legumes (e.g. *Phaseolus vulgaris*) and in sugar beet (*Beta vulgaris*), where the subunit composition changes with nitrogen nutrition and organ development. *P. vulgaris* root GS

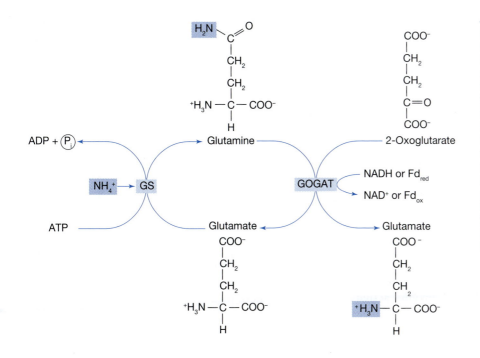

Figure 8.15 The glutamine synthetase/glutamate synthase (GS/GOGAT) cycle. Glutamine synthetase (GS) incorporates NH_4^+ into the δ-carboxyl group of glutamate to form glutamine. The reaction is driven by the hydrolysis of ATP to ADP and P_i. Glutamate synthase (GOGAT) forms two molecules of glutamate, one of which serves as the substrate for GS, while the other is available for further metabolism. Two forms of GS exist, one present in the plastid (GS2) and one in the cytosol (GS1). There are also two forms of GOGAT, both are plastidic, one is Fd-dependent and the other is NADH-dependent.

consists of α and β polypeptides, while nodule GS (Gs_{n1}), which is induced upon *Rhizobium* infection, consists of β and γ subunits.

Two isoenzymes of GS exist in most higher plants. GS1 is found in the cytosol, and GS2 in the plastids. All GS2 subunits contain an N-terminal signal peptide of 48 amino acids that target it to the plastid. The tissue and cellular distribution of GS1 and GS2 differs between species and is influenced by environmental conditions. This spatial distribution of GS isoenzymes appears to be important in determining the partitioning of nitrogen within and between different parts of the plant.

The GS genes and proteins show discrete cellular localization and different responses to light and nutrients

The plastid isoenzyme, GS2, is encoded by a single nuclear gene, *GLN2*, in most higher plant species. *GLN2* is highly expressed in photosynthetic tissue but is also found in roots and other nonphotosynthetic organs. *GLN2* expression is increased in leaves in the light, and under photorespiratory conditions, particularly in cells with functional chloroplasts. Phytochrome and the blue light receptor appear to be involved in this response in arabidopsis. Sucrose may replace the requirement for light, while amino acids counteract this effect, which indicates that the C:N balance may modulate *GLN2* expression (enhanced at high C:N and repressed at low C:N).

GS2 protein and activity is higher in photosynthetic than in nonphotosynthetic tissue. It is present in high concentrations in the mesophyll cells of leaves and has also been detected in the plastids of vascular cells of some species. The main function of GS2 is to reassimilate ammonia produced during photorespiration, as discussed later.

GLN2 is expressed in the roots of a limited number of species, and plastid GS activity can comprise up to 4% of total root GS activity. Evidence from enzyme activity and protein measurements indicate that root plastid GS may be inducible by nitrate and/or ammonium in pea and barley.

The regulation of expression of the GS1 gene family is rather complex and a complete picture has yet to emerge. GS1 is encoded by a multigene family (denoted *GLN1;1, 1;2*, etc.), ranging from three genes in rice to five in maize and arabidopsis. Studies of maize, rice, and arabidopsis all show that the *GLN1* genes have quite distinct patterns of expression, differing in cellular location and in response to nitrogen nutrition.

In leaves, GS1 genes are particularly highly expressed in the vascular cells. Immunolocalization studies of rice and barley, in particular, have confirmed that GS1 protein, while present in mesophyll cells, is particularly concentrated in the cytosol of vascular parenchyma cells of the leaf. This distribution indicates that GS1 is important in forming glutamine for transport through the vascular system, from source to sink tissue.

The most comprehensive studies of GS1 expression have been carried out in rice and arabidopsis, where measurements of enzyme activity and kinetics have been included alongside analysis of gene expression. This is an important approach to take, as gene expression should not be used as an indicator of enzyme activity, particularly where there is post-translational modification (see below and Chapter 2). In arabidopsis, the *GLN1* genes expressed in the roots (four of the five *GLN1* genes) fall into two categories. One group of

GLN1 genes encodes GS1 isoenzymes with high substrate affinity (i.e. low K_m for ammonium and glutamate), and the other with low substrate affinity. The high-affinity GS1 isoenzymes are located mainly in the outer layer of root cells, which are involved in nutrient uptake from the soil. The expression of both the genes and low-affinity isoenzymes is reduced in the presence of high concentrations of ammonium. The low-affinity GS1 proteins, in contrast, are expressed mainly within the vascular cells, which are responsible for transporting nutrients, including amino acids, from root to shoot. Expression of these low-affinity GS1 isoenzymes is increased in the presence of high concentrations of ammonium in the soil. The current hypothesis is that this distribution of GS isoenzymes serves to modulate the flow of ammonium and amino acids from cell to cell and tissue to tissue. At low concentrations of soil ammonium, the high-affinity GS1 isoenzymes are highly expressed in the surface root cells in contact with the soil, acting as scavengers for the low resources that are available. As soil ammonium concentrations rise, the low-affinity GS1 isoenzymes increase within the vascular cells, while the high-affinity peripheral GS1 isoenzymes decrease. This response is thought to enable glutamine to be formed within the vascular tissue for export to the shoot. This important study leads the way in investigating the function of the different isoforms of GS within higher plants. It is hoped that further investigations will adopt a similar approach, in measuring not only gene expression but also enzyme activity and kinetics, in order to unravel the cell-specific functions of each isoenzyme within the ammonium assimilation pathway.

Glutamine synthetase activity is regulated by metabolites and effectors, and may be modified by phosphorylation and 14-3-3 binding

GS from higher plants is regulated by a range of metabolites and there is evidence that the chloroplast isoenzyme, GS2, is particularly sensitive to sulfhydryl reagents. Dithiothreitol, for example, can activate GS2 *in vitro* and the enzyme can be inactivated by oxidation. An –SH site on the enzyme may be the target for redox regulation, with cysteine and histidine residues being potential candidates. ATP and glutamate protect GS2, in particular, from oxidative inactivation and these compounds are generally added to the isolated enzyme to help to retain its activity *in vitro*. It is some years since the kinetic properties of GS1 and GS2 were first investigated and these studies pre-date the identification of the multiple isoforms of GS1 following genomics studies. It is hoped that further attention will now be paid to the regulatory properties of each of these isoenzymes.

Another emerging hypothesis with respect to GS regulation is that it is subject to modification by phosphorylation. GS proteins are capable of binding to 14-3-3 proteins, but it is too early to determine what function this may serve *in vivo*. Nevertheless, several studies have shown that there is not always a good correlation between the level of mRNA, protein, and GS activity and it has long been suspected that GS is subject to post-transcriptional control. We await further investigation of the mechanism of GS regulation in higher plant cells.

Further evidence of the functions of glutamine synthetase isoenzymes has come from studies of mutants and transgenic plants

Attempts to overexpress GS genes in higher plants have frequently been confounded by factors that are still poorly understood. In many cases, high levels of mRNA do not result in equivalent increases in protein production nor in any increase in GS activity. These complications further point to substantial post-transcriptional regulation of GS expression. Furthermore, the complex cellular distribution of the various GS isoenzymes is frequently overlooked, yet it is important to take this into account when attempting to manipulate GS expression.

A screen for photorespiratory mutants in barley (see Chapter 5) identified plants deficient in chloroplast GS (GS2). These plants remained healthy providing they were kept under very high concentrations of CO_2 (1% by volume), which prevented photorespiration. Exposure to air caused yellowing of the leaves and eventually killed the plants. From the analysis of these GS2 mutants, it became clear that this GS isoenzyme is essential for reassimilating the ammonium released during conversion of glycine to serine in the photorespiratory pathway (Chapter 5). The fact that the plants were able to grow quite normally under nonphotorespiratory conditions, while they still retained wild-type levels of the cytosolic isoenzyme, GS1, indicates that GS1 is able to assimilate ammonium produced during primary assimilation (i.e. from nitrite reduction) at rates that are sufficient to support normal plant growth. These experiments provide persuasive evidence that photorespiratory ammonium is assimilated within the chloroplast. Yet photorespiratory ammonium is released during glycine oxidation in the mitochondria, and then, presumably, passes through the cytosol (where GS1 is located) before entering the chloroplast to be assimilated by GS2. Similarly, as nitrite reduction occurs in the plastid, ammonium has to move from there to the cytosol if it is to be assimilated by GS1. We still do not understand how this movement of ammonium is controlled, or whether the GS isoenzymes exhibit such clear, nonoverlapping functions *in vivo*. While GS1 is mainly present in the vascular cells of leaves, it is still found in the mesophyll cells in some species, where it could take part in photorespiration. We await further, cell-specific analysis of these pathways to determine the extent to which these functions might be interchangeable.

Overexpression of GS1 has produced mixed results in a number of species, with variable effects on growth. Increased leaf GS1 activity resulted in increased growth in poplar (*Populus tremula* × *P. alba*) and wheat, whereas no effect was seen in tobacco unless the plants were grown under nitrogen-deficient conditions, when they grew better than the wild type. These results have been taken to indicate that GS1 is important in nitrogen mobilization, so that its overexpression would enable plants to reassimilate ammonium produced from internal recycling of proteins during nitrogen deficiency. In tobacco plants with antisense reduction of GS1 protein in the phloem, no effects were seen unless the plants were grown on very high concentrations of ammonium. Under these conditions, proline formation appeared to be limited by a lack of glutamine and the plants were less tolerant of salt treatment. Although somewhat extreme, these treatments appear to demonstrate a requirement for GS1 in the phloem to produce glutamine for synthesis of proline, an amino acid that is known to accumulate during water stress. It remains to be seen whether overexpression of GS1 will confer increased drought tolerance to a plant.

In a recent study in rice, a T-DNA insertion mutant was identified where only one (*OsGS1;1*) of the three GS1 genes was knocked out. These plants had severely reduced growth and biomass production, and reduced glutamine content. *OsGS1;1* is normally expressed more highly in leaves than in roots, particularly in the companion and parenchyma cells of vascular tissue, where it is thought to produce glutamine for transport purposes. These results further support a role for GS1 in nitrogen mobilization, forming glutamine within the vascular cells for export to other parts of the plant. Indeed, the reduction in grain production in these knockout lines points to the importance of GS1 in providing nitrogen to the developing grain; hence this gene is becoming a target for cereal breeders.

The manipulation of GS gene expression in nitrogen-fixing legumes has also produced differing results. *Lotus corniculatus* transformed with a soybean GS1 gene showed increased growth when the gene was expressed in both roots and shoots, whereas growth was reduced when the transgene was only expressed in the roots. These results further indicate that GS localization is a critical component of the nitrogen balance within a plant and any attempt to manipulate nitrogen allocation must take this into account.

Higher plants contain two forms of GOGAT, one is ferredoxin-dependent and the other is NADH-dependent

GOGAT was first discovered by Tempest, Meers, and Brown in 1970 in the bacterium *Aerobacter aerogenes* (now called *Klebsiella aerogenes*). The enzyme was NADPH-dependent. In 1974 three papers presented evidence for the existence of GOGAT in higher plants. Dougall reported NADPH- and NADH-dependent GOGAT activity in nonphotosynthetic cultures of carrot cells. Fowler, Jessup, and Sarkissian also found evidence for NADPH- and NADH-dependent GOGAT in sycamore cell cultures and in pea roots. In the same year, Miflin and Lea provided the first evidence for the existence of GOGAT in photosynthetic tissue, reporting that the leaf form was dependent on reduced ferredoxin rather than NAD(P)H. Although NADPH-dependent GOGAT activity has been detected in a range of higher plants, no protein has been unambiguously identified and no genes have been identified within the arabidopsis genome. It is generally accepted that higher plants contain only two forms of GOGAT: Fd-GOGAT and NADH-GOGAT.

Both Fd- and NADH-GOGAT are located in the plastid and exist as monomeric proteins in most species

Unlike GS, which has both plastidic and cytosolic isoforms, both Fd- and NADH-GOGAT are located exclusively in the plastid. Fd-GOGAT is a monomeric protein, in most species, with a molecular mass ranging between 145 and 180 kDa. The exception is rice Fd-GOGAT, which exists as a dimer of two 115 kDa subunits. Fd-GOGAT is found in cyanobacteria, algae and plastids of higher plants, where reduced Fd may be generated photosynthetically, in the chloroplasts, or through NADPH and ferredoxin NADP$^+$ oxidoreductase within nonphotosynthetic plastids (see above).

Most plants appear to contain a single gene for Fd-GOGAT (*GLU1*), although an additional gene (*GLU2*) has been identified in arabidopsis. *GLU1* is expressed mainly in the leaves and, similar to the GS2 gene (*GLN2*), its expression is increased by light and sucrose, and is repressed under nonphotorespiratory conditions (e.g. high CO_2 concentrations) and may also be under phytochrome control. *GLU1* is the gene whose mutation is responsible for the Fd-GOGAT deficient arabidopsis photorespiratory mutants, as discussed below. *GLU2*, although expressed predominantly in roots, is not as highly expressed as *GLU1*, nor is its expression influenced by light or sucrose. It is possible that the low level of expression of *GLU2* has delayed its discovery in species other than arabidopsis and there may well be other *GLU2* genes yet to be identified.

NADH-GOGAT is a monomer with a molecular mass ranging from 200 to 240 kDa. It appears to be encoded by a single nuclear gene, *GLT1*. It is expressed more highly in roots than in leaves and is highly expressed in the nodules of nitrogen-fixing legumes. *GLT1* expression is increased in rice roots by both nitrate and ammonium treatment.

The tissue and cellular localization of Fd- and NADH-GOGAT provides a clue to their function in higher plants

Fd-GOGAT protein is the predominant form of GOGAT in photosynthetic tissue. It can account for up to 97% of the total GOGAT activity in leaves. Fd-GOGAT is present in high concentrations in the mesophyll cell chloroplasts of leaves, and is detectable in the vascular parenchyma plastids to a lesser extent. The main function of Fd-GOGAT in leaves appears to be the formation of glutamate to support the assimilation of photorespiratory ammonium by GS2. This function is supported by evidence from arabidopsis and barley mutants with 5% and <1%, respectively, of wild-type Fd-GOGAT activity. Both mutants are incapable of surviving under photorespiratory conditions, and the conditional-lethal mutation in the Fd-GOGAT gene (*GLU1*) causes the plants to accumulate glutamine, the substrate for GOGAT. If placed under conditions that prevent photorespiration (i.e. 1% CO_2), the mutants grow normally and no longer accumulate glutamine, thus confirming that the major, if not the only, function for Fd-GOGAT is to support photorespiratory ammonium assimilation.

Fd-GOGAT protein has also been found in the roots of a range of species, including rice, maize, barley, and pea. In rice roots, it is found in all cells within young tissue at the tip, but as its activity decreases with cell maturity it becomes confined specifically within the central cylinder of the root cortex.

NADH-GOGAT protein is most highly concentrated in immature cells of rice leaves, particularly within the vascular cells, where it coincides with the cytosolic isoform of GS. These two isoforms together provide a means of assimilating ammonium within the vascular cells, either resulting from nitrate reduction or via remobilization of proteins and amino acids. NADH-GOGAT distribution within rice root cells is highly influenced by ammonium nutrition. In nitrogen-depleted rice, NADH-GOGAT protein is present within the central cylinder and the apical meristematic cells. Ammonium causes a rapid induction of NADH-GOGAT protein within the peripheral cells (exodermis and epidermis) that are involved in ammonium uptake from the soil, which indicates that NADH-GOGAT is directly involved in primary ammonium assimilation in rice roots. The distribution of Fd-GOGAT and NADH-GOGAT within the central cylinder coincides with the location of GS1, which indicates that these three proteins are involved in producing glutamate and glutamine for transport to the shoot.

NADH-GOGAT protein is present within the plastids of legume root nodules, particularly in *Rhizobium*-infected cells. The absence of Fd-GOGAT in root nodules indicates that only NADH-GOGAT is involved in the assimilation of ammonium produced during nitrogen fixation. Further support for this function has come from the analysis of transgenic plants, as discussed in the next section.

Further evidence of the function of Fd- and NADH-GOGAT has come from the analysis of mutants and transgenic plants

Antisense lines of alfalfa (*Medicago sativa*) with a 50% reduction in NADH-GOGAT activity in the root nodule suffered from chlorosis and reduced growth due to their inability to assimilate ammonium produced during nitrogen fixation. The plants had normal levels of Fd-GOGAT, which suggests that NADH-GOGAT, and not Fd-GOGAT, is responsible for producing glutamate to support ammonium assimilation during nitrogen fixation in root nodules.

The Fd-GOGAT-deficient photorespiratory mutants described above provide the strongest evidence of the involvement of this isoform in photorespiratory ammonium assimilation. Transgenic tobacco plants with between 20 and 90% of Fd-GOGAT activity show similar effects to the mutants and further confirm this function for Fd-GOGAT.

The production of transgenic lines with altered NADH-GOGAT has only been achieved relatively recently. In lines of tobacco and rice in which NADH-GOGAT is overexpressed, there is a significant increase in growth. As the rice overexpressing lines also showed an increase in grain weight, this indicates that NADH-GOGAT might function in the remobilization of glutamine during grain filling. Hence, both GS1 (as described above) and NADH-GOGAT have become targets for crop breeders who are aiming to increase grain yield in cereals.

An NADH-GOGAT mutant has been isolated from arabidopsis. It resembles wild-type arabidopsis under normal growth conditions but shows a 20% reduction in growth and also accumulates glutamine when grown under non-photorespiratory conditions (1% CO_2). This suggests that, while Fd-GOGAT is essential for photorespiration, NADH-GOGAT is needed under nonphotorespiratory conditions.

Thus it appears that there are distinct functions for Fd- and NADH-GOGAT in higher plants and that this is probably due to the specific localization of the two isoforms. NADH-GOGAT appears to be required for nitrogen remobilization within the leaf vascular tissue, controlling levels of transported glutamate and cooperating with GS1 to metabolize ammonium. Fd-GOGAT is active in leaf mesophyll cells where its prime function is to produce glutamate to support the assimilation of photorespiratory ammonium by GS2 within the plastid.

Sulfur is an essential macronutrient but it represents only 0.1% of plant dry matter

Plants have a major role as primary producers of organic sulfur compounds. Sulfur is described as one of the most versatile elements in living organisms and is of key importance to plant life. Part of this versatility is due to the existence of sulfur in nature in a number of stable oxidation states, that is in inorganic, organic, and bioorganic forms. The oxidation state is in constant flux as it circulates through the global sulfur cycle (Figure 8.16). In the reduced form sulfur is found in the amino acids cysteine and methionine, and in peptides, glutathione and phytochelatins, proteins, iron–sulfur clusters, lipoic acid, vitamins, and other cofactors.

The sulfur-containing amino acids in plants, cysteine and methionine, perform an essential role in the structural and catalytic functions of proteins. Disulfide (S–S) bridges in proteins have important structural and regulatory roles, iron–sulfur clusters are involved in electron transport, and many enzymes and coenzymes have sulfur in their catalytic sites. In addition, a number of secondary and essential metabolites are generated from these amino acids.

The cysteine and methionine content of crop plants is important, as animals are unable to reduce sulfur and therefore need dietary sources of sulfur-containing amino acids. A number of crop plants, for example legumes, are methionine-deficient and if used as animal feeds, methionine must be added as a supplement. A key focus of biotechnological research today is to alter the amino acid composition of seed storage proteins and other polypeptides in plants to improve their nutritional index. More recently, awareness of pollution

Figure 8.16 Biogeochemical sulfur cycle.

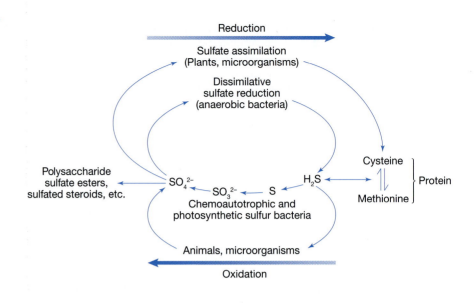

levels and its impact on the environment has led to legislative moves to decrease sulfur dioxide and other air pollutant emissions. In many industrialized nations such restriction of sulfur emissions has led to a marked decrease in both aerial dry and wet sulfur deposition. Unfortunately an additional effect of this has been that the previously adequate levels of sulfur are now suboptimal for the needs of some crop plants. Sulfur deficiency leads to decreased crop yields and specific effects on quality in terms of sulfur amino acid and protein content. Fertilizer applications are now essential to maintain crop yields and quality. For instance, in agricultural areas in northern Europe, sulfur, previously not considered to be a limiting nutrient, must now be added in the form of sulfate fertilizers.

Sulfate is relatively abundant in the environment and serves as a primary sulfur source for plants

Micromolar concentrations of sulfate (SO_4^{2-}) are absorbed by plants from the soil and concentrated in the cells to millimolar concentrations. Transport into plant cells, mediated via plasma membrane localized H^+/SO_4^{2-} cotransporters, is driven by proton motive force (PMF). The proton gradient is maintained across the membrane by a proton-pumping H^+–ATPase, with one sulfate cotransported with every three protons. Sulfate uptake is inhibited if the pH of the external medium increases or if the proton pump activity decreases. Within the cells, sulfate can be stored in the vacuole, can be found in the cytoplasm or it can enter metabolism. Once taken up sulfate is actively transported within cells and between cells by sulfate transporters.

Multiple sulfate transporters with different affinities for sulfate have been identified in plants. For example, arabidopsis has 14 sulfate transport genes and there is thought to be little redundancy. The diversity of transporters reflects the need to obtain and transport sulfate throughout the plant irrespective of availability in the soil. Protein synthesis is required for induction of sulfate transport activity, suggesting that the sulfate transporter expression is regulated. Individual transporters are expressed in specific tissues. Some are strongly regulated at the mRNA level by changes in the sulfur nutritional status of the plant. Although the precise function of all the transporters is not known, they are categorised into five groups based on their sequence similarity and function (Table 8.6).

Table 8.6 Summary of groups of sulfate transporters found in plants

Sulfate transporter	Sulfate affinity	Location	Role
Group 1	High	Root cells	Responsible for uptake of sulfate from soil solution
Group 2	Low	Xylem, parenchyma, and phloem cells of roots and leaves	Involved in translocation of sulfate within plant
Group 3	Not known	Nitrogen-fixing legume nodules	Unknown
Group 4	Not known	Tonoplast	Sulfate efflux from the vacuole
Group 5	Not known	Tonoplast	Unknown

Sulfate transport across the tonoplast is dependent on a PMF. However, in contrast to the situation in the plasma membrane, vacuolar transport is mediated by a uniporter. The precise mechanism for transport of sulfate into plastids is unclear, although a number of transporters including the triose phosphate/P_i translocator have been proposed to be involved. In addition, a plasma membrane-like proton sulfate cotransporter has also been identified. Sulfate uptake in plants is inhibited by several anions, including selenate, molybdate, and chromate, which compete with sulfate for binding to the transporters.

The assimilation of sulfate

The initial assimilation of sulfate involves activation followed by reduction to sulfide (S^{2-}), and incorporation of the sulfide into cysteine. Until recently there has been much controversy over how sulfate might be assimilated in higher plants. With the identification, cloning, and characterization of genes and enzymes involved, the route of sulfate reduction (Figure 8.17) and the subsequent assimilation into cysteine (Figure 8.31) and methionine (Figure 8.32) is now fairly clear. The whole pathway is made more complex by the finding that most enzymes involved belong to multigene families and, in most cases, with the exception of sulfite reductase (SiR), the specific gene products of families are localized in a number of individual subcellular compartments.

Higher plant cells absorb sulfate from the soil solution. Sulfate is extremely stable and is first activated by reaction with ATP to form adenosine 5′-phosphosulfate (APS) and pyrophosphate. ATP sulfurylase, an enzyme found primarily in the plastids (85–90%), and as a minor cytosolic form, catalyzes this energetically unfavorable reaction. APS is directly reduced to produce free sulfite by APS reductase catalyzing a thiol-dependent two-electron transfer. SiR is a ferredoxin-dependent enzyme that reduces the sulfite (SO_3^{2-}) to sulfide (S^{2-}). In chloroplasts the reductant, reduced ferredoxin (Fd_{red}), is generated photosynthetically by the noncyclic electron transport chain, while in nonphotosynthetic plastids the reduced ferredoxin is generated indirectly via the oxidative pentose phosphate pathway (in a similar manner to NiR, see Figure 8.12). SiR is the only enzyme in the sulfate assimilatory pathway to be represented by a single gene. It is only found in the plastids, with the chloroplasts being the primary site for sulfate reduction and assimilation. Adequate reductant and ATP are available to support these processes. Sulfide is subsequently incorporated into the β-position of the amino acid skeleton of O-acetylserine to form cysteine via the enzymes serine acetyltransferase and O-acetylserine (thiol) lyase, or cysteine synthase (see page 280).

Figure 8.17 Pathway of sulfate assimilation showing the activation of sulfate to form APS and its reduction to form sulfide.
Abbreviations: APS, adenosine 5′-phosphosulfate; Fd, Ferredoxin; GSH, glutathione (reduced); GSSG glutathione (oxidised).

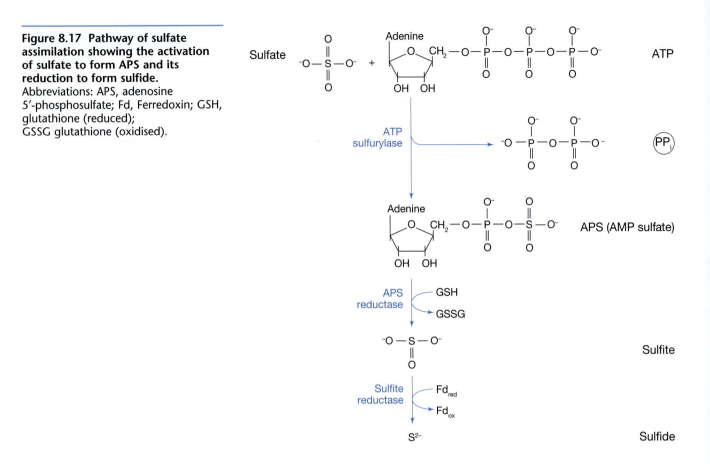

Adenosine 5′-phosphosulfate reductase is composed of two distinct domains

Adenosine 5′-phosphosulfate (APS) reductase is a dimer of 43 kDa monomer subunits joined by a disulfide bond at a conserved cysteine residue. It has two distinct domains, an APS reductase-like domain and a thioredoxin-like domain. These two domains can be expressed separately, have independent enzyme activities and catalyze distinct steps in APS reduction. It is now clear that, although the precise molecular mechanisms of the reaction are far from understood, the APS reductase reaction can be divided into three independent steps (Figure 8.18). First sulfite is transferred from APS to the active cysteine residue with AMP released. For this reaction neither the thioredoxin-like domain nor an external electron source is needed. The sulfite probably then forms an intramolecular disulfide bridge and is then released by the C-terminal domain. The electrons needed for this release are provided by APS reductase. Finally the active enzyme dimer is recovered by reaction with glutathione.

Figure 8.18 The three reaction steps in APS reduction catalyzed by APS reductase. The square represents the N-terminal domain of APS reductase and the circle the C terminal domain. S indicates the active site cysteine residue and a cysteine thought to form an intramolecular disulfide (see text for details). Abbreviations: APS, adenosine 5′-phosphosulfate; SO_3^{2-}, sulfite; GSH, glutathione (reduced); GSSG glutathione (oxidised).

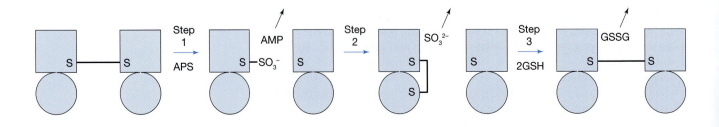

Sulfite reductase is similar in structure to nitrite reductase

SiR is a homo-oligomeric hemeprotein. It is composed of two or four identical 64–71 kDa subunits. Each subunit, like the C-terminal domain of NiR, contains a siroheme center and a 4Fe–4S cluster. Higher plant SiR has high sequence homology to NiR. Homology is greatest in the C-terminal domain region. Both SiR and NiR belong to the superfamily of anion redox enzymes.

Sulfation is an alternative minor assimilation pathway incorporating sulfate into organic compounds

Sulfation is the transfer of a sulfate group from 3′-phospho-5′-adenylsulfate (PAPS) by the cytosolic sulfotransferases to a substrate (Figure 8.19). APS can be withdrawn from sulfate assimilation in the plastids and is phosphorylated by APS kinase to produce PAPS. The reaction equilibrium for ATP sulfurylase favors the reverse reaction and the formation of ATP and sulfate. To drive this reaction in the forward direction the products APS and pyrophosphate must immediately be metabolized by the enzyme APS reductase and APS kinase pyrophosphatase. The sulfation dedicated PAPS synthesis can occur in the cytosol or the plastids. Compounds sulfated by

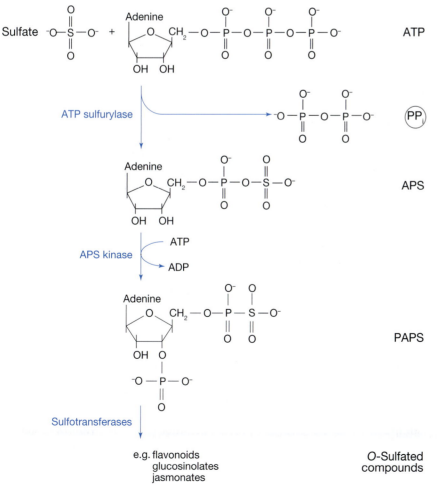

Figure 8.19 Pathway of sulfation via PAPS. Abbreviations: APS, adenosine 5′-phosphosulfate; PAPS, 3′-phospho-5′-adenylsulfate.

the pathway include flavonoids, glucosinolates, and jasmonates, which are all involved in plant defense against biotic and abiotic stress.

Amino acid biosynthesis is essential for plant growth and development

Some amino acids act primarily to assimilate and transport nitrogen within the plant. Amino acids are also the precursors of many nitrogen-containing compounds, including chlorophyll, and enzyme cofactors, including thiamine pyrophosphate (from alanine and methionine) and CoA (from valine, aspartate and cysteine). Amino acids also serve as the carbon and nitrogen source for secondary products including alkaloids (Chapter 10) and phenolic acids (Chapter 11).

Amino acid biosynthesis is a dynamic process that requires carbon and nitrogen and is controlled by a range of metabolic, environmental, and developmental factors. It is a balanced process, ensuring that multiple amino compounds are synthesized in a precise manner. Depending on the specific tissue, developmental stage, etc. many amino acids can be made by alternative synthetic pathways. In some cases, such as histidine biosynthesis, the pathway of synthesis in higher plants has only recently been characterized. Regulation of enzymes involved in amino acid biosynthesis occurs at a number of different levels.

Amino acids may be grouped together into families derived from a single head or precursor based on their biosynthetic pathway. However, as these divisions are artificial the amino acids can often be assigned to a number of families. Of the protein amino acids (i.e. those that occur mainly within proteins) nine are essential in that they cannot be synthesized by humans or monogastric animals and must therefore be obtained through diet. Approximately 50% of plant protein consumed worldwide are derived from cereal crops, but cereal grain is generally nutritionally limited in a number of essential amino acids. This has led to widespread research to try and improve the nutritional value of cereal seeds. Moreover as amino acid biosynthetic enzymes are the target for the action of a number of herbicides, the study of amino acids has important implications for both basic and applied research (Box 8.3).

Carbon flow is essential to maintain amino acid production

The metabolites, pyruvate (Figure 8.20), oxaloacetate (Figure 8.21), and 2-oxoglutarate (Figure 8.22) are directly available as carbon skeleton backbones for amino acid biosynthesis. The reactions of the GS/GOGAT cycle have the biggest drain on carbon pathways, in the form of 2-oxoglutarate. During both the reductive and oxidative pentose phosphate pathways ribose-5-phosphate (Figure 8.23, page 273) and erythrose 4-phosphate (Figure 8.24, page 273) are generated and subsequently used for aromatic amino acid biosynthesis. The Calvin cycle provides the carbon skeletons, in the form of 3-phosphoglycerate (Figure 8.25, page 274), needed for the photorespiratory synthesis of glycine and serine. In addition, phospho*enol*pyruvate (Figure 8.24) and acetyl CoA are also needed for amino acid biosynthesis. Any consumption of amino acids for protein or other synthesis withdraws carbon from these pathways. The carbon skeletons are then restored through increased glycolytic and anaplerotic activities (Chapter 6).

Box 8.3 The use of herbicides in agriculture

The yield and quality of crops are significantly affected by weeds, through competition for resources, such as light and nutrients. Also, weed populations are a source for harboring pests and diseases. As such, weeds are one of the major biotic impacts on the world crop yield available for human consumption. To control weeds, a number of herbicides or weedkillers have been developed. Many of these weedkillers are broad-spectrum herbicides, meaning that they are active against a broad spectrum of weeds. As, by definition, herbicides are generally more toxic to plants than animals, then they tend to target plant-specific biological processes, such as photosynthesis and essential amino acid synthesis. Herbicides that inhibit amino acid biosynthesis have proved to be particularly useful, in that they ultimately block the biosynthesis of proteins. As animals are unable to synthesize all the amino acids that they need for healthy development they must rely on an input from diet. This means that herbicides inhibiting amino acid biosynthesis generally have less impact on animals than other available herbicides. It is for this reason that herbicides that inhibit amino acid biosynthesis have been the focus of many research groups for new herbicide discovery.

Many herbicides act on a single enzyme or protein and are classified according to their mode of action. Although most herbicides have only one mode of action, it is clear that particular enzymes are more susceptible to herbicide activity. To date there are three enzymes that are generally targeted for commercial herbicides, which inhibit amino acid biosynthesis: glutamine synthase in ammonium assimilation, acetolactate synthase in the branched chain amino acid biosynthesis pathway, and 5-*enol*pyruvylshikimic acid 3-phosphate synthase in the shikimic pathway.

Understanding the site of action of herbicides is important for developing strategies aimed at engineering resistance. As the chemical properties of herbicides differ, then this has an impact in terms of toxicity, environmental persistence, and biodegradability. Assessment of such factors and benefits needs to be made before developing a herbicide for commercial use.

A limitation to herbicide application is that it can only be used when a crop will not be vulnerable to herbicide action. It is for this reason that herbicide-resistant crops have been developed. Such crops can be sprayed with herbicide when it is most appropriate, without damaging the crop itself. Herbicide-tolerant crops have been developed in a number of different ways. However, commercially two strategies have proved useful: altering the target gene to make the target enzyme less sensitive to the herbicide and using a metabolism enzyme that detoxifies the herbicide before it reaches the target enzyme in the plant, bypassing the target enzyme via an alternative pathway.

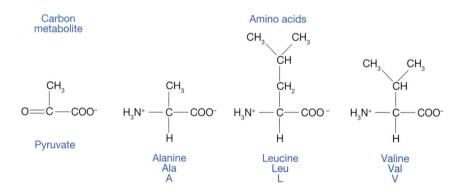

Figure 8.20 Pyruvate is a carbon skeleton for the amino acids alanine (Ala or A), leucine (Leu or L), and valine (Val or V).

The demand for carbon skeletons is affected by plant species, tissue age, time of day, and environmental stress. Depending on the situation this has an impact on carbon metabolism. For example, when wounded, plant tissues show an increase in the production of carbon skeletons necessary for the biosynthesis of aromatic amino acids (e.g. phenylalanine, tyrosine). These amino acids are essential for the generation of secondary defense metabolites, including phenylpropanoids, phenolic acids, and flavonoids (see Chapter 11).

Figure 8.21 Oxaloacetate is a carbon skeleton for the amino acids aspartate (Asp or D), asparagine (Asn or N), threonine (Thr or T), isoleucine (Ile or I), methionine (Met or M), and lysine (Lys or K).

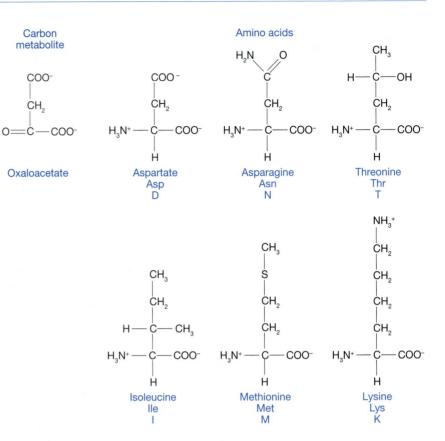

There are species differences in the form of nitrogen transported through the xylem

The form of nitrogen that is moved between roots and shoots differs between species. In wheat, for example, nitrate is transported from the roots via the xylem and assimilated to glutamine and glutamate in the leaves prior to further amino acid biosynthesis. In legumes nitrate is reduced in the roots and converted to an organic form (usually into asparagine or glutamine) prior to xylem transport. Nitrogen flow, like carbon skeletons, is also affected by age, with young leaves consuming all of the incoming nitrogen for growth. In contrast, mature leaves re-export the majority of the organic nitrogen via the phloem to the growing apex or developing fruits. Similarly, senescing leaves convert much of their proteins and other nitrogenous molecules, including chlorophyll, to transport compounds for export.

Figure 8.22 2-Oxoglutarate is a carbon skeleton for the amino acids glutamate (Glu or E), glutamine (Gln or Q), histidine (His or H), proline (Pro or P), and arginine (Arg or R).

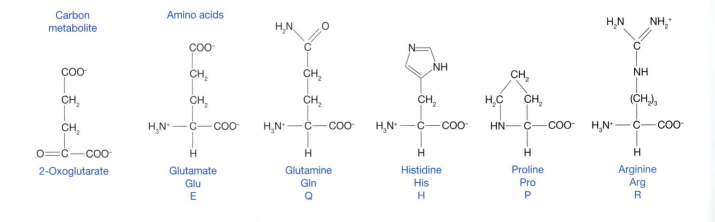

A clear relationship between carbon and nitrogen movement can be seen when nitrogen-deficient plants are transferred to a nitrogen-rich environment. Increased nitrogen assimilation leads to increased consumption of 2-oxoglutarate by GS/GOGAT (see page 259). This leads to an increased demand for the supply of carbon skeletons via respiratory metabolism of carbohydrates (Chapter 6).

Aminotransferase reactions are central to amino acid metabolism by distributing nitrogen from glutamate to other amino acids

Once nitrogen has been assimilated into glutamine and glutamate, aminotransferases, also called transaminases, are able to redistribute nitrogen to a range of amino acids. Aminotransferases, as the name suggests catalyze the transfer of an amino group from the α-carbon of an amino acid to the α-carbon of a keto acid in a freely reversible reaction (Figure 8.26). This results in the production of a new amino acid and a new keto acid. Occasionally in these reactions amines act as amino donors and aldehydes as amino acceptors. The coenzyme pyridoxal 5′-phosphate is needed for the activity of all known aminotransferases. Pyridoxal 5′-phosphate is covalently bound to the enzyme and accepts an amino group from the amino acid substrate. The keto acid produced is then released and the aminated form of the coenzyme undergoes a reversal of the process, releasing its newly acquired amino group to the keto acid substrate to produce the amino acid product. The biosynthesis of one amino acid often leads to the breakdown of another and occurs in all cells and in most subcellular compartments. Aminotransferases are ubiquitous and the reactions catalyzed are indispensable for a variety of metabolic pathways including amino acid biosynthesis and catabolism. However, they are also of fundamental importance in vitamin biosynthesis, carbon and nitrogen assimilation and transport, alkaloid biosynthesis (Chapter 10), photorespiration (Chapter 5), and gluconeogenesis (Chapter 7).

Enzymes capable of catalyzing aminotransfer of all the common amino acids have been identified. The sequencing of the entire arabidopsis genome has led

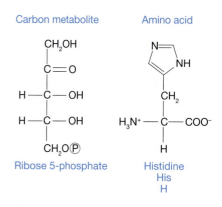

Figure 8.23 Ribose 5-phosphate is a carbon skeleton for the amino acid histidine (His or H).

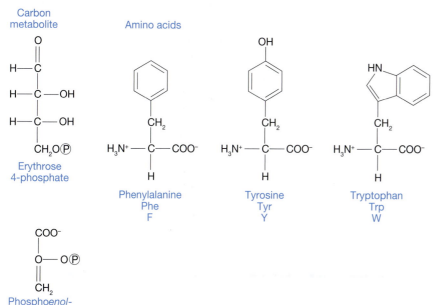

Figure 8.24 Erythose 4-phosphate or phospho*enol*pyruvate are possible carbon skeletons for the amino acids phenylalanine (Phe or F), tyrosine (Tyr or Y), and tryptophan (Trp or W).

Figure 8.25 3-Phosphoglycerate is a carbon skeleton for the amino acids serine (Ser or S), glycine (Gly or G), and cysteine (Cys or C).

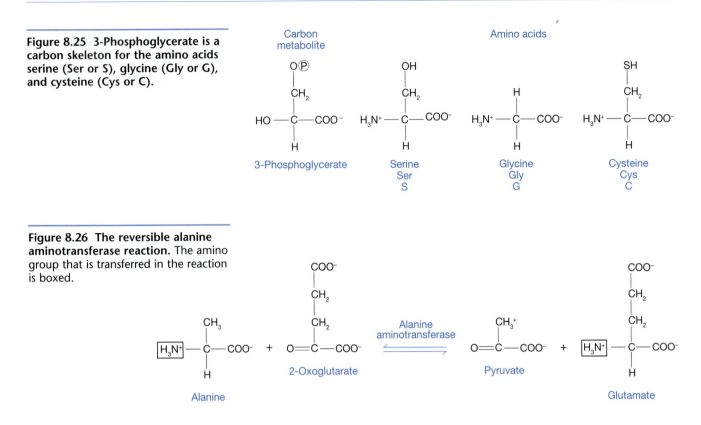

Figure 8.26 The reversible alanine aminotransferase reaction. The amino group that is transferred in the reaction is boxed.

to the identification of 60 distinct sequences encoding aminotransferases and closely related pyridoxal 5′-phosphate-dependent enzymes. As many amino acids are present in millimolar concentrations, the kinetic constants for the enzymes are very high, with the K_m values for amino acid substrates being in the range of 1–5 mM. Aminotransferases are sometimes not particularly specific, reacting with a number of amino acid or keto acid substrates. For example, an aspartate aminotransferase might also be able to use glutamate as an amino acid substrate and either 2-oxoglutarate or oxaloacetate as a keto acid substrate. Most plant nitrogen passes through glutamate and most aminotransferases can use glutamate as an amino donor to synthesize a wide range of amino acids (Figure 8.27). This widespread use of glutamate as an amino donor leads to the production of large amounts of 2-oxoglutarate, which is then reaminated by the GS/GOGAT pathway during ammonia assimilation.

Figure 8.27 Main routes of nitrogen into amino acid biosynthesis (shown by bold arrows). Amino acids generated then serve as the precursors and amino donors in the biosynthesis of other amino acids.

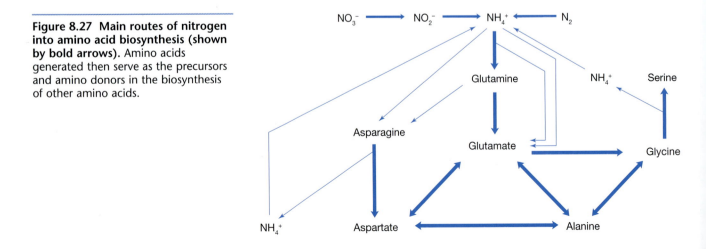

Glutamate is also an important precursor for the biosynthesis of tetrapyrroles that serve as components of NiR and SiR, chlorophyll, cytochromes, and phytochrome (Box 8.2).

Asparagine, aspartate, and alanine biosynthesis

Asparagine synthetase uses, in an ATP-dependent manner, either glutamine or ammonia as the amino donor. In most cases, however, it is the former that acts as the *in vivo* substrate (Figure 8.28). Asparagine biosynthesis, via the combined activities of glutamine synthetase and asparagine synthetase, is important in root nodules of legumes with fixed nitrogen being quickly transferred to asparagine and subsequently transported away from the nodules in the xylem. Alternatively, β cyanoalanine formed from hydrogen cyanide and cysteine may be used to synthesize asparagine by the action of β cyanoalanine synthase and β cyanoalanine hydrolase. This pathway, being dependent upon a supply of hydrogen cyanide, is restricted to plants such as lupins (*Lupinus lutea*), sorghum (*Sorghum* spp.), sweet pea (*Lathyrus odoratus*), and asparagus (*Asparagus* spp). Although glutamine and asparagine are amides, which differ only in carbon chain length, they are distinct in their biochemical activities and cellular roles. For example, asparagine is more soluble and less reactive than glutamine. As such, asparagine is well suited as a transport and storage compound.

Aspartate dehydrogenase aminates oxaloacetate with ammonia to aspartate. Similarly, aspartase catalyzes the addition of ammonia to fumarate to produce aspartate. However, neither of these enzymes is considered to make a marked contribution to ammonia assimilation.

Asparagine has several important roles in plants. During the day its synthesis from aspartate is important in supporting protein synthesis. It also serves as an important molecule for nitrogen storage during the night, when photosynthesis is inactive. Asparagine also acts as the nitrogen transport form between source and sink tissue. Metabolic channeling of aspartate into asparagine or the aspartate family of amino acids is regulated by the expression of the genes encoding asparagine synthetase and aspartate kinase. In plants there are two forms of asparagine synthetase genes. One form is induced by light and sucrose enabling asparagine biosynthesis during the day. The other form is repressed by light and sucrose and induced during the night. So, during the day aspartate is channeled into asparagine and into the aspartate family pathway to facilitate synthesis of all its end products of amino acids. At night, the aspartate family pathway is blocked due to repression of expression of the aspartate kinase genes, channeling aspartate preferentially into asparagine. Asparagine levels are much higher and lysine levels are much lower during the night than the daytime.

Figure 8.28 The pathway for the biosynthesis of asparagine from glutamine and aspartate via asparagine synthetase.

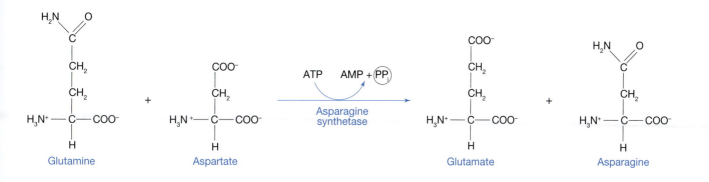

Whether nitrogen arrives at a tissue in the form of nitrate, glutamate, or amino acids, it is rapidly distributed within pools of amino acids with different subcellular compartments. Transamination between alanine and oxaloacetate transfers amino groups between alanine and aspartate. All the aminotransferases involved are present in the chloroplast and the cytosol. In addition, some are also found in mitochondria.

In developing legume seeds and fruits much of the available nitrogen is in the form of asparagine and comes via the xylem or phloem. Asparaginase is a cytosolic enzyme that also sythesizes aspartate by the hydrolysis of asparagine. Its levels vary during development and during the diurnal cycle, increasing in the light and decreasing in the dark. This diurnal variation of the enzyme is essential as it is involved in the production of ammonia that is toxic to the plant. By functioning in the light, asparaginase ensures there is ATP and reductant supply available to support the operation of the GS/GOGAT pathway.

Glycine and serine biosynthesis

Serine and glycine are generated during photorespiration. While this is the major route, there is evidence of an additional 3-phosphoglycerate dependent synthesis of serine in chloroplasts.

The aspartate family of amino acids: lysine, threonine, isoleucine, and methionine

The conversion of aspartate to aspartic semi-aldehyde catalyzed by aspartate kinase and aspartate semialdehyde dehydrogenase is a key reaction in the biosynthesis of a number of amino acids (Figure 8.29). Aspartate kinase phosphorylates aspartate to produce β-aspartylphosphate. There are at least two distinct forms of aspartate kinase in plants, the major form is a lysine-sensitive monofunctional isozyme, which is involved in the overall regulation of the pathway. This form can also be synergistically inhibited by S-adenosyl methionine. The other form of aspartate kinase is a bifunctional threonine-sensitive aspartate kinase and homoserine dehydrogenase. Homoserine dehydrogenase is the first committed enzyme in threonine biosynthesis and is also present as a separate threonine-insensitive enzyme with an unknown physiological function. Overexpression of the plant aspartate kinase–homoserine dehydrogenase in *Escherichia coli* demonstrated that the enzyme contains a regulatory domain with two similar subdomains, each containing two nonequivalent threonine-binding sites. Interaction with threonine at the first subdomain inhibits the aspartate kinase activity and also facilitates the binding of a threonine to the second subdomain resulting in the inhibition of the homoserine dehydrogenase activity.

Aspartate semialdehyde is situated at a branchpoint within the aspartate metabolic pathway, where subsequent steps lead to either the biosynthesis of lysine, or of threonine and isoleucine (Figure 8.29). Aspartate semialdehyde is synthesized from β-aspartylphosphate in an NADPH-dependent reaction catalyzed by aspartate semialdehyde dehydrogenase. To date there has been little characterization of this enzyme. Aspartate semialdehyde is subsequently condensed with pyruvate to form dihydropicolinate. Dihydrodipicolinate synthase catalyzes this reaction and is extremely sensitive to inhibition by low concentrations of lysine. As a result of lysine binding to the dimers the tetramer formed is rearranged. Feeding studies and isolation of intermediates have enabled the pathway of lysine synthesis to be deduced. It involves six

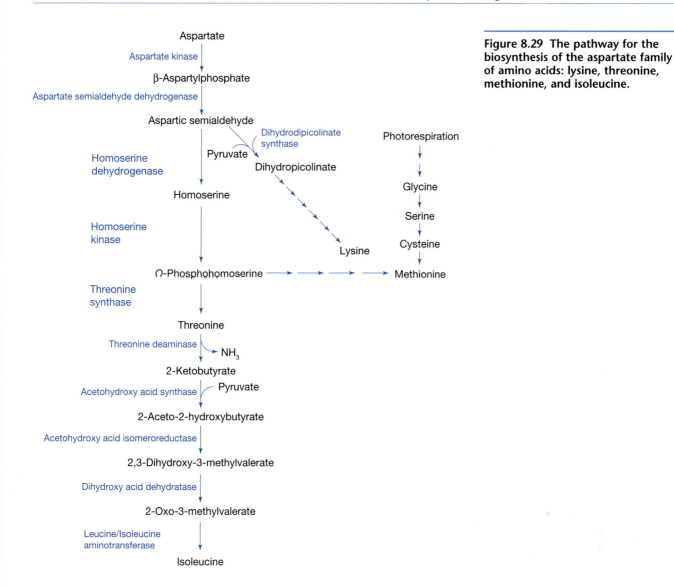

Figure 8.29 The pathway for the biosynthesis of the aspartate family of amino acids: lysine, threonine, methionine, and isoleucine.

steps, including acylation, transamination, and decarboxylation; however, there is only limited information on a number of the enzymes involved.

The homoserine generated via homoserine dehydrogenase is phosphorylated to O-phosphohomoserine in an ATP-dependent manner by homoserine kinase. O-Phosphohomoserine is an important intermediate of the aspartate pathway as it is at the branchpoint where threonine and methionine synthesis diverge. In the case of threonine synthesis the allosteric threonine synthase converts O-phosphohomoserine to threonine. From *in vitro* studies threonine synthase has 250–500-fold higher activity for the substrate O-phosphohomoserine than the enzyme involved in methionine synthesis (cystathione γ-synthase, see Figure 8.32, page 283) suggesting that carbon from aspartate is preferentially channeled to the synthesis of threonine rather than methionine. Threonine synthase has been highly studied in plants and appears to exist in a single isoform.

The only enzyme that is exclusive to the isoleucine biosynthetic pathway is the first enzyme, threonine deaminase. This enzyme catalyzes the deamination and dehydration of threonine to produce 2-ketobutyrate and ammonium. One biosynthetic threonine deaminase isoenzyme is present mainly in young developing tissues and is inhibited by isoleucine. A second biodegradable threonine deaminase isoenzyme is found in older senescing tissue and is insensitive to isoleucine inhibition. The biosynthetic threonine deaminase in arabidopsis is a tetramer composed of identical 59.6 kDa monomer subunits. Isoleucine inhibits the enzyme by inducing dimerization upon binding, while tetramerization is restored by the addition of high concentrations of valine. Kinetic and binding experiment studies on the arabidopsis enzyme have demonstrated that each regulatory domain of the threonine deaminase monomer has two effector binding sites containing tyrosine 449 and tyrosine 543. The tyrosine 449 site belongs to a high-affinity binding site and interaction with an isoleucine molecule induces conformational modification. This results in a protein with an increased ability to bind a second isoleucine at the lower affinity binding site containing tyrosine 543. The resulting conformational modification leads to inhibition of the enzyme. The tyrosine 449 also interacts with valine, leading to a conformational modification that reverses the isoleucine binding and therefore the inhibition.

Acetohydroxy acid synthase carries out two distinct reactions; one that forms 2-aceto-2-hydroxybutyrate within the isoleucine pathway (Figure 8.29), and another that provides intermediates for leucine and valine synthesis (Figure 8.30). Acetohydroxy acid synthase is inhibited by leucine, isoleucine, and valine, and is subjected to additional inhibition when valine and leucine are both present. In the case of arabidopsis it has been proposed that the regulatory subunit contains two domains created by an internal duplication, with one domain binding isoleucine or valine and the other leucine.

Acetohydroxy acid isomeroreductase isomerizes and then reduces the two acetohydroxyacids (either 2-aceto-2-hydroxybutyrate in isoleucine synthesis or acetolactate in valine and leucine synthesis, see Figures 8.29 and 8.30) to produce dihydroxy acids. The enzyme is thought to exist as a dimer. Either

Figure 8.30 The pathway for the biosynthesis of the leucine and valine family of amino acids.

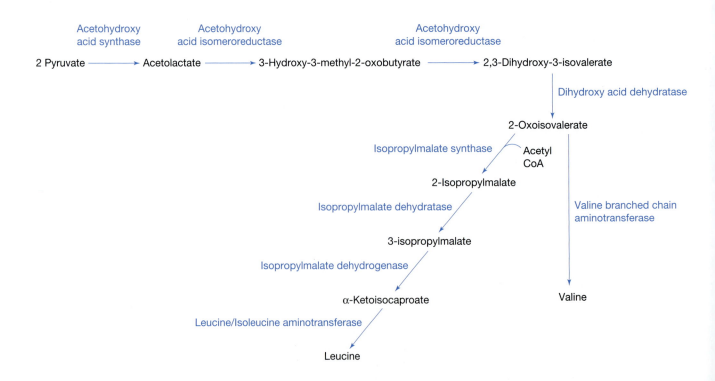

substrate competes for the same binding site with a high affinity. However, the V_{max} of either enzyme is six to 11 times higher with 2-aceto-2-hydroxybutyrate. Upon binding of Mg^{2+} ions and NADPH to acetohydroxy acid isomeroreductase there is a conformational change at the interface of the two domains of each monomer. The binding of the two cofactors alters the structure of the active site promoting or inhibiting substrate binding. Acetohydroxy acid isomeroreductase has a very high affinity, one of the strongest ever recorded between an enzyme and a metal ion, towards Mg^{2+}.

Dihydroxy acid dehydratase carries out the reaction where, in the case of isoleucine synthesis 2,3-dihydroxy-3-methylvalerate is dehydrated to form 2-oxo-3-methylvalerate (while in the case of valine and leucine synthesis 2,3-dehydroxy-3-isovalerate is converted to 2-oxoisovalerate). Although this enzyme is essential for branched chain amino acid production there is currently little information about the properties and regulation of this enzyme.

The final metabolic step in the production of isoleucine (and leucine and valine via this pathway) is a transamination reaction catalyzed by the branched chain amino acid aminotransferase. In plants two spinach aminotransferases have been identified, one, termed the leucine/isoleucine aminotransferase, which uses 2-oxo-3-methylvalerate or α-ketoisocaproate as a substrate to generate isoleucine or leucine, respectively (see later), the other, the valine aminotransferase, uses 2-oxoisovalerate as a substrate to generate valine (Figure 8.30).

The branched chain amino acids valine and leucine

Although isoleucine is also a branched chain amino acid, with a similar structure to leucine, it derives its carbon from aspartate. A unique feature of the biosynthesis of branched chain amino acids, valine, leucine, and isoleucine, is that the same four enzymes, that convert 2-ketobutyrate to isoleucine (Figure 8.29), also convert pyruvate to valine in a set of parallel reactions but with different substrates (Figure 8.30). These four enzymes are acetohydroxy acid synthase, acetohydroxy acid isomeroreductase, dihydroxyacid dehydratase, and branched chain aminotransferase. Although this is a parallel pathway, sometimes called the common enzyme pathway, it is quite distinct with no common intermediates shared between the pathway forming valine and leucine and the sequence of reactions forming isoleucine (see earlier for details including regulation of the different enzymes involved).

In the reaction producing valine and leucine, acetohydroxy acid synthase catalyzes condensation of two molecules of pyruvate to form acetolactate in the presence of thiamine pyrophosphate, FAD, and a divalent cation. Acetohydroxy acid synthase catalyzes a two-step reaction involving the isomerization of 2-acetolactate to 3-hydroxy-3-methyl-2-oxobutyrate in a Mg^{2+}-dependent reaction, and then the reduction to the dihydroxyacid 2,3-dihydroxy-3-isovalerate. This latter reduction requires NADPH and Mg^{2+}. 2,3-Dihydroxy-3-isovalerate is converted to 2-oxoisovalerate by dihydroxy acid dehydratase. 2-Oxoisovalerate is a branchpoint and is either directly transaminated to valine via the valine branched chain aminotransferase, or acts as the initial substrate in the branch towards leucine synthesis. In the case of leucine synthesis it is condensed with the methyl group of acetyl CoA to form 2-isopropylmalate, this reaction is catalyzed by isopropylmalate synthase. 2-Isopropylmalate is then isomerized to 3-isopropylmalate by the poorly characterized enzyme isopropylmalate dehydratase. The subsequent oxidative decarboxylation to α-ketoisocaproate is catalyzed by isopropylmalate dehydrogenase, prior to transamination to leucine by the leucine/isoleucine aminotransferase (Figure 8.30). On the basis of the results of feeding experiments with $^{14}CO_2$ and localization studies, these pathways appear to be in the chloroplast.

Sulfur-containing amino acids cysteine and methionine

As animals and humans lack the ability to reduce sulfate, then they rely on the provision of sulfur via cysteine and methionine in their diet. Plants are a major source of these amino acids. Cysteine residues are important in proteins for stabilizing, through disulfide bridges, the tertiary and quaternary conformation of proteins. The sulfur atoms in thiol groups of cysteine are the sites for catalytic activities. Similarly, both protein-associated and free thiols are necessary for the binding of metals and reaction with reactive oxygen, while the reversible conversion of free thiol groups to disulfide bridges is the basis for redox modulation of various proteins that regulate a number of key metabolic processes (see Chapters 2 and 7). Cysteine is also involved in the biosynthesis of a number of biomolecules including antioxidants, vitamins, and cofactors, where the catalytic mechanism is due to the reactivity of the thiol group arising from the relative weakness of the C–S bond present.

Methionine is involved in initiating mRNA translation. It is also, through S-adenosylmethionine, an important intermediate in the biosynthesis of a number of essential metabolites including alkaloids (Chapter 10), polyamines (Chapter 10), vitamins, cofactors, and osmoprotectants. Methionine is also the precursor for the biosynthesis of the plant growth regulator ethylene (Box 8.4). The role of methionine in protein structure relates to its hydrophobic properties for protein folding. Also, in the presence of chemicals or oxidants, the oxidation of exposed methionine residues within proteins leads to alterations in protein folding and catalysis.

Most sulfate is reduced to sulfide prior to incorporation into cysteine in a multistep process (see page 267). The sulfide becomes the thiol group in cysteine, the cysteine may then be further metabolized to a range of compounds. Hence, cysteine is the central intermediate form from which most sulfur compounds are synthesized. The concentration of free cysteine in a cell is very low (less than 10 μM). However, the flux through cysteine is very high because of its rapid utilization for synthesis of methionine, proteins, and glutathione (Box 8.5). The complete pathway for cysteine synthesis is present in the cytosol, plastids, and mitochondria (Figure 8.31). It has been suggested that the triple location of these enzymes reflects an inability to transport cysteine across membranes. O-Acetylserine is synthesized by the acetylation of serine with acetyl CoA catalyzed by serine acetyltransferase. O-Acetylserine (thiol) lyase then converts O-acetylserine to cysteine (Figure 8.31). Serine acetyltransferase and O-acetylserine (thiol) lyase are associated through protein–protein interactions to form the multienzyme complex of cysteine synthase. The interaction of these two enzymes plays a central role in the regulation of cysteine synthesis (Box 8.6).

Figure 8.31 The pathway for the biosynthesis of cysteine and methionine from serine. Abbreviations: OAS, O-acetylserine.

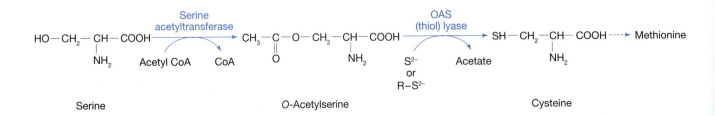

Serine   O-Acetylserine   Cysteine

The reduced sulfur found in cysteine is the thiol group that is strongly nucleophilic. It is this property that defines the function of cysteine in all its forms. Oxidizing the thiol groups of cysteines can lead to the formation of a stable disulfide bond, which is important as a covalent linkage that is necessary to establish tertiary, and sometimes even quaternary, protein structures. Reduction of disulfide bonds breaks them and restores the thiol groups. It is this process that controls the reduction state of thiol metabolites and proteins. Thiols readily react with electrophilic compounds. Reduction may be necessary for catalytic activity, for example, in the case of metalloenzymes where cysteine acts as a metal ligand. But when metal ions are present in excess or in the incorrect form or the cells are exposed to poisons, the reaction with protein thiols can lead to enzyme inactivation, highlighting the vulnerability of protein thiols to toxins.

Methionine occupies a central position in cellular metabolism. Methionine is essential for mammals and must be derived entirely from the diet. Methionine

Box 8.4 Methionine is the precursor for the biosynthesis of the plant hormone ethylene

Ethylene (or ethene) is a gaseous hormone involved in the regulation of plant growth, development and responses to stress and pathogen attack in plants. Virtually all plant tissues have the capacity to make ethylene, although in most cases the amount produced is relatively low. Ethylene is synthesized from the amino acid methionine via the intermediate S-adenosyl-L-methionine (Figure 1). The latter is first converted to 1-aminocyclopropane-1-carboxylate (ACC) and 5'-methylthioadenosine (MTA) by the enzyme ACC synthase (ACS). ACC is then converted to ethylene by ACC oxidase (ACO). In the majority of plant species both ACS and ACO are encoded by multigene families. Recent studies have identified that differential

transcription of the gene family members of ACS and also regulation of ACS protein stability are important factors in controlling the production of ethylene in response to different stimuli.

Genetic modification of the ethylene biosynthesis pathway has resulted in the production of new tomato cultivars that are slower to ripen and retain more flavor. The UK-based company, Zeneca, produced a transgenic tomato with reduced ACC synthase activity (through antisense) so that the fruits produced less ethylene and therefore took longer to ripen. These tomatoes were on sale in the UK between 1996 and 1999.

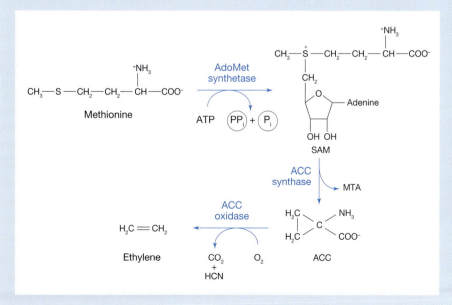

Figure 1 The pathway of ethylene biosynthesis.
Abbreviatins: AdoMet synthetase, S-adenosyl-L-methionine (SAM) synthetase; ACC, 1-aminocyclopropane-1-carboxylate; MTA, 5'-methylthioadenosine.

Box 8.5 Glutathione metabolism

Glutathione, an important low molecular weight nonprotein thiol in plants, is a key end product of the reductive sulfate assimilatory pathway. In higher plants the glutathione pool is estimated to be up to 10 mM. Glutathione and its derivatives are involved in both the storage and long distance transport of reduced sulfur for a number of processes (Table 1). Glutathione is an enzymatically synthesized tripeptide, made up of three amino acids, glutamate, cysteine, and glycine. The γ-carboxyl group of glutamate forms a peptide link with the amino group of cysteine. Glutathione has been described as a redox buffer. Upon oxidation one reduced glutathione (GSH) can react with another giving the disulfide form (GSSG). GSH is restored by the activity of NADH-glutathione reductase. Of the total intracellular glutathione content, over 90% is in the form GSH. However, during development and in response to biotic and abiotic stimuli, large and transient changes in the total glutathione pool and in the GSH:GSSG ratio occur. Glutathione homeostasis is maintained by synthesis, transport, storage, oxidation/reduction, continued metabolism, and catabolism.

In addition to acting as a redox buffer, glutathione has a number of critical functions including acting as a signal of stress, a trigger for development and a detoxifier of xenobiotics, for example, herbicide detoxification mediated by glutathione-*S*-transferases. Glutathione is also a precursor of phytochelatins, which are important in the detoxification of heavy metals. It is also a storage and transport form of reduced sulfur and a signal for the regulation of sulfur assimilation.

Table 1 Roles for glutathione in higher plants

Processes in which glutathione and its derivatives are involved
Signal transduction
Scavenging H_2O_2
Scavenging reactive O_2 species
Detoxifying xenobiotics
Activating and conjugating phenylpropanoids
Activating and conjugating hormones
Substrate for phyochelatin biosynthesis

is synthesized from cysteine and *O*-phosphohomoserine (Figure 8.32). Methionine is derived from three convergent pathways, aspartate gives the carbon backbone via *O*-phosphohomoserine (Figure 8.29), the β carbon of serine gives the methyl group (Figure 8.31), and the sulfur atom comes from cysteine (Figure 8.32). In higher plants the reactions are unique to methionine biosynthesis and catalyzed by the pyridoxal 5′-phosphate-dependent enzymes cystathione γ-synthase and cystathione β-lyase. The terminal step is a methylation of the thiol group of homocysteine, catalyzed by methionine synthase (Figure 8.32). Methionine may be withdrawn from amino acid pools for the biosynthesis of a number of compounds, including the polyamines spermine and spermidine (Chapter 10). Much of the methionine synthesized in the plant cell is converted into *S*-adenosylmethionine, which is essential for metabolism. Methionine regulates its own synthesis through *S*-adenosylmethionine at the level of cystathione γ-synthase.

Glutamine, arginine, and proline biosynthesis

For arginine biosynthesis, glutamate is first converted via a series of acetylated intermediates, to the nonprotein amino acid ornithine (Figure 8.33). It is the use of these acetylated intermediates that is thought to act to prevent competition between arginine and proline biosynthesis. In addition, arginine has been found to inhibit the phosphorylation of acetylglutamate. Ornithine

Box 8.6 Cysteine synthase is a multienzyme complex comprising SAT and OASTL

Free *O*-acetylserine (thiol) lyase (OASTL) is active and synthesizes cysteine efficiently, whereas when associated with serine acetyltransferase (SAT) in the cysteine synthase complex it is nearly inactive. The formation of the complex alters the kinetic behavior of SAT, changing it from displaying Michaelis–Menten type kinetics in the free form to positive cooperativity with respect to the substrate when associated in the complex. Where there is limited acetyl CoA concentration in the plastids this positive cooperation is probably very significant. In the chloroplast the ratio of OASTL to SAT is about 300:1, so the majority of OASTL is in the free form. OAS formation is regulated in response to the activity of the sulfate reduction pathway. Complex formation is promoted by sulfide, while OAS disrupts complex formation (Figure 1).

A model has been proposed for the cysteine synthase multienzyme complex based on *in vitro* enzyme data. Under conditions of sulfate starvation the concentration of OAS increases, the complex disassembles and production of OAS is slowed down. When production of OAS does not keep pace with sulfate reduction, sulfide accumulates, which stimulates complex formation and leads to a stimulation of OAS production. In transgenic tobacco plants where chloroplast OASTL was overexpressed, this led to no effect on cysteine content. In contrast, when OAS was fed to isolated chloroplasts, cysteine was overproduced. Although the enzyme complex is also localized in the cytosol and mitochondria, the ratios may vary markedly. The role of extra chloroplastic isoforms in overall cysteine biosynthesis is not known.

Figure 1 Regulation of cysteine biosynthesis by sulfide (S^{2-}) and *O*-acetylserine (OAS).

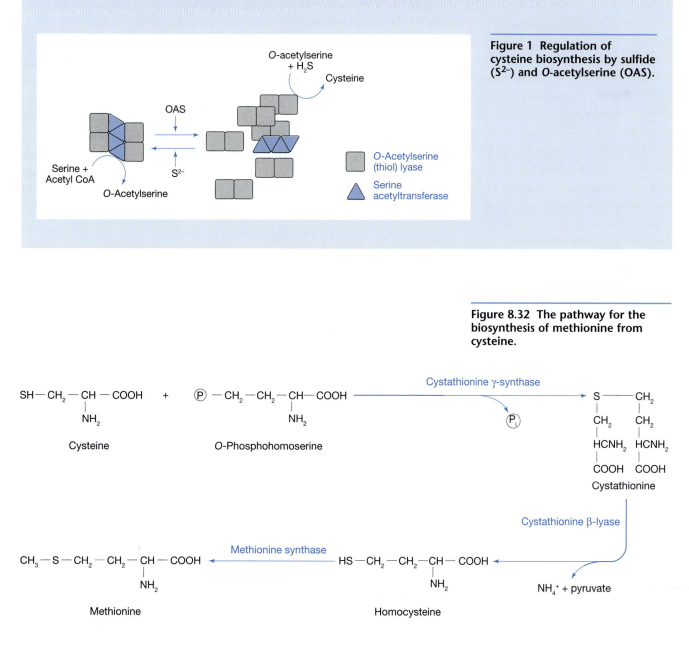

Figure 8.32 The pathway for the biosynthesis of methionine from cysteine.

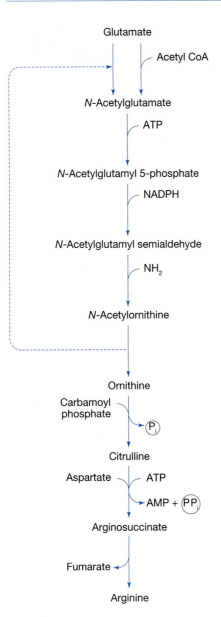

is converted into arginine via the input of two additional amino groups from carbamoyl phosphate and aspartate (Figure 8.33).

In proline biosynthesis, the first step involves the reduction of glutamate to glutamyl-5'-semialdehyde via glutamyl 5'-phosphate, an enzyme-bound intermediate (Figure 8.34). The semialdehyde spontaneously cyclizes to form pyrroline 5'-carboxylic acid. This is then reduced by NADPH to proline by the enzyme pyrroline 5'-carboxylate reductase. Proline biosynthesis most likely occurs in the cytosol. In the mitochondrion it appears that the same series of reactions operates in the opposite direction to form glutamate. It is thought that, although this reverse pathway has little significance under normal conditions, under stressful conditions it may be important for linking proline and arginine metabolism. Ornithine may be converted to glutamate semialdehyde directly by transamination or via proline through a series of reactions and the reverse of the proline synthetic pathway.

In early experiments it was shown that proline biosynthesis was regulated by end-product inhibition in unstressed tobacco and barley leaves. It was also shown that there was an increased synthesis of proline during water stress, which involved a loss of feedback regulation of the pathway leading to the formation of pyrroline 5'-carboxylic acid. Although the mechanism underlying this latter process has not been identified, it is thought that under osmotic stress the enzyme pyrroline 5'-carboxylate reductase becomes less sensitive to feedback inhibition by proline. Proline accumulation has also been proposed to be controlled by the level of the mitochondrial enzyme proline dehydrogenase whose activity decreases during salt stress.

Proline levels may increase up to 200-fold under stress conditions and proline has been shown to have a special protective role against desiccation damage in leaves. Under conditions such as drought or salt stress, which lead to water stress, many plants accumulate very high concentrations of proline in the leaves, to the order of several times that of other amino acids, resulting in a significant lowering of the cellular water potential.

Figure 8.33 The pathway for the biosynthesis of arginine.

The biosynthesis of the aromatic amino acids: phenylalanine, tyrosine, and tryptophan

As discussed elsewhere (see Chapter 11) the shikimic acid pathway involves the condensation of erythrose 4-phosphate (formed via the oxidative pentose phosphate pathway or the Calvin cycle) with phospho*enol*pyruvate (generated

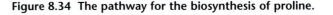

Figure 8.34 The pathway for the biosynthesis of proline.

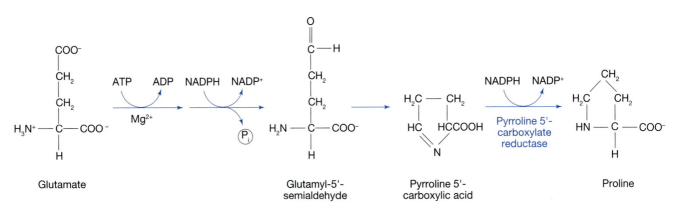

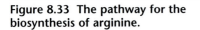

via glycolysis) to produce 3-Deoxy-D-*arabino*-heptulosonic acid 7-phosphate (Figure 8.35). Following a series of reactions shikimic acid is generated, which is then phosphorylated to shikimate phosphate via shikimic kinase. Upon combination with phospho*enol*pyruvate, 3-*enol*pyruvylshikimic acid 5-phosphate is formed. This is then converted to chorismic acid. Chorismic acid is a branchpoint in the pathway and may either be converted to tryptophan or phenylalanine and tyrosine (Figure 8.35). Tryptophan can serve as a precursor for auxin (e.g. indolyl acetic acid) biosynthesis. However, there are multiple routes for auxin biosynthesis in plants, including a number of tryptophan-independent routes, which remain to be characterized in any detail due to the paucity of appropriate mutants (see Further Reading).

Histidine biosynthesis

Histidine is synthesized in a linear pathway from phosphoribosylpyrophosphate via a number of imidazole intermediates (Figure 8.36). In contrast to the extensive gene redundancy seen in other plant amino acid biosynthesis pathways, many of the enzymatic steps of histidine biosynthesis are represented by single genes. This apparent simplicity makes histidine biosynthesis an attractive target for herbicide discovery (especially as histidine biosynthesis does not occur in animals). Histidine plays an important part in regulating the biosynthesis of a number of other unrelated amino acids, in chelation and transport of metal ions, and in plant reproduction and growth.

Large amounts of nitrogen can be present in nonprotein amino acids

A large number of amino acids exist that are not normally constituents of proteins. They represent a very diverse and often complex group of compounds

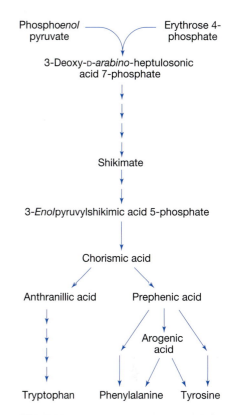

Figure 8.35 The pathway for the biosynthesis of the aromatic family of amino acids.

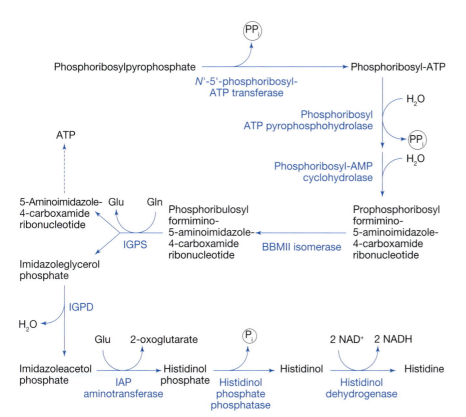

Figure 8.36 The pathway for the biosynthesis of histidine. Abbreviations: BBMII isomerase, N'-[(5'phosphoribosyl)-formimino]5-aminoimidazole-4-carboxamide-ribonucleotide; IAP, imidazoleacetol phosphate; IGPD, imidazoleglycerol-phosphate dehydratase; IGPS, imidazoleglycerol phosphate synthase; Glu, glutamate; Gln, glutamine.

termed nonprotein amino acids. Several hundred different types exist in plants, in particular in seeds and they can accumulate to high levels. Nonprotein amino acids are found in all plant tissues as intermediates in the biosynthesis of protein amino acids. As discussed earlier homoserine is a nonprotein amino acid intermediate in the biosynthesis of threonine, isoleucine, and methionine and also serving as a transport metabolite in some legumes. Similarly diaminopimelic acid is involved in the biosynthesis of lysine. Nonprotein amino acids usually exist in unconjugated forms and often have toxic properties. The highest concentration and most diverse range of nonprotein amino acids is found in legumes. A major role of nonprotein amino acids is that they form a chemical defense system against predation and disease. Nonprotein amino acids are classified based on their structural relationship with amino acids or with their structural analogy with, for example, aromatic amino acids or sulfur-containing amino acids. The toxic action of individual nonprotein amino acids occurs through a range of mechanisms, including inhibition of key enzymes and competition with essential amino acids for transport and protein synthesis.

An example of a four-carbon nonprotein amino acid is γ-aminobutyric acid (GABA). GABA is rapidly produced in response to biotic and abiotic stress. It is primarily metabolized via the GABA shunt, a conserved pathway in eukaryotes and prokaryotes. This pathway is composed of three enzymes: the cytosolic glutamate decarboxylase, and the mitochondrial GABA transaminase and succinic semialdehyde dehydrogenase (Figure 8.37). Glutamate decarboxylase catalyzes the irreversible decarboxylation of glutamate to produce GABA. GABA is transported into the mitochondria where, using either 2-oxoglutarate or pyruvate as amino group acceptors, it is converted into succinic semialdehyde by GABA aminotransferase. Succinic semialdehyde is then reduced by succinic semialdehyde dehydrogenase to form succinate, which enters the tricarboxylic acid cycle but bypasses the two steps involving succinyl CoA ligase and 2-oxoglutarate dehydrogenase. Although the role of GABA in plants is still to be totally confirmed, there is increasing evidence that it is important in different aspects of plant development, metabolism, and response to stress.

Plant storage proteins: why do plants store proteins and what sort of proteins do they store?

Storage proteins are found in seeds, tubers, stems, roots, and other organs whose function is to enable plant growth to resume after a period of dormancy. These proteins are broken down and mobilized to yield essential organic building materials for new stems and leaves.

Figure 8.37 The γ-aminobutyric acid (GABA) shunt. The cytosolic enzyme glutamate decarboxylase catalyzes the irreversible decarboxylation of glutamate to GABA. GABA is transported into the mitochondria and converted into succinic semialdehyde by GABA aminotransferases using either 2-oxoglutarate (by GABA-TK) or pyruvate (by GABA-TP) as amino acid acceptors. Succinic semialdehyde is then reduced to succinate by succinic semialdehyde dehydrogenase. (From *Biochemistry and Molecular Biology of Plants*, edited by Russell L. Jones, Bob B. Buchanan, Wilhelm Gruissem, 2000. Copyright John Wiley & Sons Limited. Reproduced with permission.)

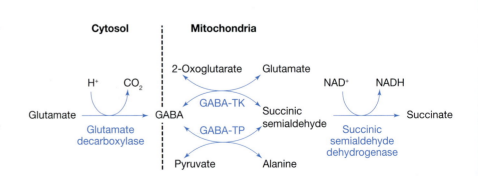

In most, but not all cases, the genes encoding these proteins have evolved to specifically serve this storage function. Storage proteins can be recognized by a number of criteria. They generally do not have enzymatic activity and they are resistant to the normal proteolytic turnover of cellular proteins. Despite this, they do have cleavage sites that are susceptible to attack by specific proteinases. Storage proteins are accumulated late in the maturation of a storage organ, and the level of their synthesis is markedly affected by the availability of nitrogen. Under nitrogen-rich conditions synthesis of storage proteins is enhanced, while it is depressed in nutrient-poor conditions.

Originally storage proteins were classified according to their solubility in successively stronger solvents ranging from water, to salt solutions, to alcohol solutions, and to acids and alkalis. Albumins are water soluble, but globulins require dilute salt solutions to dissolve them. This approach separated prolamins, soluble in alcohol solutions, from glutelins, soluble only in acidic or basic solutions. Modern analysis of these two proteins shows that in fact they are very similar to each other, except that glutelins have disulfide bridges conferring extra stability to the molecule. Molecular analysis has revealed families of proteins descended from ancestral precursors, so it is now more meaningful to classify proteins according to their family relationships, rather than their solubility characteristics.

Across the plant kingdom there are remarkably few different types of storage proteins. Globulins are found in nearly all dicotyledonous seeds. These are usually referred to by their sedimentation coefficients with names first coined from legume seed extracts: 7S vicilins and 11S legumins. Small 2S albumins are also found in seeds. Seeds of grasses and cereals (Gramineae) are important for the nutrition of humans and other mammals, birds, and reptiles. Barley, wheat, and maize contain predominantly prolamins, whereas oats and rice contain only small amounts of prolamin, but mainly contain glutelin-type 11S globulin.

Although globulins and prolamins are the main storage proteins found in plants, other types of proteins are found in specific nonseed (vegetative) organs, such as tubers and other underground stems and roots. These organs store diverse proteins which, unlike seed storage proteins, are often cellular enzymes.

From a plant's perspective, the precise amino acid composition of storage proteins is not of great significance, as plants are able to synthesize all their own amino compounds. Animals have come to rely on plants as sources of amino acids, to a greater or lesser extent, so the precise amino acid composition of these proteins is of considerable importance. For example, 70% of human demand for protein is met by consumption of seeds. As we shall see, some seeds are poor sources of certain critical amino acids for humans and other animals, so plant breeding programs have been designed to tailor seed proteins of crop plants to meet the demands of animals.

Because storage proteins only represent a source of carbon skeletons, nitrogen, and sulfur to the plant, there has been little evolutionary pressure to conserve particular amino acid sequences. This has led to the formation of a very large number of variations of a small number of proteins that are slightly divergent from each other. So long as the basic shape requirements and proteolytic susceptibility criteria are met, there is no pressure to conserve a particular amino acid sequence. In this chapter we shall review the main protein storage types, and also provide accounts of the proteins in three particular plants as examples of the range of information available.

Vicilins and legumins are the main storage proteins in many dicotyledonous plants

7S and 11S globulins are related storage proteins that occur widely in seed plants. The following is a general account of these storage proteins, more specifically pea seed globulins are described in detail in Box 8.7. Vicilins are synthesized as a polypeptide that is split into fragments by cleavage at one or more proteolytic sites. This generates proteins of different sizes that are later glycosylated, before final storage in the cotyledons of developing seeds.

11S legumins are synthesized as proproteins that fold to form a hair pin shape (Figure 8.38). This fold is generated and maintained by two conserved cysteine disulfide bridges. Cleavage of this proprotein at an asparagine site at the fold generates two polypeptide chains. These are a larger α-subunit, and a smaller β-subunit, linked by disulfide bridges. Cleavage also allows rearrangement of

Figure 8.38 Synthesis of 11S globulin. (From *Biochemistry and Molecular Biology of Plants*, edited by Russell L. Jones, Bob B. Buchanan, Wilhelm Gruissem, 2000. Copyright John Wiley & Sons Limited. Reproduced with permission.)

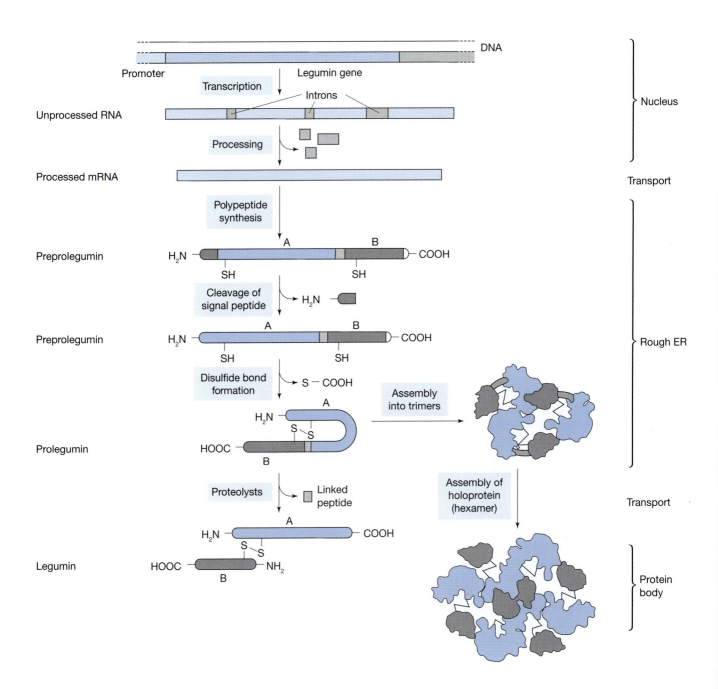

Box 8.7 Pea seed globulins

Legumes provide a major nutritional resource for humans and animal feedstuffs across the world. Their seeds are rich in proteins, so the amino acid content and storage form of the proteins are of major importance. In this box we will describe the globulins found in seeds of peas, *Pisum sativum*.

The main storage proteins are vicilin, a trimeric 7S globulin protein and legumin, a hexameric 11–12S globulin protein. Pea globulins contain relatively low levels of methionine and cysteine, with vicilin being especially low. There is considerable variation in amino acid composition between different genotypes, and between the products of individual genes in any given genotype. Legumin typically contains 1–1.5 moles% sulfur amino acids.

The polypeptide composition of these proteins is somewhat variable, but they share common structural features. Vicilin is a mixture of about nine different polypeptides; the major ones are 50 kDa and 30–33 kDa. The 50 kDa protein consists of three segments, α, β, and γ. The smaller proteins are derived from the 50 kDa polypeptide as a result of processing at one or both cleavage sites located between each segment pair. Cysteine is mostly absent from these proteins, so there is an absence of disulfide bonds.

Vicilin is glycosylated at a single site close to the C-terminus, processing then yields three glycosylated polypeptides: 50, 25–30, and 12–16 kDa.

A related 70–75 kDa protein, convicilin, is also found in peas. This protein is homologous to vicilin, but with additional sequences inserted into the protein.

As with vicilins from other species, pea vicilin exists as a trimer of three polypeptides associated together. Each polypeptide has two domains. Convicilin can form trimers on its own, but exists in the native state as mixed trimers with vicilin.

Legumin is found as six subunits arranged as two trimers. Each subunit consists of a pair of disulfide-bonded polypeptides, α and β. The α polypeptides are larger (40 kDa) and have an acidic isoelectric point (pI), while the β polypeptides are smaller (20 kDa) and have an alkaline pI. These are synthesized from larger precursor proteins: 60 kDa preproteins are cleaved to give the 40 and 20 kDa polypeptides. Five different pairs have been identified, called L1–L5.

There are four major legumin genes, all coding for similar polypeptides, and five to six minor genes that make rather different proteins. These represent the A and B subfamilies that are also found in *Vicia faba* and *Glycine max*.

the interdomain linker sequence from the face to the side of the disc-shaped molecule. This removes the steric hindrance to face to face aggregation. Three such discs associate by their edges to form a circular trimer. Two trimers can now aggregate face to face to form a large hexamer, 350–400 kDa in size. In general, legumins, unlike the vicilins, are not glycosylated.

The primary and tertiary structures of vicilins and legumins share some similarities, suggesting that they have evolved from a common ancestor. This ancestor can be traced back to early land plants, mosses, and ferns, and to prokaryotic organisms. Bacteria form cupins, so named because their tertiary structure contains a barrel-shaped fold of β-polypeptide ribbons. These are small thermostable and desiccation-resistant proteins found in both prokaryotes and eukaryotes. The cupin family contains a variety of different enzymes, some of which contain a range of transition metal ions as cofactors at the active site. Duplication and evolution of the cupin gene has led to the formation of a variety of homologous proteins that can be traced up through the evolutionary tree to angiosperms. For instance, moss spores contain a related protein called germin, which has homologs in barley seeds. Vicilin-like proteins, that are ancestral to the vicilins and legumins of seed plants, are found in fern spores. All these proteins exist as hexameric aggregates, or similar structures based on trimers of two-domain proteins. The barrel-shaped tertiary structures are conserved, despite extensive changes to the primary amino acid sequences.

All these proteins exhibit marked stability and desiccation tolerance, which favored their adoption as storage proteins. The compact structure of these proteins makes them resistant to protease attack; germins are especially resistant. Seed globulins have extensions to the surface loops to provide proteolysis attack points, which allow them to be degraded during germination.

Prolamins are major storage proteins in cereals and grasses

Seeds of grasses and cereals contain two related storage proteins, the prolamins and glutelin-type prolamins. Glutelin-type prolamins resemble prolamins, but are much more stable owing to the presence of disulfide bridges. Prolamins are the major storage proteins in most of these grasses, but in some, such as oats and rice, globulins predominate. All these proteins are stored in the endosperm, which is a triploid tissue retained into grain maturity.

In barley, wheat (see Box 8.8 for details of wheat prolamins), and rye (all members of the Triticeae) three major classes of prolamins are recognized, sulfur-rich prolamins, sulfur-poor prolamins, and high molecular weight (HMW) prolamins (should more properly be high molecular mass). Sulfur-rich prolamins are the major storage protein, accounting for up to 90% of the total. As their name implies, they are rich in the sulfur-containing amino acids, cysteine and methionine, and have molecular masses in the range 30,000–40,000. Sulfur-poor prolamins constitute 10–20% of storage protein;

Box 8.8 Wheat seed proteins

Species of wheat provide a major source of human nutrition, with the grains containing both starch and proteins. The proteins have been extracted by washing out the starch grains and cell walls, leaving a glutinous mass of protein called gluten, which has been the subject of analyses for over 100 years. The properties of gluten vary across wheat species and varieties, as do the properties of their starch grains. This imparts differing physicochemical characteristics to the flour derived from each species. Hence the flour from *Triticum durum* (hard durum wheat, a tetraploid species of two genomes, AABB) is favored for pasta making, while bread making flour comes from varieties of *Triticum aestivum* (a hexaploid species of three genomes, AABBDD). Pastries, noodles and other food products also employ flour from specific sources.

Gluten was divided into different fractions according to the classical system based on solubility properties. Alcohol-soluble glutens were termed gliadins, which are monomeric proteins that act as plasticizers in doughs. Alcohol-insoluble glutens are called glutenins. These are polymeric owing to the presence of disulfide bonds, and impart viscoelastic properties to doughs. In this account we shall concentrate on storage proteins of hexaploid wheats. The modern classification follows that given in the main text: sulfur-poor prolamins (10–20% of total), sulfur-rich prolamins (70–80% of total), and HMW-prolamins

(5–10% of total). With three genomes (A, B, and D) each contributing several copies of each gene for these proteins, there is a considerable variety of proteins produced, even from a single cultivar. Different members of each group can sometimes be distinguished from one another by the tripeptide sequence located at the N-terminal end of each molecule (RQL, ARQ, TRQ in *T. monococcum* and ARE in *T. aestivum*).

The presence of these proteins in wheat flour, a basic carbohydrate source for a significant proportion of the world's population, can give rise to allergic reactions in some people. Albumins and globulins can elicit production of IgE antibodies in sensitive individuals. Celiac disease is hereditary and is due to production of IgA and IgG antibodies in response to consuming gluten. These responses also occur with food products derived from rye and barley grains. Oats were thought to cause similar problems, but probably reaction to oats is due to contamination of the oat flour with wheat flour from the same mill. Rice and corn products do not cause the same allergic reactions.

The main sulfur-poor polypeptides are called ω-gliadins, which are encoded by genes on the A and D genomes. The repeat sequence motifs are rich in Q (glutamine), P (proline), and F (phenylalanine), with the N-terminal SRL-types having a higher proportion of glutamine and

Box 8.8 Wheat seed proteins (continued)

glutamate (50% of total) than the other sulfur-poor pro-lamins (Figure 1). This is at the expense of proline, which is lower than others at 20% of the total.

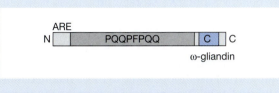

Figure 1 The structure of wheat gliadin. The repeat sequence is rich in proline (P), glutamine (Q), and phenylalanine (F). The N-terminal of this polypeptide is characterized by the amino acid sequence ARE (alanine, arginine, glutamic acid).

ω-Gliadins have proved difficult to characterize, so there is no complete sequence or agreed molecular mass value available.

The sulfur-rich prolamins of wheat are very varied, with considerable polymorphism within any given genotype. They have characteristic A, B, and C segments at the C-end of the polypeptide, which contain conserved cysteine residues. Between these segments and between the repetitive domain and segment A there are four interspersed regions (I₁–I₄) which are conserved between members of the same S-rich prolamin type, but differ between different types.

The ancestral prolamin is γ-gliadin, which is very widespread and makes up the major portion of sulfur-rich prolamins. The N-terminal repetitive domain has repeating PQQPFPQ motifs (Figure 2).

α-Gliadin is similar to γ-gliadin, but there are only six conserved cysteine residues in the A, B, and C segments (Figure 2). The repetitive domain sequences differ slightly, being based on PQPQPFP and PQQPY motifs. They also

contain two polyglutamine (QQQQ–) regions, one between the repetitive domain and I₁ and the other within the B region. α-Types are only found in wheat, they do not occur in other triticeae cereals such as barley and rye.

A further group of sulfur-rich prolamins are the B-group LMW (low molecular weight) proteins of glutenin. These are disulfide-bridge stabilized polymers consisting of two different polypeptides recognized by their N-terminal sequences, either MET– (LMWm) or SHI– (LMWs) (Figure 2).

HMW prolamins are glutenins. Each group 1 chromosome contains two HMW genes, one called x-type (larger, 83–89 kDa) and one y-type (smaller, 65–67 kDa). However, not all these genes are transcribed. HMW is the only prolamin for which the genes have been identified. They exhibit gene silencing, with different cultivars containing three, four, or five HMW gene transcripts. All cultivars make 1D and 1B x-types and 1D y-type. Some also have 1Ax and/or 1By. 1Ay is only found in two bread wheat lines. Tetraploid pasta wheat seeds contain one or two HMW proteins coded for by chromosome 1B, but some of these wheats also contain transcripts from a 1A gene.

Ten HMW genes have been sequenced, coding for polypeptides with amino acid lengths ranging from 627 to 827 residues. The N-terminal repeats range from 81 to 89 residues in x-type proteins to 104 in y-type ones. The repeat sequences contain hexapeptide (PGQGQQ, G is glycine) and nonapeptide repeat motifs. The C-terminal 42 amino acids are the same in all members of the group. The N-terminal domains contain three cysteine residues in x-type and five in y-type.

Wheat prolamins are synthesized exclusively in the endosperm during mid-late grain development. The levels of deposition are dependent on the availability of soil nitrogen and sulfur. Unusually there is no evidence of post-translational processing of polypeptides. All the observed polymorphisms (e.g. in electrophoretic mobility) are based on variations in the gene nucleotide sequences, not on variations in glycosylation patterns. Regulation of transcription is effected by sequences that are similar for sulfur-rich and sulfur-poor proteins, but different for HMW proteins. Transport of the proteins is initially to endoplasmic reticulum-derived protein bodies, but some are subsequently absorbed into the vacuole system by autophagy.

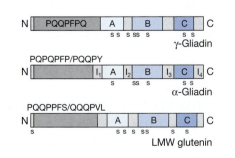

Figure 2 Structure of wheat S-rich prolamins. γ-Gliadin is shown with the main repeat sequence motif and the approximate positions of the eight cysteine residue sulfurs (S). α-Gliadin is shown with the repeat sequence motifs and the interspersed regions (I). Note that only six of the eight cysteine residues are found in this protein, while in the low-molecular weight protein (LMW glutenin) there are eight residues of which six are conserved from the ancestral γ-gliadin.

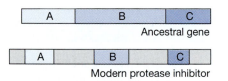

Figure 8.39 The ancestral gene for prolamins consisted of three distinct segments. These were protease inhibitors and the same basic structure can be recognized in modern day proteins, such as amylase and trypsin inhibitors.

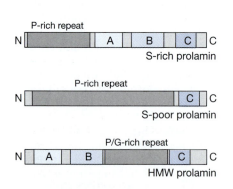

Figure 8.40 The three main classes of storage proteins found in cereals. These are related to the ancestral gene shown in Figure 8.39. The repeat sequences are rich in either proline (P) or proline and glycine (G).

they are poor in sulfur-containing amino acids, but rich in glutamine and proline and have molecular masses of 40,000–80,000. HMW prolamins are unusually rich in glycine (15–20%, compared with 0.5–3% in the other prolamins) and have molecular masses of 90,000–150,000.

Prolamins appear to have evolved from an ancestral gene by a series of gene duplication events followed by insertion and deletion of gene fragments. The original gene was duplicated to provide the basis for protease inhibitors. Such a property would clearly benefit a putative storage protein as it would be resistant to proteolysis. One element of the duplicated gene (or the original gene) was triplicated to provide three sequences that then diverged from each other during evolution. These gave rise to three segments of the one polypeptide, called A, B, and C. Gene duplications provided additional copies of this gene that underwent further evolution. Some of these retained most of their original structure and their inhibitory properties, forming modern protease and α-amylase inhibitors (Figure 8.39).

The sulfur-rich A, B, C prolamin proteins have a proline-rich repeat sequence added at the N-terminal end of the polypeptide (Figure 8.40). The sulfur-poor proteins have a much larger version of this repeated sequence, but have lost most of the A, B, and C sequences at the C-end of the molecule. The HMW proteins retained the A, B, and C sequences, but also have an extensive proline- and glycine-rich insert between the B and C sequences.

The main storage protein of maize, α-zein is related to ancestral prolamin sequences, but bears little resemblance to the prolamins found in the Triticeae (Box 8.9). However, the minor components, β- and γ-zein do contain the A, B, and C motifs, with the addition of a methionine-rich region between B and C.

2S albumins are important but minor components of seed proteins

Albumins are relatively small proteins as indicated by their sedimentation coefficient, 2S. They occur in seeds of many angiosperms, both dicotyledons and monocotyledons. Larger proproteins are synthesized and then cleaved to yield the initial albumin, which folds with the formation of disulfide bridges. A further cut creates two unequal-sized fragments that are held together by the bridges forming two subunits of a dimer.

2S albumins have evolved from the same ancestral gene that gave rise to the prolamins. They are homologous with the A, B, and C segments of sulfur-rich prolamins and HMW prolamins. The proprotein cleavage site lies between the A and B segments of the polypeptide. Albumins generally have a higher content of sulfur amino acids than other seed proteins, in particular brazil-nut albumin has a methionine content of 17%, and sunflower albumin has a methionine content of 3%. These seeds represent a valuable source of essential amino acids for those trying to achieve a balanced diet from purely vegetable sources.

Where are seed proteins synthesized and how do they reach their storage compartment?

Storage proteins in seeds are usually found within the cytoplasm in distinctive membrane-bound organelles. The origin of these has been the subject of much debate, partly because of differences between different plants, and because of differences within the same plant at different stages of seed development

Box 8.9 Maize seed proteins

Maize is a major world crop, for both human consumption and animal feedstocks. While starch is the principal component of the seed (kernel), storage proteins are present and are critical as an amino acid source for the consumers. As we shall see, these storage proteins are very deficient in lysine and tryptophan, essential amino acids for humans. While the storage proteins of maize are categorized as prolamins, they bear little relationship to wheat prolamins described in Box 8.8. *Zea mays* is in the family Maydeae, order Panicoideae, whereas the Triticeae are in the order Festucoideae. This evolutionary divergence has led to the development of very different storage proteins in the two cereals.

In maize, the main storage protein is a prolamin called zein, which is laid down mainly in the outer endosperm cells. In contrast the center of the endosperm is taken up with starch-storing cells, which have much lower levels of these proteins. At maturity, the center is soft, granular, and opaque, while the outer shell of protein-storing cells is hard, glassy, and translucent. In the case of popcorn rapid heating of this structure leads to a build up of internal pressure and explosive bursting of the outer shell. The release of pressure instantly converts the internal superheated water into rapidly expanding steam, causing the kernel to pop.

A feature of the cell biology of endosperm tissue is that the DNA is repeatedly duplicated in S-phase, without the nucleus going through mitosis. This endoreduplication results in nuclei containing C-levels of DNA of 90–100, so greatly increasing the copy number of genes available for transcription.

Maize storage proteins can be divided, on the basis of alcohol solubility, into two main fractions after solubilization in a reducing buffer (to break S–S bonds). The major fraction is alcohol-soluble prolamin, called zein. The minor fraction is alcohol insoluble, and is a mixture of albumins, globulins, and some glutelin-type proteins. Zeins are divided into four classes according to molecular structure and size (Table 1).

α-Zein is the major storage protein in maize, accounting for about 70% of the stored protein. Although there are up to 40 different α-zein polypeptides, they can be divided into two classes on the basis of size, either 19 kDa or 22 kDa. The size difference is due to the protein structure. α-Zein consists of a series of 20 amino acid sequences that are repeated either nine times in the case of the 19 kDa proteins or 10 times in the 22 kDa ones. The repeats are rich in glutamine and proline, and also alanine and leucine. Each repeat forms an α helix and the helices are linked by glutamine residues. Within each class, there is a high degree of homology between the repeat units (90%), but between classes there is only 60% homology. Notable absences from the spectrum of amino acids in α-zein are lysine and tryptophan.

γ-Zein is, in some respects, similar to α-zein in that it lacks lysine and tryptophan and has a series of repeated peptide structures, but γ-zein is cysteine-rich, with a high level of disulfide bridges in the molecules. This makes the protein very stable and hard to dissolve without reducing agents. There are three types of γ-zeins, 50 kDa, 27 kDa, and 16 kDa. The 27 kDa zein has eight tandem repeat sequences in the N-terminal part of the molecule. These consist of a repeating hexapeptide motif PPPVHL (V is valine, H is histidine). The repeating section is followed by a 22 amino acid linker of proline alternating with other amino acids. The smallest type (16 kDa) has only two degenerate copies of this hexapeptide, which partly accounts for its smaller size. The C-terminal end is rich in glutamine and cysteine.

β-Zein is a minor component of total zein. It has a molecular mass of 15 kDa and is rich in cysteine, and hence contains S–S bridges that stabilize the molecule. Like α- and γ-zein it also lacks lysine and tryptophan.

δ-Zein is quite different from the other zeins. It is rich in methionine and is found as two size classes, 18 and 10 kDa. The larger one contains a duplicated sequence from the methionine-rich core of the 10 kDa protein. This larger class also contains one lysine and two tryptophan residues in each polypeptide, the only zein to contain these important amino acids.

Table 1 Maize zein classification

Protein	M_r	Structure
α-Zein	19/22,000	9/10 copies of 20 aa repeats
γ-Zein	27/16,000	8 tandem repeats/2 repeats
β-Zein	15,000	β-strand, cysteine rich
δ-Zein	10/18,000	Methionone-rich center, duplicated

continued

Box 8.9 Maize seed proteins (continued)

Synthesis of maize storage proteins takes place on the endoplasmic reticulum and the proteins are retained in the lumen. The endoplasmic reticulum retention signals have not yet been identified, but they are not the usual tetrapeptides HDEL/KDEL found on endoplasmic reticulum-specific proteins. Removing part of the N-terminal sequence allows the zeins to proceed to the Golgi and be secreted to the cell walls. Immunolocalization studies have revealed that β- and γ-zeins are the first to be made and accumulate into small, densely staining protein bodies. Later α-zeins accumulate in the interior, along with δ-zeins, to form a large central core that stains less intensely. If zeins are expressed in other plants, protein bodies only form when multiple zeins are synthesized together. Single zeins cannot form protein bodies. β- and γ-zeins are needed to initiate protein body formation so that α-zeins can accumulate later.

α-zein genes are present in large multigene families totaling 70–100 gene copies. Polypeptide analysis shows that many of these genes are not expressed. Mutations in the synthesis and processing of zeins lead to distinctive kernel phenotypes. Those lacking protein lose the translucent property of the outer shell, so the white starchy core predominates. Such mutants are called opaque (O). For example, opaque mutants *o2* and *o7* lack α-zein 22 kDa and 19 kDa, respectively.

Quality Protein Maize

As emphasized above, most of the zein proteins are deficient in lysine and tryptophan. This has serious consequences for a significant proportion of the human population who are dependent on maize for their own nutrition and that of their farm animals. This deficiency cannot simply be rectified by increasing the cereal intake. The unbalanced amino acid supply means that the body has to dispose of the excess amino acids, principally leucine, tyrosine, and valine. The result is that in regions where the human population, and their livestock, depend on maize

corn as the principal component of their diet, both can be malnourished.

In the 1960s it was recognized that the *opaque2* (*o2*) mutant offered a solution to this problem due to its better balance of amino acids. In this mutant the synthesis of α-zeins, especially the 22 kDa α-zeins, which contain no lysine or tryptophan, is reduced. There is a compensatory increase in the synthesis of other endosperm proteins, including storage globulins, which do contain these amino acids. Scientists at the International Maize and Wheat Improvement Center, Mexico (CIMMYT) worked for 20 years to produce Quality Protein Maize from *opaque 2* mutants. There were significant problems to overcome because the *opaque 2* mutants produced chalky grains with a low yield. The plants had high levels of ear rot, and ears which were slow to dry and very susceptible to pests on storage. The aim was to produce Quality Protein Maize that would perform at least as well as the best common varieties in the field across a wide part of the globe. This involved breeding for improved kernel and ear characteristics, and improved digestibility, while retaining the increased levels of lysine and tryptophan in the storage proteins.

In 1993, 33 tropical and 22 subtropical lines were released for testing internationally. In 1997, four countries were growing Quality Protein Maize on a commercial scale; this had risen to 27 countries 5 years later. Local breeding programs have been set up, assisted by CIMMYT, to provide varieties adapted to local conditions, such as short growing seasons in the foothills of the Himalayas. In general, these new varieties yield 110–120% of the best local varieties. More recently developed strains are achieving yields that are 30–50% higher again. Nutritionally, the minimum daily intake requirement of Quality Protein Maize for humans is about 250 g day^{-1} compared with 547 g day^{-1} for the traditional varieties. The Quality Protein Maize story is a major achievement of modern plant biochemistry and plant breeding.

(early versus late). Only a very small number of crop species have been examined in detail, and these are not representative of angiosperms in general. It is now known that seeds of different plants can store proteins in one of two different compartments and their derivatives. These are the endoplasmic reticulum (ER) and the protein storage vacuole (PSV). In all cases storage proteins have signal sequences that ensure that they are synthesized on the rough ER. The initial translation product is referred to as a preproprotein, which carries all the signal sequences. Usually the ER-targeting sequence is removed immediately to yield the proprotein. Further targeting and processing sequences are cleaved as the protein is transported to its final destination, giving the mature protein molecule.

In some plants, for example maize, the proteins remain in the ER where they aggregate into spherical masses forming globular extensions of the ER. These may eventually pinch off to form discrete protein bodies that are bounded by a membrane derived from the ER. The mechanisms that ensure that these proteins remain in the ER, but are not degraded by the normal protein quality control systems, are unknown. Maize storage proteins do not carry the normal ER retention signals, such as KDEL or HDEL (see Box 8.9).

In other plants, such as legumes and cereals, storage proteins are processed through the normal secretory route to the Golgi and then transported to a specific class of vacuoles, the PSVs, rather than the normal lytic vacuoles found in all plant cells (Plate 8.1 and Figure 8.41). The two vacuole types can be distinguished at an early stage, as normal plant cell vacuoles have acidic contents, but PSVs have pH neutral contents. The signals involved in targeting storage proteins to these vacuoles are being discovered. To date, one tetrapeptide sequence (NPIR) at the N-terminal end has been identified that targets vegetative storage proteins to acidic lytic vacuoles. C-terminal signals appear to be important for the transport of seed proteins to PSVs, but these seem to be less well defined, at least at the primary sequence level. The signal may well involve a large part of the polypeptide chain. Deletion of a C-terminal tetrapeptide sequence (AFVY) from phaseolin, the 7S globulin in the common bean, *Phaseolus vulgaris*, leads to it being secreted from the cell. It has not yet been demonstrated that this sequence alone is sufficient to direct a nonstorage protein to the PSV compartment. Defense hydrolases, chitinases, and toxic proteins are also targeted to the PSVs of developing seeds. In contrast, proteins accumulating in nonseed (vegetative) storage organs are stored in lytic vacuoles that differentiate from existing vacuole structures.

During normal secretion, lytic vacuole-targeted proteins bind to 80 kDa glycoprotein (BP80) receptors in the forming of CCVs at the rims of Golgi cisternae

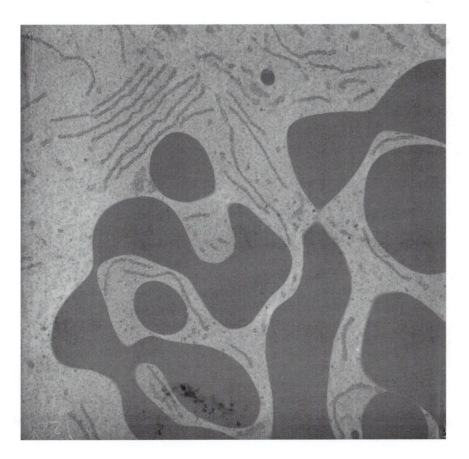

Figure 8.41 Protein storage vacuoles in the cytoplasm of developing *Vicia faba* cotyledon cells. (Reprinted with permission from Gunning BES, Plant Cell Biology on DVD, 2007.)

(Figure 8.42). BP80 is a large transmembrane protein that binds clathrin on the outside (cytoplasmic side) of the vesicle and has protein-binding domains on the inside (luminal side) of the vesicle. Work on castor beans shows that this conventional route via CCVs is utilized in the transport of storage protein from the Golgi to the PSV (Figure 8.42). This includes binding of the protein to BP80 receptors on the inner face of the CCVs. In arabidopsis deletion of one of the genes for BP80 leads to secretion of storage proteins at the cell surface, suggesting that the normal sorting mechanism for lytic vacuole proteins also applies to storage proteins in this plant. In contrast, there are no known receptors for the PSV-directed proteins that are transported in dense vesicles found in pea. The protein products appear to condense on to the inner face of the rims of Golgi cisternae (from *cis* to *trans* face) and form large aggregates that are very hydrophobic in nature. These aggregates are then pinched off forming dense vesicles, which eventually fuse with the PSVs. The mechanism for sorting pea storage proteins into the dense aggregates is unknown. These examples serve to emphasize the problems faced by researchers in trying to build general models for storage protein deposition in seeds.

Protein stores are degraded and mobilized during seed germination

Storage proteins are synthesized, processed, and stored for mobilization at some future date (germination in the case of seeds). During the synthesis and accumulation phases, these proteins have to be protected from proteases that are also being stored in the same compartments at the same time. Also, some proteins are subject to specific proteolytic cleavage as part of their maturation process. However, these same molecules must be protected from unwanted digestion by the same enzymes. How are these conflicting requirements handled?

Two general mechanisms have evolved to provide a means of storing and mobilizing proteins. One is the compartmentation of these storage proteins at synthesis, in the ER, and their transport and storage in membrane bound compartments that protect the proteins from general cytoplasmic proteases. The compartment barrier is preserved during the later degradation phase, so that complete mobilization can be achieved without exposing the cytoplasm to intensive proteolytic activity.

Figure 8.42 Summary diagram of possible transport routes for storage proteins from the Golgi to protein storage vacuoles. LV, lytic vacuole; PSV, protein storage vacuole; CCV, clathrin-coated vesicles. Normal secretory vesicles are targeted to either the cell wall or to the lytic vacuoles. Clathrin-coated vesicles may also be targeted to the lytic vacuoles. In developing seeds, there is evidence for transfer of storage proteins to PSV from either the *trans*-face via CCVs in some plants, or in other plants from the *cis*-face via dense vesicles.

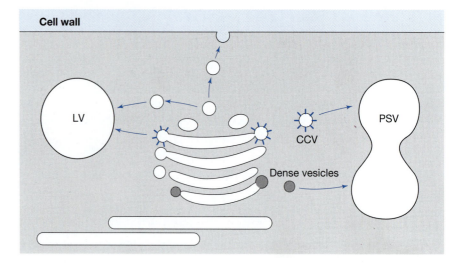

Second, in seeds where storage proteins and proteases coexist within the same compartment, there has been extensive coevolution of their respective structures. Highly specific proteases have been developed alongside storage proteins with very restricted cleavage sites. For example, legumin maturation is achieved through cleavage by an asparagine-specific vacuolar processing enzyme, called legumain. This only cleaves the peptide bond between an asparagine residue on the N-terminal side and any other amino acid on the C-terminal side. Hence there has been considerable evolutionary pressure to ensure that the polypeptide chain of the storage protein has an asparagine residue in the appropriate position. For example, in prolegumin there is a single exposed asparagine residue in the loop between the A and B domains that allows cleavage to form two disulfide-bridged subunits.

Conversely, the α-chain and β-sheet extensions, away from the proteolysis-resistant β-sheet barrel, do not contain asparagine residues, or if they do, these are located at sites that are hidden from protease attack. Aggregation of legumin trimers into hexamers is a further protection against legumain attack, so both can be stored in the same PSVs.

Some protein storage compartments do contain low specificity papain-type proteases at maturity. These are not active in proteolysis until germination, for reasons that are not thoroughly understood. They may be held in some different part of the vacuole, or the storage proteins may have condensed sufficiently to avoid destruction before these proteases are added to the vacuole.

Mobilization appears to be triggered by a destabilization of the tertiary structure of the storage proteins. This is achieved by an initial limited proteolysis by legumain. Complete degradation is aided by broad-spectrum enzymes, such as papain-like proteases, which are synthesized *de novo* in the cytoplasm and imported into the storage organelle. The relative importance of these two enzymes varies across species. In some, for example in *Phaseolus vulgaris*, a simultaneous attack is needed to break down phaseolin, while in others, such as *Vicia faba*, legumin is degraded by the papain-type enzyme alone.

In summary, temporary protection of susceptible sites allows deposition and storage, while limited proteolysis during early mobilization paves the way for complete degradation.

Vegetative organs store proteins, which are very different from seed proteins

Apart from storage in reproductive structures, proteins may be stored in a variety of vegetative organs. These enable plants to survive periods of inhospitable growth conditions, and they also may act as vegetative propagules. In temperate climates most people are familiar with potatoes, but in the tropics sweet potato, yams, cassava, and taro among others are more common. These storage organs are derived from a variety of plant structures, stems, roots, rhizomes, corms, etc. Alongside this diversity is a corresponding diversity of proteins that have been adapted to a storage function.

The proteins stored in vegetative organs barely fall within the definition of storage proteins. Unlike seed proteins, vegetative-storage proteins are much more closely allied to the normal enzyme complement of the tissues, and are present in relatively small quantities compared with those found in seeds. They do not accumulate late in the maturation of the storage organ, nor is their level of accumulation greatly influenced by the level of nitrogen available. Under low nitrogen conditions the storage organs are simply reduced in size.

The potato, a major temperate-climate crop

In potato, *Solanum tuberosum*, the main storage protein is patatin, which is a 40 kDa glycoprotein that accounts for 20–40% of the total soluble protein in tubers. The tubers are swollen underground stems that contain the protein in vacuoles. Synthesis of patatin occurs on the ER and the molecule is glycosylated at two sites (asparagines 60 and 90) with glycans composed of xylose, fucose, mannose, and *N*-acetylglucosamine in the ratio 1:1:3:2.

Cultivated potato is a tetraploid plant and synthesizes two slightly different patatins, class I and class II. Most of the protein, 99%, is class I. These have N-terminal signal sequences of about 23 residues, as expected for a protein targeted to the endomembrane system.

Patatins have a number of properties that may enable them to act as plant defense agents against insect and fungal attack. They can act as deacylating enzymes against a number of lipids, discouraging herbivorous insects. Fungal pathogens may be adversely affected by the glucanase activity of patatin, which could digest the hyphal cell walls.

Patatin can invoke an allergic response in susceptible people, especially when handled in the uncooked state. Also, there is an homology between patatin and a protein in rubber latex, so allergic individuals react to both, and to similar proteins in other members of the Solanaceae, such as tomatoes.

Tropical roots and tubers, sweet potato, yams, taro, and cassava

Sweet potato is a member of the Convolvulaceae (*Ipomoea batatus*), unlike potatoes that are members of the Solanaceae. The plants produce tubers on the roots that contain up to 10% protein. The main storage protein is sporamin, which accounts for 70–85% of the tuber protein content. Molecular analysis reveals the presence of two closely related proteins, with twice as much sporamin A present as sporamin B. Both contain four cysteine residues at conserved positions, which form two intra-chain (within the molecule) disulfide bridges.

Sporamins are related to Kunitz-type trypsin inhibitors, and also inhibit these digestive enzymes. They are therefore defense proteins against attack by insects. The trypsin inhibitors prevent the insects from digesting their food, so they starve. As with patatin, sporamin is targeted to the ER during synthesis, and transported to the vacuole. Sporamin is classified as a storage protein, as it is only synthesized in the sweet potato tubers. When leaves and petioles are subjected to experimentally induced high sucrose conditions, they synthesize sporamin. This suggests that it is high local sucrose levels, such as would be found in developing tubers, that induce sporamin synthesis and limit it to the tuber tissues.

Yams (Dioscoreaceae) develop at the stem–root junction, the hypocotyl, and contain only a low level of protein, up to about 5%. The main protein component (about 70% of the total) of yams is called dioscorin, which is a 30 kDa glycoprotein found in vacuoles. Two types of dioscorin have been characterized, class A and class B. These each have two conserved cysteine residues, which form an intra-chain disulfide bridge, and a third nonconserved cysteine. The amino acid sequences show similarities with α-carbonic anhydrase enzymes from arabidopsis, mice, and human genomes. The class B polypeptide is a maltose-binding lectin, the only plant lectin known to have an exclusive binding specificity for maltose. In the native state it is associated with two class A subunits. Of the remaining 30% of total protein, more than half is a 10 kDa lectin, which binds mannose. The remainder, about 10% of the total protein, is a chitinase, which, combined with the lectin content, probably accounts for the high resistance of yams to insect attack.

Taro is obtained from corms or other enlarged underground stems of four different genera in the family Araceae (aroids). They generally contain very low levels of protein, less than 1%, but higher values have been recorded. The main protein components are two globulins (G) that account for 80% of the protein content. About half of these are 28 kDa G1 proteins called tarins. These are mannose-binding lectins that serve a defensive function for the corms, as well as a storage function. The remaining half of the taro proteins are two related proteins of masses 22 and 24 kDa, called G2a and G2b. These resemble sporamin protease inhibitors. Both G1 and G2 proteins are stored in the vacuole.

In cassava the storage organ is a swollen root, which yields tapioca (manioc or yucca). It is mostly consumed for its starch content. The protein content is very low, 1–2%, and no specific storage proteins have yet been characterized from this crop.

Despite their diversity, storage proteins share common characteristics

We have seen that a great variety of proteins are stored in different seed and vegetative tissues, from the cotyledons and endosperm of seeds, to the parenchyma of stems and tubers. The proteins are all either protease resistant, or are synthesized and stored in protected compartments. The mechanisms involved are varied, implying that storage of proteins has been developed many times in plant evolution. Functionally, there does not appear to be any pressure to store particular amino acids for future growth, reflecting the biochemical versatility of plant tissues. Shape considerations, such as a tightly folded structure more resistant to proteolysis and dehydration, are more important. Disulfide bridges act to cross-link and stabilize the proteins, hiding potential proteolytic sites and acting as a store of sulfur. Placement of cleavage sites is important for degradation at a future time when mobilization is to occur.

Some animals rely on seed storage proteins as an integral part of their diet, including many animals adapted for agriculture. Many of these animals have lost the ability to synthesize certain amino acids, because they were readily available when the species first evolved. Changing habitats and availability of food sources have now brought these animals into a situation where their diet may be deficient. Human populations have expanded rapidly across the planet over geologically and evolutionary recent times. Many people now find themselves in a situation where food supplies are scarce and seed proteins from local crops are insufficient for their well being because they lack the appropriate balance of amino acids. This has driven research programs aimed at introducing new seed crops with a diversity of amino acids that is more suitable for a human diet. Crops such as Quality Protein Maize are now becoming available owing to recent advances in our knowledge of storage proteins (Box 8.9).

Summary

Nitrogen and sulfur are crucial nutrients for the plant in terms of growth and development. The variety of sources of these elements available to plants in the environment is reflected by the metabolic routes in which these nutrients are used metabolically. These processes require a combination of energy, carbon skeletons, and reductant. Nitrogen is mainly used for structural macromolecules, while sulfur has a key role in catalytic and electrochemical functions of biomolecules. Both nitrogen and sulfur are essential for the biosynthesis of amino acids and again this is dependent upon a supply of carbon skeletons, energy, and reductant. These processes need to occur in a coordinated manner to support plant growth and development. Although beyond the scope of this book it is becoming increasingly clear that nitrogen and sulfur metabolism directly impact upon each other. This requires coordination of metabolism through a number of regulatory routes. Integration is at a number of levels from the control of gene expression through to the level of proteins necessary to support different processes. Amino acids are used throughout the plant for general plant development, but storage proteins also play an important part in fueling the plant to resume growth and development following dormancy.

Further Reading

Nitrogen fixation

Igarashi RY & Seefeldt LC (2003) Nitrogen fixation: the mechanism of the Mo-dependent nitrogenase. *Crit. Rev. Biochem. Mol. Biol.* 38, 351–384.

Considers recent developments in providing an understanding of the structures of the active site and substrate interactions with the active site.

Nitrate transporters

Miller AJ, Fan X, Orsel M, Smith SJ & Wells DM (2007) Nitrate transport and signaling. *J. Exp. Bot.* 58, 2297–2306.

Ammonium transporters

Howitt SM & Udvardi MK (2000) Structure, function and regulation of ammonium transporters in plants. *Biochim. Biophys. Acta* 1465, 152–170.

Nitrogen assimilation

Andrews M, Lea PJ, Raven JA & Lindsey K (2004) Can genetic manipulation of plant nitrogen assimilation enzymes result in increased crop yield and greater N-use efficiency? An assessment. *Ann. Appl. Biol.* 145, 25–40.

This paper provides an overview on the impact that genetic manipulation of components of nitrogen assimilation has had on plant yield.

Kaiser W, Weiner H & Huber SC (1999) Nitrate reductase in higher plants: a case study for transduction of environmental stimuli into control of catalytic activity. *Physiol. Plantarum* 105, 385–390.

Mini review discussing the ways in which NR is regulated in response to the changing environment and the impact this has on metabolism.

Nitric oxide signaling

Lamotte O, Courtois C, Barnavon L, Pugin A & Wendehenne D (2005) Nitric oxide in plants: the biosynthesis and cell signalling properties of a fascinating molecule. *Planta* 221, 1–4.

Yamasaki H (2005) The NO world for plants: achieving balance in an open system. *Plant Cell Environ.* 28, 78–84.

These two papers provide summaries of the importance and implications of NO signaling in plants.

GS/GOGAT

Cren M & Hirel B (1999) Glutamine synthetase in higher plants: regulation of gene and protein expression from the organ to the cell. *Pl. Cell Physiol.* 40, 1187–1193.

Ishiyama K, Inoue E, Watanabe-Takahashi A, Obara M, Yamaya T & Takahashi H (2004) Kinetic properties and ammonium-dependent regulation of cytosolic isoenzymes of glutamine synthetase in *Arabidopsis. J. Biol. Chem.* 279, 16598–16605.

Lancien M, Martin M, Hseih M-H, Leustek T, Goodman H & Coruzzi GM (2002) Arabidopsis glt1-T mutant defines a role for NADH-GOGAT in the non-photorespiratory ammonium assimilatory pathway. *Pl. J.* 29, 347–358.

Lea PJ & Miflin BJ (2003) Glutamate synthase and the synthesis of glutamate in plants. *Pl. Physiol.Biochem.* 41, 555–564.

Miflin BJ & Habash DZ (2002) The role of glutamine synthetase and glutamate dehydrogenase in nitrogen assimilation and possibilities for improvement in the nitrogen utilization of crops. *J. Exp. Bot.* 53, 979–987.

Tobin AK & Yamaya T (2001) Cellular compartmentation of ammonium assimilation in rice and barley *J. Exp. Bot.* 51, 591–604.

Yamaya T, Obara M, Nakajima H, Sasaki S, Hayakawa, T & Sato T (2002) Genetic manipulation and quantitative-trait loci mapping for nitrogen recycling in rice. *J. Exp. Bot.* 53, 917–925.

Sulfate assimilation

Hawkesford MJ & De Kok LJ (2006) Managing sulphur metabolism in plants. *Pl. Cell Environ.* 29, 382–395.

Kopriva S (2006) Regulation of sulfate assimilation in arabidopsis and beyond. *Ann. Bot.* 97, 479–495.

Saito K (2004) Sulfur assimilatory metabolism. The Long and Smelling Road. *Pl. Physiol.* 136, 2443–2450.

Recently there has been marked progress in understanding more completely the pathway of sulfate assimilation in higher plants. These papers summarize the new developments and current understanding of regulation.

Rausch T & Wachter A (2005) Sulfur metabolism: a versatile platform for launching defence operations. *Trends Pl. Sci.* 10, 503–509.

This paper highlights the importance of sulfur-containing defense compounds (including H$_2$S, glutathione and secondary metabolites) for the survival of plants under biotic and abiotic stress.

Amino acid biosynthesis

Liepman AH & Olsen LJ (2004) Genomic analysis of aminotransferases in Arabidopsis thaliana. *Crit. Rev. Pl. Sci.* 23, 73–89.

Recent advances in the identification and characterization of aminotransferase sequences from arabidopsis and other plants are discussed in this paper, particularly with reference to their importance in plant proteins and metabolic pathways.

Binder S, Knill T & Schuster J (2007) Branched-chain amino acid metabolism in higher plants. *Physiol. Plantarum* 129, 68–78.

Singh BK (ed.) (1999) Plant Amino Acids. Biochemistry and Biotechnology. Marcel Decker, Inc.

Nikiforova VJ, Bielecka M, Gakie're B, Krueger S, Rinder J, Kempa S, Morcuende R, Scheible W-R, Hesse H & Hoefgen R (2006) Effect of sulfur availability on the integrity of amino acid biosynthesis in plants. *Amino Acids* 30, 173–183.

Wirtz M & Droux M (2005) Synthesis of the sulfur amino acids: cysteine and methionine. *Photosynthesis Res.* 86, 345–362.

Tan S, Evans R & Singh B (2006) Herbicidal inhibitors of amino acid biosynthesis and herbicide-tolerant crops. *Amino Acids* 30, 195–204.

This paper discusses targeting of amino acid biosynthesis pathways for identifying effective herbicides and also herbicide tolerant crops.

Tanaka R & Tanaka A (2007) Tetrapyrrole biosynthesis in higher plants. *Annu. Rev. Pl. Biol.* 38, 321–346.

This review covers current knowledge of the biosynthesis of tetrapyrrole biosynthesis, including localization and regulation, and interactions with other pathways.

Woodward AW & Bartel B (2005) Auxin: regulation, action, and interaction. *Ann. Bot.* 95, 707–735.

A comprehensive review with good coverage of the multiple pathways for auxin biosynthesis.

Chae HS & Kieber JJ (2005) Eto Brute? Role of ACS turnover in regulating ethylene biosynthesis. *Trends Pl. Sci.* 10, 291–296.

Chen Y-F, Etheridge N, Schaller GE (2005) Ethylene signal transduction. *Ann. Bot.* 95, 901–915.

Reviews considering the post-transcriptional control of ethylene biosynthesis and the impact of signal transduction on ethylene signaling.

Characterization of seed storage proteins

Shewry PR & Casey R (1999) Seed Proteins. Kluwer Academic Publishers.

Dunwell JM (2003) Structure, function and evolution of vicilin and legumin seed storage proteins. In Biopolymers, Vol. 8: Polyamines and Complex Proteinaceous Materials (Steinbüchel AF, Fahnestock SR eds), pp. 223–253. Wiley-VCH.

Shewry PR (2003) Tuber storage proteins. *Ann. Bot.* 91, 755–769.

These are the main references for accessing descriptions of the proteins.

Synthesis of seed proteins

Shutov AD, Bäumlein H, Blattner FR & Müntz K (2003) Storage and mobilization as antagonistic functional constraints on seed storage globulin evolution. *J. Exp. Bot.* 54, 1645–1654.

Vitale A & Hinz G (2005) Sorting of proteins to storage vacuoles: how many mechanisms? *Trends Pl. Sci.* 10, 316–323.

These reviews deal with the targeting of proteins to storage vacuoles and the conflicting requirements for the construction of efficient storage proteins.

Mobilization of seed storage proteins

Müntz K, Belozersky MA, Dunaevsky YE, Schlereth A & Tiedemann J (2001) Stored proteinases and the initiation of storage protein mobilization in seeds during germination and seedling growth. *J. Exp. Bot.* 5, 1741–1752.

Müntz K & Shutov AD (2002) Legumains and their functions in plants. *Trends Pl. Sci.* 7, 340–344.

This paper and review provide a good insight into the main features of proteinase degradation of seed proteins.

Evolution of seed storage proteins

Dunwell JM, Culham A, Carter CE, Sosa-Aguirre CR & Goodenough PW (2001) Evolution of functional diversity in the cupin superfamily. *Trends Biochem. Sci.* 26, 740–746.

This review traces the origins of seed storage proteins from bacteria to higher plants.

Lipid Biosynthesis

9

Key concepts

- Fatty acid biosynthesis occurs in the plastids through the sequential addition of two carbon units.

- There are two pathways of glycerolipid biosynthesis, one in the plastid (the prokaryotic pathway) and one in the endoplasmic reticulum (the eukaryotic pathway).

- Phosphatidic acid produced in the plastids or endoplasmic reticulum is a central intermediate in glycerolipid biosynthesis.

- Oxylipins are metabolites produced as a result of the oxidation of lipids.

- Triacylglycerols are essentially a storage form of carbon and chemical energy in seeds, fruits, and pollen grains.

- Fatty acid oxidation at the β-carbon and the subsequent removal of two carbon units provides biosynthetic precursors from stored lipids.

Overview of lipids

Lipids are a structurally diverse group of chemicals that are essential for the structure and function of living cells. Lipids are not classified on the basis of their chemical structure but instead by their solubility properties. Lipid is a term used by chemists to identify a chemically heterogeneous group of substances with the common property that they are insoluble in water but soluble in non-polar organic solvents, such as chloroform. Lipids may also be classed, based on their reactivity with strong bases, as saponifiable and non-saponifiable. Saponification is the process used to produce soaps by reacting lipids with a strong base. A saponifiable lipid is made of a glycerol backbone and fatty acids (Figure 9.1). The non-saponifiable lipids, including the fat-soluble vitamins A and E, and cholesterol will not be discussed further in this chapter.

It is clear that lipids are made up of a range of structures that may be aliphatic or aromatic, acyclic or cyclic, straight or branched, saturated or unsaturated, flexible or rigid. In addition most lipids also have some polar or hydrophilic features that make them amphipathic (containing both polar and nonpolar domains) or amphiphilic (having a polar, water soluble group, attached to a nonpolar, water insoluble, hydrocarbon chain) molecules with hydrophobic as well as hydrophilic portions. The fatty acid composition of lipids can be determined by using gas chromatography, high pressure liquid chromatography or electron capture mass spectrometry to separate the

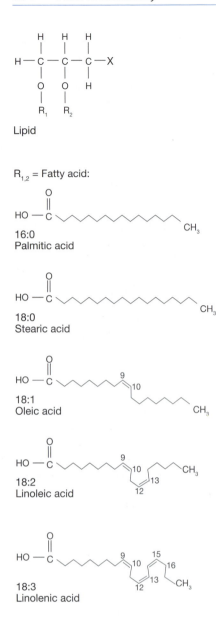

Lipid

R$_{1,2}$ = Fatty acid:

16:0
Palmitic acid

18:0
Stearic acid

18:1
Oleic acid

18:2
Linoleic acid

18:3
Linolenic acid

Figure 9.1 Examples of fatty acids found in plants. The top structure is a lipid comprised of a glycerol backbone with two fatty acid groups attached at R$_1$ and R$_2$. X varies depending upon the type of lipid. For example, in the case of diacylglycerol X = H; triacylglycerol X = R$_3$; phospholipid, X = $\circledP$; galactolipid X = galactose (see Table 9.1 for more details). Fatty acid structures are denoted by a shorthand notation, for example in the case of 18:2, the number before the colon, 18, represents the chain length, while the number after the colon, 2, indicates the number of double bonds. Sometimes the position of the double bond in relation to the terminal methyl group, or ω carbon, is also denoted. For instance, a ω3, or n3 fatty acid has a double bond three carbons from its methyl end.

methylated derivatives of the fatty acids. Generally, the purification of a particular type of lipid from a plant extract gives a complex mixture. Lipidomics is an emerging technology to collect data on lipids using a systems biology approach. Although in the early stages of development, in the future lipidomics should provide a highly sensitive approach for the complete characterization of the lipids present in a sample (see Further Reading for more details).

Fats, oils, some hormones and pigments, and most nonprotein membrane components are lipids. The basic classes of saponifiable lipids are listed in Table 9.1. In this introductory section a brief overview of the structures of different fatty acids and lipids will be provided. Fatty acids are carboxylic acids with long chain hydrocarbon side groups. They usually occur in an esterified form as the major component of different lipids or as thioesters. The naming system of fatty acids can be quite complex and the basic terminology used for identifying chain lengths, presence of unsaturation (double bonds), and of substituent groups will be discussed (Table 9.2). This information will form a basis for the remainder of the chapter where a more complete discussion relating to their formation and functionality in the plant cell will be made.

Fatty acids are long chain monocarboxylic acids typically 12–24 carbons long. As their biosynthesis results from the concatenation (link in a chain) of two carbon units, the carbon chain length is almost always an even number of carbons. Rather than stating that the chain length is of, for example, 16 carbon atoms, the terminology used will be to refer to a 16C fatty acid. Numbering of fatty acid carbon atoms is from the carboxyl end of the chain with carboxyl carbon as C1. The predominant fatty acid residues are those of 16C and 18C species, palmitic, stearic, oleic, linoleic, and linolenic acids (Table 9.2). Fatty acids of less than 14C or more than 20C are uncommon. A saturated fatty acid contains no C=C double bonds and an unsaturated fatty acid contains 1, 2, 3, or more (polyunsaturated) double bonds. The shorthand nomenclature to show this is that two numbers are separated by a colon, where the number before the colon is the number of carbon atoms and that after represents the number of double bonds. For example, the polyunsaturated fatty acids linoleic acid and linolenic acid, representing the major proportion of fatty acids in plants, are described as 18:2 and 18:3 fatty acids, which means they are 18 carbons long and have two and three double bonds,

Table 9.1 Basic classes of saponifiable lipids	
Class	**Feature**
Fatty acid	CH$_3$(CH$_2$)$_n$COOH
saturated	No double bonds
unsaturated	Double bonds
Glycerolipids	Fatty acids esterified to glycerol
triacylglycerols	3 fatty acids esterified to glycerol
phospholipids	Esterification of 2 fatty acids to glycerol and a phosphate group
galactolipids	Esterification of 2 fatty acids to glycerol and a galactosyl group
sulfolipid	Esterification of two fatty acids to glycerol and a sulfoquinovosyl group

Table 9.2 Some naturally occurring straight chain saturated fatty acids, mono-unsaturated fatty acids, and polyunsaturated fatty acids found in plants

	Number of C atoms	Systematic name	Common name
Straight chain saturated acids	2	*n*-Ethanoic	Acetic
(No double bonds)	8	*n*-Octanoic	Caprylic
	10	*n*-Decanoic	Capric
	12	*n*-Dodecanoic	Lauric
	16	*n*-Hexadecanoic	Palmitic (16:0)
	18	*n*-Octadecanoic	Stearic (18:0)
	22	*n*-Docosanoic	Behenic
	24	*n*-Tetracosanoic	Lignoceric
	26	*n*-Hexacosanoic	Cerotic
	28	*n*-Octacosanoic	Montanic
Mono-unsaturated fatty acids	16	*trans*-3-hexadecenoic	
(1 double bond)	16	*cis*-9-hexadecenoic	Palmitoleic (16:1)
	18	*cis*-9-octadecenoic	Oleic (18:1)
	20	*cis*-11-eicosenoic	Gadoleic (20:1)
	22	*cis*-13-docosenoic	Erucic (22:1)
Polyunsaturated fatty acids	Dienoic acids		
(>1 double bond C=C)	18	*cis,cis*-9,12-octadecadienoic	Linoleic (18:2)
	Trienoic acids		
	18	All-*cis*-6,9,12-octadecatrienoic	γ-Linolenic (18:3)
		All-*cis*-9,12,15-octadecatrienoic	α-Linolenic (18:3)
	Trienoic acids (conjugated)		
	18	*cis*-9,*trans*-11,*trans*-13-octadecatrienoic	Eleostearic (18:3)

respectively (Figure 9.1; Table 9.2). Saturated fatty acids are fully reduced or saturated with hydrogen and are very flexible molecules with relatively free rotation around each C–C bond meaning they can assume a wide range of conformations. Over half the fatty acids of plant lipids are unsaturated and often polyunsaturated. Unsaturated fatty acid isomers may be either positional (structural) or geometric. In the case of positional fatty acids the double bond is located in different positions in the carbon chain. The disadvantage of this shorthand system is that it does not give more precise information relating to the position and configuration of the double bonds. Instead this information is sometimes also incorporated into a shorthand nomenclature with systematic names. So linoleic acid can also be described as *cis*(Δ-)9, *cis* (Δ-)12-18:2, or (*cis,cis*)9,12-octadecadienoic acid, meaning it is an 18C fatty acid with a *cis* double bond at 9 and 12 carbons from the carboxyl end. The overwhelming majority of unsaturated fatty acids in plants have the *cis* configuration.

An exception is *trans* Δ3 hexadecenoic acid, found in the chloroplast lipid phosphatidylglycerol.

Geometric isomerism refers to the possibility that the configuration at the double bond can be either *cis* or *trans*. If a biological membrane contained only saturated or *trans*-unsaturated fatty acids then the hydrophobic tails would give a semi-crystalline gel, which would impair the permeability barrier and interfere with the mobility of membrane components. However, the *cis* configuration is the most naturally occurring fatty acid with double bonds and puts a 30° bend in the hydrocarbon chain. This allows the unsaturated fatty acids to pack together much less closely than saturated fatty acids. With increasing unsaturation van der Waals interaction decreases and there is a lowering of melting points. For example, desaturation of stearic acid, an 18C fatty acid with no double bonds, to oleic acid, an 18C fatty acid with one double bond, decreases the melting temperature from 69°C to 13°C. The fluidity of lipids containing fatty acid residues also increases with the degree of unsaturation of fatty acids.

A saponifiable lipid or glycerolipid is made of a glycerol backbone and fatty acids (Figure 9.1). The type of fatty acid that is made, in terms of chain length and degree of saturation, will determine the type of lipid produced. Similarly, the selection of fatty acid incorporated into the glycerol backbone influences the type of lipid. Finally, the location where the fatty acids are built into lipids will influence the combination of fatty acids and the type of lipid produced. There are four main types of glycerolipids in plants, triacylglycerols (TAGs), phospholipids, galactolipids, and a sulfolipid (Table 9.1).

The TAGs have three fatty acids esterified to glycerol, and they differ according to the identity and placement of the three fatty acid residues. Many TAGs contain two or three different types of fatty acid residues and are named according to their placement of the glycerol moiety (see Figure 9.17). They are often called neutral lipids due to their nonpolar nature. Fats and oils (which only differ in that fats are solid and oils are liquid at room temperature) are complex mixtures of TAGs, which, depending on the organism that has produced them, have varying fatty acid composition. The TAGs represent the major storage lipids in plants.

Phospholipids are amphipathic in nature containing a glycerol core linked to two hydrophobic fatty acid chains by an ester linkage, and a hydrophilic headgroup linked by a phosphate ester linkage. Phospholipids are complex lipids that are essential structural elements of membranes. They are formed in the plastids and the endoplasmic reticulum (ER) as a result of the operation of the so-called prokaryotic or eukaryotic pathway, respectively (Figure 9.9). Phosphatidic acid (PA) is the first phospholipid formed in both the plastids and ER. In the plastids further esterification of the polar headgroup of the PA can give rise to phosphatidylglycerol via the prokaryotic pathway. Phosphatidylcholine, phosphatidylethanolamine, phosphatidylserine, and phosphatidylinositol are generated in the ER via the eukaryotic pathway (Figure 9.10).

In the galactolipids, monogalactosyl diacylglycerol and digalactosyl diacylglycerol, the glycerol residue carries a sugar (galactosyl group) instead of a phosphate (Table 9.1). Sulfolipid sulfoquinovosyl diacylglycerol has a sulfoquinovosyl group replacing the phosphate group. Galactolipids are found in plastid membranes. They are synthesized in the plastid from plastidic PA via the prokaryotic pathway (Figure 9.10).

Perhaps the most obvious role of lipids in the majority of plant cells, in particular phospholipids and galactolipids, is as the major component forming the hydrophobic barrier of biological membranes (Chapter 3). Fatty acids and fatty acid derived molecules are highly reduced organic molecules and therefore can act as storage reserves that can be metabolized to release energy. Lipids also include a number of secondary compounds and pigments. These latter compounds are metabolically unrelated to fatty acid metabolism and will not be considered in this chapter. (Carotenoid synthesis is dealt with in Chapter 12.) Instead this chapter will focus on fatty acid biosynthesis, the lipids arising as a result of this and the subsequent release of energy from lipids when they are metabolized. Fatty acid and lipid metabolism in plants have a number of features in common with other organisms. However, the added complexity of cellular compartmentation found in plants means that lipid pools must also be moved within the plant cells to meet demands.

Fatty acid biosynthesis occurs through the sequential addition of two carbon units

Over 200 different fatty acids are known to accumulate in plants. However, only six or seven structures with chain lengths of 16 or 18 carbons and one to three double bonds are commonly found (Table 9.2, Figure 9.1). In plants, *de novo* fatty acid biosynthesis occurs in the plastids where acyl groups, attached to a soluble acyl carrier protein (ACP), are elongated by the sequential addition of 2C units. As the site of fatty acid biosynthesis, the plastid must supply membrane fatty acids not only for itself, but also for the rest of the plant cell. Plastids of most (if not all) tissues are able to synthesize fatty acids. For example, the plastids of developing seeds play a major part in the synthesis of fatty acids for TAGs. An ACP has also been found in the mitochondria where it participates in lipoic acid biosynthesis. However, the contribution of these organelles to fatty acid biosynthesis still needs to be fully determined and is undoubtedly very minor when compared with plastids.

Fatty acid biosynthesis uses acetyl CoA as the initial carbon precursor and the building block for assembly of long chain 16C and 18C fatty acids (Figure 9.2). Acetyl CoA is supplied to the pathway mainly by the action of plastid pyruvate dehydrogenase (Figure 9.3, page 309). In chloroplasts, pyruvate can be provided via 3-phosphoglycerate produced by the Calvin cycle reactions (see Chapter 5). In nonphotosynthesizing tissues, pyruvate comes from the glycolytic pathway (see Chapter 6). In isolated oilseed plastids there is sufficient pyruvate dehydrogenase activity to support the *in vivo* rates of fatty acid biosynthesis. It has been suggested that in some plastids acetate is generated outside the plastid, by, for instance, a mitochondrial pyruvate dehydrogenase or a cytosolic ATP citrate lyase. In this case the acetate would be taken into the plastids and activated in the stroma by acetyl CoA synthase. Plastids have a very active acetyl CoA synthase, with isolated chloroplasts or developing seed plastids rapidly incorporating radiolabeled acetate into fatty acids. However, most recent research suggests that this route only plays a minor role and that in most plastids acetyl CoA is generated by pyruvate dehydrogenase, which in contrast to the mitochondrial pyruvate dehydrogenase is not down-regulated by protein phosphorylation (Chapter 6).

The condensation of nine two-carbon units is necessary for the assembly of an 18C fatty acid

Although the two-carbon units for fatty acid biosynthesis are derived from acetyl CoA their actual assembly first requires the acetyl group to be further activated. This activation occurs when acetyl CoA is carboxylated by acetyl CoA carboxylase (ACCase) to form malonyl CoA (Figure 9.2). ACCase, at least

Figure 9.2 Simplified pathway of fatty acid biosynthesis. Malonyl-ACP also acts as the co-substrate and two-carbon donor for each condensing reaction labeled c–e. The enzymes involved are (a) acetyl CoA carboxylase; (b) malonyl CoA:ACP transacylase; (c) 3-ketoacyl-ACP synthase III; (d) 3-ketoacyl-ACP synthase I; (e) 3-ketoacyl-ACP reductase; (f) 3-hydroxyacyl-ACP dehydratase; (g) enoyl-ACP reductase; (h) 3-ketoacyl-ACP synthase II; (i) stearoyl-ACP desaturase; (j) oleoyl-ACP thioesterase; (k) medium-chain acyl-ACP thioesterase; (l) acyl CoA synthetase. Enzymes a–h are all part of the type II fatty acid synthase complex. Abbreviations: ACP, acyl carrier protein.

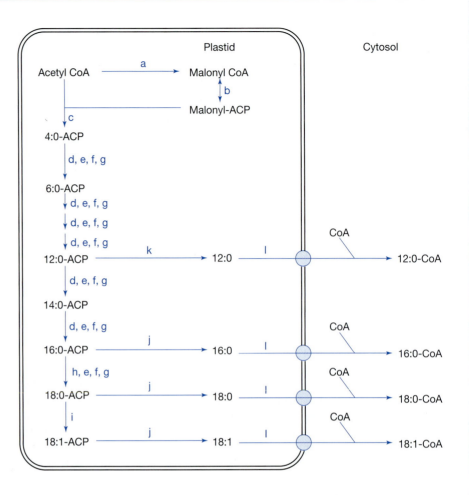

in non-Gramineae species in plants (Box 9.1), exists in two specific forms, a heteromeric plastidic form and a homomeric cytosolic form (Figure 9.4). These forms differ in the quaternary organization of their structural domain. In the heteromeric type, the ACCase is made up of four types of smaller subunits organized into a 700 kDa complex. The three domains occur on separate dissociable proteins. In the homomeric type, the complete ACCase, consists of a very large (>200 kDa) multifunctional polypeptide with three functional domains (Figure 9.4). These domains are ordered sequentially on a single polypeptide. As membranes are impermeable to acyl CoA, the two ACCases generate physically isolated malonyl CoA pools. This compartmentalization provides a means of regulating the supply of malonyl CoA for a range of biosynthetic processes: the plastidic malonyl CoA pool is the precursor for *de novo* fatty acid biosynthesis, producing 16C and 18C fatty acids; the cytosolic malonyl CoA pool elongates fatty acids to 20C and longer. Cytosolic malonyl CoA is also used in the biosynthesis of flavonoids and stillbenes (Chapter 11), as well as malonyl derivatives. For example many D-amino acids and other secondary metabolites react to form malonyl derivatives.

ACCase is a soluble class 1 biotin-containing enzyme. It catalyzes a two-step reaction requiring the three conserved structural domains (Figure 9.4) and a non-catalytic biotin-containing component. The biotin prosthetic group is covalently bound to the subunit and, as the intermediate carrier of the carboxyl group that carboxylates acetyl CoA, is essential for enzymatic function of ACCase (Figure 9.5). CO_2 (in the form of HCO_3^-) is activated and attached to a nitrogen in the biotin ring of the biotin carboxyl carrier protein via the action of the biotin carboxylase domain in an ATP-dependent reaction.

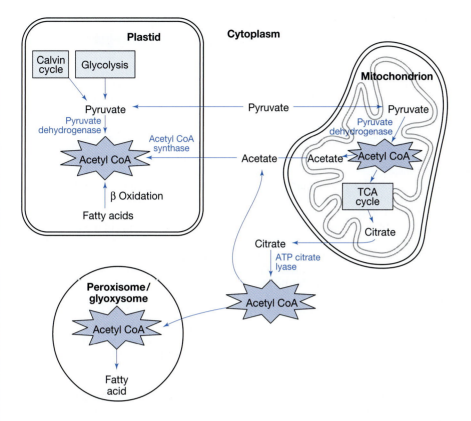

Figure 9.3 Potential routes for generating acetyl CoA in plant metabolism. Acetyl CoA is not thought to be able to cross membranes and must therefore be produced in the compartment where it is required. The major routes for acetyl CoA production are glycolysis, via pyruvate dehydrogenase, and fatty acid oxidation.

Following the formation of the biotin carboxyl carrier protein, the CO_2 is transferred to acetyl CoA by the carboxyltransferase, forming malonyl CoA (Figure 9.5).

The three subunits of multisubunit ACCase are encoded by single genes in arabidopsis, while two cDNAs encoding the biotin carboxyl carrier protein have been identified, one of which is constitutively expressed and the other predominantly expressed in flowers and developing seeds. This multisubunit form of ACCase must be able to associate easily and immunoprecipitation

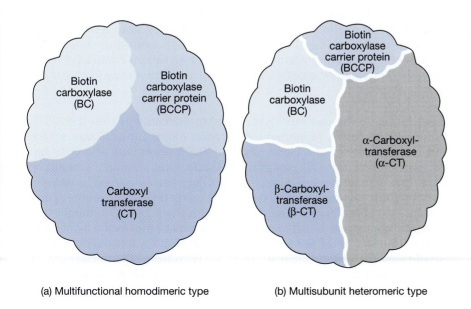

(a) Multifunctional homodimeric type   (b) Multisubunit heteromeric type

Figure 9.4 The two different forms of acetyl CoA carboxylase (ACCase) found in plants.

Box 9.1 The sensitivity of acetyl CoA carboxylase (ACCase) to selective herbicides

For many years the structure of plant ACCase was the subject of much controversy. A major impetus to solving the structure was one of commercial exploitation of the sensitivity of certain plant ACCases to selective herbicides. The type of ACCase structure present in a plant depends on the plant species and subcellular localization. For example, dicots have a multisubunit form in their plastids and a multifunctional structure in their cytoplasm (see Figure 9.4). This enzyme is different in the grass family where similar homomeric forms with large, multifunctional polypeptides are found in both the plastid and cytosol. In addition the biotin carboxylase transferase gene is partially or completely missing from the grass chloroplast genome and none of the heteromeric ACCase subunits can be detected. Although the reason for these differences is not clear, at the practical level it is an extremely important observation. This difference between the grass family and other plants means that the herbicides can be specifically targeted to the homomeric form of ACCase but not the heteromeric form. This inhibition *in vivo* results in the blocking of fatty acid biosynthesis and kills the plant.

Two commercially important groups of selective herbicides have ACCase as their primary targets, aryloxyphenoxypropionic acids (AOPPs/FOPs) and the cyclohexanediones (CHDs/DIMs) and act specifically on grass weeds. In 1998 these two herbicides represented 10% of the global herbicide market. It is now clear that the carboxyltransferase partial reaction on the plastid-located multidomain ACCase is the site of action. The different structure of AOPP and CHD give double inhibition on kinetic analysis. That is, the binding of a member of one herbicide class stops the binding of a herbicide on a second class. The

selectivity of these herbicides in broad-leaved crops is due to the fact that they do not affect the multisubunit plastidic ACCase of dicots or the multidomain cytosolic form of ACCase present in all plants.

Apart from a few grass species that can metabolize specific graminicides, resistance in the whole plant is generally correlated with an insensitive plastid isoform of ACCase. Some grasses (*Poa* or *Festuca* spp.) show inherent resistance to at least some AOPPs and CHDs. Acquired resistance is an increasing problem for agriculture and is associated with the presence of insensitive forms of the enzyme. Inheritance studies have shown that ACCase resistance to both types of graminicide is governed by a single dominant or semi-dominant nuclear gene and resistance to different herbicides is coded at the same locus.

Experiments with chimeric genes containing wheat plastid ACCase that complemented a yeast ACC1 mutation located the herbicide sensitivity section in a 400 amino acid fragment. When compared with other sequence information it was possible to locate the fragment on the carboxyltransferase domain. Further work identified that an isoleucine to leucine substitution was critical for herbicide interaction. Although other residues most likely play a role in binding specificity, the identified isoleucine to leucine substitution has been shown in acquired resistance biotypes of black grass. Interestingly the leucine found in the plastidic ACCase of resistant grass weeds is also found in the multisubunit plastidic enzyme of dicots and the cytosolic multidomain enzymes, which are naturally resistant to these herbicides. This suggests that the herbicides selectivity also occurs at this site.

experiments have confirmed that the two subunits of the carboxyltransferase have a tight association. The whole multisubunit of ACCase is strongly associated with the chloroplast envelope through non-ionic interactions between the carboxyltransferase units. Such associations could have important functional significance by allowing substrate channeling in plastid fatty acid biosynthesis.

When biotin carboxylase expression levels were decreased to less than 25% of the wild-type levels in transgenic tobacco plants, growth was severely retarded under low light. Plants also had a 26% lower fatty acid content in comparison with the wild-type plants. No effect was seen on the biotin carboxyl carrier protein levels. Similar results have been observed in arabidopsis transgenic plants; however, the mRNA levels for biotin carboxyl carrier protein and biotin carboxylase together with the α- and β-subunit of carboxyltransferase accumulated at a constant molar ratio. As the β-subunit of carboxyltransferase of the heteromeric form of ACCase is encoded by the plastid genome and the other subunits are encoded by the nuclear genome then this indicates that there is communication between the nuclear and plastid genomes.

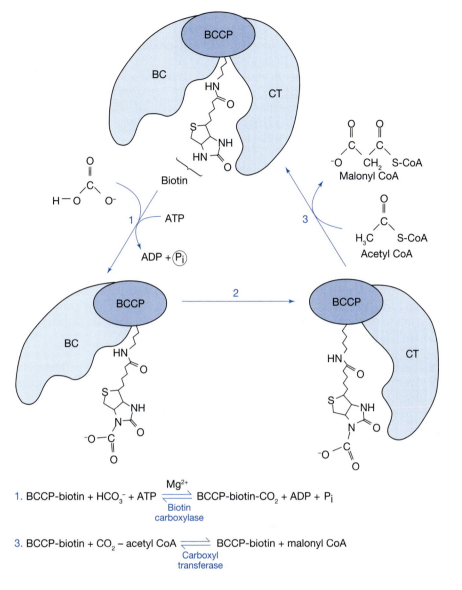

Figure 9.5 Acetyl CoA carboxylase (ACCase) reaction. The three functional components of ACCase facilitate the carboxylation of acetyl CoA to malonyl CoA. (1) Biotin carboxylase (BC) activates HCO_3^- by attaching it to a nitrogen in the biotin ring of the biotin carboxyl carrier protein (BCCP) in an ATP-dependent reaction. (2) The flexible biotin arm of BCCP moves, in a pendular movement, the activated CO_2 to the carboxyltransferase (CT) site. (3) Activated CO_2 is transferred by the transcarboxylase from biotin to acetyl CoA to form malonyl CoA.

1. BCCP-biotin + HCO_3^- + ATP $\underset{\text{Biotin carboxylase}}{\overset{Mg^{2+}}{\rightleftharpoons}}$ BCCP-biotin-CO_2 + ADP + P_i

3. BCCP-biotin + CO_2 – acetyl CoA $\underset{\text{Carboxyl transferase}}{\rightleftharpoons}$ BCCP-biotin + malonyl CoA

Experiments measuring the pool size of acyl thioesters following marked changes in fatty acid formation were used to initially determine whether ACCase had a regulatory role in fatty acid biosynthesis. Upon exposing dark grown leaves to light the malonyl-thioesters were detectable while the acetyl CoA levels remained stable. Stimulation of fatty acid biosynthesis with the detergent Triton-X100 led to a decrease in the acetyl CoA pool. The availability of specific inhibitors for ACCase in grasses meant that experiments could be undertaken to measure flux control coefficients (see Chapter 2). Under these experimental conditions ACCase exerted 45–61% of the total control of flux to lipids, indicating the importance of this enzyme in regulating fatty acid biosynthesis. A number of regulatory mechanisms have now been proposed, including the effect on the enzyme of changing stromal properties, acyl thioester concentration and CoA levels (see Ohlrogge and Jaworski, 1997 in Further Reading for more details). The presence of reducing agents has also been shown to increase ACCase activity in spinach chloroplasts and partially purified enzyme preparations. Studies of the chloroplast ACCase showed that carboxyltransferase was influenced by the reducing agent dithiothreitol, whereas biotin carboxylase was not. *In vitro* studies using site-directed mutagenesis of recombinant carboxyltransferase identified two cysteines, which, as a result of redox regulation, can form an intermolecular disulfide bridge between the two carboxyltransferase units. Recent work suggests that this disulfide–dithiol exchange could be relevant during light–dark changes *in vivo*. Current research should confirm whether ACCase redox regulation *in planta* is of any physiological relevance. Finally, extra-chloroplastic fatty acyl CoAs have been shown to significantly decrease fatty acid biosynthesis and might be a means of communicating the demand for cytosolic fatty acids.

The acyl group of malonyl CoA is linked to acetyl CoA and then transferred to acyl carrier protein via a malonyl CoA:ACP transacylase:

Malonyl CoA + ACP $\rightleftharpoons$ Malonyl-ACP + CoASH

ACPs are small acidic proteins that are about 80 amino acids long, which carry the nascent acyl chains during the synthesis of 16C and 18C acyl groups. ACP has a 4′-phosphopantetheine prosthetic group covalently attached to a serine residue near the middle of the protein (Figure 9.6). The prosthetic group has a similar structure and function to CoA. A sulfhydryl group at the end of the pantetheine can, in the presence of ATP, join with the carboxyl carbon of a fatty acid forming a thioester.

ACP is at the center of what is described as a type II fatty acid synthase. The acyl residue, bound as a thioester to ACP located in the center of the complex, is attached to a flexible chain and subsequently transferred from enzyme to enzyme during the reaction cycle.

For the assembly of an 18C fatty acid from acetyl CoA using type II fatty acid synthase, 48 reactions are necessary and at least 12 different proteins involved

Based on experimental estimations the concentrations of substrates give the appearance of being insufficient to support the observed *in vivo* rates of fatty acid biosynthesis. Type II fatty acid synthase is a complex allowing the operation of a tightly coupled pathway. Type II fatty acid synthase found in plants functions as a metabolic pathway, where each enzyme activity is located on an individual protein that can be readily separated from the other components involved in fatty acid biosynthesis (Figure 9.2).

The combined action of the four enzymes, 3-ketoacyl-ACP synthase, 3-ketoacyl-ACP reductase, 3-hydroxyacyl-ACP dehydratase, and enoyl-ACP reductase,

Figure 9.6 Structure of the central region of a acyl carrier protein (ACP). ACP has a prosthetic group, 4′-phosphopantetheine, which is covalently attached to the hydroxyl group of a serine residue near the middle of the polypeptide chain.

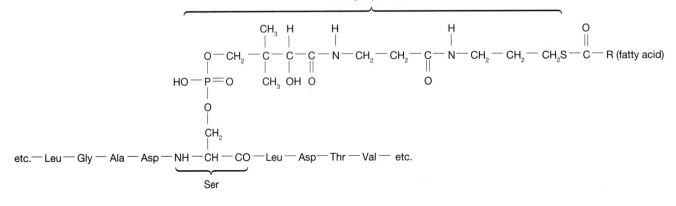

leads to the lengthening of the 2C acetic acid to the 4C butyric acid, which remains attached to ACP as a thioester (Figure 9.2):

- 3-Ketoacyl-ACP synthase catalyzes the condensation reaction between acetyl CoA and malonyl-ACP forming a new C–C bond.

  In the first instance condensation of acetyl and malonyl groups by 3-ketoacyl-ACP synthase forms the four-carbon intermediate acetoacetyl-ACP (Figure 9.2). This results in the formation of a C–C bond and in the release of the CO_2 that was previously added by the ACCase reaction. The removal of CO_2 drives the reaction in the forward direction, making it essentially irreversible:

  Acetyl CoA + malonyl-ACP $\longrightarrow$ acetoacetyl-ACP + CO_2

- The acetoacetyl-ACP is then reduced at the carbonyl group by the enzyme 3-ketoacyl-ACP reductase, which uses NADPH as the electron donor:

  Acetoacetyl-ACP + NADPH $\longrightarrow$ D-3-hydroxybutyryl-ACP + $NADP^+$

- 3-Hydroxyacyl-ACP dehydratase removes a water molecule from hydroxyl-ACP to yield *trans*-2-acyl-ACP:

  D-3-Hydroxybutyryl-ACP $\longrightarrow$ *trans*-2-butanoyl-ACP + H_2O

- The enzyme enoyl-ACP reductase uses NAD(P)H to reduce the *trans*-2 double bond and form a saturated fatty acid, completing one round of fatty acid synthesis:

  trans-2-Butanoyl-ACP + NAD(P)H $\longrightarrow$ butyryl-ACP + $NAD(P)^+$

  The major form of this enzyme is NADH-specific and is required for the reduction of growing fatty acid chains up to 10 carbons. Another form of the enzyme can use NADH or NADPH to perform the final reduction of chains of 10 or more carbons bound to ACP. The source of the required NADH is uncertain, but it may be the NADH produced by plastidic pyruvate dehydrogenase complex (see page 315).

The condensation reaction is repeated with additional malonyl-ACP followed again by the keto reduction, dehydration, and enoyl reduction steps. This four-step cycle continues to be repeated until the fatty acid chain length is 16C or 18C long.

Single isozymes of the 3-ketoacyl-ACP reductases and 3-hydroxyacyl-ACP dehydratase can act on acyl-ACP substrates of all chain lengths; however, at least three separate 3-ketoacyl-ACP synthase enzymes are necessary to produce an 18C chain. 3-Ketoacyl-ACP synthase III carries out the first condensation forming the four-carbon product. 3-Ketoacyl-ACP synthase I is most active with 4C–16C acyl-ACPs. 3-Ketoacyl-ACP synthase II is required for the elongation of 16C palmitoyl-ACP to 18C stearoyl-ACP.

Each fatty acid synthesis cycle adds two carbons to the acyl chain. For common fatty acid formation, the growing acyl chain is terminated when it is 16C or 18C long. This is presumably controlled to a large degree by the presence of the specific 3-ketoacyl-ACP synthase isoforms. Synthesis ends by either hydrolysis of the acyl moiety from ACP by a thioesterase, transfer of the acyl moiety from ACP directly on to a glycerolipid by an acyltransferase, or the formation of a double bond on the acyl moiety by an acyl-ACP desaturase.

Type II fatty acid synthases generally form 16C and 18C fatty acids. However, plants have a requirement for longer fatty acids in, for example, waxes of 26–34C fatty acids or sphingolipids of 22–24C fatty acids. There is a specialized elongase system for extension beyond 18C. Each elongase uses the reaction series that condenses two carbons at a time from malonyl CoA to an acyl primer followed by reduction, dehydration, and the final reduction. Elongases do not require ACP and are localized outside the plastid and are membrane bound.

Once the saturated fatty acid is produced, the next stage is to introduce a double bond into an acyl chain with a molecule of dioxygen completely reduced to water. Double bonds have a large effect on the melting temperature of fatty acids and also lipids that contain them (see page 306). The production of an unsaturated fatty acid through the introduction of double bonds is catalyzed by desaturase enzymes. There are two evolutionarily unrelated classes of desaturases that give a complete array of unsaturated lipids, soluble desaturases, and integral membrane desaturases. This has been confirmed by studies of arabidopsis mutant lines that were analyzed for lipid content in leaves and seeds. At least eight desaturase genes have been identified, differing in substrate specificity, subcellular localization, tissue location, and mode of regulation. It should also be noted that in addition to unsaturated fatty acids many plants also contain unusual fatty acids, which have a wide variety of functional substituents. Many of these fatty acids are generated through modification of conventional fatty acids by enzymes closely related to fatty acid desaturases.

The formation of each double bond requires two electrons to support the desaturation. In the plastid these are supplied by reduced ferredoxin generated either photosynthetically (via the noncyclic electron transport scheme) or nonphotosynthetically (via the oxidative pentose phosphate pathway). In the ER, reducing equivalents are supplied by NADH via cytochrome b_5 reductase and the heme protein cytochrome b_5.

Acyl-ACP utilization in the plastid

Acyl-ACP formed through fatty acid biosynthesis in the plastids has two important functions. First, the acyl-ACP may be used directly by plastid localized acyltransferases for the biosynthesis of prokaryotic lipids within the plastids. Second, the acyl-ACP can be exported to the cytoplasm where it is primarily incorporated into glycerolipids at the ER and via the eukaryotic pathway (Figure 9.7). In leaves of plants such as arabidopsis and spinach both of these pathways contribute equally to the biosynthesis of leaf glycerolipids. In nonphotosynthetic tissue and in the leaves of plants such as pea, the eukaryotic pathway is the major route for glycerolipid biosynthesis. For this latter process to occur acyl-ACP hydrolysis occurs in the plastid stroma by the action of soluble acyl-ACP thioesterases, possibly on the inner envelope. The free fatty acids generated are then transferred to the outer envelope of the plastid where they are reactivated to acyl CoA by acyl CoA synthetase for utilization in cytosolic glycerolipid biosynthesis (Figure 9.7). Plastid thioesterases hydrolyze primarily 16:0 (palmitate) and 18:1 (oleate)-acyl ACP, and to a lesser extent 18:0 (stearate)-acyl-ACP. Therefore plastids mainly provide CoA esters with acyl residues of 18:1 (oleate) and 16:0 (palmitate) for lipid metabolism outside the plastids (Figure 9.7).

Regulation of fatty acid formation

A metabolon is an organization framework for a number of soluble metabolic enzymes (see Chapter 2) and has been proposed to exist for fatty acid formation in the chloroplast. Such a multienzyme complex would channel acetate into long chain acids. This would require specific interaction and organization

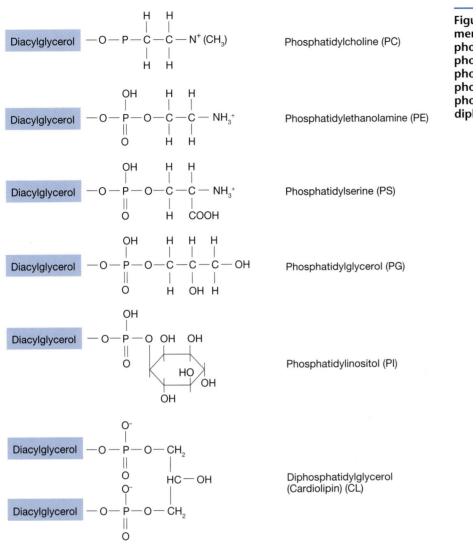

Figure 9.12 The structures of the membrane lipids phosphatidylcholine, phosphatidylethanolamine, phosphatidylserine, phosphatidylglycerol, phosphatidylinositol, and diphosphatidylglycerol.

position 3, 4, or 5 of the inositol ring (Figure 9.13) either alone or in combination. Their synthesis or hydrolysis increases within minutes of a plant being subjected to drought, salinity, temperature stress, and pathogen signaling. This has been suggested to be indicative of a potential role in rapid regulatory processes such as cell signaling events. In recent years a number of PPI-binding domains have been identified (over 70 in the arabidopsis genome). These lipid-binding domains are thought to function as effectors of lipid signaling by either recruiting proteins to specific membranes where specific PPIs are synthesized, or by activating proteins via a conformational change, induced upon ligand binding. The ratio between PPI and any interacting ligands can be very large, allowing a high number of different proteins to be targeted to a specific membrane site without saturating the available number of binding sites. In addition, the specific isomeric conformation of the phosphorylated myo-inositol head-group combined with a rapid monoester phosphate turnover rate allows the phosphphoinositide head-group moieties and different ligands to interact. All these factors are thought to be important properties contributing to PPI being a good messenger or precursor for messenger molecules.

Figure 9.13 The classic phosphatidylinositol response where phosphatidylinositol 4,5-bisphosphate is converted to inositol 1,4,5-triphosphate (IP₃) and diacylglycerol (DAG) by phospholipase C.

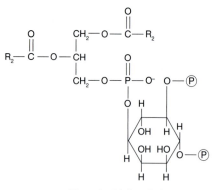

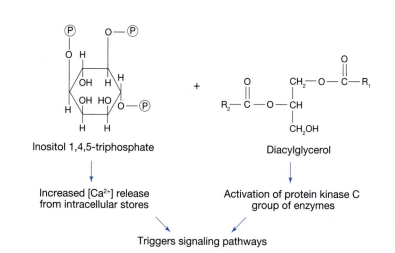

Phosphatidylinositol 4,5-bisphosphate

Phospholipase C — $-H_2O$

Inositol 1,4,5-triphosphate + Diacylglycerol

Increased [Ca²⁺] release from intracellular stores

Activation of protein kinase C group of enzymes

Triggers signaling pathways

The products of the oxidation of lipids and the resulting metabolites are collectively known as oxylipins

During plant development lipid composition often changes throughout the plant. In addition to the turnover of fatty acids within lipids, the formation of oxidized polyunsaturated fatty acids is an extremely important process in lipid metabolism. The products of the oxidation of lipids and the resulting metabolites are collectively known as oxylipins. Most oxylipins are members of families of closely related molecules. Their formation is usually the result of serial modifications leading to the synthesis of structurally related molecules carrying one or more functional oxygen-containing groups (keto groups, hydroxyl groups, etc.). Many oxylipins are generated from more than one fatty acid substrate. Polyunsaturated fatty acid oxidation is thought to play a regulatory role in plant development and adjustment by the plant to diverse and variable environmental conditions (Figure 9.14). A number of environmental stress-induced and developmental cues regulate oxylipin biosynthesis. A key component in this regulation is thought to be phospholipases and their release of fatty acid precursors from membrane lipids. To date much of the oxylipin research has focused on the jasmonate family of molecules, including jasmonic acid and its methyl ester, methyl jasmonate. There

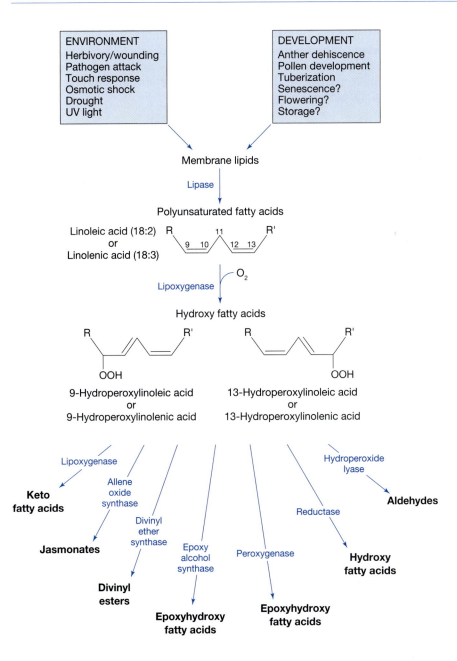

Figure 9.14 Biosynthesis of oxylipins in response to environmental stress and developmental cues.
Polyunsaturated fatty acids (PUFAs) are released from membrane lipids via lipases prior to the lipoxygenase-catalyzed addition of molecular oxygen. Hydroperoxy fatty acids are then metabolized through the action of different enzyme systems. These then form the substrates for a wide range of oxylipins (shown in bold). R, R′ = fatty acid chain.

is much less information available on the biosynthesis and function of other oxylipins, including aldehyde, ketol, epoxy, and hydroxyl derivatives.

Initially, oxylipins are formed by autoxidation or via enzymes such as lipoxygenases (LOXs) or α-dioxygenases. Autoxidation is an unspecific process leading to complex mixtures of fatty acids. The LOX-catalyzed step and subsequent reactions make up the LOX pathway and it is this latter pathway and its role in distinct stress- and developmentally-regulated pathways that has been the focus of much recent research.

In general, plant LOXs preferentially react with free fatty acids as substrates. There are a range of LOX isoforms that are distinguishable based on their subcellular localization, expression pattern, and substrate utilization. LOX is a nonheme iron dioxygenase that adds molecular oxygen to either the 9 or the 13 position of the 18C chain of linoleic and linolenic acids. 13-LOX and 9-LOX represent enzymes that generate predominantly 13- or 9-hydroperoxy fatty acids, respectively. These hydroperoxy fatty acids can then act as the substrate for a number of different enzyme families all forming different branches of the LOX pathway.

The CYP74 family is one of the best characterized members of the LOX pathway, made up of at least three different enzyme subfamilies: allene oxide synthase, hydroperoxide lyase, and divinyl ether synthase. All these enzymes share an intermediate epoxy allylic carbocation formed from the acyl hydroperoxide. Allene oxide synthase and divinyl ether synthase deprotonate the carbocation at different positions to give stable derivatives. Hydroperoxide lyases catalyze a rearrangement of the positive charge leading to a fragmentation of the molecule into two aldehydes. Allene oxide synthase, hydroperoxide lyase, and divinyl ether synthase have specificity for either 9- or 13-hydroperoxides, supporting the idea that oxylipin metabolism is organized into discrete 9-LOX and 13-LOX pathways.

A main product of the allene oxide synthase pathway is jasmonic acid. The biosynthesis of this phytohormone occurs in the chloroplast at the level of the free fatty acid derivative 13-LOX. Following reduction and three steps of β-oxidation, jasmonic acid is formed. Jasmonic acid has a role in defense and wound signaling. It is thought that upon wounding linolenic acid is released from membranes by the action of phospholipase A and subsequently converted to jasmonic acid. By acting with other compounds, including abscisic acid (ABA) and ethylene, jasmonic acid induces gene expression following wounding.

A number of studies characterizing the 9-LOX pathway have suggested that this is important in plant defense against microbial pathogens. For example, antisense expression of a 9-LOX gene in tobacco enhanced the plant's susceptibility to both compatible and incompatible plant pathogens.

A waxy cuticle coats all land plants

The cuticle, consisting of soluble and polymerized lipids, covers the epidermal cells of many aerial plant surfaces. Its role is essentially one of protection against water loss and provides some resistance to pathogens and insects. Among different plant species and even between different organs of a single species, or within one organ during development, the structure and chemical composition of the cuticle varies. The lipid polyester, cutin, is the major component of the cuticle making up 40–80% of cuticular weight. It is insoluble in organic solvents, consisting of oxygenated 16C or 18C fatty acids cross-linked by ester bonds so that the carboxyl group of one fatty acid is linked to a primary or secondary hydroxyl group of another. This leads to a fairly inelastic hydrophobic network, which gives some rigidity to the turgid plant tissues. Since to penetrate the cutin layer, pathogens probably need cutinase enzymes to hydrolyze the ester linkages, then the cuticle may act as a limited defense against pathogen attack. As the pore size of the cutin network is quite large then it does not particularly provide a barrier against water loss, instead cuticular waxes are embedded in the cutin matrix. Cuticular waxes are complex mixtures of very-long-chain fatty acid (VLCFA) derivatives, although they may also include triterpenoids (Chapter 11) and minor secondary metabolites such as sterols, alkaloids (Chapter 10), phenylpropanoids, and flavonoids (Chapter 12). During plant development the epidermis synthesizes and secretes the cuticle.

In epidermal cells there is an important control point resulting in a large flow of fatty acids into the production of cutin and wax. Although the mechanism for facilitating the movement of fatty acid precursors into different biosynthetic pathways is unclear, activities and specificities of a wide range of enzymes, including acyl-ACP acyltransferases, fatty, and elongase enzymes, have been identified. Fatty acid availability due to compartmentation or metabolic channeling has been proposed as a means of affecting *in vivo* distribution. However, this is an area that needs further investigation.

Acyl chains are released from the ACP by a thioesterase and exported from the plastid to the ER (Figure 9.7). The 16:0 and 18:1 CoAs are then used as precursors for cutin biosynthesis, while 16:0 or 18:0 acyl groups are elongated further in wax biosynthesis to generate VLCFA wax precursors of between 26C and 34C in length (Figure 9.15).

FatA- and FatB-type acyl-ACP thioesterases with different substrate preferences have been identified. The former, with 18:1 ACP as its preferred substrate, is thought to be the key acyltransferase involved in the export of 18:1 acyl

Figure 9.15 The pathways for elongation of long-chain fatty acids producing carbon skeletons for cuticular lipid. Elongation of long-chain fatty acid pathways for generating carbon skeletons for cuticular lipids. Cutin monomers are derived from 16C and 18C fatty acids without further elongation. Very-long-chain fatty acids (20C–34C) are produced by the fatty acid elongation pathway and are the substrates of modifying wax biosynthetic enzymes, whilst the β-diketone elongation pathway produces β-diketones. In all cases sequential 2C additions are made to the growing chains by enzymes of fatty acid biosynthesis: (1) 3-ketoacyl-ACP synthase; (2) 3-ketoacyl-ACP reductase; (3) 3-hydroxyacyl-ACP dehydratase; (4) enoyl-ACP reductase.

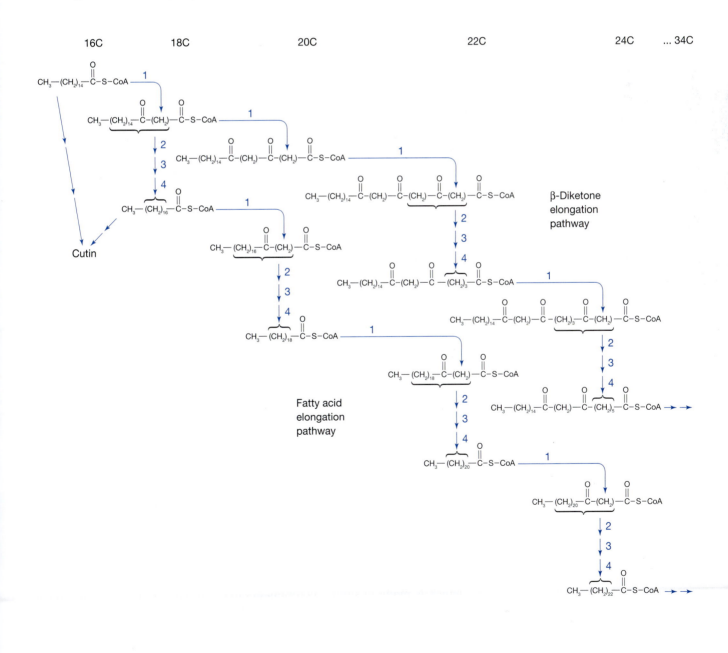

groups for 18C cutin monomer production, while the latter appears to have a preference for saturated acyl-ACP substrates and is thought to be involved in supplying 16:0 for the production of 16C cutin monomers. Arabidopsis mutants deficient in FatB activity have a 20% decrease in total wax load in leaves and 50% decrease in stems, suggesting the FatB type thioesterases are involved in providing fatty acyl substrates for wax biosynthesis.

Little is known about the biosynthesis of cutin, its transport to the plant surface, and monomer polymerization or of the enzymes involved in the whole process. Originally a pathway using the 16:0 CoA and 18:1 CoA was proposed. Here the acyl groups undergo hydroxylation and epoxidation reactions catalyzed by a number of cytochrome P450-dependent enzymes. Although cytochrome P450 enzymes that are able to catalyze in-chain and ω-hydroxylation have been identified, there is no biological confirmation of the role of these enzymes in cutin formation. More recently an alternative LOX–peroxygenase–epoxide hydrolase pathway for the formation of 18C cutin monomers has been identified. Here a peroxygenase catalyzes the epoxidation of oleic acid in the presence of unsaturated fatty acid hydroperoxides, generated from linoleic acid or linolenic acid via LOX, which act as co-substrates.

Biosynthesis of very-long-chain fatty acid wax precursors

The first step in wax biosynthesis is the elongation of 16:0 and 18:0 plastid produced fatty acids to generate VLCFA wax precursors with chain lengths of between 20C and 34C (Figure 9.15). The composition of the wax varies from plant species to plant species, though long chain hydrocarbons, acids, primary and secondary alcohols, ketones, aldehydes, and alkanes all contribute. Interestingly some pathogens and herbivorous insects are attracted to or repelled by surfaces covered with specific wax compositions.

An extraplastidial membrane-associated multienzyme complex fatty acid elongase catalyzes the fatty acid elongation via four enzyme reactions that are analogous to type II fatty acid synthase. The initial elongation reaction, catalyzed by a β-ketoacyl CoA synthase, adds 2C from malonyl CoA to a CoA-esterified fatty acid substrate. Subsequently, β-ketoacyl CoA is reduced to β-hydroxy CoA, then dehydrated to enoyl CoA prior to a second reduction yielding acyl CoA extended by two carbons (Figure 9.16).

Experiments following the incorporation of radiolabeled precursors into wax components of different chain lengths in the presence of inhibitors, together with the analyses of mutants defective in fatty acid elongation have been used to show that a number of fatty acid elongases are necessary for the production of VLCFA precursors. To meet this requirement for fatty acyl precursors of various chain lengths in different plant tissues during plant development, there is a family of 21 elongase condensing enzyme-like sequences in the arabidopsis genome. How many of these putative condensing enzymes are wax-specific and how many separate condensing enzymes are required for the elongation of an 18C to 34C fatty acid are not yet known, although it is thought that some elongase condensing enzymes may catalyze multiple elongation steps.

Role of suberin as a hydrophobic layer

Epidermal cells of root tissues are not covered in cutin to the same degree as aerial tissue. However, they do have an internal hydrophobic suberin layer, which is made of similar components to cutin. The VLCFAs, fatty alcohols, diacids, and phenolics are the main components of suberin. Unlike cutin the

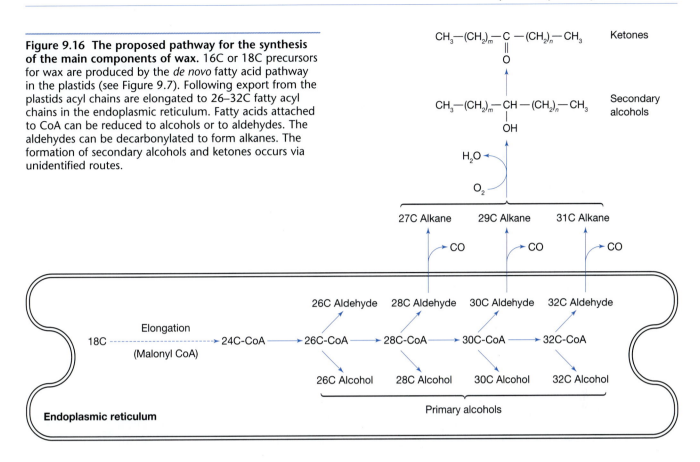

Figure 9.16 The proposed pathway for the synthesis of the main components of wax. 16C or 18C precursors for wax are produced by the *de novo* fatty acid pathway in the plastids (see Figure 9.7). Following export from the plastids acyl chains are elongated to 26–32C fatty acyl chains in the endoplasmic reticulum. Fatty acids attached to CoA can be reduced to alcohols or to aldehydes. The aldehydes can be decarbonylated to form alkanes. The formation of secondary alcohols and ketones occurs via unidentified routes.

fatty acids in suberin are usually longer than 18C and do not have epoxy or secondary alcohol groups. Suberin also has a high phenolic content. All these factors contribute to suberin being extremely impermeable to aqueous solutions meaning that materials moving across the endodermis and into the vascular tissue must do so by symplastic transport. Suberin also surrounds bundle sheath cells in C_4 species, preventing any CO_2 produced by decarboxylation reactions from diffusing out of the bundle sheath cells (Chapter 5).

Storage lipids are primarily a storage form of carbon and chemical energy

Plant storage lipids are a highly reduced form of carbon, which is available as an energy source during germination and early seedling development. Storage lipids found in oleaginous crops are also a source of edible oils for human consumption (Boxes 9.2 and 9.3). The major storage lipids found in plants are triacylglycerols (TAGs) and the liquid wax of the jojoba (*Simmondsia chinensis*) (Box 9.4). TAGs consist of a glycerol backbone, with fatty acids esterified to all three carbons (Figure 9.17). Because glycerol lacks rotational symmetry the carbon atoms can be distinguished from one another. In labeling the three positions on the glycerol derivative stereospecific numbering (*sn*) can be used with position 1 at the top and 3 at the bottom. Analysis of seed oils from a variety of sources has shown that generally saturated fatty acids occupy position 1, unsaturated fatty acids occupy position 2 and there is more variation in the fatty acid composition found at position 3. The liquid wax of the jojoba seed is made up of esters of very long chain monosaturated fatty acids and alcohols (Box 9.4).

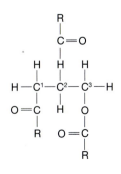

Figure 9.17 Triacylglycerol structure. The position at which a fatty acid (R) is esterified to the glycerol backbone of glycerolipids.

Box 9.2 Manipulating fatty acids for industrial applications

Fatty acids represent the most abundant form of reduced carbon chains available in living things. They have a diverse range of dietary and industrial uses and are a major commercial resource. In 2006 the world vegetable oil production was estimated at over 100 million metric tonnes and had a market value of 50 billion US dollars. Soybean, oil palm, rapeseed, and sunflower are the four most important oilseed crops accounting for 75% of current worldwide production. The fatty acids produced in these oilseeds are primarily the saturated palmitic (C16:0) and stearic (C18:0) acids, monounsaturated oleic acid (C18:1), and the polyunsaturated linoleic (C18:2) and α-linolenic (C18:3) acids. Although the demand for vegetable oils continues to rise, production capacity is easily able to meet these demands. As such vegetable oils have been maintained at approximately $500 per tonne.

Most vegetable oil is used directly for human dietary consumption. In developed countries up to 25% of human calorific intake comes from plant fatty acids. Plant fatty acids are also a key component of soaps, detergents, lubricants, biofuels, cosmetics, and paints. Up to 20% of plant oil is used for nonfood applications. Vegetable oils are seen as a possible means to meet renewable alternatives to petroleum-derived chemical feedstocks. To be economically viable in terms of industrial applications, plant oils must be enriched in a particular fatty acid with either double bonds or a functional group in a particular position. Also some industrial applications need oils with high oxidative stability (e.g. lubricants) or that will be readily oxidized (e.g. drying oils). These properties are usually absent from the major vegetable oils. However, the typical mix of 16C and 18C fatty acids making up the common vegetable oils has chemical and physical properties of only limited value for most industrial uses. For example, soybean is composed of a complex mixture of five fatty acids with only the high oleic soyoil being used industrially as a lubricant for printing inks.

Box 9.3 Improvement of oil quality by plant breeders

In plants belonging to the Brassicaceae family, such as rapeseed, there tends to be a high proportion of very long chain (20–24C) monounsaturated fatty acids. These are synthesized by chain elongation of 18:1. Studies on laboratory animals have indicated that erucic acid (22:1), which makes up over 50% of the fatty acids present in rapeseed oil, leads to an elevated incidence of heart disease. Although this problem makes rapeseed oil of little commercial value as an edible oil, erucic acid is particularly useful for industrial applications. Work on the genetic analysis of rapeseed plants identified that erucic acid (22:1) was formed by the sequential elongation of oleic acid via eicosenoic acid. Subsequently, experiments monitoring acetate incorporation patterns in immature rapeseed pods confirmed that the elongation pathway forming erucic acid was being controlled by only a few genes. A selective breeding approach identifying naturally occurring mutant varieties of rapeseed with a virtually abolished fatty elongation pathway resulted in the high value crop renamed Canola, where the plants contained instead the high-quality oleic edible oil. Canola is now the major Canadian Prairie crop and is also used extensively in the USA and worldwide with an annual value of more than $6 billion. It was later demonstrated that there was a single point mutation in two genes encoding isoforms of β-ketoacyl CoA synthase that is responsible for the generation of this low erucic acid phenotype of Canola.

The biosynthesis of TAG was characterized in the 1960s as the Kennedy pathway. This involves the sequential transfer of fatty acids to position 1 and 2 of glycerol 3-phosphate forming PA (Figure 9.18). Acyl CoA-dependent and acyl CoA-independent steps involve microsomal enzymes and are known to produce TAG; however, the extent that the different stages of reactions occur in different oilseeds at different stages of development changes. Glycerol 3-phosphate and acyl CoA is first converted to lysophosphatidic acid (LPA) by glycerol 3-phosphate acyltransferase. Studies from a range of oil-forming tissues indicate that microsomal glycerol 3-phosphate acyltransferases utilize a range of acyl CoA species, although many oilseed species have enhanced specificity and selectivity for 16:0 CoA. In many cases it is for this reason that the sn-1 position of TAG tends to be occupied by saturated fatty acids. Acyl-ACP is the natural substrate for plastidial glycerol 3-phosphate

Box 9.4 Biosynthesis of liquid wax esters

The only angiosperm that is known to produce liquid wax esters of 40–42 carbon atoms as an energy source is the jojoba desert plant (*Simmondsia chinensis*). Triacylglycerols are totally absent even as an intermediate. In the developing jojoba seed plastids, and in a manner equivalent to that seen in oilseed crops, oleic acid is produced. Oleoyl CoA is then converted into liquid wax through the enzymes fatty acid elongase, acyl CoA reductase, and wax synthase. In the first step of the microsomal elongation pathway malonyl CoA is initially condensed with long chain acyl CoA through the action of the enzyme β-ketoacyl CoA synthase. This initially leads to the formation of eicosenoic acid (20:1), 22:1 and nervonic acid (24:1). Acyl CoA reductase then catalyzes the formation of the equivalent CoAs before the final step where linear esters and fatty alcohols are produced through the action of wax synthase.

Commercially the jojoba shrub oil is obtained from the fruit which is a nut. About 50% of the nut is oil and approximately 2000 tonnes of jojoba oil are used per annum. Jojoba oil is primarily used in the cosmetics industry, although it also has limited additional applications as a lubricant, currently representing a market of 100 tonnes per annum. It is the high price of this oil when compared with other lubricant oils, which has been a limiting factor to its more widespread application. A reason for interest in jojoba oil as a commercial lubricant is that unlike many other lubricants it has a very tough molecular structure, making it stable even at high temperatures and pressures. Cloning of the cDNAs encoding β-ketoacyl CoA synthase, acyl CoA reductase, and wax synthase have been accomplished. Jojoba acyl CoA reductase and β-ketoacyl CoA synthase have been co-expressed in arabidopsis, resulting in seed oils with wax levels up to 70% during seed development, while expression of jojoba acyl CoA reductase in the high eruic acid form of *Brassica napus* led to the synthesis of small amounts of wax ester.

acyltransferase, although the enzyme can also accept acyl CoAs. Purified plastidial glycerol 3-phosphate acyltransferase was selective for 18:1 moieties over 16:0 moieties. LPA acyltransferase (LPAAT) then catalyzes a second acyl CoA dependent acylation to form PA. In general, low activities of the microsomal LPAAT are observed with acyl CoAs consisting of saturated fatty acids. In the seed of high-erucic acid *Brassica napus* (canola/oil seed rape) microsomal LPAAT does not effectively utilize erucoyl (22:1) CoA or lauroyl (12:0) CoA *in vitro*. In selected high-erucic acid *B. oleracea* over 23% of 22:1 is at the *sn*-2 position of TAG. In contrast microsomal LPAAT from palm (*Butia capitata* and *Syagrus cocides*) effectively utilizes 12:0 CoA, while the enzyme from California Bay (*Limnanthes alba*) can utilize 22:1 CoA. LPAAT specificity is also affected by the molecular species of the acyl acceptor LPA; i.e. microsomal LPAAT from *Zea mays* and *B. napus* have both been shown to have a preference for LPA containing 18:1 acyl chains compared with 12:0 acyl chains.

Acyl-ACP:LPAAT is associated with the plastid envelope and shows an increased preference for 16:0 ACP while discriminating against 18:1 ACP. Such selectivity of the plastidial LPAAT is consistent with the selectivity of the prokaryotic pathway incorporating 16:0 as the major species of fatty acid at the *sn*-2 position of plastidial glycerolipids. Site-directed mutagenesis studies have identified two conserved motifs responsible for enzyme activity.

Dephosphorylation of the PA, catalyzed by PA phosphatase, produces DAG and then a final fatty acyl group is transferred to the empty position 3 by diacylglycerol acyltransferase (DGAT) to produce TAG. The DAG can also be used for the biosynthesis of phosphatidylcholine in a reaction catalyzed by CDP-choline:1,2-diacylglycerol cholinephosphotransferase (Figure 9.18). This latter reaction is reversible and can potentially regenerate DAG for use by DGAT for TAG biosynthesis. Studies of PA phosphatase from a variety of sources have shown that this enzyme is a fairly promiscuous enzyme accepting a wide range of molecular species of PA, meaning that it probably has little

Figure 9.18 Simplified scheme for the pathway leading to the biosynthesis of fatty acids in oil seeds and their assembly into triacylglycerols.

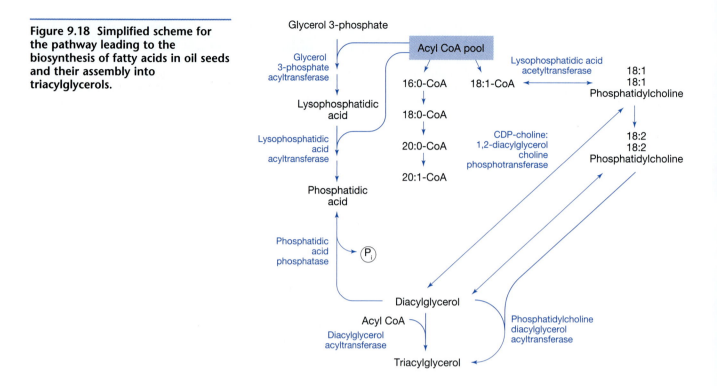

effect on the fatty acid composition of TAG. To date there are mixed results in assessing the importance of DGAT on the flow of carbon into seed oil.

Recently, alternative reactions for TAG biosynthesis have been identified involving the enzyme phospholipid:diacylglycerol acyltransferase. This enzyme moves fatty acids directly from phospholipids into the TAG pool in non-acyl CoA-dependent synthesis of TAG (Figure 9.18). Based on evidence to date it would appear that this pathway is involved in the accumulation of particular groups of fatty acids in TAG, for example, ricinoleic acid and vernolic acid in castor bean.

Small quantities of TAG are found in leaves and in certain fruit such as avocados and olives. However, the presence of TAG in seeds has been the focus of much research. Anywhere from 1% to 60% of the total dry weight of the seed can be oil. In the mature seed TAGs are stored in oil bodies. In contrast to the conserved fatty acid composition of membrane glycolipids, there can be over 200 different fatty acids in seed triglcerols. Although the reason for such diversity is unclear, many of the properties of these fatty acids are important in commercial terms (Table 9.3, see also Further Reading). The availability of a wide range of naturally diverse fatty acids has also been of interest to researchers and industry interested in exploiting their potential commercial applications (Box 9.5).

Release of fatty acids from acyl lipids

After germination the TAG reserves in oil-storing seeds begin to be broken down by the action of lipases. Lipase activity rapidly increases after imbibition of seeds and catalyzes the hydrolysis of fatty acids from the glycerol backbone. The hydrophobicity and insolubility in water of TAG means it segregates into lipid droplets and does not impact on the osmolarity of the cytosol. Being relatively inert means TAGs can be stored in high quantities with little risk of undesired chemical reactions with other cellular components. Because lipids

Peroxisomes are the site of fatty acid oxidation in leaf tissue, while gly-oxysomes are the site in germinating seeds. β-Oxidation provides biosyn-thetic precursors from stored lipids (Figure 9.19). Within the peroxisome the catabolism of TAGs by β-oxidation to acetyl CoA and the subsequent conver-sion of acetyl CoA to succinate via the glyoxylate cycle (Figure 9.20) provide germinating seeds with carbon skeletons and energy before they develop photosynthetic capacity. Acetyl CoA is converted via the glyoxylate cycle to 4-carbon precursors for gluconeogenesis.

The core pathway of peroxisomal β-oxidation involves the repeated cleavage of acetate units from the thiol end of the fatty acyl CoA (Figure 9.19). The major end products are H_2O_2, NADH, and acetyl CoA. The continued opera-tion of the cycle is dependent on removing the toxic H_2O_2 and regenerating NAD and CoASH. Glyoxysomes and peroxisomes both contain high activities

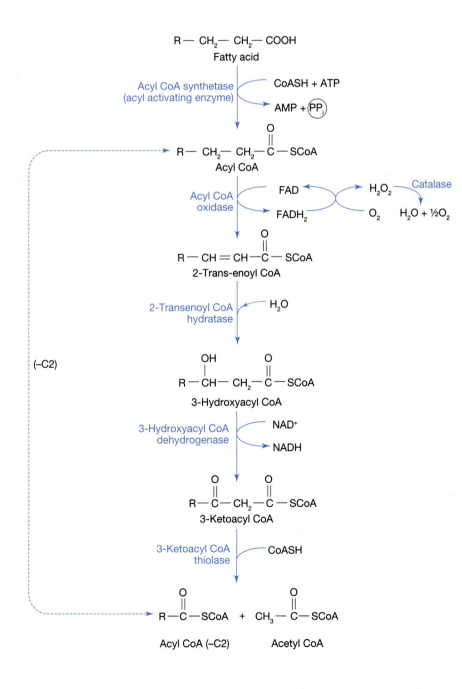

Figure 9.19 The β-oxidation pathway.

of catalase. Catalase converts any H_2O_2 produced by β-oxidation to water and oxygen. α-Oxidation of fatty acids to produce, for example, 2-OH 16:0 has been documented in a limited range of plant tissues and can be a component of plant cerebrosides (which are abundant in extraplastidic membranes such as plasma membranes and tonoplasts).

Recently, the coordinated expression of key genes and their corresponding enzyme activities involved in storage lipid mobilization during germination in arabidopsis have been demonstrated; i.e. acyl CoA oxidase and 3-ketoacyl CoA thiolase of β-oxidation, isocitrate lyase (catalyzing the formation of glyoxylate and succinate from isocitrate), malate synthase (catalyzes the formation of malate from glyoxylate and acetyl CoA) of the glyoxylate cycle, and

Figure 9.20 The mobilization of storage lipids for the synthesis of hexoses during germination.

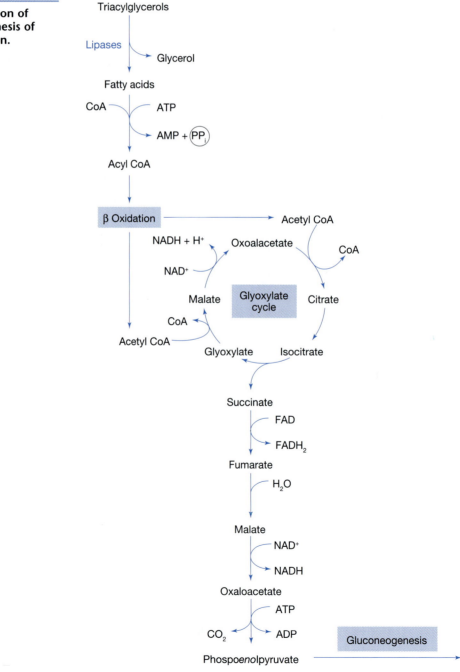

phospho*enol*pyruvate carboxykinase (*PCK*, catalyzes the conversion of oxaloacetate to phospho*enol*pyruvate in gluconeogenesis) of gluconeogenesis, are all induced in a similar manner. The latter activity has been demonstrated to be essential for seedling establishment in arabidopsis. Antisense *PCK1* arabidopsis plants have reduced ability to use both storage lipid and storage protein through gluconeogenesis to produce soluble carbohydrate.

Summary

Lipids play an essential part in the integrity of cells and organelles by providing a hydrophobic barrier of membranes. Lipids are also a major form of storage of chemical energy, particularly in seeds. As most lipids contain fatty acids esterified to glycerol, their synthesis is dependent upon the biosynthesis of fatty acids and the subsequent synthesis of lipids after esterification of fatty acids to produce PA. The *de novo* biosynthesis of fatty acids occurs in the plastids and involves the initial carboxylation of acetyl CoA to malonyl CoA followed by a series of reactions in which two carbons are added at a time to a growing carbon chain. Fatty acid biosynthesis is terminated by a number of different reactions. Five or six fatty acids are commonly found in all higher plants and these are 16C or 18C long. However, over 200 unusual fatty acids have also been identified, which are almost all located in seed oils. It is not clear why seed oils should have such diverse fatty acid constituents, although a function of defense has been proposed. Such variation may reflect random variation during the evolutionary process in plants. These novel fatty acids may be of little selective advantage for different plants, but by being sequestered in the oil bodies, which have no structural function, they do not cause any problems for the plants.

The biosynthesis of membrane glycerolipids occurs in either the plastids or the ER via the prokaryotic and eukaryotic pathways, respectively. In the case of the former, the glycerolipid components of the plastid membrane are formed through the sequential acylation of glycerol 3-phosphate using 18:1-ACP and 16:0-ACP. The main structural lipids of all other membranes are synthesized through the eukaryotic pathway. 16:0 and 18:1 fatty acids are first exported from the plastid to the ER as acyl CoAs, where they are then incorporated into various phospholipids.

Most storage lipids or TAGs accumulate in oil bodies. The mobilization and catabolism of TAGs, via the β-oxidation pathway, provides the energy and carbon needed for germination and pollination.

Further Reading

Lipids

Gurr MI, Harwood JL & Frayn KN (2002) Lipid Biochemistry (5th ed). Blackwell Science Ltd.

Murphy D (2004) Plant Lipids—Biology, Utilisation and Manipulation. Blackwell Science Ltd.

General textbooks on lipid biochemistry

Germana JB, Gilliesa LA, Smilowitza JT, Zivkovica AM & Watkins SM (2007) Lipidomics and lipid profiling in metabolomics. *Curr. Opin. Lipidol.* 18, 66–71.

Fatty acid synthesis

Ohlrogge JB & Jaworski JG (1997) Regulation of fatty acid synthesis. *Annu. Rev. Pl. Physiol. Pl. Mol. Biol.* 48, 109–36.

Provides a good overview of fatty acid synthesis in higher plants.

Bao X, Focke M, Pollard M & Ohlrogge J (2000) Understanding *in vivo* carbon precursor supply for fatty acid synthesis in leaf tissue. *Pl. J.* 22, 39–50.

Considers the relationship in the leaves between the rate of fatty acid synthesis and the total pool of acetate.

Triacylglycerols

Voelker T & Kinney AJ (2001) Variations in the biosynthesis of seed storage lipids. *Annu. Rev. Pl. Physiol. Pl. Mol. Biol.* 52, 335–361.

Provides an overview of triacylglycerol synthesis and the diversity of TAGs found in plants.

β oxidation

Baker A, Graham IA, Holdsworth M, Smith SM & Theodoulou FL (2006) Chewing the fat: β-oxidation in signalling and development. *Trends Pl. Sci.* 11, 124–132.

Discusses the information generated as a result of available mutants.

Industrial applications of fatty acids

Miller AA, Smith MA & Kunst L (2000) All fatty acids are not equal: discrimination in plant membrane lipids. *Trends Pl. Sci.* 5, 95–101.

Review discusses the more unusual fatty acids often accumulating in seed oils and why these are usually excluded from the membrane lipids of plants.

Thelen JJ & Ohlrogge JB (2002) Metabolic engineering of fatty acid biosynthesis in plants. *Metab. Eng.* 4, 12–21.

Review presents examples of some successful approaches to modifying plant fatty acid biosynthesis.

Jaworski J & Cahoon EB (2003) Industrial oils from transgenic plants. *Curr. Opin. Pl. Biol.* 6, 178–184.

Consider the difficulties that need to be overcome if plants are to be generated that overproduce the more unusual fatty acids, which often have a wide range of potential industrial applications.

Domergue F, Abbadi A & Heinz E (2006) Relief for fish stocks: oceanic fatty acids in transgenic oilseeds. *Trends Pl. Sci.* 10, 112–116.

Review discusses the recent advances in the production of very long chain polyunsaturated fatty acids in transgenic plants as a means to providing sustainable fatty acid sources for human and animal food.

Shanklin J & Cahoon EB (1998) Desaturation and related modifications of fatty acids. *Annu. Rev. Pl. Biol.* 49, 611–641.

Review summarizes understanding of enzymes and processes involved in desaturation of fatty acids.

Voelker T & Kinney AJ (2001) Variations in the biosynthesis of seed storage lipids. *Annu. Rev. Pl. Biol.* 52, 335–361.

Oxylipins

Feussner I & Wasternack C (2002) The lipoxygenase pathway. *Annu. Rev. Pl. Biol.* 53, 275–297.

Roles and regulation of lipoxygenases reviewed together with the involvement of the lipoxygenase pathway in signaling.

Howe GA & Schilmiller AL (2002) Oxylipin metabolism in response to stress. *Curr. Opin. Pl. Biol.* 5, 230–236.

Van Leeuwen W, Okresz L, Bogre L & Munnik T (2004) Learning the lipid language of plant signaling. *Trends Pl. Sci.* 9, 378–384.

Discusses structural and functional diversity of oxylipins and their potential role in response to stress.

Creelman RA & Mullet JE (1997) Biosynthesis and action of jasmonates in plants. *Annu. Rev. Pl. Physiol. Pl. Mol. Biol.* 48, 355–381.

Alkaloids

10

Key concepts

- Alkaloids are a diverse group of chemicals that are mostly synthesized from amino acid precursors.

- Alkaloids are very effective poisons and some animals can sequester them for their own defense.

- Many alkaloids have medicinal properties.

- The alkaloid biosynthesis pathways are relatively complex, with a number of branch points and a high degree of tissue and subcellular compartmentation.

- Genetic engineering approaches have provided new insights into the complexity and regulation of the alkaloid biosynthetic pathways.

- The diversity of alkaloid compounds is thought to have evolved, in part, through gene duplication and mutation.

Plants produce a vast array of chemicals that deter or attract other organisms

Plants interact with other living organisms in a number of ways, most obviously by providing food to herbivores and by acting as hosts to bacterial and fungal pathogens. Chemical feeding deterrents and poisons serve to reduce these negative interactions, which would otherwise result in damage to the plant. However, as plants are immobile, most angiosperms need to attract animals for successful reproduction and seed dispersal. The fine balance between deterring herbivores and pathogens, and attracting pollinators is maintained by the production of a vast range of specialized chemicals, traditionally referred to as secondary metabolites. Because herbivores or pathogens evolve resistance to these chemicals, plants evolve the capacity to produce new ones, in a continual arms race that adds to the chemical diversity and complexity of biosynthesis of these plant products. Over 100,000 different secondary metabolites have been identified in plants, with nearly half of them in the three major groups: the alkaloids (12,000), phenolics (10,000), and terpenoids (25,000). There are undoubtedly many more yet to be discovered, with the prospect of finding novel pharmaceuticals providing a commercial incentive in the search; a quarter of all prescribed drugs in industrialized countries contain compounds derived from plants, either directly or after further synthetic modification. In developing countries, 80% of the population relies on plant extracts for the treatment of illnesses.

In this and the next two chapters we will describe the pathways leading to the production of the major plant products, the alkaloids (this chapter), phenolics (Chapter 11), and terpenoids (Chapter 12). As we shall see, all of these groups contain poisons, while the terpenoids and phenolics also include compounds that provide colors and scents, both of which are important in attracting pollinators and repelling unwanted visitors.

Alkaloids, a chemically diverse group that all contain nitrogen along with a number of carbon rings

Almost all of the alkaloids are alkaline in solution. Their name originates from the Arabic *al-qali* (meaning ashes; the first alkaline salts were isolated from the ashes of plant material).

Alkaloids are defined as organic molecules consisting of several ring structures containing one or more nitrogen atoms, usually, but not always, located within a carbon ring. It is the structure of this carbon ring that is usually used to classify them, with the major alkaloid groups being the protopine, piperidine, pyrrolidine, pyridine, quinoline, quinolizidine, quinazoline, imidazole, steroidal, tropane, pyrrolizidine, purine, isoquinoline, and terpenoid indole alkaloids.

Functions of alkaloids in plants and animals

Alkaloids serve a range of functions in plants: as poisons, feeding deterrents, antimicrobial defenses, and germination inhibitors. As poisons they can be very effective, being particularly toxic to vertebrates, and therefore serve as defensive chemicals and are widespread within the plant kingdom, being found in about 20% of all plant species. Many alkaloids have antimicrobial properties and protect the plant from bacterial and fungal infection. Some alkaloids function as germination inhibitors and reduce competition from other plants in a process termed allelopathy, whereby a chemical released from a plant alters the growth or development of another plant.

Herbivores may learn to avoid alkaloid-containing plants because alkaloids tend to be bitter tasting, so they serve as feeding deterrents that repel the herbivore before it acquires a toxic dose.

Some animals also use alkaloids for protection against predators, although there is very little evidence for alkaloid biosynthesis within the animal kingdom. Instead, those animals that accumulate alkaloids appear to acquire them in their diet, either directly from plants or indirectly from other herbivores. This means of defense is more commonly used by invertebrates, particularly within the Lepidoptera, and is rarely used by vertebrates (a notable example being the tree frog, Plate 10.1 and Box 10.1 Alkaloids in animals).

The challenges and complexity of alkaloid biosynthetic pathways

The complexity of structures, the diversity of intermediates and the frequency with which branch points occur together present quite a challenge to anyone attempting to analyze the alkaloid biosynthesis pathways. It can take a very long time for a pathway to be unraveled. A good example of this is morphine biosynthesis. Morphine was first extracted from seeds of opium poppy (*Papaver somniferum*) in 1806, but it took until 1952 for its chemical structure to be determined, and another 50 years to identify fully the enzymes and reactions involved in its biosynthesis. Recent progress has, of course, been faster than this, benefiting from new analytical techniques for identifying intermediates, as well as from genetic approaches that will be described during the course of this chapter. Nevertheless, there is still much to learn about alkaloid biosynthesis, and many reactions and pathways remain to be characterized.

Box 10.1 Alkaloids in animals

A number of animals obtain alkaloids from their diet and store them in their body to act as feeding deterrents against potential predators.

The poison dart frogs (see Plate 10.1; *Dendrobates*, *Phyllobates* species) contain batrachotoxins in their skin. The batrachotoxins are steroidal alkaloids that act as very powerful neurotoxins, causing swift paralysis, either through ingestion or skin absorption. South American Indians coat arrowheads with the skin secretions from these frogs, finding the batrachotoxins to be even more potent than curare, which is extracted from plants of the genera *Strychnos* and *Curarea*. The poison dart frogs derive the toxins from their diet of ants and other alkaloid-containing arthropods, which ingest them from plants.

The red imported fire ant (*Solenopsis invicta*) is native to South America and they were accidentally imported from Brazil into the USA in the 1930s and into Australia and parts of Europe in the last few years. Currently, over 150 million hectares of land in the southeastern states of the USA are infested with these ants. *S. invicta* carries a potent sting that is extremely painful. It is rarely fatal to humans but it can kill smaller animals, such as birds and rodents. The venom consists mostly (>95%) of piperidine alkaloids, called solenopsins (Figure 1) that cause inflammation and swelling to the skin. Multiple stings can be serious as the piperidine alkaloids are thought to interfere with nitric oxide synthesis in humans, which can result in neurological problems.

As many as 40 species of moth are known to accumulate pyrrolizidine alkaloids from their diet and sequester them in their bodies for defense. Perhaps the most famous example is of the Monarch butterfly (*Danaus plexippus*), which migrates in huge numbers each year from winter refuges in central Mexico to breed at sites within the eastern USA and southern Canada. The Monarch butterfly lays its eggs exclusively on milkweed plants (*Asclepia* spp.), which contain pyrrolizidine alkaloids (see Figure 10.1 for structures and Figure 10.10 for biosynthesis). The eggs hatch and the larvae accumulate pyrrolizidine alkaloids as they feed on the milkweed leaves. The alkaloids remain in the moth right through to the adult stage, where their bitter taste makes them unpalatable to predatory birds. While the bitter taste serves as an effective feeding deterrent, the alkaloids will also cause birds to vomit so that they learn to avoid feeding on these brightly colored moths.

Other examples of insects that accumulate pyrrolizidine alkaloids include the tiger moth (*Arctia caja*) and cinnabar moth (*Tyria jacobaeae*). Both species feed exclusively on the pyrrolizidine alkaloid-containing *Senecio* species, such as *Senecio vulgaris* and *Senecio jacobeae* (see Plate 10.2). The senecio alkaloids appear to defend the moths throughout all stages of their life cycle, as they are accumulated in the eggs, caterpillars, and adult moths.

The moth, *Cosmosoma myrodora* (Arctiidae; Plate 10.3) uses pyrrolizidine alkaloids in a seemingly unique courtship ritual. The males acquire pyrrolizidine alkaloids by feeding on fluid exudates from plants such as *Eupatorium capillifolium* and store these throughout their bodies as protection against spiders. When they mate, the males discharge alkaloids from abdominal pouches, covering the female with the protective alkaloids. In addition to this, the males can excrete alkaloids in their seminal fluid so that the fertilized eggs are also protected against predators.

The ability of certain insects to sequester pyrrolizidine alkaloids without suffering any ill effects is remarkable, given that these alkaloids are highly toxic to mammals and birds (Box 10.2). The explanation lies in the fact that it is the breakdown products, rather than the pyrrolizidine alkaloids themselves, that cause harm. While mammals and birds are poisoned by the highly reactive pyrroles that are formed in their livers as the pyrrolizidine alkaloids are metabolized, the moths remain tolerant because they are unable to metabolize pyrrolizidine alkaloids.

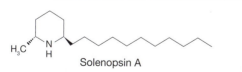

Solenopsin A

Figure 1 Solenopsin A, a piperidine alkaloid, is the main alkaloid present in the venom of red imported fire ants (*S. invicta*).

With the promise of identifying new pharmaceuticals and of producing novel crops and products, such as naturally produced decaffeinated coffee (see Box 10.3), the elucidation of the alkaloid biosynthesis pathways is an exciting and very active field of research.

Amino acids as precursors in the biosynthesis of alkaloids

Common to almost all alkaloids, with some important exceptions (e.g. purine alkaloids; see page 355), is that they are synthesized from one of a small number of L-amino acids—notably arginine, lysine, phenylalanine, tryptophan, and tyrosine acting as primary precursors. In many cases, another common feature of alkaloid biosynthesis is the generation of a central intermediate that serves as a precursor for all of the alkaloids produced within a particular pathway.

Figure 10.1 shows the five classes of alkaloids that will be described in this chapter, along with the amino acid and central intermediate from which they are made. We have chosen to focus on these five alkaloid groups—terpenoid indole, isoquinoline, tropane, pyrrolizidine, and purine alkaloids—of the 14 listed on page 336, because they represent the most abundant alkaloids produced by plants. These five groups also serve to illustrate the biochemical and taxonomic diversity of the alkaloid family. The biochemical pathways, as far as they are known, are presented in the accompanying figures with detailed legends to enable you to follow each reaction. In the text, we shall highlight the important steps and discuss current knowledge about their regulation. We shall also mention some of the other classes of alkaloids where relevant. Finally, we will identify the key biotransformations required to produce alkaloids and briefly consider the evolutionary origins of these fascinating plant products.

Terpenoid indole alkaloids are made from tryptamine and the terpenoid secologanin

There are about 3000 different types of terpenoid indole alkaloids. Within this group there are a number of effective mammalian poisons, some of which have medicinal properties if taken in appropriate doses (as summarized in Box 10.2).

Common to all terpenoid indole alkaloids is a terpenoid group, derived from isopentenyl diphosphate (produced via the deoxyxylulose phosphate pathway; see Chapter 12) via geraniol and secologanin, and an indole group (see structures in Figure 10.1) made from tryptamine. Strictosidine is the central intermediate from which all of these terpenoid indole alkaloids are made (Figure 10.2). Strictosidine synthesis marks the junction between two biochemical branches, the terpenoid branch from pyruvate and the indole branch from chorismate (via tryptamine).

Given the enormous diversity of terpenoid indole alkaloids produced by plants, there is still much to learn about how they are made. Because of their commercial value, we know most about those with pharmaceutical properties

Figure 10.1 Representatives of the alkaloid groups described in this chapter, their amino acid precursors and the central intermediate in their biosyntheses. The alkaloid groups are defined by the structure of their heterocyclic rings—these are highlighted in blue.

Alkaloid group	Amino acid precursor	Central intermediate(s)

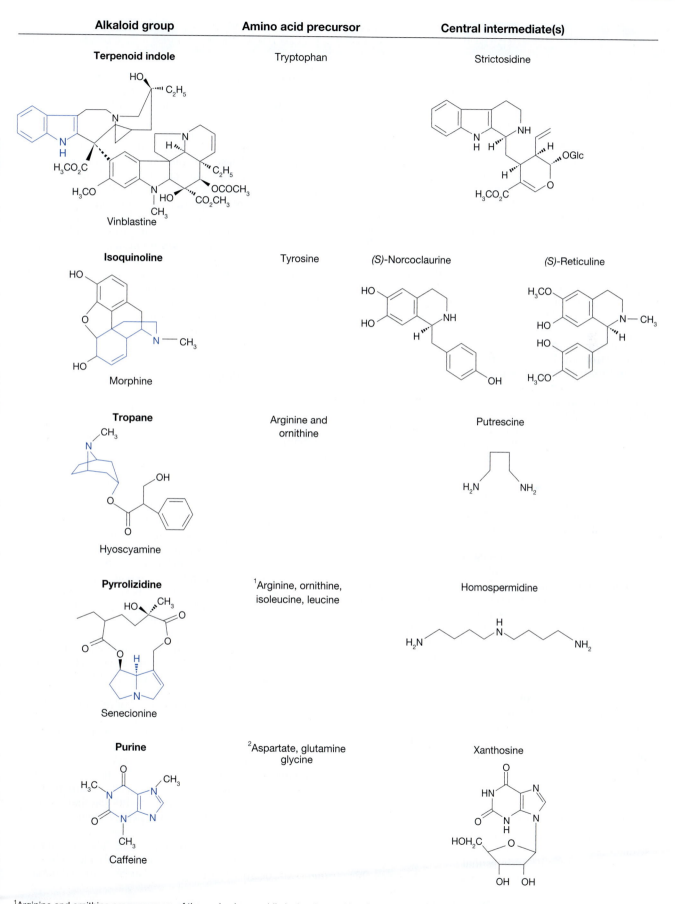

Terpenoid indole — Tryptophan — Strictosidine

Vinblastine

Isoquinoline — Tyrosine — (S)-Norcoclaurine — (S)-Reticuline

Morphine

Tropane — Arginine and ornithine — Putrescine

Hyoscyamine

Pyrrolizidine — [1]Arginine, ornithine, isoleucine, leucine — Homospermidine

Senecionine

Purine — [2]Aspartate, glutamine glycine — Xanthosine

Caffeine

[1]Arginine and ornithine are precursors of the necine base, while isoleucine and leucine are precursors of the necic acid.
[2]Xanthosine is formed from adenosine precursors, which are synthesized in turn from aspartate, glutamine, and glycine.

Box 10.2 Toxicity and medicinal properties of alkaloids

Alkaloids are nearly always poisonous but in appropriate doses many of them have valuable medicinal properties.

Terpenoid indole alkaloids

Strychnine (from *Strychnos nux-vomica*; see Figure 10.2 for structure and biosynthesis) interferes with neurotransmission in the spinal cord and medulla, causing unchecked stimulation of motor neurons. This results in muscle spasms, and death occurs through lack of oxygen as a consequence of respiratory failure and exhaustion. Strychnine is used as a rat poison, although it is now banned in some countries because of the risk of accidental poisoning. It is therefore surprising to find it in current use as a homeopathic medicine. Nux vomica is prepared from the seeds of *Strychnos nux-vomica* and is still marketed as a natural (and potentially very dangerous!) homeopathic remedy for everything from colds to oversensitivity and irritability.

Quinine (see Figure 10.2 for structure and biosynthesis), obtained from the bark of *Cinchona officinalis*, is an effective antimalarial drug. The plant is named after Ana de Osorio, Countess of Chinchon, wife of the Spanish Viceroy of Peru, who was purported to have been cured, in 1630, of a near-fatal attack of malaria by drinking a tea prepared from Cinchona bark. The story featured in a popular nineteenth century novel and the legend became established, with the Countess being credited with bringing the cure back to Europe. There is, however, no documentary evidence to support this romantic story and there is better evidence that it was a group of Jesuit priests who learned of the bark from indigenous Peruvian Indians and who sent samples back to Europe from 1620 onwards.

Quinine kills the malaria-causing parasite, *Plasmodium*, by preventing it from metabolizing the heme moiety of hemoglobin. The parasite dies from heme toxicity. Quinine also affects the mammalian heart, acting as a cardiac Na^+ channel inhibitor and causing a slowing of the heartbeat. It is therefore an effective treatment for irregular heartbeat but an overdose can lead to heart failure. Two other alkaloids in this group, ajmaline and ajmalicine (from *Rauvolfia serpentina*), have the same effect on the heart.

The terpenoid indole alkaloids, vincristine and vinblastine (extracted from *Catharanthus roseus*; see Figure 10.2 for structures) have anticancer properties, being particularly effective against childhood leukemia and Hodgkin's disease. These alkaloids are thought to interfere with mitosis by inhibiting mitotic spindle assembly, and so reducing tumor growth.

Isoquinoline alkaloids

Toxic effects of the isoquinolines (see Figure 10.4 for structures and 10.5 for the biosynthetic pathway) include disturbances of the gastrointestinal system and interference with the central nervous system. In humans this results in vomiting, drowsiness and loss of co-ordination, followed by seizures and coma. Poisoning of livestock is more common as the noxious taste tends to deter accidental poisoning in humans.

In appropriate doses, many chemicals in this group have medicinal uses. These include the morphinan alkaloid pain-killers morphine and codeine, from the opium poppy, *Papaver somniferum*; the antibacterial and antiplaque reagent, sanguinarine (a benzophenanthridine alkaloid), from *Eschscholzia californica*, which is sometimes added to toothpastes; the mitotic inhibitor, colchicine, from *Colchicum autumnale*, a gout treatment in humans and a cancer treatment in domestic pets; an antimicrobial, berberine, from *Coptis japonica*, a remedy for eye infections; the anti-amoebic, emetine, from *Uragoga ipecacuanha*, and the muscle relaxant (+)-tubocurarine, from *Chondodendron tomentosum*, used by South American Indians as a component of the arrow poison, curare.

Tropane and nicotine alkaloids

Toxicity of the tropane alkaloids (see Figure 10.7 for structures) is due mainly to their anticholinergic properties, i.e. they inhibit the transmission of parasympathetic nerve impulses by interfering with the binding of acetylcholine to its receptor in the nerve ending. Effects on humans include raised body temperature due to reduced sweating, decreased saliva production, relaxation of the bladder and uterus, irregular and/or rapid heartbeat, stomach cramps, and hallucinations. The effects can be lethal. At low doses some of the tropane alkaloids have medicinal uses. Extracts of *Atropa belladonna* (deadly nightshade) were used as eye drops to dilate the pupils, both for cosmetic (hence belladonna—beautiful lady) and medicinal purposes until it was replaced by the active principal atropine. Atropine sulfate is also injected as an antidote to organophosphorus poisoning resulting from exposure to insecticides or nerve gas.

Scopolamine, which has stronger anticholinergic properties than atropine, also acts to depress the central nervous system, producing a sedative effect that atropine does not. Scopolamine was used in the early 1900s as a means of inducing twilight sleep (a high level of relaxation) during childbirth. It was also in common use as a pre-anesthetic up until the 1960s because it reduced pre-operative anxiety and postoperative nausea, but it was prone to causing hallucinations and disorientation, particularly among the elderly. Scopolamine is still in use in the form of impregnated skin patches, both as a treatment for Ménière's disease (a disorder of the middle ear that causes vertigo and nausea) and as a travel sickness remedy.

The hallucinogenic effects of the tropane alkaloids have made them subjects of abuse, often with fatal consequences. Cocaine, from *Erythroxylon coca*, has valuable

Box 10.2 Toxicity and medicinal properties of alkaloids (continued)

effects as a topical anesthetic and central nervous system stimulant at controlled doses, and potentially toxic effects when abused. Several fatalities a year occur as a result of smoking leaves of *Datura stramonium* (Thornapple), which contains both scopolamine and hyoscyamine. Disreputable horse-dealers used to insert rolled-up leaves of this plant into the rectum of a horse before taking it to auction because it made the animal livelier. Mandrake (*Mandragora officinarum*) contains hyoscyamine, scopolamine, and atropine, and was once used as a herbal soporific. This plant was believed to have magical properties because its roots resemble the human form. During the Middle Ages it was feared that anyone uprooting Mandrake would go mad from the screams issuing from the plant, a story perpetuated in the Harry Potter novels of JK Rowling.

Nicotine has quite different toxic effects compared with the other alkaloids within this group. It acts on the vegetative ganglia, initially as a stimulant that causes raised blood pressure, and then as a blocker resulting in convulsions and respiratory paralysis. Other symptoms include nausea, vomiting, diarrhea, and mental confusion. Death can be rapid at high doses, especially as nicotine is absorbed readily through the skin. This property has led to fatalities among smugglers who have wrapped tobacco leaves around their bodies and, as a consequence, absorbed a lethal dose of nicotine. Sheep and cattle seem to be less sensitive to nicotine poisoning, being able to ingest as much as 50 times the lethal dose to humans without any ill effect. Nicotine can have teratogenic effects (i.e. harms the fetus) in both cattle and humans, causing defects in the palate and digits. Nicotine is also a very effective insecticide. Several ecological studies have

provided evidence of its action as an inducible feeding deterrent, responding systemically (i.e. throughout the whole plant) to herbivore damage. Commercial growers have exploited this knowledge by decapitating tobacco plants, as wounding results in an increase in the nicotine content of the remaining leaves.

Pyrrolizidine alkaloids

The pyrrolizidine alkaloids are only toxic when broken down to pyrroles. This reaction occurs in the liver, where the highly reactive pyrroles form cross-links with DNA, causing fibrous lesions that lead to death from liver failure. There can also be effects on the lungs, heart, and kidneys. While an acute intake can cause rapid death, the effects can also be cumulative, resulting in chronic liver damage over a period of time. Damage can also be carried over into the fetus, resulting in deformities of the palate and digits.

The mixture of pyrrolizidine alkaloids, such as senecionine (Figure 10.1 for structure and Figure 10.10 for biosynthesis), found in species of *Senecio* (Compositae), *Senecio vulgaris* (groundsel), and *Senecio jacobaea* (ragwort) is a common cause of poisoning and birth defects in cattle and horses (see Plate 10.4). Although the bitter taste of these plants is usually enough to deter grazing, this effect decreases once the plant is harvested, and contaminated hay used for animal feed is often the source of the problem. As these alkaloids are also highly toxic to humans, their presence in comfrey (*Symphytum officinale*, Boraginaceae) should be noted because this is still found in herbal remedies and natural foods in some countries.

for example, the anticancer drugs vincristine and vinblastine, produced from vindoline in *Catharanthus roseus*. The pathway for vindoline biosynthesis is shown in Figure 10.3. Here we will use the vindoline biosynthesis pathway to illustrate the key steps in the regulation of terpenoid indole alkaloid production.

The terpenoid indole alkaloid biosynthesis pathway is highly compartmentalized

Plants make strictosidine as the starting point for all terpenoid indole alkaloids (Figure 10.2). Its synthesis requires enzymes present in three different parts of the cell: (1) tryptophan and the terpenoid geraniol (see Chapter 12) are produced in the plastid; (2) tryptophan is converted to tryptamine in the cytoplasm; and (3) strictosidine is formed from tryptamine and secologanin in the vacuole. As well as this compartmentation within cells, the pathway for vindoline production requires at least three different types of cell. Genes encoding enzymes involved in the early stages, from tryptophan to strictosidine, are expressed in epidermal cells while genes coding for the last two enzymes in the pathway, i.e. deacetylvindoline-4-*O*-acetyltransferase and

Figure 10.2 Strictosidine as the central intermediate in terpenoid indole alkaloid biosynthesis. Abbreviations: STR, strictosidine synthase; TDC, tryptophan decarboxylase; IPP, isopentenyl diphosphate; Glc, glucosyl residue.

deacetylvindoline-4-hydroxylase, are expressed in specialized cells, such as laticifers and idioblasts of leaves, stems, and flower buds. This means that transport of biosynthetic intermediates across membranes, both within and between cells, is likely to contribute to the regulation of this pathway. The requirement for cell specialization also helps to explain why vinblastine and vincristine cannot be produced by undifferentiated cells of *C. roseus* grown in culture.

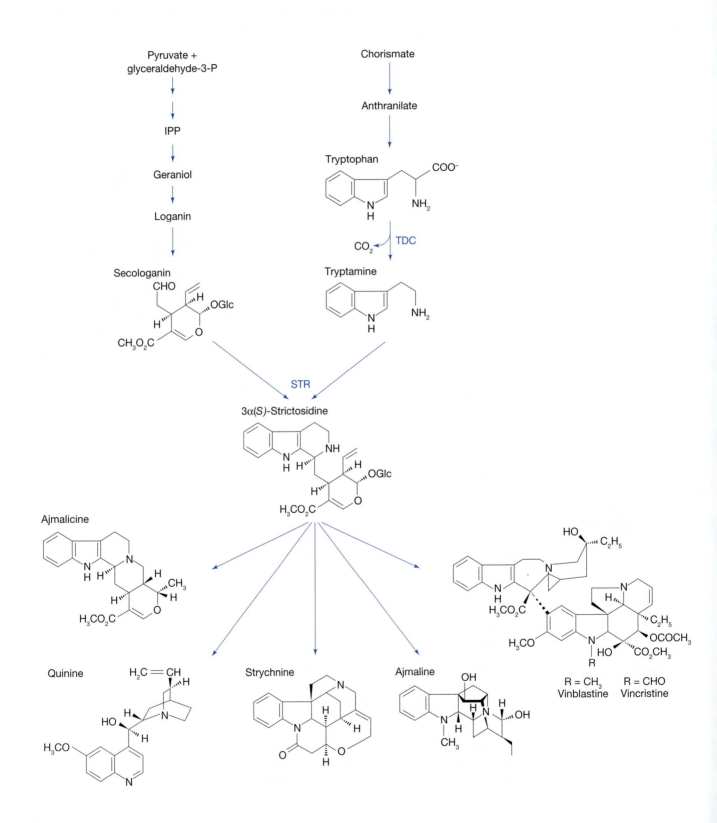

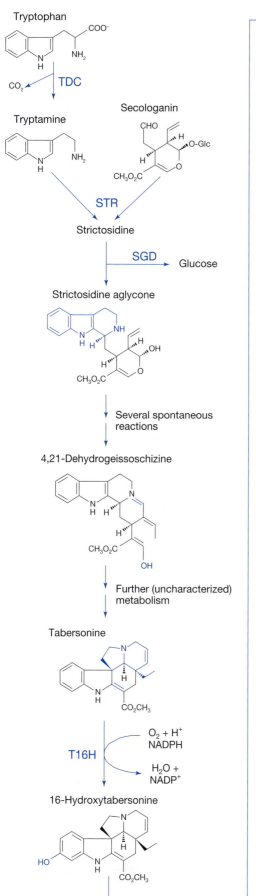

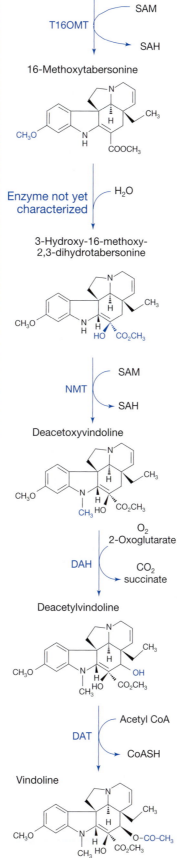

Figure 10.3 Vindoline biosynthesis, as an example of the terpenoid indole alkaloid biosynthesis pathway. The first reaction in this pathway, which results in the decarboxylation of an amino acid (in this case of tryptophan), is a common first step in many of the alkaloid biosynthetic pathways. Other frequently occurring reactions of alkaloid biosynthesis include: deglycosylation (where glucose is removed) as in the reaction catalyzed by SGD; hydroxylation via cytochrome P450 monooxygenases such as T16H; O-methylation (where a methyl group is added to an oxygen atom) as in the reaction catalyzed by T16OMT; N-methylation (where a methyl group is added to a nitrogen atom) as catalyzed by NMT, and an O-acetylation (where an acetyl group is added to an oxygen atom), as in the final step, which is catalyzed by DAT. Some of the reactions of this pathway still remain to be characterized. The alkaloids vincristine and vinblastine are produced from vindoline by dimerization with catheranthine (not shown).

Abbreviations: TDC, tryptophan decarboxylase; STR, strictosidine synthase; SGD, strictosidine β-D-glucosidase; T16H, tabersonine 16-hydroxylase; T16OMT, 16-hydroxytabersonine 16-O-methyltransferase; NMT, 16-methoxy 2,3-dihydro-3-hydroxytabersonine-N-methyltransferase; DAT, deacetylvindoline-4-O-acetyltransferase; D4H, deacetylvindoline-4-hydroxylase; SAM, S-adenosylmethionine; SAH, S-adenosylhomocysteine.

Environmental signals might increase terpenoid indole alkaloid production

There is some evidence that terpenoid indole alkaloid production increases in response to stress. Stress-responsive regions have been identified in the promoter of the tryptophan decarboxylase (*TDC*) gene of *Catharanthus roseus* and both excess light and fungal pathogens appear to increase its expression. The gene for strictosidine synthase (*STR*) is induced by jasmonate, a well-known signal of wounding and pathogen attack. It is too early to say how widespread these responses are, or whether they help the plant to tolerate stress or to resist pathogens. Attempts to mimic the response by overexpressing both *TDC* and *STR* have so far failed to increase the production of terpenoid indole alkaloids, despite an increase in accumulation of their respective intermediates. This experiment might indicate that later steps in the pathway exert more control over end-product formation, but it also certainly highlights large gaps in our understanding of how the pathway is regulated.

Isoquinoline alkaloids are produced from tyrosine and include many valuable drugs such as morphine and codeine

Isoquinoline alkaloids are produced from tyrosine. They are a diverse group of alkaloids with many subgroups, including the berberine and morphinan alkaloids (Figure 10.4). Isoquinoline alkaloids occur in a range of species, including *Papaver, Eschscholzia, Berberis, Coptis,* and *Sanguinaria* (poppy, California poppy, barberry, goldenthread, and bloodroot, respectively). Their toxic and medicinal effects are described in Box 10.2. There is no doubt that the various toxic properties of these alkaloids make them effective herbivore deterrents, but they might also benefit the plant in other ways. For example, there is evidence that some of these alkaloids provide protection against microbial and fungal pathogens: berberine has antimicrobial properties, and

Figure 10.4 Examples of isoquinoline alkaloids.

Morphinan alkaloids

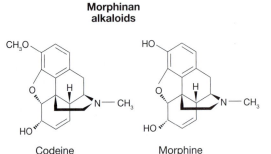

Codeine

Morphine

Benzophenanthridine alkaloids

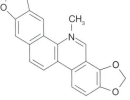

Sanguinarine

Berberine alkaloids

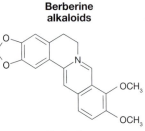

Berberine

Protopine alkaloids

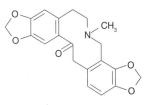

Protopine

a number of enzymes in the isoquinoline pathway (like those of the terpenoid indole alkaloids described on page 344) are inducible by fungal elicitors and by the wounding- and pathogen-induced signal, methyl jasmonate. For those plants that produce isoquinoline alkaloids, such as the opiates (e.g. morphine, codeine, and derivatives such as heroin), there is an interesting, and less direct, benefit. The addictive property of the opiates has resulted in the mass cultivation of species such as the opium poppy (*Papaver somniferum*). Hence, the effect of these particular alkaloids on human behavior has led to the proliferation of opiate-producing plants.

Norcoclaurine is the central intermediate in the pathway for isoquinoline alkaloid biosynthesis

The pathway for isoquinoline alkaloid biosynthesis is shown in detail in Figure 10.5. Here we highlight the key reactions in the pathway and explain what is known about its regulation.

Two molecules of the amino acid tyrosine are the starting point for isoquinoline alkaloid biosynthesis. A series of hydroxylations, deaminations, and decarboxylations converts one tyrosine to dopamine and the other to 4-hydroxyphenylacetaldehyde. Dopamine and 4-hydroxyphenylacetaldehyde are condensed together by norcoclaurine synthase to form (*S*)-norcoclaurine, the precursor to all isoquinoline alkaloids. Norcoclaurine is a very important plant chemical, as 50% of all known alkaloids are produced either from norcoclaurine or from strictosidine, the terpenoid indole alkaloid precursor (see Figure 10.2).

Another key intermediate in the isoquinoline pathway is (*S*)-reticuline, which is positioned at a branch point. One branch takes (*S*)-reticuline through to (*S*)-scoulerine and the formation of three subgroups, the berberine, protopine, and benzophenanthridine alkaloids. Another branch leads to the formation of morphinan alkaloids such as morphine and codeine. These routes are species-specific so it is not strictly correct to think of them as branches of the same pathway in a single organism. Berberine alkaloids, such as berberine, accumulate, for example, in *Berberis* and *Coptis* (Ranunculaceae) species, while the morphinans occur predominantly in *Papaver* and the benzophenanthridines, such as sanguinarine, are found in *Eschscholzia californica* and *Sanguinaria canadensis*. Crucial to this product specificity is the way in which (*S*)-reticuline is metabolized. It is converted to (*S*)-scoulerine by the berberine bridge enzyme, which uniquely catalyzes the direct linkage of two carbon atoms. This reaction has never been achieved by artificial chemical synthesis, nor have any other enzymes been discovered that have the capability to form these carbon–carbon linkages. The berberine bridge enzyme, and its reaction, is therefore of considerable interest to both chemists and biologists who are trying to identify the reaction mechanism with the aim of generating novel biochemical products.

Tissue culture and genetic manipulation studies provide insights into the regulation of the isoquinoline alkaloid pathway

The isoquinoline alkaloids, unlike many of the other alkaloid groups, can be produced by cell cultures. This feature has been of immense value for studying the biosynthetic pathway, as it is very much easier to supply substrates and inhibitors to a cell suspension than to an intact plant. Cell cultures also offer the promise of commercial-scale production of valuable alkaloid products. However, while sanguinarine (in *Papaver somniferum* cell cultures) and berberine (in *Coptis japonica* cell cultures) have been produced at relatively

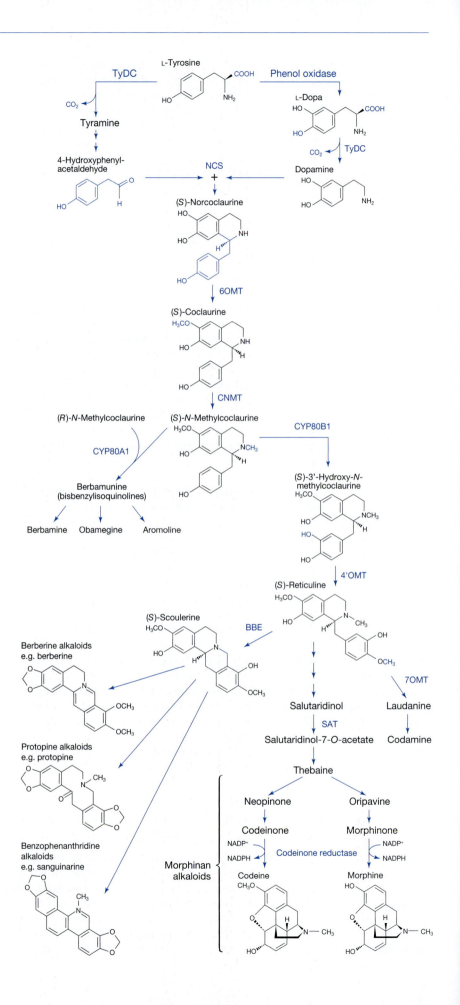

Figure 10.5 Isoquinoline alkaloid biosynthesis pathway.
Note the central intermediates (*S*)-norcoclaurine and
(*S*)-reticuline, which serve as precursors for synthesis of all of the
alkaloids in this group. Note also that the pathway is branched
in several places, resulting in the production of the various
subgroups, including the berberine, protopine,
benzophenanthridine alkaloids, the morphinans, and the
bisbenzylisoquinolines. These branches do not all occur in the
same species, it is more usual for only one subgroup of
isoquinoline alkaloids to be produced within a single plant
species. The amino acid tyrosine is the starting point for this
pathway. Two tyrosine molecules are used, one is
decarboxylated to the amine, tyramine, the other is converted to
dopamine via a phenol oxidase and a decarboxylation reaction.

The central intermediate, (*S*)-norcoclaurine is formed from
dopamine and 4-hydroxyphenylacetaldehyde (produced from
tyramine) in a reaction catalyzed by NCS.

Two methylation reactions (via 6OMT and CNMT) convert
(*S*)-norcoclaurine to (*S*)-*N*-methylcoclaurine, which has two
potential fates, depending on species. (*S*)-*N*-Methylcoclaurine
can react with its isomer (*R*)-*N*-methylcoclaurine to form
berbamunine. The reaction is catalyzed by the cytochrome
P450 hydroxylase berbamunine synthase (CYP80A1).
Berbamunine is the precursor for the bisbenzylisoquinoline
alkaloids such as berbamine, obamegine, and aromoline.
Alternatively, (*S*)-*N*-methylcoclaurine can be converted to
(*S*)-reticuline via a hydroxylation reaction (catalyzed by a
cytochrome P450 hydroxylase, CYP80B1) and a methylation
reaction (catalyzed by 4'OMT).

(*S*)-Reticuline serves as the precursor for a further range of
alkaloids. It can be used to form (*S*)-scoulerine (reaction

catalyzed by BBE), from which the berberine (e.g. berberine),
protopine (e.g. protopine), and benzophenanthridine alkaloids
(e.g. sanguinarine) may be formed via a series of oxidations,
hydroxylations, and methylations (see Figure 10.6).
Alternatively, (*S*)-reticuline can be used to form the morphinan
alkaloids, a class that includes the important drugs morphine
and codeine.

The morphinan alkaloids are produced in high concentrations
in the opium poppy, *Papaver somniferum*. (*S*)-Reticuline is
converted to salutaridinol by four successive enzyme steps,
involving an oxidase and three consecutive NADPH-dependent
reductases (not shown in detail). Salutaridinol
O-acetyltransferase (SAT) adds an acetyl group (from acetyl
CoA) to form salutaridinol 7-*O*-acetate, which spontaneously
degrades to thebaine, losing the acetate group in the process.
Thebaine sits at another branch point leading either to
codeine or to morphine production. The final step in the
synthesis of both of these alkaloids is an NADPH-dependent
reduction, catalyzed by codeinone reductase. Opium poppy
also accumulates laudanine and codamine, produced via a
side reaction of *S*-reticuline catalyzed by (*R,S*)-reticuline
7-*O*-methyltransferase (7OMT).

Abbreviations: TyDC, Tyrosine decarboxylase; 6OMT,
(*R,S*)-norcoclaurine-6-*O*-methyltransferase; CNMT,
(*S*)-coclaurine *N*-methyltransferase; CYP80A1, berbamunine
synthase (*N*-methylcoclaurine, NADPH:oxygen oxidoreductase
[carbon-oxygen phenol coupling]); CYP80B1,
(*S*)-*N*-methylcoclaurine-3'-hydroxylase; 4'OMT,
(*S*)-3'-hydroxy-*N*-methylcoclaurine-4'-*O*-methyltransferase;
7OMT, (*R,S*)-reticuline 7-*O*-methyltransferase; BBE, berberine
bridge enzyme; SAT, salutaridinol 7-*O*-acetyltransferase; NCS,
norcoclaurine synthase.

high concentrations in the laboratory, the yields are still not sufficient for
commercial exploitation.

The manipulation of isoquinoline alkaloid production by genetic engineering
has provided further information about pathway regulation. The branch
point at (S)-scoulerine is of particular interest because it offers scope to
change the type of alkaloids that a plant might make. For example,
Eschscholzia californica makes benzophenanthridine alkaloids, such as sanguinarine, but cannot make berberine alkaloids because it lacks the enzyme
(S)-scoulerine 9-*O*-methyltransferase (SMT). Introducing the *SMT* gene from
Coptis japonica into *Eschscholzia californica* cells resulted in a reduction in
sanguinarine production and an accumulation of columbamine, which is
formed from (S)-tetrahydrocolumbamine, the product of the SMT reaction
(Figure 10.6). Although no berberine alkaloids were produced, as the further
steps from columbamine to berberine were lacking, this experiment demonstrates that a single genetic modification can introduce a new branch point
into an existing alkaloid pathway.

A recent success story has been the genetic modification of the opium poppy,
Papaver somniferum, to produce plants that no longer make any of the morphinan alkaloids, such as morphine and codeine. This result was achieved by
using RNA interference (RNAi; see Chapter 2, Box 2.3) to silence the multigene family encoding codeinone reductase. The enzyme, codeinone reductase, forms codeine from codeinone and also produces morphine from morphinone (Figure 10.5). Hence, the loss of codeinone reductase activity might

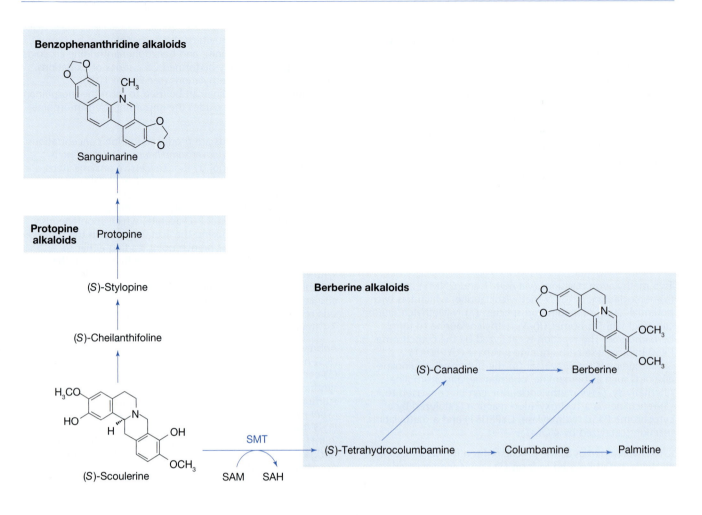

Figure 10.6 Reactions from (S)-scoulerine in the isoquinoline alkaloid pathway. (S)-Scoulerine is the primary product of the berberine bridge enzyme (as shown in Figure 10.5) and can be biotransformed to berberine, protopine or benzophenanthridine (e.g. sanguinarine) alkaloids. When 9-O-methylated by SMT, (S)-scoulerine is committed to berberine synthesis. SMT is absent from the Californian poppy (*Eschscholzia californica*) hence, while it can make protopine and benzophenanthridine alkaloids it cannot produce berberine alkaloids. Introduction of the *SMT* gene from *Coptis japonica* into transgenic *E. californica* cells resulted in columbamine production, and a decrease in sanguinarine formation. The plants failed to produce berberine as other steps from columbamine to berberine were lacking. Abbreviations: SMT, S-scoulerine 9-O-methyltransferase; SAM, S-adenosylmethionine; SAH, S-adenosylhomocysteine.

be expected to lead to an accumulation of codeinone and morphinone in the transgenic plants. The results were somewhat unexpected. The transgenic plants did not simply lose the capacity to produce morphine and codeine, they were incapable of producing any metabolites from (S)-reticuline onwards, i.e. seven enzymatic steps before the missing codeinone reductase reaction (Figure 10.5). This unexpected result indicates that either one or both of the products of the codeinone reductase reaction, codeine and morphine, might be regulating the flux of (S)-reticuline into the morphinone branch of the pathway; perhaps via a positive feedback regulation of one or several of the intervening enzyme steps. Consequently, (S)-reticuline would accumulate in the transgenic plants as the positive feedback can no longer function in the absence of morphine or codeine production. Although there is no direct evidence for such a feedback regulation, these results serve to highlight the complexity of regulation of such a highly branched pathway. The technical breakthrough that enabled the production of transgenic poppy will open up new approaches to understanding this regulation. It also offers a new use for a crop that has often been grown for illegal drug production. While the transgenic poppy cannot be used for heroin production, because it lacks the morphinan alkaloids that are the precursors for heroin synthesis, these plants may have legitimate commercial uses. For example, (S)-reticuline has been found to stimulate the growth of cultured hair cells of mice and it is currently being tested as a potential treatment for baldness. (S)-reticuline might also be of use as the precursor for the synthetic production of other pharmaceutical products. Consequently, there are a number of patents in place that relate to the production and use of these transgenic poppy plants.

Tropane alkaloids and nicotine are found mainly in the Solanaceae

Tropane alkaloids are esters of carboxylic acids and tropine. There are about 150 different types, found mainly in the Solanaceae, especially in species of *Atropa*, *Datura*, *Duboisia*, *Hyoscyamus*, and *Mandragora* (nightshades, jimsonweed, corkwood tree, henbane, and mandrake, respectively). The main tropane alkaloids are L-hyoscyamine and L-scopolamine. Atropine is also found in plant extracts. Atropine is an optically inactive form of L-hyoscyamine (i.e. a racemic mixture of DL-hyoscyamine) that can form through racemization during drying. Hence it might be an artifact of the extraction process rather than a natural component of the plant.

While nicotine is not a tropane alkaloid its biosynthesis requires precursors from the tropane alkaloid pathway; it is for this reason best considered in this section. Nicotine, like the tropane alkaloids, is found in a number of species within the Solanaceae (including potato, tomato, and eggplant—also known as aubergine (*Solanum melongena*)) but it only accumulates to appreciable concentrations in tobacco. Of the 50 or more species of tobacco, *Nictotiana tabacum* is the species that is usually cultivated for cigarette production.

The toxicity and medicinal properties of some of the tropane alkaloids and nicotine are described in Box 10.2.

The pathway for biosynthesis of the tropane alkaloids is highly branched and shares precursors with the pathways for polyamine, nicotine, and pyrrolizidine alkaloid biosynthesis

The pathway for tropane alkaloid biosynthesis is highly branched and its precursors are also used in nicotine biosynthesis, as well as in the formation of pyrrolizidine alkaloids (see page 354). Putrescine is not only an intermediate in the biosynthesis of all of these alkaloids, it is also a precursor in the synthesis of polyamines, such as spermidine and spermine (Figure 10.7). Polyamines are ubiquitous, evolutionarily ancient chemicals that occur in plants, animals, fungi, and microorganisms where they serve a range of fundamental functions, including the control of cell proliferation. Consequently, the capacity to produce putrescine was most likely present in a range of organisms before the alkaloid pathways evolved. Indeed, there is evidence that the gene encoding a key enzyme in the tropane and nicotine alkaloid pathway evolved from one of the genes of the polyamine pathway, as discussed below.

In plants and bacteria, both arginine and ornithine can be used to synthesize putrescine (Figure 10.7), whereas all other organisms, which lack arginine decarboxylase, can only use ornithine. One interesting anomaly is arabidopsis, which lacks ornithine decarboxylase and can only use arginine. We do not know whether arabidopsis is unique in this respect, nor do we understand why ornithine decarboxylase appears to have been lost from this species.

The first committed step in the pathway leading to the formation of tropane alkaloids is the conversion of putrescine to *N*-methylputrescine, in a reaction catalyzed by putrescine *N*-methyltransferase (PMT). This enzyme is only found in plants that produce either tropane or nicotine alkaloids. In contrast, the capacity to convert putrescine to spermidine, by the enzyme spermidine synthase (SPDS), is present in plants, animals, microorganisms, and fungi as this is a key step in polyamine biosynthesis. There is compelling evidence that the gene encoding PMT evolved from the gene for SPDS. The two proteins share a high degree of identity, with the amino acid sequence of plant SPDS

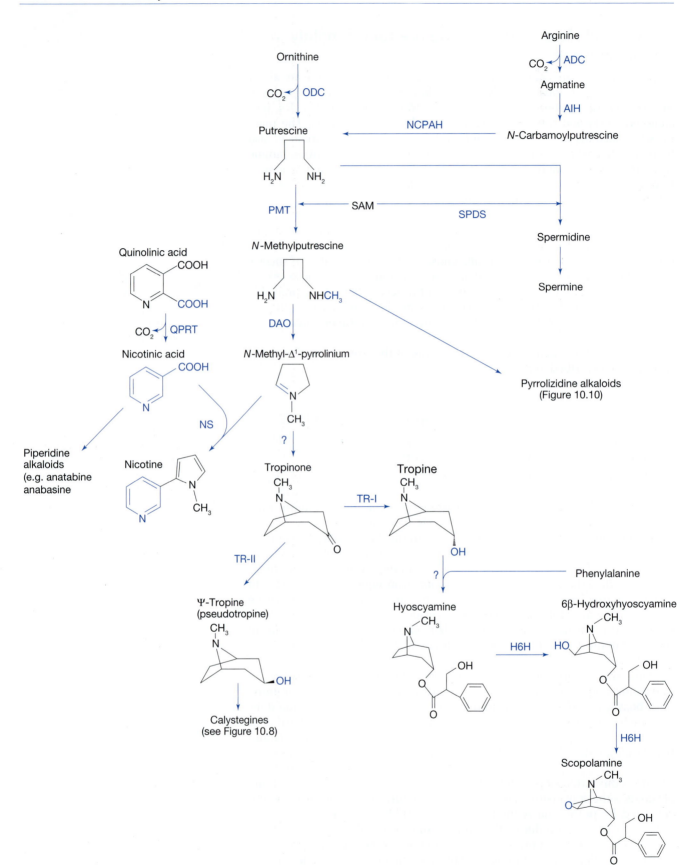

Figure 10.7 Biosynthesis of tropane and nicotine alkaloids and branch points to the calystegines and the pyrrolizidine and piperidine alkaloids. The amino acids arginine and ornithine can be used by most plants as precursors for the biosynthesis of these alkaloids. In both cases, the precursor amino acid is decarboxylated (via ADC or ODC), with putrescine being formed either directly from ODC, or in two further reactions following ADC.

Putrescine can be used to produce the two most abundant polyamines, spermidine and spermine, and these steps mark a branch point away from the alkaloid biosynthesis pathway. Spermidine formation from putrescine is catalyzed by SPDS. Putrescine also serves as a precursor for the biosynthesis of pyrrolizidine alkaloids (Figure 10.10).

The first committed step in tropane alkaloid production is catalyzed by PMT, which carries out an N-methylation using SAM as the methyl donor, to produce N-methylputrescine.

N-methylputrescine is converted to N-methyl-Δ^1-pyrrolinium, the common precursor for all tropane alkaloids. The reaction is catalyzed by a DAO (a putative N-methylputrescine oxidase with broad specificity). The enzyme remains to be fully characterized. DAO would appear to convert N-methylputrescine to the aldehyde, 4-methylaminobutanal which then spontaneously cyclizes to yield the N-methyl-Δ^1-pyrrolinium ion. This intermediate marks another branch point in the pathway, whereby N-methyl-Δ^1-pyrrolinium can be used to produce nicotine, or further metabolized to tropinone.

The reaction that converts N-methyl-Δ^1-pyrrolinium to tropinone remains to be characterized. The subsequent formation of tropine from tropinone is rather better understood. This reaction is mediated by TR-I, which converts the 3-keto group of tropinone to the 3α-hydroxyl of tropine. A second isoenzyme, TR-II, also reduces tropinone but with a stereospecificity opposite to that of TR-I, converting the 3α-hydroxyl of tropinone instead to the 3β-hydroxyl of

Ψ-tropine (also known as pseudotropine) producing another branch point, in this case to the production of calystegines, which are classed as polyhydroxynortropine alkaloids (see Figure 10.8).

The formation of hyoscyamine from tropine is another step that remains to be elucidated. Hyoscyamine is thought to result from the condensation of tropine with a tropic acid derivate of phenylalanine but no enzymes have been isolated as yet. Hyoscyamine to scopolamine conversion is catalyzed by H6H, a 2-oxoglutarate-dependent enzyme. This conversion occurs via two reactions, both catalyzed by H6H (see Figure 10.9).

Nicotine biosynthesis shares the steps to N-methyl-Δ^1-pyrrolinium with the tropane alkaloid pathway. Nicotine is composed of a pyridine ring and an N-methylated pyrrolidine ring (Figure 10.1). The pyridine ring is synthesized from quinolinic acid, which is formed from glyceraldehyde 3-phosphate and L-aspartic acid (not shown). Quinolinic acid is decarboxylated by QPRT to produce nicotinic acid. The subsequent reactions remain uncharacterized but are thought to involve an enzyme, nicotine synthase (that has yet to be purified) that reacts nicotinic acid with N-methyl-Δ^1-pyrrolinium (which provides the pyrrolidine ring) to form nicotine.

Nicotinic acid is a precursor in the biosynthesis of another group of alkaloids, the piperidines; anabasine is formed from the reaction between the amino acid lysine and nicotinic acid, while anatabine is formed from the condensation of two molecules of nicotinic acid.

Abbreviations: AIH, agmatine iminohydrolase; ADC, arginine decarboxylase; NCPAH, N-carbamoylputrescine amidohydrolase; PMT, putrescine N-methyltransferase; DAO, diamine oxidase; TR-I, tropinone reductase I; TR-II, tropinone reductase II; H6H, hyoscyamine 6β-hydroxylase; ODC, ornithine decarboxylase; QPRT, quinolinic acid phosphoribosyltransferase: SPDS, spermidine synthase; NS, nicotine synthase; ? uncharacterized reaction.

proteins being about 64–68% identical to PMT proteins from plants. This homology is higher than that shared between plant SPDS and animal SPDS proteins (47–50% identical). Consequently, the plant *PMT* gene is thought to have evolved from the plant *SPDS* gene after plant and animal *SPDS* genes diverged. Thus, the capacity to produce tropane and nicotine alkaloids appears to have evolved from a fundamental and highly conserved pathway (i.e. polyamine synthesis) that is essential for the growth and survival of virtually all organisms.

Species that produce pyrrolizidine, rather than tropane alkaloids, do not possess the *PMT* gene and putrescine is reacted with spermidine in the first committed step (as discussed on page 354). Hence, the retention of SPDS without acquiring PMT activity appears to have resulted in pyrrolizidine, rather than tropane alkaloid production in this group of plants.

Because of its position at the start of the tropane alkaloid pathway, PMT has been a target enzyme for genetic engineering experiments aimed at increasing alkaloid production. The results have been mixed, depending on the chosen species. Increased PMT expression in *Atropa belladonna* and in *Duboisia* had no effect on the production of tropane alkaloids, whereas it increased hyoscyamine and scopolamine in *Datura metel*, and hyoscyamine in *Hyoscyamus niger*. PMT activity seems to be closely correlated with nicotine production in *Nicotiana*, where overexpression results in increased nicotine formation, while gene silencing (to block *PMT* expression) has the opposite effect.

Another key metabolite is N-methyl-Δ^1-pyrrolinium. It can either form tropinone, in the tropane alkaloid pathway, or react with nicotinic acid to yield nicotine (see next section and Figure 10.7). Once formed, tropinone can also be metabolized by one of two different reactions, leading either to tropine and the production of the tropane alkaloids scopolamine and hyoscyamine, or to Ψ-tropine (pseudotropine), the starting point for the formation of calystegines. These two reactions, catalyzed by isoforms of tropinone reductase, have been overexpressed in transgenic *Atropa belladonna* root cultures. Overexpression of tropinone reductase-I, which forms tropine, resulted in increased hyoscyamine and scopolamine accumulation, while overexpression of tropinone reductase-II, which forms Ψ-tropine, led to increased calystegine pools.

The calystegines differ from other tropane alkaloids in that they lack the methyl group of the bridge nitrogen and possess between three and five hydroxyl groups on the tropane ring (Figure 10.8). They are structurally similar to monosaccharides, making them strong inhibitors of glucosidases, which causes disruption to carbohydrate metabolism in animals that ingest them. Calystegines also have antiviral properties and they are currently being screened for efficacy against diseases such as HIV.

Hyoscyamine, formed from tropine, is converted to scopolamine in a two-step reaction catalyzed by hyoscyamine 6β-hydroxylase (H6H; Figure 10.9).

Figure 10.8 Comparison of the structures of some typical calystegines and monosaccharides. Calystegines are structurally similar to monosaccharides. Consequently calystegines can interfere with carbohydrate metabolism by acting as competitive inhibitors of enzymes, such as glucosidases, that react with sugars. The calystegines are synthesized from tropinone and share much of their biosynthetic pathway with that of the tropane alkaloids (Figure 10.7), although calystegines are classed as polyhydroxynortropines, rather than tropane alkaloids. Calystegines occur in many plants of the Convolvulaceae, Solanaceae, and Moraceae families.

Examples of calystegines found in plants

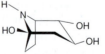

Calystegine A$_3$

Calystegine A$_5$

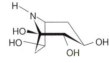

Calystegine B$_1$

Calystegine B$_2$

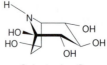

Calystegine C$_1$

Examples of monosaccharides found in plants

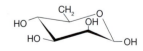

D-Glucose

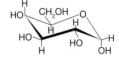

D-Rhamnose

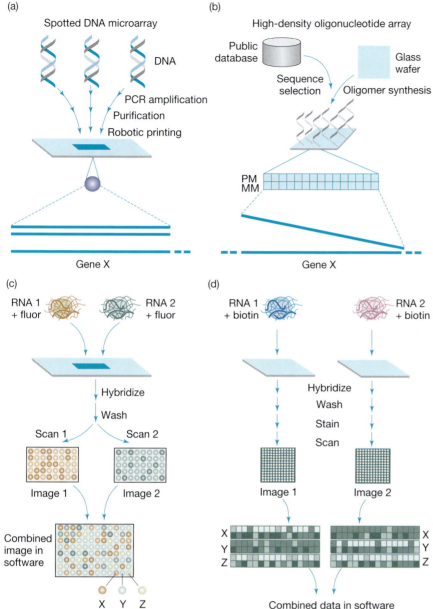

(a) Spotted DNA microarray

DNA

PCR amplification
Purification
Robotic printing

Gene X

(b) High-density oligonucleotide array

Public database

Sequence selection

Glass wafer

Oligomer synthesis

PM
MM

Gene X

(c)

RNA 1 + fluor

RNA 2 + fluor

Hybridize
Wash

Scan 1

Scan 2

Image 1

Image 2

Combined image in software

X Y Z

(d)

RNA 1 + biotin

RNA 2 + biotin

Hybridize
Wash
Stain
Scan

Image 1

Image 2

X
Y
Z

X
Y
Z

Combined data in software

Plate 2.1 DNA microarrays to study expression analysis. (a) Robotic printing of amplified cDNA molecules on to glass slides produces spotted microarrays. Each spot corresponds to a contiguous gene fragment of hundreds of base pairs. (b) A process of light-directed combinatorial chemical synthesis is used to produce high-density oligonucleotide chips consisting of different sequences in a highly ordered array. Genes are represented by 15–20 different oligonucleotide pairs. PM, perfectly matched; MM, mismatched. (c) Comparative expression assays on spotted microarrays can be carried out by using different flurophores to differentially label two mRNA or cDNA samples. Shaded spots labelled x, y and z represent transcripts present at increased levels in sample 1 (x) or sample 2 (y), and similar levels in samples 1 and 2 (z). (d) With Affymetrix GeneChips, biotinylated cRNA is hybridized to the array, stained with a flurophore conjugated to avidin, and the signal detected by laser scanning. Sets of paired oligonucleotides for hypothetical genes present at increased levels in sample 1 (x) or sample 2 (y), and similar levels in samples 1 and 2 (z) are shown.

Plate 4.1 Photosystem I polypeptides represented as a ribbon model, with Lhca2 (yellow) and Lhca4 (light blue) on the right, alongside PsaF (blue–green), and a docked plastocyanin molecule (dark purple) below (the orientation is the same as in Box 4.4, Figure 1, page 81). Note that the negatively charged amino acids of plastocyanin (orange) bind to the positively charged lysine amino acids on PsaF (dark blue). The copper atom of plastocyanin (black dot) lies close to the tryptophans on PsaA and PsaB, which feed electrons up to P700 (red), chlorophylls (A0, green), phylloquinones (PQ, blue), and finally to F_X, F_A, and F_B (yellow/red spheres). (From *Nature*, "The structure of a plant photosystem I supercomplex at 3.4Å...resolution." by Alexey Amunts, Omri Drory, Nathan Nelson, 2007. Reprinted with permission from Nature Publishing Group.)

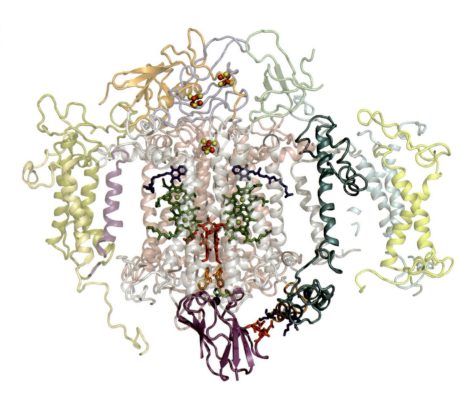

Plate 5.1 Model of Rubisco (side view) with the small subunits in yellow and the large subunits in blue and green. (Original image courtesy of Professor Inger Andersson, Professor of Plant Biochemistry, Swedish University of Agricultural Sciences, Department of Molecular Biology, Uppsala Biomedical Centre, Sweden.)

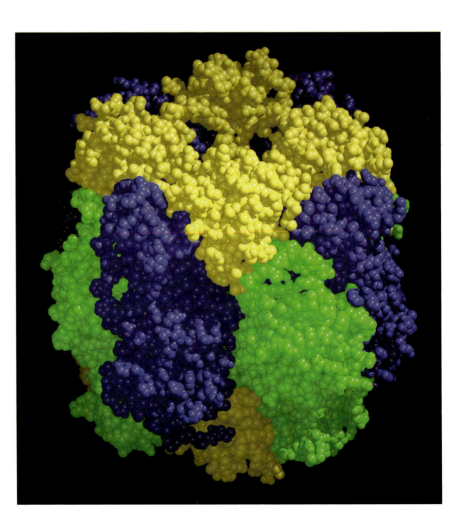

Plate 6.1

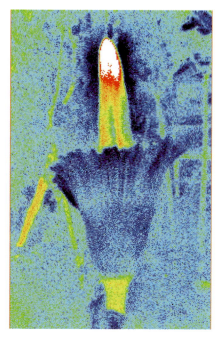

Plate 6.2

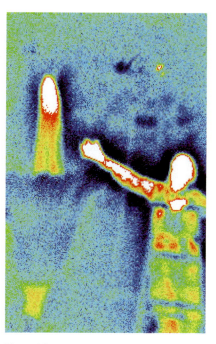

Plate 6.3

Plates 6.1–6.3 *Amorphophallus titanum.* The titan arum, produces a huge flowering structure (inflorescence) up to 3 m tall. The thermal images (plates 6.2 and 6.3) were taken at full bloom and show the spadix temperature to be close to human body temperature. (Plates 6.1, 6.2, and 6.3 reprinted with permission from http://news.wisc.edu/titanarum/facts.html, Kandis Elliot, 2006, University of Wisconsin, Wisconsin.)

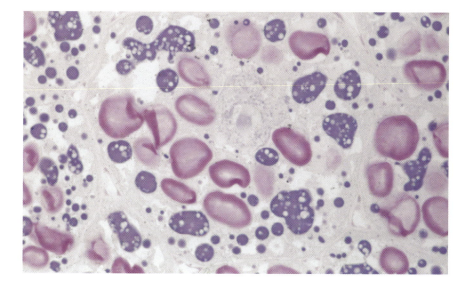

Plate 8.1 Developing protein storage vacuoles bodies (protein bodies, stained dark blue with Coomassie blue) in young cotyledon cells of *Vicia faba.* Note the clear lytic vacuoles in the cytoplasm. The nucleus is in the center of this cell and the amyloplasts contain single large starch grains (stained pink with periodic acid-Schiff). (Reprinted with permission from Gunning BES, Plant Cell Biology on DVD, 2007.)

Plate 10.1 A harlequin poison dart frog (*Dendrobates histrionicus*), which accumulates batrachotoxins (steroidal alkaloids) in its skin as a means of deterring predators (see Box 10.1, Chapter 10). (Original image courtesy of Dr. Zoltan Takacs through website http://zoltantakacs.com)

Plate 10.2 Caterpillars of the cinnabar moth (*Tyria jacobaeae*) feeding on ragwort (*Senecio jacobaea*). The caterpillar accumulates pyrrolizidine alkaloids from the plant to protect it from predators (see Box 10.1, Chapter 10). (Original image courtesy of Christopher Sheehan, © Christopher Sheehan/Ark Gallery.)

Plate 10.3 The scarlet-bodied wasp moth (*Cosmosoma myrodora*), which accumulates pyrrolizidine alkaloids as a protection against spiders. The males also use these alkaloids during mating (see Box 10.1, Chapter 10). (Original image courtesy of Ronald Gaubert, Prairieville, Louisiana.)

Plate 10.5 Fruit of the coffee plant (*Coffea arabica*), which contains the purine alkaloid, caffeine (see Box 10.3, Chapter 10). (Original image courtesy of David A. Anderson, Associate Professor, Department of Economics, Centre College, Danville, Kentucky, USA.)

Plate 10.4 The yellow-flowered ragwort (*Senecio jacobaea*) contains pyrrolizidine alkaloids that are toxic to horses, which usually avoid eating these plants due to their bitter taste (see Box 10.2, Chapter 10). (Original image courtesy of Kit Houghton Photography, Bridgwater, Somerset, UK.)

Plate 11.1 Giant hogweed (*Heracleum mategazzianum*), which contains phototoxic furanocoumarins that cause large blisters to form on the skin when exposed to sunlight. These plants can grow to a height of 5 meters and a single plant can produce 30,000–50,000 viable seeds per year. It was introduced into the UK in 1893 as an ornamental plant and has spread rapidly along water courses as its seeds can be dispersed through water. (Reprinted with permission from Alex Tobin.)

Plate 11.2 Lupin (*Lupinus podophyllus*) flowers, whose orange, pink, and red colors are provided by pelargonidin anthocyanins. (Reprinted with permission from Alex Tobin.)

Plate 11.3 Field poppy (*Papaver rhaeas*) flowers whose red color is due to the presence of cyanidin anthocyanins. (Reprinted with permission from Alex Tobin.)

Plate 11.4 Delphinium (*Delphinium occidentale*) flowers, whose blue color is due to the presence of delphinidin anthocyanins. (Reprinted with permission from Alex Tobin.)

Plate 11.5 *Potentilla anserina* flowers photographed under white light (yellow flower) and ultraviolet light, showing the ultraviolet-absorbing flavonoid pigments that provide nectar guides to attract pollinators. (Original image courtesy of Samfoto Photo Agency, Oslo, Norway.)

Plate 11.6 Hydrangea (*Hydrangea macrophylla*) flowers. The blue coloration occurs in plants growing on acid soils, while the pink coloration occurs on alkaline soils, as explained in Chapter 11. (Reprinted with permission from Alyson Tobin.)

Plate 12.1 Taxol, a diterpene that is an effective treatment for breast cancer, was first discovered in bark extracts from the Pacific yew (*Taxus brevifolia*). (Original image courtesy of George Young, Professor of Meteorology, Pennsylvania State University, Pennsylvania.)

This enzyme has also been manipulated in transgenic plants in an attempt to modify the alkaloids they produce. *Atropa belladonna* normally accumulates over 90% of its alkaloids as hyoscyamine. When the *H6H* gene from *Hyoscyamus niger* was overexpressed in *Atropa belladonna*, however, it switched from hyoscyamine to produce virtually 100% of its alkaloids as scopolamine. The overexpression of both *PMT* and *H6H* genes together resulted in an even greater increase in alkaloid production in transgenic *H. niger*. In this case, scopolamine accumulation increased ninefold.

These experiments demonstrate that both the quality and quantity of tropane alkaloids can be successfully manipulated through genetic engineering.

Nicotine is formed from nicotinic acid and N-methyl-Δ^1-pyrrolinium. A similar reaction forms a different class of alkaloids, the piperidine alkaloids

Nicotine contains both a pyrrolidine ring, derived from *N*-methyl-Δ^1-pyrrolinium in the tropane alkaloid pathway, and a pyridine ring derived from nicotinic acid (Figure 10.7). The reaction between nicotinic acid and *N*-methyl-Δ^1-pyrrolinium is a condensation reaction catalyzed by nicotine synthase, an enzyme that is still rather poorly characterized.

As well as being the substrate for nicotine formation, nicotinic acid can be used to produce anatabine and anabasine, two piperidine alkaloids that are also found in tobacco. Anatabine is formed from two molecules of nicotinic acid, while anabasine is formed from the reaction of lysine, which produces the piperidine ring, and nicotinic acid, which forms the pyridine ring (Figure 10.7).

Tropane and nicotine alkaloids are synthesized in the roots and accumulate in the shoots

Tropane alkaloids are produced near to the root apex and accumulate in the shoots of the plant. PMT, which catalyzes an early step in the pathway, and H6H, which operates at a late stage, are both found in the root pericycle. Tropinone reductase-I, however, which acts at a stage in between these two reactions, is found in the endodermis and outer cortex. This means that intermediates have to be moved between cells, with the final products accumulating in the shoot.

Although nicotine is produced in the roots of *Nicotiana* species, damage to the leaves by grazing herbivores can result in increased nicotine production, leading to increased nicotine accumulation in the leaves. This response, which is an example of herbivore-induced chemical defense, can be mimicked by artificial

Figure 10.9 The two-step hyoscyamine 6β-hydroxylase (H6H) reaction. H6H catalyzes the conversion of hyoscyamine to scopolamine in a two-step reaction, with 6β-hydroxyhyoscyamine as the common intermediate. In the first step, H6H adds an atom of oxygen to the 6β position of hyoscyamine to form 6β-hydroxyhyoscyamine, and uses a second oxygen atom to oxidize 2-oxoglutarate to succinate and CO_2. In the second step H6H removes the 7β-hydrogen, to bring about intramolecular epoxide formation, which results in the formation of scopolamine.

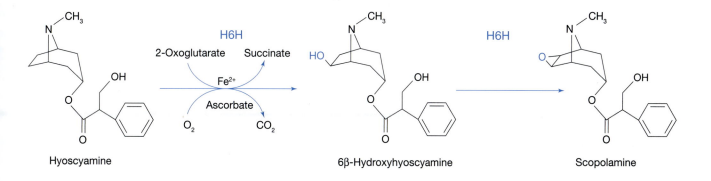

Hyoscyamine   6β-Hydroxyhyoscyamine   Scopolamine

wounding of the leaves and by decapitation of the shoot apical meristem, a technique that has been used routinely to increase the nicotine content of tobacco crops. Leaf damage is thought to induce nicotine synthesis in the roots via a systemic signaling mechanism that involves ethylene and methyl jasmonate movement from the shoot to the root. The roots appear to be sensing the removal of the shoot apical meristem by detecting a decrease in auxin concentration (this plant growth regulator is synthesized in the shoot apex, see Chapter 11). In support of this interpretation, the induction of root nicotine biosynthesis can be blocked by adding synthetic auxin (1-naphthylacetic acid) to the wounded shoot tissue.

Pyrollizidine alkaloids are found in four main families

Of the 500 or so different types of pyrrolizidine alkaloids more than 95% are found in just four families; the Asteraceae, Boraginaceae, Fabaceae, and Orchidaceae. They are scattered around other families, sometimes occurring in a single genus. This taxonomic distribution indicates that the biosynthetic pathway may have evolved several times.

The pyrrolizidine alkaloids are highly toxic to mammals, in particular, and serve as effective feeding deterrents to most herbivores, yet many species of moth accumulate them without any ill effect and use them to deter predators (see Box 10.1).

Pyrrolizidine alkaloids are synthesized from putrescine and spermidine

The pathway for the biosynthesis of a common pyrrolizidine alkaloid, senecionine *N*-oxide, is shown in Figure 10.10. The branch point from putrescine to the pyrrolizidine alkaloids is also indicated in Figure 10.7. Several of the component enzymes remain to be identified and the pathway, as shown, has been predicted from radiolabeling experiments and the analysis of intermediates. Indeed, it is one of the least characterized of all of the alkaloid pathways.

Pyrrolizidine alkaloids all contain a necine base that is attached, through esterification, to a necic acid residue. The necine base is produced from the polyamines spermidine and putrescine, while the necic acid residue is derived from either isoleucine or leucine (Figure 10.10).

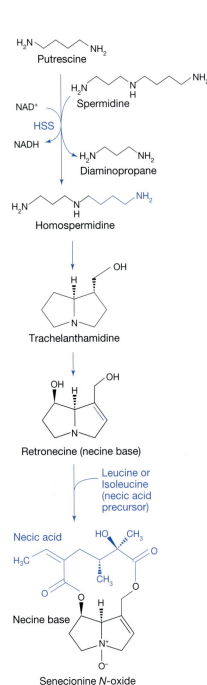

Figure 10.10 Pyrrolizidine alkaloid biosynthesis. The example given here is of senecionine biosynthesis, one of the pyrrolizidine alkaloids found in *Senecio* species (see Figure 10.1 and Box 10.2). Pyrrolizidine alkaloids are composed of two structures, a necine base and a necic acid moiety, that are esterified together in reactions that are still largely uncharacterized. The necine base is derived from homospermidine, while the necic acid residue can be formed from either leucine or isoleucine. Homospermidine is formed in an NAD-dependent reaction catalyzed by homospermidine synthase (HSS), which transfers an aminobutyl group from spermidine to putrescine. Homospermidine is converted into the necine base, retronecine, via the intermediate trachelanthamidine. Either leucine or isoleucine may form the necic acid moiety. The necine base and necic acid esterify together to produce the pyrrolizidine alkaloid skeleton typical of all alkaloids within this group. Evidence for these reactions has come from radiolabeling studies and none of the enzymes that catalyze the reactions from homospermidine onwards have been characterized yet. Homospermidine synthase has been studied in some detail because of its pivotal role in this pathway and its interesting evolutionary origin, as explained in the text.

The first committed step in the pathway is catalyzed by homospermidine synthase (HSS). This enzyme carries out an NAD-dependent reaction whereby an amino butyl group is transferred from spermidine to putrescine to form homospermidine (Figure 10.10). The amino acid sequence of HSS closely resembles (up to 80% identity) that of a fundamental enzyme of primary metabolism, deoxyhypusine synthase (DHS). The reactions catalyzed by these two enzymes are also very similar. Both are NAD-dependent and both remove an amino butyl group from spermidine. However, while HSS transfers this amino butyl group to putrescine, DHS adds it to a eukaryotic translation initiation factor, eIF5A, as part of a fundamental mechanism that initiates cell growth. While HSS is unique to plants that produce pyrrolizidine alkaloids, DHS is ubiquitous, being found in all eukaryotes. It now seems that HSS evolved from DHS through gene duplication, and a small mutation in the duplicate gene enabled the resulting enzyme to produce homospermidine. The mutation survived under the selection pressure from herbivores, and so the first step in the pyrrolizidine pathway became established. This is an important example of the recruitment of an enzyme essential for cell function (i.e. DHS) for use in a metabolic pathway (i.e. HSS) that occurs in only a few Angiosperm families.

Purine alkaloids as popular stimulants in beverages, and as poisons and feeding deterrents against herbivores

The purine alkaloids caffeine and theobromine are found in a limited number of species. They have been studied mostly in species of *Camellia* and *Coffea* because of their value in the drinks industry. Caffeine is the principal alkaloid present in tea (*Camellia sinensis*) and coffee (*Coffea arabica*; Arabica coffee), while theobromine is the dominant purine alkaloid in cocoa beans (*Theobroma cacao*). Other species with significant concentrations of purine alkaloids include *Ilex paraguariensis* (maté), which contains both caffeine and theobromine, and *Cola nitida* (cola) and *Paulliania cupana* (guaraná), which contain high concentrations of caffeine. Extracts from cola and guaraná are popular additives in high energy drinks because of the stimulatory effect of caffeine on the central nervous system. Less desirable is their diuretic effect and a tendency for mild addiction to the caffeine contained in these drinks.

Purine alkaloids are effective insecticides. Caffeine and theobromine are particularly toxic to insect larvae and can cause sterility in some species of beetles. The coffee plant, *Coffea arabica*, times its production of caffeine to coincide with the months when the plant is most susceptible to herbivory. The caffeine concentration is at its highest in soft, young tissue, such as the developing fruit, and immature leaves, and declines as leaves mature, allowing for recycling of alkaloids into developing tissue.

Caffeine is an effective germination inhibitor, and being water-soluble it can leach from leaves into the soil where it acts as an allelopathic agent (a plant chemical that affects the growth of another plant) to reduce competition from other plants. However, because soil microorganisms break it down very slowly, caffeine can accumulate in soils where it can lead to self-poisoning in coffee plantations if there is insufficient crop rotation.

Xanthosine provides the purine group for the purine alkaloids

The detailed biosynthetic pathway for the purine alkaloids theobromine and caffeine is shown in Figure 10.11. It begins with xanthosine, which provides the purine region of these alkaloids. Most of the reactions are well characterized, as a result of radioisotope labeling and chemical analysis together with the isolation and identification of the component enzymes and genes. Part of the reason for the interest in this pathway is because of the high commercial value of caffeine. Ironically, given that early research was driven by the need to produce high caffeine yields, recent success has come from the production of transgenic coffee plants that no longer produce caffeine—offering the potential to produce naturally decaffeinated coffee beans (Box 10.3).

The diversity of alkaloids has arisen through evolution driven by herbivore pressure

In any organism that uses chemicals as a means of defense there is always a chance that its predators will evolve resistance. Consequently, new chemical defenses are produced and the arms race continues. In plants, herbivore pressure drives the evolution of biosynthesis of increasingly diverse defense chemicals. Alkaloids are a good example of this diversity, with over 12,000 known compounds. As we have seen, these almost all originate from just a few amino acid precursors. How, then, do plants produce such an impressive variety of alkaloids?

The important first step is to produce a set of chemicals, or central intermediates, that can be modified in such a way as to create a seemingly infinite

Box 10.3 Producing decaffeinated coffee

There is a significant and growing worldwide market, at present about 10% of total sales, for decaffeinated tea and coffee. Commercial production of decaffeinated drinks relies on chemical extraction of caffeine, using organic solvents, but this also removes some of the flavor and can leave behind solvent residues. Breeding and genetic modification experiments are currently aimed at producing plants without caffeine, so removing the need for chemical extraction. There has been some success already in the search for naturally decaffeinated coffee.

Breeding approach

Naturally occurring decaffeinated wild coffee species were discovered in Madagascar some time ago. Reproductive barriers prevent these wild species from being crossed with *Coffea arabica*, the species used commercially because of its superior quality (see Plate 10.5); *C. arabica* is a tetraploid that is largely self-pollinating, while all other coffee species are diploid and out-crossing. The recent discovery of a number of wild, caffeine-free *C. arabica* cultivars in Ethiopia is promising because these plants can be readily crossed with commercial varieties with the aim of producing high-quality coffee beans that are naturally decaffeinated. The mutation in the Ethiopian cultivars seems to have occurred in the gene coding for caffeine synthase (see pathway in Figure 10.11). This mutation has caused theobromine to accumulate instead of caffeine. The plants are still at the

research stage and it will be some time before they reach the commercial market.

Genetic engineering approaches

Transgenic approaches have concentrated on the *N*-methyltransferase steps in the pathway (see Figure 10.11). Genes have been isolated for each of these steps and transgenic plants have been produced with reduced expression of theobromine synthase. This result was achieved by RNAi (see Chapter 2), i.e. by introducing double-stranded RNA coding for theobromine synthase (TS in Figure 10.11, gene denoted *CaMXMT1* as the enzyme is also called MXMT (7-methylxanthine methyltransferase)). Two species have now been successfully transformed, *Coffea canephora* and the more commercially valuable *Coffea arabica*. Although these transformations have resulted in reductions in both the theobromine and caffeine content of the young plants and embryonic material, it is too early to tell whether there is any effect on the coffee beans. However, as caffeine has been reduced by as much as 70% in the leaves, it would seem reasonable to expect the beans to have significant reductions too. Given that caffeine and theobromine are produced as a defense against herbivores, the ultimate aim would be to produce plants with these alkaloids in all tissues except for the beans, otherwise the resulting crops would be more susceptible to insect damage.

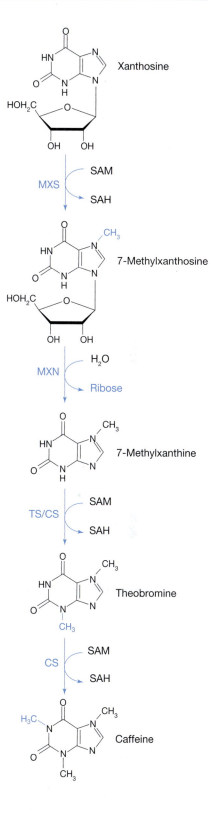

Figure 10.11 Pathway for biosynthesis of the purine alkaloids caffeine and theobromine. Unlike most other alkaloids, which are formed from amino acid precursors, the purine alkaloids are produced from xanthosine, which is a purine ribonucleoside (i.e. a purine group attached to ribose). The ribose group is removed in the second reaction of the pathway, which is catalyzed by MXN.

Three methyltransferase reactions feature in this pathway, each one adding a methyl group; the first (MXS) converts xanthosine to 7-methylxanthosine, the second converts 7-methylxanthine to theobromine (catalyzed by TS in some plants and CS in others), while the third produces caffeine from theobromine (catalyzed by CS). Each of these methyltransferase reactions uses SAM as the methyl donor. Abbreviations: MXS, 7-methylxanthosine synthase; MXN, methylxanthine nucleosidase; TS, theobromine synthase; CS, caffeine synthase; SAM, S-adenosyl-L-methionine; SAH, S-adenosyl-L-homocysteine.

number of end products. The central intermediates in the alkaloid pathways described in this chapter are shown in Figure 10.1. As we have already seen, two of them, strictosidine and norcoclaurine, cater for the formation of over 50% of all known alkaloids. These central intermediates almost all originate from the formation of amines by decarboxylation of amino acid precursors. In joining these amines to different chemical partners, plants produce central intermediates with a chemical versatility that allows further structural diversity to be introduced by highly substrate-specific enzymes. The evolution of these enzymes is a second crucial step in the expansion of alkaloid diversity.

Gene duplication followed by mutation is thought to be a major factor in the evolution of the alkaloid biosynthesis pathways

Among the most important reactions that introduce chemical variety into the alkaloids are: oxidation (via cytochrome P450 monooxygenases and dioxygenases), acylation (via acyltransferases), glucosylation (via glucosyltransferases), and methylation (via methyltransferases). Examples of these reactions are given in Box 10.4. None of these reactions is unique to alkaloid-producing plants. For example, there are orthologs of genes for cytochromes P450 and methyltransferases in unrelated organisms such as bacteria and animals that do not make alkaloids. What is unique about the enzymes present in alkaloid-producing plants is their substrate specificity, and this appears to have been acquired through mutation of existing genes. The large gene families such as the cytochrome P450 monooxygenases, *N*- and *O*-methyltransferases, which are widespread and abundant in many unrelated organisms, may be seen as a gene pool from which secondary metabolism has been recruited. The first event would be gene duplication, where an existing gene is copied and the original is retained (Figure 10.12, page 360). Duplication is important if the original gene is involved in a fundamental process, for example, primary metabolism. Mutations within the copied gene might then provide it with a completely new function, perhaps enabling the resulting enzyme to react with a different substrate, for example. In many cases, duplicated genes share considerable sequence similarity with each other, as only small changes in the sequence can be sufficient to introduce a completely different function. We have already seen an example of this in the pyrrolizidine alkaloid pathway, where duplication and mutation of the *DHS* gene has produced *HSS*, which still shares considerable homology (>80%) with its ancestor, yet its function is entirely different. DHS is essential for cell growth while HSS provides the first step in the pathway for pyrrolizidine alkaloid biosynthesis. Thus, gene duplication followed by mutation and selection by herbivore pressure seems to have been the essential driving force for the evolution of secondary metabolism and the acquisition of chemical defenses in plants.

Box 10.4 Five key reactions that account for much of the diversity of alkaloids

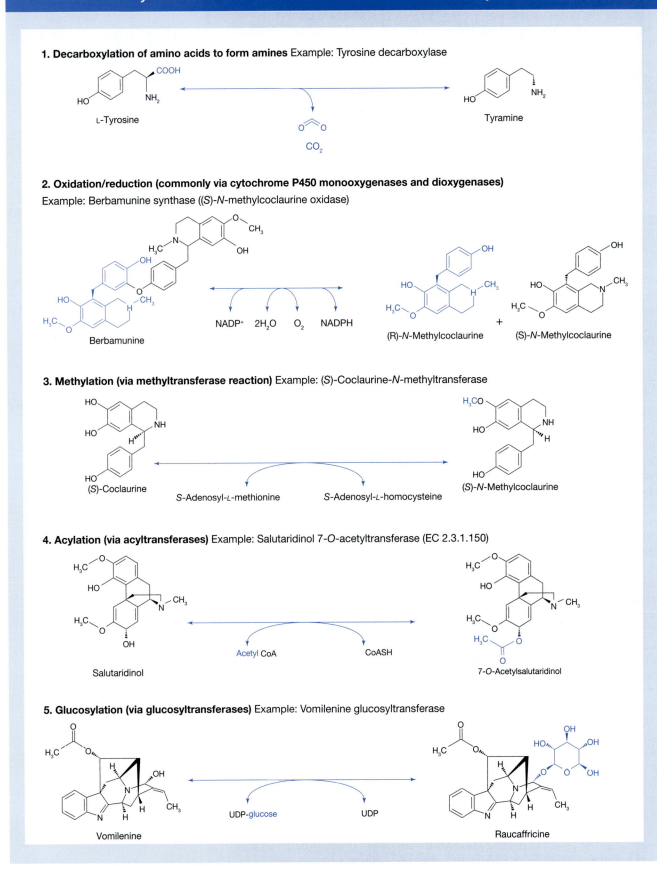

1. Decarboxylation of amino acids to form amines Example: Tyrosine decarboxylase

L-Tyrosine → Tyramine

CO_2

2. Oxidation/reduction (commonly via cytochrome P450 monooxygenases and dioxygenases)
Example: Berbamunine synthase ((S)-N-methylcoclaurine oxidase)

Berbamunine

$NADP^+$ $2H_2O$ O_2 $NADPH$

(R)-N-Methylcoclaurine   +   (S)-N-Methylcoclaurine

3. Methylation (via methyltransferase reaction) Example: (S)-Coclaurine-N-methyltransferase

(S)-Coclaurine

S-Adenosyl-L-methionine   S-Adenosyl-L-homocysteine

(S)-N-Methylcoclaurine

4. Acylation (via acyltransferases) Example: Salutaridinol 7-O-acetyltransferase (EC 2.3.1.150)

Salutaridinol

Acetyl CoA   CoASH

7-O-Acetylsalutaridinol

5. Glucosylation (via glucosyltransferases) Example: Vomilenine glucosyltransferase

Vomilenine

UDP-glucose   UDP

Raucaffricine

Box 10.4 Five key reactions that account for much of the diversity of alkaloids (continued)

The chemical cores for a diverse range of alkaloid structures can be produced by these five enzyme-catalyzed reactions.

1. The **decarboxylation** step converts the amino acid precursor to an amine. This reaction is often the first step in the pathway. The example shown here, tyrosine decarboxylase, is the starting point for isoquinoline alkaloid biosynthesis (Figure 10.5).

2. **Cytochrome P450-dependent oxidases**, such as berbamunine synthase shown here, carry out many (but not all) of the oxidation reactions in the alkaloid biosynthetic pathways. There is a large gene family of cytochrome P450-dependent oxidases in plants and most are involved in secondary, rather than primary metabolism. Berbamunine synthase functions within the pathway for isoquinoline alkaloids, forming berbamunine, the precursor for the bisbenzylisoquinolines (Figure 10.5). It is found in species of poppy. Cytochrome P450-dependent enzymes can catalyze both oxidation and reduction reactions.

3. **Methyltransferases** transfer a methyl ($-CH_3$) group from one molecule to another. Typical reactions are O-and N-methylations where methyl groups are added to an atom of oxygen or nitrogen, respectively. The example

shown here, (S)-coclaurine-N-methyltransferase, functions in the isoquinoline alkaloid pathway and transfers a methyl group from S-adenosyl-L-methionine (SAM) to form (S)-N-methylcoclaurine (Figure 10.5).

4. **Acyltransferases** transfer acyl groups (any group or radical of the form RC=O where R is an organic group) from one molecule to another. The example given here, salutaridinol 7-O-acetyltransferase (SAT), transfers an acetyl group from acetyl CoA to salutaridinol to form 7-O-acetyl-salutaridinol, which can spontaneously convert to thebaine, the precursor for the morphinan alkaloids in opium poppy (Figure 10.5). Acyltransferases are widespread in primary as well as in secondary metabolism.

5. **Glycosyltransferases** add sugars, such as glucose, galactose or rhamnose, to produce a glycosylated molecule. The addition of a sugar often reduces the toxicity so that there is a reduced risk of self-poisoning. Tissue damage, by grazing herbivores, can release glucosidase enzymes from the vacuole, which cleave the sugar and restore toxicity. Glycosylation is also a means of facilitating transport, as the sugar moiety makes the complex more water-soluble. The example here, vomilenine glucosyltransferase, transfers glucose from UDP-glucose to vomilenine to form the terpenoid indole alkaloid raucaffricine.

The distribution of enzymes between different cell types allows for further chemical diversity

Cell specialization is another important aspect of alkaloid biosynthesis, and the failure of undifferentiated cell cultures to produce some classes of alkaloids is a possible indicator of this requirement. Many of the alkaloid biosynthetic pathways described in this chapter show high levels of compartmentation both within and between cells. The need to transport intermediates and end products from cell to cell, and often from root to shoot, adds another layer of complexity to these pathways. It also introduces another means of creating metabolic diversity, by keeping competing pathways apart, and localizing substrates within specialized alkaloid-producing cells.

There is no simple taxonomic relationship in the distribution of different classes of alkaloids

The metabolic pathways that lead to the production of different types of alkaloid are not always found in taxonomically related species. Plants seem to have acquired these pathways by diverse means, with most alkaloid groups appearing to be distributed among unrelated species. For example, the majority of pyrrolizidine alkaloids (about 95% of structures) occur in three unrelated families, the Asteraceae, Boraginaceae and in a single genus of the Fabaceae. Isolated occurrences have been found in single species of some additional families. This situation could have occurred by two different processes; either the genes were recruited very early in angiosperm evolution and they were subsequently lost, or they were recruited several times

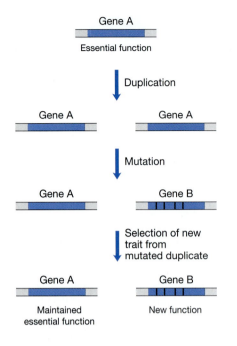

Figure 10.12 Gene duplication followed by mutation can result in gain of function. In this example gene A encodes a protein that serves an essential function. If a mutation occurred in gene A that led to a loss of this essential function then the mutation would be lethal. However, if gene A is duplicated the essential function can be retained in the event that a mutation occurs in the duplicate gene. If the mutation results in a gain of function that is beneficial to the organism, the new gene (B) is retained through natural selection. This process is thought to have driven the evolution of diversity of secondary plant products through the emergence of novel enzymes.

independently. Either process would result in the pathway being found in unrelated species. Although we still do not know how the entire pyrrolizidine pathway arose, it seems that the gene encoding the first enzyme, HSS, did evolve independently on several occasions from its *DSS* ancestor.

There are examples where particular alkaloids show strong taxonomic grouping, and yet they are rarely entirely specific to a single family. The tropane alkaloids, such as hyoscyamine and scopolamine for example, are widespread among the Solanaceae family and yet they occur in other unrelated families as well, such as the Euphorbiaceae. Another example is nicotine, which is a major alkaloid in all species within the genus *Nicotiana*. Even so, it is also found in unrelated genera within the Solanaceae, and in low concentrations in completely unrelated families.

Alkaloids cannot, therefore, be relied upon as taxonomic markers, and we are left with the conclusion that their evolutionary origin is complex. Their erratic distribution is attributable to the independent evolution of the pathways on a number of occasions, or else to the loss of pathways during the course of evolution. Because alkaloids play a vital role as defense chemicals, herbivory exerts a strong selection pressure, and, providing this continues, the diversity of alkaloid production will continue to expand, so maintaining the chemical arms race.

Summary

Alkaloids serve a range of functions in plants, as poisons, feeding deterrents, antibiotics, and allelopathic agents. As poisons, they serve a major function in deterring herbivores, and the chemical arms race between plants and herbivores has driven the evolution of an enormous range of alkaloid structures. Many alkaloids are of commercial value, primarily as pharmaceuticals (e.g. in cancer treatment) but also as food and drinks additives (e.g. caffeine). The metabolic pathways leading to alkaloid biosynthesis are highly complex and many of the enzymes and intermediates remain to be identified. Virtually all of the pathways begin with the decarboxylation of an amino acid, with the subsequent formation of a central intermediate that is further modified to generate a diverse array of alkaloid structures. Genetic engineering, to alter single or multiple reaction steps, is providing a valuable experimental approach that is serving to improve our understanding of these biosynthetic pathways. This approach has already resulted in the production of new products and plants, such as opium poppy that accumulates a potential treatment for baldness, as well as decaffeinated coffee plants.

Further Reading

Review articles

Asihara H & Crozier A (2001) Caffeine: a well known but little mentioned compound in plant science. *Trends Pl. Sci.* 6, 407–413.

This review covers both the biosynthesis and degradation of caffeine.

De Luca V & St Pierre B (2000) The cell and developmental biology of alkaloid biosynthesis. *Trends Pl. Sci.* 5, 168–173.

A review of the tissue, cellular and subcellular distribution of alkaloids and their biosynthetic pathways.

Facchini PJ (2001) Alkaloid biosynthesis in plants: biochemistry, cell biology, molecular biology and metabolic engineering applications. *Annu. Rev. Pl. Physiol. Pl. Mol. Biol.* 52, 29–66.

A very thorough review of the pathways of alkaloid biosynthesis.

Facchini PJ & St-Pierre B (2005) Synthesis and trafficking of alkaloid biosynthetic enzymes. *Curr. Opin. Pl. Biol.* 8, 657–666.

A detailed review that concentrates on the tissue and cellular compartmentation of the pathways for benzylisoquinoline, monoterpenoid indole, and quinolizidine alkaloid biosynthesis.

Kessler A & Baldwin IT (2002) Plant responses to insect herbivory: the emerging molecular analysis. Scavenging deleterious oxygen radicals. *Annu. Rev. Pl. Biol.* 53, 299–328.

A review of inducible chemical defenses in plants from the authors who first discovered wound-induced nicotine biosynthesis in tobacco.

Ober D (2005) Seeing double: gene duplication and diversification in plant secondary metabolism. *Trends Pl. Sci.* 10, 444–449.

A very useful review of the evolution of alkaloid (and general secondary metabolite) biosynthesis pathways.

Raskin I, Ribnicky DM, Komarnytsky S, Ilic N, Poulev A, Bortsjuk N, Brinker A, Moreno DA, Ripoli C, Yakoby N, O'Neal JM, Cornwell T, Pastor I & Fridlender B (2002) *Trends Biotechnol.* 20, 522–531.

A wide-ranging review that discusses scientific, political, and economic aspects of using plants and plant products for medicinal purposes.

Manipulation of alkaloid biosynthesis by genetic engineering and breeding

Allen SA, Millgate AG, Chitty JA, Thisleton J, Miller JAC, Fist AJ, Gerlach WL & Larkin PJ (2004) RNAi-mediated replacement of morphine with the non-narcotic alkaloid reticuline in opium poppy. *Nat. Biotechnol.* 22, 1559–1566.

The paper that describes the production of transgenic opium poppy that produces reticuline instead of morphine.

Moyano E, Jouhikainen K, Tammela P, Palazón J, Cusidó RM, Piñol T, Teen TH & Oksman-Caldentey K-M (2003) Effect of *PMT* gene overexpression on tropane alkaloid production in transformed root cultures of *Datura metel* and *Hyoscyamus muticus*. *J. Exp. Bot.* 54, 203–211.

The overexpression of putrescine *N*-methyltransferase results in an increase in tropane alkaloid production—in contrast to the work of Rothe *et al.* indicating specifies differences in the response.

Ogita S, Uefuji H, Yamaguchi Y, Koizumi N & Sano H (2003) Producing decaffeinated coffee plants. *Nature* 423, 823.

A short report on the production of transgenic coffee plants with a reduced concentration of caffeine, due to antisense repression of a methyltransferase gene.

Rothe G, Hachiya A, Yamada Y, Hashimoto T & Dräger B (2003) Alkaloids in plants and root cultures of *Atropa belladonna* overexpressing putrescine *N*-methyltransferase. *J. Exp. Bot.* 54, 2065–2070.

This paper describes the production and analysis of transgenic *Atropa belladonna* plants where overexpression of putrescine *N*-methyltransferase fails to lead to an increase in tropane alkaloid production.

Richter U, Rothe G, Fabian A-K, Rahfeld B & Dräger B (2005) Overexpression of tropinone reductases alters alkaloid composition in *Atropa belladonna* root cultures. *J. Exp. Bot.* 56, 645–654.

The overexpression of two tropinone reductases resulted in an increase in scopolamine and hyoscyamine production.

Silvarolla MB, Mazzafera P & Fazuoli LC (2004) A naturally decaffeinated Arabica coffee. *Nature* 429, 826.

This paper describes the identification of a number of low caffeine coffee plants among a collection from Ethiopia, offering promise for a breeding-based approach to the production of low caffeine coffee plants.

Phenolics

11

Key concepts

- The main classes of plant phenolic compounds are the simple phenylpropanoids, flavonoids, lignin, and tannins.

- The different classes of phenolics serve a variety of functions within plants, from poisons to antibacterial compounds and plant pigments and scents.

- Phenylalanine is the precursor for plant phenolic biosynthesis.

- The shikimic acid pathway produces phenylalanine.

- The phenylpropanoid pathway is central to the formation of plant phenolic compounds.

- The addition of carbon from malonyl CoA to the phenylpropanoid structure provides the basis for flavonoid biosynthesis.

- The pathways for the biosynthesis of the more complex phenolics, such as the flavonoids and lignin, still remain to be fully characterized.

- Breeding and genetic engineering are being used to manipulate the biosynthetic pathways to yeild useful phenolic products.

Plant phenolic compounds are a diverse group with a common aromatic ring structure and a range of biological functions

Plants produce a very large range of phenolic compounds with diverse structures and properties. About 10,000 different plant phenolics have been identified to date. All of them have a common component, an aromatic hydrocarbon ring (phenyl, or benzyl ring) that is usually attached to at least one hydroxyl group. The simplest form is the phenol molecule itself (Figure 11.1) and although free phenol is never found in plants, this structure can usually be recognized somewhere within a plant phenolic molecule. As most plant phenolics are synthesized from products of the phenylpropanoid pathway (see below) they are frequently referred to as phenylpropanoids. The basic structure of a phenylpropanoid is a phenyl ring with a three-carbon side chain attached (C_6–C_3; Figure 11.2 a and b).

The biological functions of plant phenolics are many and varied, ranging from scents and pigments, poisons and feeding deterrents, allelopathic compounds, signaling molecules, structural components, and antifungal and antimicrobial agents (Box 11.1).

Phenyl ring

OH   Hydroxyl or phenolic group

Figure 11.1 Structure of phenol.
A hydroxyl, or phenolic group, is attached to a 6-carbon phenyl ring. Phenols are acidic, as the phenyl ring is tightly bound to the –O of the hydroxyl group while the relatively loose bonding between the –O and –H allows a H$^+$ to dissociate in solution to leave a negatively charged phenolate ion.

The phenolics may be grouped into different classes according to their basic structure. In this chapter we will describe the biological functions and biochemical pathways leading to the synthesis of the four most abundant plant phenolic groups: simple phenolics, flavonoids, lignin, and tannins.

The simple phenolics

The simple phenolics are a mixed group of phenolic compounds with a range of biological functions (summarized in Box 11.1). The three main groups are the simple phenylpropanoids, coumarins, and benzoic acid derivatives.

The simple phenylpropanoids all share the basic phenylpropanoid structure, with a three-carbon side chain attached to a six-carbon phenyl ring (Figure 11.2a). The side chain is linear. Phenylpropanoids are the central units from which almost all phenolic compounds are made. Examples of simple phenylpropanoids include caffeic, ferulic (Figure 11.2a), cinnamic, and *p*-coumaric acids (see Figure 11.5).

The coumarins also have the basic C$_6$–C$_3$ phenylpropanoid skeleton (Figure 11.2b). They differ from the simple phenylpropanoids in that the side chain is

Box 11.1 Biological functions of plant phenolics

Plant phenolics serve a range of biological functions. They can provide visual signals, scents, and flavors that attract pollinators and deter herbivores. Some are toxic to mammals and insects, while others provide defense against fungal and bacterial pathogens. Some serve plant-specific functions, as germination inhibitors, ultraviolet-protectants, and signaling compounds.

Plant phenolics are involved in animal–plant interactions

As well as their important function as floral pigments (described in Box 11.2), many plant phenolics provide scents, flavors, and poisons. The volatile benzoic acid derivatives are present in the floral scents of more than 100 species of 30 different families and function as pollinator attractants. Vanillin, for example, is an attractive scent present in the flowers of vanilla orchids (*Vanilla planifolia*, *V. pompona*, and *V. tahitiensis*) while its flavor also aids with fruit dispersal in a range of species.

The flavors produced by the phenolics range from the pleasant-tasting vanillins, through to the pungent or astringent flavors of ginger (gingerols), capsicum peppers (capsaicin), and the tannins. Bitter tastes tend to deter most herbivores. Cattle, for example, will avoid eating plants with high tannin content. Some of the aromatic phenolics are used as food flavorings, for example cloves (eugenol), cinnamon (cinnamaldehyde), and nutmeg (myristicin).

The indigestibility of lignin makes it a very effective feeding deterrent. Many herbivores avoid lignified tissue because it is simply too tough to bite into.

The furanocoumarins are toxic only when they are activated by ultraviolet A light (320–400 nm wavelength). When activated in this way they can bind to the pyrimidine bases (cytosine and thymine) of DNA, blocking transcription and leading ultimately to cell death. These phototoxic furanocoumarins are particularly abundant in the Umbelliferae. Examples include the food crops celery, parsley, and parsnip as well as the giant hogweed (*Heracleum mantegazzianum*; Plate 11.1). The photosensitive reaction can cause severe burns to the skin, for example in agricultural workers when picking celery. Some caterpillars are able to feed on furanocoumarin-containing plants by rolling themselves up inside a leaf in order to shade themselves from sunlight until digestion is complete.

Isoflavonoids, found in high concentrations in some cultivars of clover, have strong anti-estrogenic properties and can cause infertility in sheep. This effect is due to the steroidal-like structure of the isoflavonoids that enables them to bind to estrogen receptors.

The tannins (both hydrolyzable tannins and proanthocyanidins), apart from serving as feeding deterrents, are toxic towards many insects, mammals, and birds. In binding to proteins, tannins can cause severe digestive problems in monogastrics by inhibiting gut enzymes. In contrast, ruminants benefit from having a low concentration of tannins in their diet because they form complexes with soluble proteins in the rumen and permit subsequent absorption of amino acids in the lower digestive tract.

Box 11.1 Biological functions of plant phenolics (continued)

The antibiotic role of phenolics

Many of the phenolic compounds are important in protecting plants against bacterial and fungal pathogens. These phenolics are classed as phytoalexins (defined as a plant product with antifungal or antibacterial properties that is synthesized *de novo* as a response to attack). Phenolic phytoalexins include: tannins (both hydrolyzable and proanthocyanidins), which can inhibit fungal growth in trees; isoflavonoids that are induced in legume roots in response to fungal infection; and the stilbenes, such as resveratrol, that have strong antifungal properties. Lignin can also protect a plant from fungal growth due to its mechanical, rather than its toxic properties. Lignification can sometimes be induced in response to wounding, as a defense against further penetration of the pathogen.

These phytoalexin properties of the phenolics make them targets for breeding and genetic engineering, with some success. For example, introduction of the gene encoding stilbene synthase, from grape, into tobacco and alfalfa enabled these plants to produce stilbenes and resulted in increased resistance to their respective fungal pathogens, *Botrytis cinerea* and *Phoma medicaginis*. Increased production of the isoflavonoid, medicarpin, in alfalfa, as a result of over-expression of the gene encoding isoflavone *O*-methyltransferase also resulted in increased resistance to *P. medicaginis*.

Some phenolics are involved in plant signaling and development

Many of the simple phenolics are involved in biochemical interactions between plants (known as allelopathy). Examples include syringic, caffeic, and ferulic acids, which, as water-soluble compounds, are readily leached out from leaves and roots into the surrounding soil where they function as effective germination inhibitors. Bracken (*Pteridium aquilinum*) is particularly rich in these allelopathic phenolics and this enables bracken to dominate large areas of land. Interestingly, there is evidence of self-poisoning, as bracken will begin to degenerate after a number of years growing in the same site. After a resting period, during which the phenolics are leeched from the soil, the bracken returns to recolonize the area.

Salicylic acid is a key regulator of plant development and of plant defense responses. It reaches high concentrations prior to the development of cyanide-resistant respiration in Arum lilies, for example (see Chapter 6) and can induce this response if applied to immature Arum flowers. Salicylic acid appears to be a widespread signal for the systemic induction of pathogen-related protein genes and for the oxidative burst that precedes cell death during pathogen attack.

Flavonoids provide a means of signaling between plant hosts and bacterial symbionts. They are released from legume roots into the soil where they induce the expression of nodule-inducing genes in *Rhizobium* bacteria. The response appears to be highly species-specific, both in terms of the type of flavonoid released from the plant, and the effect that it has on different species and individual strains of *Rhizobium*. The isoflavonoids, daidzein and genistein, for example, will induce the expression of *nod* genes in particular strains of *Rhizobium* while inhibiting their expression in others. Hence, the specific release of flavonoids serves an important role in establishing the correct symbiosis between plant and bacterium (see Chapter 8).

Some phenolics can protect plants from ultraviolet B radiation

The ultraviolet-absorbing properties of flavonoids provide the plant with protection against the potential damaging effects of this short-wavelength radiation. Quercetin, lutonarin, and kaempferol are all induced in epidermal cells in response to increased ultraviolet B (wavelengths of 280–320 nm) radiation in a range of plants. Mutants with reduced flavonoid content have increased susceptibility to the DNA-damaging effects of ultraviolet B radiation. The flavonoids appear to serve two protective functions. First, they reduce the amount of ultraviolet B radiation penetrating into the mesophyll cells. Second, they function as antioxidants, protecting cells from free radicals generated during exposure to ultraviolet B.

---

cyclized to form a ring and, consequently, they may also be classed as phenylpropanoid lactones. Examples of coumarins include coumarin, umbelliferone (Figure 11.2b), esculetin, and scopoletin. The furanocoumarins are a subgroup within the coumarins that contain a furan ring (shown in blue, Figure 11.2b) attached to the phenyl group. Examples include psoralen, methoxypsoralen (Figure 11.2b), xanthotoxin, bergapten, and sphondin. Furanocoumarins are particularly abundant in species within the Umbelliferae, such as the giant hogweed (*Heracleum mantegazzianum*) (see Box 11.1 and Plate 11.1).

A third group of simple phenolics is the benzoic acid derivatives (Figure 11.2c). These compounds differ from the others in this group in having a

C_6–C_1, rather than a C_6–C_3 structure. Examples include vanillin and salicylic acid (Figure 11.2c; Box 11.1) as well as gallic acid and erusic acid that are components of the hydrolyzable tannins (Figure 11.2g).

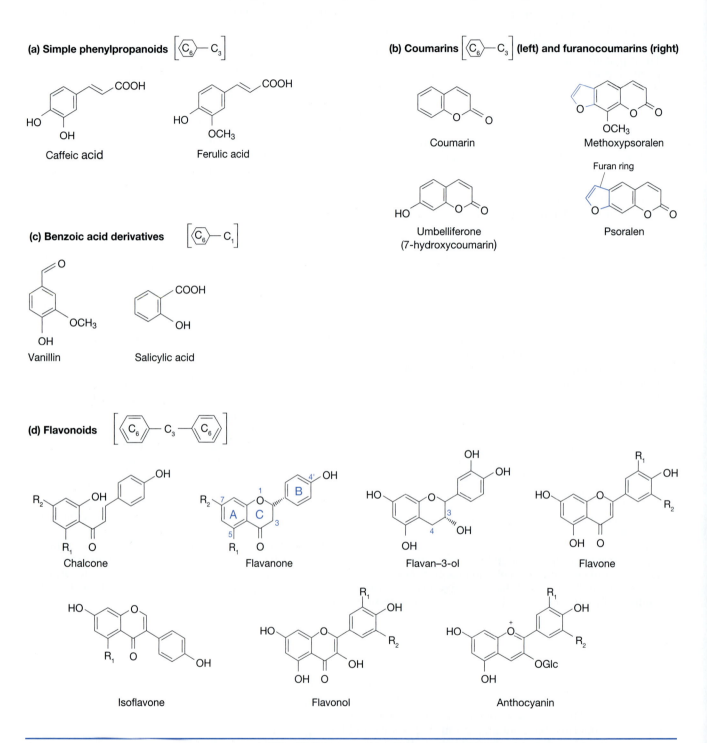

(a) Simple phenylpropanoids $[\langle C_6 \rangle - C_3]$

Caffeic acid

Ferulic acid

(b) Coumarins $[\langle C_6 \rangle - C_3]$ **(left) and furanocoumarins (right)**

Coumarin

Methoxypsoralen

Furan ring

Umbelliferone
(7-hydroxycoumarin)

Psoralen

(c) Benzoic acid derivatives $[\langle C_6 \rangle - C_1]$

Vanillin

Salicylic acid

(d) Flavonoids $[\langle C_6 \rangle - C_3 - \langle C_6 \rangle]$

Chalcone

Flavanone

Flavan–3-ol

Flavone

Isoflavone

Flavonol

Anthocyanin

Figure 11.2 a to g Examples of the major phenolic groups that occur in plants. Chalcones, R_1= OH R_2 = OH tetrahydroxychalcone; R_1=H R_2=OH isoliquiritigenin. **Flavanones,** R_1= OH R_2 = OH naringenin; R_1= H R_2 = OH liquiritigenin; **Flavan–3-ols,** (-)-epicatechin has R-stereochemistry at C3 (i.e. OH group enters the plane, as shown); (+)-catechin has S-stereochemistry at C3 (i.e. OH group comes out from the plane). **Flavones,** R_1= H R_2 = H apigenin; R_1= OH R_2 = H luteolin. **Isoflavones** R_1− OH genistein; R_1= H daidzein. **Flavonols** R_1= H R_2 = H kaempferol; R_1= OH R_2 = OH myricetin; R_1= H R_2 = OH quercetin. **Anthocyanins** R_1= OH R_2 = H cyanidin; R_1= OH R_2 = OH delphinidin; R_1= H R_2 = H pelargonidin. (From *Plant Physiology, 2nd edition,* edited by Lincoln Taiz & Eduardo Zeiger, Figure 8.14, 2006. Reprinted with permission from Sinauer Associates Inc.)

The more complex phenolics include the flavonoids, which have a characteristic three-membered A, B, C ring structure

The flavonoids are a very large and diverse group of phenolics with over 6000 known types. They have a range of biological functions, from pigments to poisons and antibiotics (Box 11.1).

All flavonoids have the same basic skeleton, consisting of two six-carbon rings (ring A and ring B) linked by a three-carbon bridge that usually forms a

(e) Stilbenes
e.g. Pinosylvin [R_1, R_2 = H]; Resveratrol [R_1 = OH, R_2 = H]

Figure 11.2 a to g Examples of the major phenolic groups that occur in plants.

(f) Lignin $\left[\langle C_6 \rangle - C_3\right]_n$ (partial structure of a lignin molecule from beech)

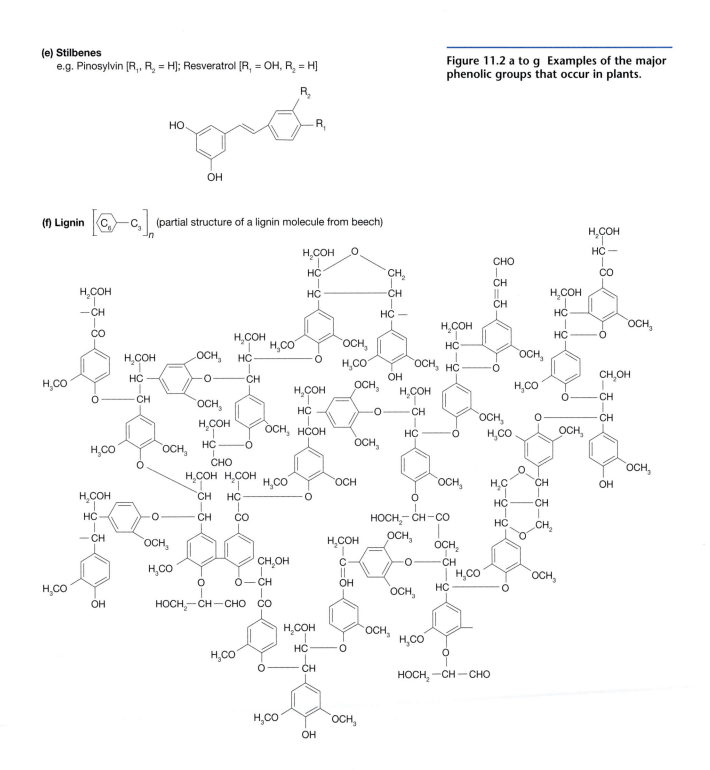

third ring (ring C, as shown for flavanone; Figure 11.2d). Modifications to these three rings, in particular the degree of oxidation of ring C, form the basis of further classification of the flavonoids into various chemical groups. These groups include: chalcones, flavanones, flavonols, flavones, isoflavones, flavan-3-ols, and anthocyanins (Figure 11.2d) as well as the flavonoid-based polymers, the phlobaphenes and proanthocyanidins (condensed tannins; Figure 11.2g). The structure of anthocyanin is described later in this chapter and in Box 11.2. Chalcones are distinguished from other flavonoids in that the C ring is left open (Figure 11.2d). Flavanones are isomers of chalcones, with the C ring closed as in other flavonoids. Flavones differ from flavanones in having a C=C double bond within ring C. Flavones are isomeric with

Figure 11.2 a to g (continued) Examples of the major phenolic groups that occur in plants. (From *Plant Physiology, 2nd edition*, edited by Lincoln Taiz & Eduardo Zeiger, Figure 8.14, 2006. Reprinted with permission from Sinauer Associates Inc.)

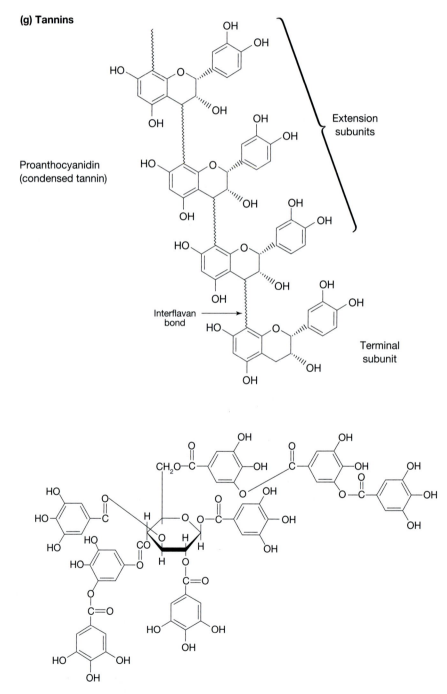

(g) Tannins

Extension subunits

Proanthocyanidin (condensed tannin)

Interflavan bond

Terminal subunit

Hydrolyzable tannin

isoflavones, with the key difference between these two groups being the positioning of ring B. In flavones, as in all other flavonoid groups, ring B is attached to ring C at the C-2 position. In isoflavones, ring B is attached to ring C at the C-3 position (Figure 11.2d). Flavonols differ from flavones in having a hydroxyl group at the C-3 position within the C-ring.

Flavonoid groups can polymerize to form proanthocyanidins, which consist of flavan-3-ols (both (+)-catechin and (–)-epicatechin, which are stereoisomers; see Figure 11.2d) joined by interflavan bonds, usually forming between the 4-position of the upper unit and the 8-position of the lower unit (Figure 11.2g). There may be between two and 50 flavan-3-ol units, with an overall molecular mass that can exceed 20,000. The structure varies according to the types of flavan-3-ols present, the positioning of the bonds between them, and the degree of hydroxylation of the subunits. In rare cases, there may be components other than flavan-3-ols in the polymer. Gallic acid, for example, is frequently incorporated into the flavan-3-ol units of grape seed proanthocyanidins, while a flavan (butiniflavan) occurs within the proanthocyanidins in the bark of the tropical legume, *Senna petersiana*. Reports of colored anthocyanidins being found in proanthocyandins are open to question, however, as these can be formed by hydrolysis during the extraction process. Indeed, proanthocyanidins are so called because they can be hydrolyzed to anthocyanidins by strong acid treatment.

The stilbenes, while not strictly flavonoids, share common precursors and have similar properties to some flavonoids and they are classed as phenolic compounds. Stilbenes consist of two phenyl rings linked by an ethene bridge (Figure 11.2e). They are found in a limited range of unrelated species and have a range of biological functions (Box 11.1). Resveratrol is a stilbene that is found in particularly high concentrations in grapes and, consequently also in red wines.

Lignin is a complex polymer formed mainly from monolignol units

Lignin is a very complex and heterogeneous polymer formed mainly from three different monolignols (*p*-coumaryl, coniferyl, and sinapyl alcohols; Figure 11.2f). These monolignols differ only in the number and positioning of methoxyl ($CH_3O–$) groups attached to the phenyl ring (see also Figure 11.17). When incorporated into the lignin polymer the monolignols *p*-coumaryl, coniferyl, and sinapyl alcohols form their respective phenylpropanoid subunits: *p*-hydroxyphenyl (H), guaiacyl (G), and syringyl (S). Lignin may also contain small amounts of intermediates of monolignol biosynthesis as well as some phenylpropanoids such as *p*-coumarates, ferulates, and *p*-hydroxybenzoates.

The subunit composition of lignin varies between taxa and can differ from one cell to another and even within different regions of the same cell. In most angiosperm dicots, lignin contains guaiacyl and syringyl monolignols (forming G and S lignins), while the monocots also incorporate *p*-hydroxyphenyl units (H lignins). In gymnosperms, lignin is mostly composed of guaiacyl with small amounts of *p*-hydroxyphenyl monolignols (G and H lignins) and S lignins are absent. Gymnosperm lignin, being rich in G units, contains more C–C bonds than the relatively S-rich lignin of angiosperms because the G units are able to form bonds at the C-4 and C-5 position while S units only link between C-4 (see Figure 11.17). These differences in lignin composition affect not just the structural properties of the woods, but also their suitability for commercial applications. Gymnosperm woods require more chemical processing than angiosperm woods during paper production because the additional C–C bonds that form between adjacent G units render the lignin more resilient to chemical degradation. There has been some success with genetic engineering of poplar trees so that they produce lignins more suitable for pulping and requiring less chemical treatment, but there has yet to be any breakthrough in manipulating gymnosperm lignin (see Further Reading).

The tannins are phenolic polymers that form complexes with proteins

The tannins, like lignin, are phenolic polymers. There are two different categories, condensed and hydrolyzable tannins. Although they share similar biological functions, they are structurally and biosynthetically quite distinct.

The condensed tannins (Figure 11.2g) are more frequently referred to as proanthocyanidins. As they are flavonoid polymers, their structure has been described further in the section relating to flavonoids (page 369).

Hydrolyzable tannins are mixed polymers of phenolic acids, usually gallic (gallotannins) or ellagic acids (ellagitannins), together with simple sugars, usually D-glucose (Figure 11.2g). They are more readily hydrolyzed than the proanthocyanidins, degrading to their component carbohydrates and phenolic acids in the presence of dilute, rather than concentrated, acids as well as by hot water treatment. The hydrolyzable tannins are smaller than the proanthocyanidins, with molecular weights ranging between 600 and 3000. Although structurally quite distinct from one another, both the hydrolyzable tannins and proanthocyanidins will bind to proteins. This property is the source of the name, tannin, as it was first used to describe chemicals that could be used to turn raw animal hides into leather in the tanning process. Tannins and proteins can bind in two different ways: hydrogen bonds may form between the hydroxyl groups on the phenyl ring and electronegative sites on a protein; alternatively, covalent bonds may form between the same groups in a reaction catalyzed by a polyphenol oxidase. While these protein–tannin complexes are beneficial to the tanning process, making the leather more water resistant, they can cause problems to the digestive system of herbivores if present in large amounts in forage (Box 11.1).

Most plant phenolics are synthesized from phenylpropanoids

Almost all plant phenolic compounds are synthesized from phenylpropanoid precursors. The exceptions include: the isocoumarin, 6-methoxymellein, found in carrots, and tetrahydrocannabinol, the psychoactive chemical of *Cannabis sativa*, which are both products of polyketide synthesis. Some doubt also remains over the biosynthetic route to the benzoic acid derivatives, such as salicylic acid and the gallic acid and erusic acid units of the hydrolyzable tannins, which may form directly from chorismic acid, rather than via the phenylpropanoid pathway, as discussed later.

For the vast majority of plant phenolics that are derived from phenylpropanoid precursors, the starting point for their biosynthesis is the shikimic acid pathway. As we shall see, this pathway produces the aromatic amino acid, phenylalanine, from which the first phenylpropanoid, cinnamic acid, can be formed.

A second pathway, the malonic acid/acetate pathway, is also required for the production of some plant phenolics. It is used in two ways. First, it may be used in polyketide synthesis where acetyl CoA units are successively condensed together to form phenolic products. Although these reactions are important in fungi and bacteria, they play a lesser role in the biosynthesis of plant phenolics, being responsible for the formation of only a few products, such as the isocoumarin, 6-methoxymellein, described above. The second, and far more frequent, use for the malonic acid/acetate pathway is to provide the six-carbon A-ring of flavonoids, by condensation of three molecules of malonyl CoA.

The inter-relationships between the shikimic acid and malonic acid/acetate pathways, and the various groups of phenolic compounds are summarized in Figure 11.3. The pathways will be described in more detail in the following sections.

The shikimic acid pathway provides the aromatic amino acid, phenylalanine, from which the phenylpropanoids are all derived

The initial substrates for the shikimic acid pathway are formed from carbohydrate breakdown; erythrose 4-phosphate is a product of the oxidative pentose phosphate pathway, while phospho*enol*pyruvate is produced in glycolysis (see Chapter 6). The shikimic acid pathway is shown in detail in Figure 11.4.

The first seven reactions of the shikimic acid pathway result in the formation of chorismic acid, which is sometimes defined as the end-point of the pathway. However, subsequent metabolism of chorismic acid to the aromatic amino acids, tyrosine and phenylalanine, has been included here as an essential part of the pathway that leads to phenylpropanoids (Figure 11.4). The shikimic acid pathway is present in plants, fungi, and bacteria, but absent from animals, which therefore require a dietary intake of aromatic amino acids.

The entire shikimic acid pathway, including the final steps leading to phenylalanine and tyrosine, takes place in the plastids except for the reaction catalyzed by chorismate mutase (as discussed below).

Figure 11.3 Overview of the pathways leading to the production of the major plant phenolic groups. Most plant phenolics are derived from phenylalanine, produced in the shikimic acid pathway. Phenylalanine provides the essential 6-carbon ring and 3-carbon side chain that is central to all phenylpropanoids (shorthand used here shows the 6-carbon phenyl ring (see Figure 11.1) and the various chain-lengths of carbon to which it is joined (e.g. C₆-C₃). Phenylalanine then enters the core phenylpropanoid pathway, generating intermediates (e.g. cinnamic acid) that can be used to form simple phenolics and benzoic acid derivatives, and the more complex phenolics such as lignin. While phenylpropanoids are also incorporated into the structure of flavonoids and proanthocyanidins, additional carbon units are provided by the malonic acid/acetate pathway. This pathway produces malonyl CoA that is condensed to form an additional phenyl ring for the flavonoid and proanthocyanidin molecules. (From *Plant Physiology, 2nd edition*, edited by Lincoln Taiz & Eduardo Zeiger, Figure 8.14, 2006. Reprinted with permission from Sinauer Associates Inc.)

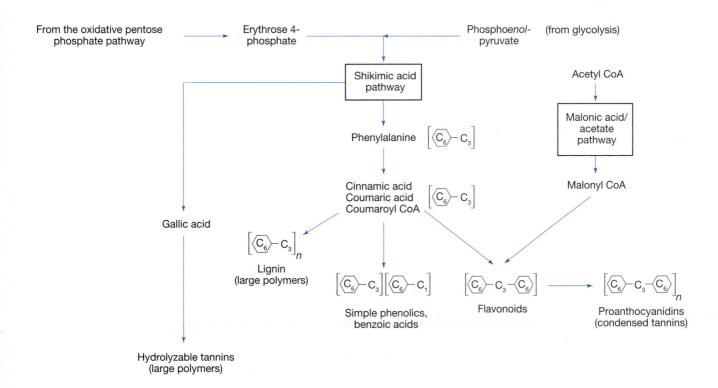

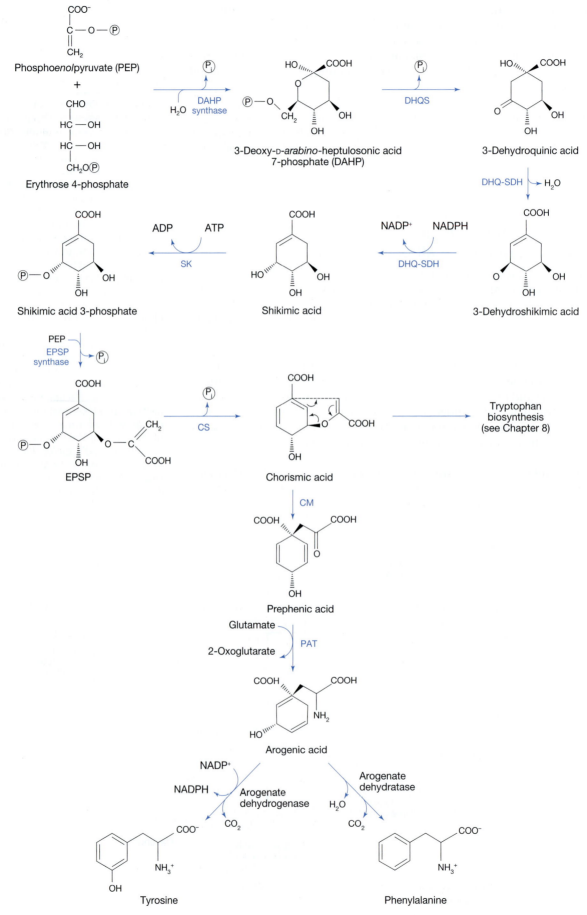

Phosphoenolpyruvate (PEP)

Erythrose 4-phosphate

3-Deoxy-D-arabino-heptulosonic acid 7-phosphate (DAHP)

3-Dehydroquinic acid

Shikimic acid 3-phosphate

Shikimic acid

3-Dehydroshikimic acid

EPSP

Chorismic acid

Tryptophan biosynthesis (see Chapter 8)

Prephenic acid

Arogenic acid

Tyrosine

Phenylalanine

Figure 11.4 The shikimic acid pathway. Some versions of this pathway show chorismic acid as the end-point. However, we have extended the pathway to include the formation of the aromatic amino acids, tyrosine and phenylalanine, as phenylalanine is the precursor for the majority of phenylpropanoid products. Note that the conversion of chorismic acid to prephenic acid, by chorismate mutase, involves an intramolecular rearrangement where the *enol*pyruvyl side-chain (i.e. –CCH$_2$COO$^-$) migrates to a different position on the phenyl ring in a reaction classed as a Claisen rearrangement. Note also that the subsequent structures for arogenic acid, tyrosine, and phenylalanine are slightly re-orientated in order to conform to the structures used elsewhere in this chapter.

Abbreviations: PEP, phospho*enol*pyruvate; DAHP, 3-deoxy-D-*arabino*heptulosonate-7-phosphate; DAHP synthase, 3-deoxy-D-*arabino*heptulosonate-7-phosphate synthase; DHQS, 3-dehydroquinate synthase; DHQ-SDH, dehydroquinate dehydratase-shikimate dehydrogenase; SK, shikimate kinase; EPSP synthase, 5-*enol*pyruvylshikimic acid 3-phosphate synthase; EPSP, 5-*enol*pyruvylshikimic acid 3-phosphate; CS, chorismate synthase; CM, chorismate mutase; PAT, prephenate aminotransferase.

The shikimic acid pathway begins with a condensation reaction between erythrose 4-phosphate and phospho*enol*pyruvate, catalyzed by 3-deoxy-D-arabino-heptulosonate 7-phosphate (DAHP) synthase. The plant and bacterial enzymes share only 20% amino acid identity and have very different regulatory properties. While the bacterial enzyme is strongly inhibited by the end-products of the pathway, tyrosine and phenylalanine, these amino acids have no effect on the plant enzyme. It seems that the plant DAHP synthase is regulated mainly at the level of gene expression, with light, wounding, and pathogens all influencing its expression. As we shall see, stress-induced expression is a fairly common feature of genes that encode enzymes within the phenolic biosynthetic pathways. As many phenolic compounds provide protection against pests, pathogens, and other stresses, this stress-induced response is likely to serve a valuable defensive function.

The remaining reactions, leading to chorismic acid formation, show further significant differences between organisms. In prokaryotes, these reactions are catalyzed by separate enzymes, whereas in eukaryotic microbes, such as *Aspergillus nidulans*, *Neurospora crassa*, and *Saccharomyces cerevisiae*, a single multi-enzyme AROM complex carries out all of these reactions. In higher plants, the third and fourth reactions (3-dehydroquinate dehydratase and shikimate dehydrogenase) are catalyzed by a single bifunctional enzyme that dehydrates 3-dehydroquinic acid to 3-dehydroshikimic acid and then reduces this to shikimic acid. The enzyme, dehydroquinate dehydratase-shikimate dehydrogenase, a single 59-kDa polypeptide, has been purified from a number of plant species. The dehydratase activity is found in the amino terminal half, while the dehydrogenase is in the carboxy-terminal half. In plants, the remaining steps in the pathway require separate enzymes that are structurally similar to their prokaryotic counterparts.

The most fully characterized of all enzymes in the shikimic acid pathway is 5-*enol*pyruvylshikimic acid 3-phosphate (EPSP) synthase, which catalyzes the reaction that precedes chorismic acid formation. The reason why this enzyme is particularly well characterized is because it is inhibited by glyphosate, which is the world's best-selling herbicide (marketed as Roundup™). Glyphosate binds to EPSP synthase uncompetitively with respect to one of its substrates, shikimic acid 3-phosphate, and competitively with respect to its co-substrate, phospho*enol*pyruvate. Glyphosate is highly specific and does not affect any other phospho*enol*pyruvate-dependent enzymes. Glyphosate-resistant crops have been produced by introducing a gene from a naturally occurring *Agrobacterium tumefasciens* strain (CP4) that encodes a glyphosate-insensitive EPSP synthase. A number of glyphosate-resistant crops, including cotton, soybean, and canola (oilseed rape), have been grown commercially for several years, particularly in North America.

Following the EPSP synthase reaction, chorismate synthase forms chorismic acid from EPSP. Chorismate synthase requires reduced FMN (flavin mononucleotide) as a cofactor, although the reaction does not involve a net redox change, which is unusual for a reaction that requires a reductant.

Chorismic acid is situated at a branch point in the pathway leading to the synthesis of aromatic amino acids. One branch leads to the formation of phenylalanine and tyrosine, while the other branch results in tryptophan formation (Figure 11.4, also see Chapter 8). The reaction that commits chorismic acid to the phenylalanine and tyrosine branch is catalyzed by chorismate mutase, which converts chorismic to prephenic acid. Chorismate mutase is the only enzyme in the entire pathway that is not exclusive to the plastid, as it exists in two isoforms, one plastidic (CM1) and the other cytosolic (CM2). The plastidic form, CM1, is regulated in a manner that is consistent with its role in aromatic amino acid biosynthesis. For example, CM1 is feedback-inhibited by phenylalanine and tyrosine and activated by tryptophan. This regulation would serve to divert chorismic acid into the tryptophan-forming branch when there is either an accumulation of phenylalanine and/or tyrosine and a deficiency of tryptophan. Similarly, an excess of tryptophan or deficiency of phenylalanine and/or tyrosine would stimulate CM1 and divert chorismic acid towards phenylalanine and tyrosine production. These regulatory properties of CM1, together with its location in the plastid along with that of all other enzymes of the shikimic acid pathway, are consistent with the conclusion that CM1 is involved in the biosynthesis of phenylalanine and tyrosine. In contrast, the cytosolic location of CM2, together with its apparent lack of responsiveness to changes in the concentrations of phenylalanine, tyrosine, and tryptophan, indicate that this isoform is unlikely to be involved in amino acid biosynthesis. The function of CM2 in plants has yet to be determined. Given that the genes for CM1 and CM2 have been cloned in a number of organisms, including arabidopsis, it should be possible to investigate the functional differences between these isoforms by the use of genetic and biochemical approaches (see Chapter 2).

Nitrogen is introduced into the shikimic acid pathway at the point where prephenate aminotransferase transfers an amino group on to prephenic acid to form arogenic acid. Glutamate is the preferred amino donor, although aspartate may also be used at a slower rate.

Phenylalanine, the precursor for phenylpropanoid biosynthesis, is formed from arogenic acid (Figure 11.4). The same substrate can be used to form another aromatic amino acid, tyrosine. Two different enzymes are involved. Arogenate dehydrogenase catalyzes the NADP-dependent oxidative decarboxylation of arogenic acid to produce tyrosine. Arogenate dehydrogenase is strongly inhibited by its product, tyrosine. Phenylalanine is produced by arogenate dehydratase, which also decarboxylates arogenic acid. This reaction does not require NADP and instead relies on C–O cleavage by elimination of water (dehydratase reaction). Arogenate dehydratase is also inhibited by its product, phenylalanine.

The shikimic acid pathway is regulated by end-product inhibition and is affected by wounding and pathogen attack

The shikimic acid pathway is a major biosynthetic pathway in plants. It provides precursors for a number of plant products, notably the flavonoids and lignin that can accumulate in significant quantities (up to 50% of the dry weight in some tissue). It is also the pathway that forms the precursors for the aromatic amino acids (phenylalanine, tyrosine, and tryptophan; Chapter 8).

Consequently, the shikimic acid pathway must be regulated in order to meet the varying demands for phenylpropanoid precursors, without compromising the fundamental requirement for amino acid biosynthesis.

One form of regulation of the shikimic acid pathway is feedback inhibition. An example of this regulation is the inhibition of arogenate dehydratase by its product, phenylalanine. The feedback works because phenylalanine will accumulate and inhibit its own synthesis when there is a reduced demand for phenylalanine for phenylpropanoid production.

Another form of regulation of the shikimic acid pathway operates at the level of gene expression. Genes encoding DAHP synthase, shikimate kinase, EPSP synthase, and chorismate synthase are all induced in response to the same environmental signals, i.e. wounding and pathogen attack, that result in increased phenylpropanoid biosynthesis. Furthermore, induction of these genes coincides with increased expression of the gene encoding phenylalanine ammonia lyase (PAL), the first committed enzyme of the phenylpropanoid pathway. Hence, plants appear to be able to regulate the supply of aromatic amino acids and phenylpropanoids according to demand and in response to wounding and pathogen attack.

The core phenylpropanoid pathway provides the basic phenylpropanoid units that are used to make most of the phenolic compounds in plants

The core phenylpropanoid pathway converts phenylalanine to the simple phenylpropanoids *trans*-cinnamic acid, *p*-coumaric acid, and *p*-coumaroyl CoA. Its central role in the production of plant phenolics is illustrated within the overview in Figure 11.3 and in detail in Figure 11.5. Several branches radiate from these core reactions. Early branch points lead to the formation of benzoic acid derivatives, simple phenylpropanoids, coumarins and the precursors for lignin formation and stilbenes. The later incorporation of three malonyl CoA molecules leads to the formation of the chalcones, from which all of the various flavonoids can be produced (as described below).

The first step in the core phenylpropanoid pathway is catalyzed by PAL. The reaction catalyzed by PAL results in the release of ammonia and the formation of a C=C double bond in the product, *trans*-cinnamic acid (Figure 11.5). The ammonia released from phenylalanine is reassimilated by glutamine synthetase (see Chapter 8). PAL is a tetramer and its subunits (ranging between 75 and 83 kDa depending on the species) are encoded by a multigene family in most species. The different *PAL* genes are expressed in different cells and tissues and are subject to regulation by environmental signals, including wounding, light (both visible and ultraviolet B radiation), and pathogen attack, which all lead to rapid induction of one or more *PAL* genes. PAL is a key enzyme in regulating flux into the phenylpropanoid pathway and its induction in response to pathogen attack is an important mechanism that leads to enhanced synthesis of phenolic compounds, such as phytoalexins, that help to defend plants from further attack.

Cinnamate 4-hydroxylase (C4H) introduces a hydroxyl group into the phenyl group of *trans*-cinnamic acid, thus forming the first phenolic product, in the form of *p*-coumaric acid. The enzyme is a cytochrome P450 monooxygenase, characterized by the presence of a heme group that absorbs light at a wavelength of 450 nm. Most cytochrome P450 monooxygenases are membrane-bound, usually within the endoplasmic reticulum, but sometimes associated with the mitochondria. The cytochrome P450 monooxygenase is responsible for transferring electrons from O_2 to the substrate, which in this case is *trans*-cinnamic acid. Although molecular oxygen is used, only one atom is transferred to the substrate, hence this is a monooxygenase reaction. The other atom of oxygen is transferred to NADPH, thus making the general monooxygenase reaction:

$$NADPH + H^+ + substrate + O_2 \longrightarrow NADP^+ + substrate\text{-}OH + H_2O$$

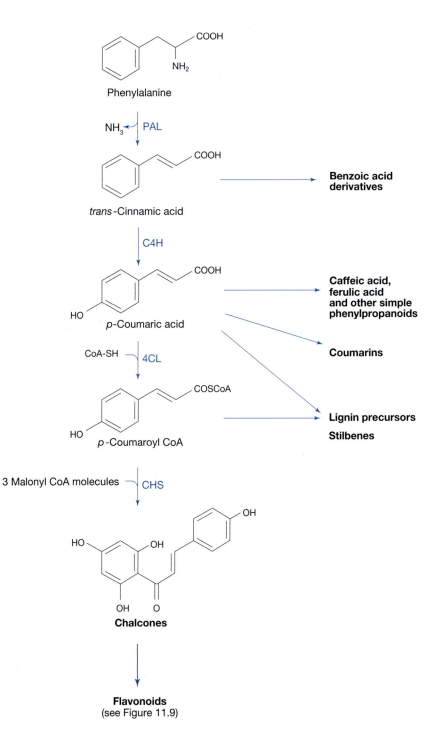

Figure 11.5 The formation of the major phenolic groups from phenylalanine and the core phenylpropanoid pathway. As an expanded version of Figure 11.3, this figure shows the central role of the core phenylpropanoid pathway and the branch-points leading to the synthesis of the major phenolic groups, along with their characteristic structures. Abbreviations: PAL, phenylalanine ammonia lyase; C4H, cinnamate 4-hydroxylase; 4CL, 4-coumarate:CoA ligase; CHS, chalcone synthase.

The final step in the core phenylpropanoid pathway is the formation of *p*-coumaroyl CoA from coumaric acid. This reaction is catalyzed by 4-coumarate: CoA ligase (4CL):

$$ATP + p\text{-coumaric acid} + CoASH \longrightarrow AMP + PP_i + p\text{-coumaroyl CoA}$$

4CL will also react with caffeic, ferulic, and sinapic acids to form their respective CoA esters, as discussed later. 4CL is an important enzyme as it sits at the branch point between the flavonoid and lignin biosynthetic pathways (Figure 11.5). It has been successfully manipulated to produce trees with reduced lignin content and more rapid growth rates (see Further Reading). 4CL exists in several isoforms, encoded by a multi-gene family in a range of species. These 4CL families can be subdivided into two classes, I and II, that appear to have distinct metabolic functions. Class I 4CLs are associated with lignin biosynthesis, while class II 4CLs are involved in flavonoid biosynthesis.

The hydroxycinnamic acids, ferulic, caffeic, and sinapic acid, are produced from *p*-coumaric acid (Figure 11.5) by the introduction of hydroxyl and methoxyl groups on to the phenyl ring. Ferulic acid is an important component of the primary cell wall of many monocots, particularly the gramineae, where it forms an ester with the arabinosyl residues within arabinoxylan polysaccharides. Further ferulic acid groups can be attached to the feruloyl moiety to form ferulate dimers and trimers, via oxidative reactions within the Golgi body. The feruloyl groups can form cross-links between adjacent polysaccharide chains and also between polysaccharides and lignins. These cross-links provide structural rigidity to the cell wall and are an important means of restricting cell expansion. Cereal grains are a particularly rich source of ferulic acid, which is a valuable antioxidant that is often added to food, both for human and animal consumption.

The coumarins (e.g. umbelliferone and coumarin; Figure 11.2b) are thought to be formed from *p*-coumaric acid, via hydroxylation followed by isomerization and ring closure, although the component enzymes have yet to be identified. The recent identification of a range of coumarins in arabidopsis offers the promise of using molecular genetic approaches to identify the biosynthetic pathway. The furanocoumarins differ from coumarins in having a furan ring attached (Figure 11.2b). The reactions leading to furanocoumarins also originate from *p*-coumaric acid. Some of the component enzymes have been identified in a limited range of plants. An overview of the relationship between the precursor, *p*-coumaric acid, and the coumarins and furanocoumarins is shown in Figure 11.6 along with some of the reactions thought to be involved.

Figure 11.6 Some of the proposed reactions involved in coumarin and furanocoumarin biosynthesis. *p*-Coumaric acid, formed in the core phenylpropanoid pathway (Figure 11.5) serves as the common precursor for the formation of coumarins (e.g. scopoletin and 7-hydroxycoumarin) and furanocoumarins (e.g. psoralen). The pathway to the coumarins is based on identification of some of the intermediates, as none of the enzymes have been identified. However, the enzymes involved in furanocoumarin formation are known to include: **1** Umbelliferone:*O*-dimethylallyl transferase. This enzyme carries out a prenyltransferase reaction, where a hydrocarbon group from dimethylallyl diphosphate (DMAPP) is transferred to the C-6 of umbelliferone. This type of reaction features in the terpenoid biosynthesis pathway, where DMAPP is an important precursor (see Chapter 12). **2** Marmesin synthase, a cytochrome P450 NADPH-dependent oxidase. **3** Psoralen synthase, which is also a cytochrome P450 NADPH-dependent oxidase. Note that reaction 1 subsequently leads to the formation of the furan ring of the furanocoumarins (as shown in Figure 11.2b, psoralen structure).

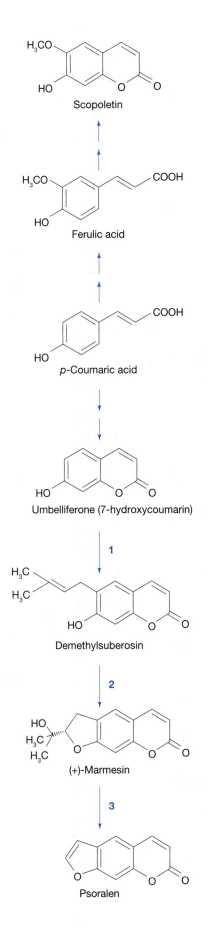

The pathways for biosynthesis of the benzoic acid derivatives (e.g. vanillin and salicylic acid; Figure 11.2c) are still not resolved. There is evidence to support two different routes, either from the core phenylpropanoid pathway, via phenylalanine (Figure 11.7a), or from the shikimic acid pathway (Figure 11.7b). Three different routes have been proposed for the formation of benzoic acids from the core phenylpropanoid pathway, via phenylalanine, to benzoic acid (Figure 11.7a). In the first of these, the CoA-dependent β-oxidation route (Figure 11.7a, route I), the sequence of reactions resembles those of the fatty acid β-oxidation pathway (Chapter 9). Here, cinnamoyl CoA is hydrated to form β-hydroxyphenylpropionyl CoA, which is oxidized to β-oxophenylpropionyl CoA and then cleaved in a CoA-dependent reaction to form benzyl CoA and then benzoic acid. In the most recently proposed of the three routes, the CoA-dependent non-oxidative pathway (Figure 11.7a, route II), cinnamoyl CoA is thought to be converted to benzaldehyde and then to benzoic acid. Finally, the CoA-independent, non-oxidative pathway is thought to involve hydration of the free acid (*trans*-cinnamic acid), removal of an acetyl group, followed by oxidation of benzaldehyde to form benzoic acid (Figure 11.7a, route III). None of the component enzymes or genes has been identified for any of these three proposed routes, with the exception of PAL and 4CL. Evidence has come from radioisotope-feeding experiments (e.g. [14]C precursors, see Chapter 2), using either cell-free extracts or plant cell cultures.

The synthesis of benzoic acids from the shikimic acid pathway derivative, chorismic acid (Figure 11.7b) was first discovered in bacteria. Recent experiments with arabidopsis provide strong support for the involvement of this pathway in the synthesis of the benzoic acid derivative, salicylic acid. Arabidopsis mutants that lack isochorismate synthase (*ics1* mutants), the enzyme responsible for forming isochorismic from chorismic acid (Figure 11.7b), were found to have also lost the capacity to synthesize salicylic acid in response to pathogen attack. While this study provides the first genetic evidence for the formation of a benzoic acid derivative via the shikimic acid route (Figure 11.7b), the enzymes that convert isochorismic acid to salicylic acid have yet to be characterized in higher plants.

Figure 11.7 Proposed routes for the biosynthesis of benzoic acids (a) from the core phenylpropanoid pathway. Abbreviations: 4CL, 4-coumarate:CoA ligase; PAL, phenylalanine ammonia lyase.

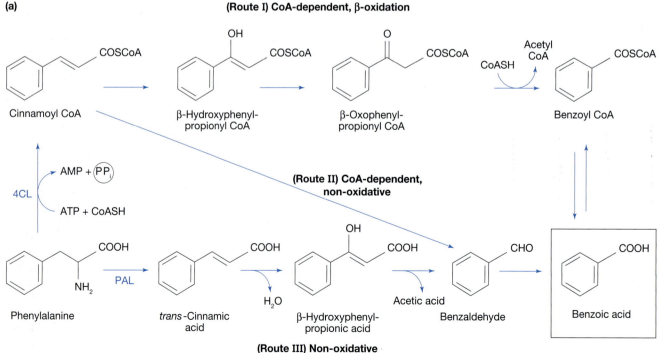

(a)

(Route I) CoA-dependent, β-oxidation

Cinnamoyl CoA   β-Hydroxyphenyl-propionyl CoA   β-Oxophenyl-propionyl CoA   Benzoyl CoA

(Route II) CoA-dependent, non-oxidative

Phenylalanine   *trans*-Cinnamic acid   β-Hydroxyphenyl-propionic acid   Benzaldehyde   Benzoic acid

(Route III) Non-oxidative

We are left to conclude, at this stage, that there are a number of possible reactions that could result in the production of benzoic acid derivatives in plants. Until the component enzymes and genes have been isolated, and thorough labeling experiments have been completed, we are unable to conclude with any certainty which of these routes occurs in plants. It is quite possible that all of the reactions described above serve a role in benzoic acid biosynthesis and that there are species- and tissue-specific differences in the expression of each alternative route.

Flavonoids are produced from chalcones, formed from the condensation of *p*-coumaroyl CoA and malonyl CoA

The basic C-6–C-3–C-6 structure, common to all flavonoids, is derived from two separate pathways (Figure 11.8). The core phenylpropanoid pathway produces *p*-coumaroyl CoA, which provides the skeleton for synthesis of ring B, as well as the three-carbon bridge that forms ring C. The malonic acid/acetate pathway provides the three molecules of malonyl CoA that condense together to form the six-carbon A-ring (Figure 11.8).

Because there are so many different groups of flavonoids, we shall first present an overview of the processes involved in generating the major flavonoid groups (as summarized in Figure 11.9) before providing a more detailed account of the pathways.

The three-ringed flavonoid structure is established in the first committed reaction of the flavonoid biosynthesis pathway, where three molecules of malonyl CoA are condensed with a molecule of *p*-coumaroyl CoA to form a chalcone, in a reaction catalyzed by chalcone synthase (CHS). Chalcones may also be formed by a CHS/chalcone reductase complex in some species, as discussed below. The stilbenes are also formed from *p*-coumaroyl CoA and malonyl CoA, in a reaction catalyzed by stilbene synthase (STS) (Figure 11.9).

Once formed, chalcones serve as the precursors for all of the various groups of flavonoids, as outlined in Figure 11.9, and discussed in detail in the next

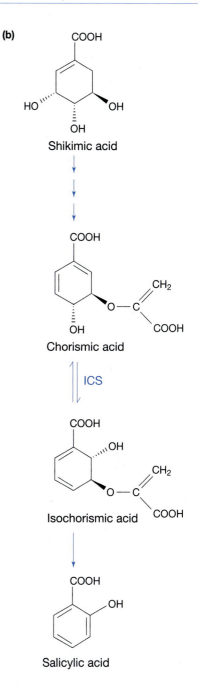

(b)

Shikimic acid

Chorismic acid

ICS

Isochorismic acid

Salicylic acid

Figure 11.7 (continued) Proposed routes for the biosynthesis of benzoic acids (b) from the shikimic acid pathway. Abbreviations: ICS, isochorismate synthase. (Reprinted with permission from the *Annual Review of Plant Biology*, Volume 54, © 2003 by Annual Reviews, www.annualreviews.org)

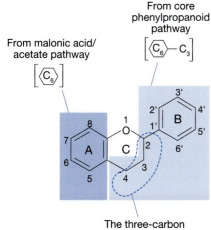

From core phenylpropanoid pathway

From malonic acid/acetate pathway

The three-carbon bridge

Figure 11.8 Basic flavonoid skeleton showing its origins in the shikimic acid and malonic acid/acetate pathways, and the numbering of the carbon atoms on the three rings A, B and C.

Figure 11.9 Overview of the pathways and branch-points leading to the biosynthesis of the major flavonoid groups and related compounds. The key reactions that lead to the formation of the major flavonoid groups (in bold). Further detail of the reactions and structures is provided in Figures 11.10 to 11.15 as indicated here. The first committed step in flavonoid biosynthesis is catalyzed by chalcone synthase, which condenses *p*-coumaroyl CoA with three molecules of malonyl CoA to produce the chalcone, tetrahydroxychalcone. Note that malonyl CoA is produced by acetyl CoA carboxylase, the same enzyme that features in fatty acid biosynthesis (Chapter 9). In some species, malonyl CoA and *p*-coumaroyl CoA are converted to the chalcone isoliquiritigenin by a complex of chalcone synthase and chalcone reductase. Abbreviations: CHS/CHR, chalcone synthase/chalcone reductase; ACC, acetyl CoA carboxylase; STS, stilbene synthase.

section. Chalcones are converted to flavanones by isomerization. The flavanones are important intermediates, as they can be used to form several other major flavonoid groups. The flavones, for example, are formed from flavanones by a dehydration reaction, while flavanones may also be isomerized to produce isoflavones, which are subsequently used to produce isoflavonoids in some species. Flavanones may also be reduced to flavan-4-ols, which serve as precursors for the formation of the phlobaphene polymers. Finally, hydroxylation reactions convert flavanones to dihydroflavonols, which are converted to flavonols by a desaturation reaction (i.e. introduction of a double bond).

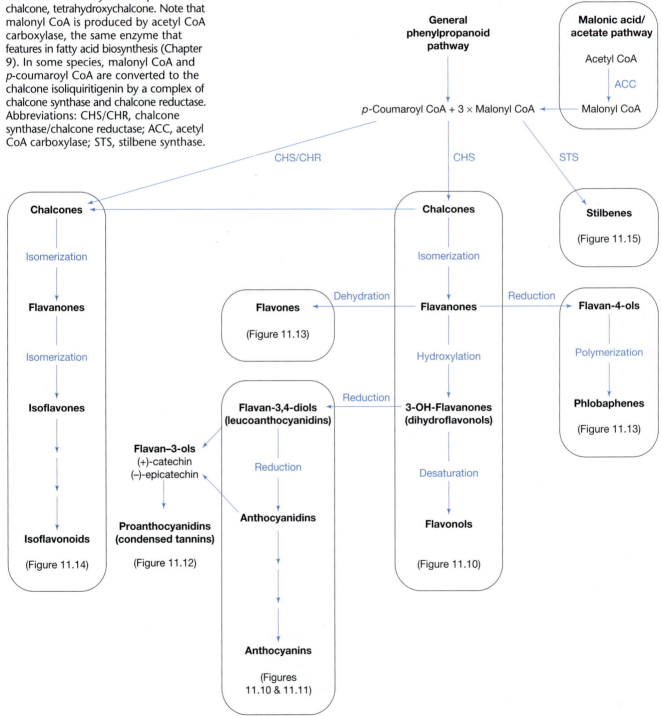

The dihydroflavonols (also called 3-OH-flavanones) are precursors for the synthesis of the colored anthocyanin pigments. The dihydroflavonols are reduced to form the colorless leucoanthocyanidins (also called flavan-3,4-diols), which are converted to anthocyanidins and then to the anthocyanins. The leucoanthocyanidins also serve as precursors for the biosynthesis of the proanthocyanidins, which consist of polymers of leucoanthocyanidins, together with anthocyanidins in some plants. The detailed reactions involved in the formation of these various flavonoid groups are described in the sections that follow.

The reaction catalyzed by chalcone synthase is an important regulatory step in flavonoid biosynthesis

CHS forms the chalcone tetrahydroxychalcone by condensing three molecules of malonyl CoA together with *p*-coumaroyl CoA (Figure 11.10). CHS is an important enzyme as it is rapidly induced in response to a range of environmental stresses, and appears to exert a major control over the flux through the flavonoid pathway.

Environmental stresses such as wounding, pathogen attack and ultraviolet B radiation can all induce CHS gene expression, and CHS protein begins to accumulate within an hour of exposure. This rapid response is accompanied by an increased rate of flavonoid production, which provides the plant with protection against the various stresses. Ultraviolet B-absorbing flavonoids, for example, accumulate within the epidermal cells where they provide a protective sun-screen, while some flavonoids function as phytoalexins whose antimicrobial and antifungal properties help to restrict the invasive growth of pathogens through the wounded plant tissue.

Chalcones may also be formed from malonyl CoA and *p*-coumaroyl CoA in some legume species by an enzyme complex formed by the association of CHS with an NADPH-dependent chalcone reductase. This complex produces the chalcone, isoliquiritigenin, which can be used as a precursor in the biosynthesis of the isoflavonoids, which are unique to the leguminosae, as discussed on page 388.

The flavanones are formed from chalcones and are precursors for the biosynthesis of a range of flavonoids

The flavanones are produced from chalcones by an isomerase reaction that forms a closed ring (ring C) between the two phenolic rings (ring A and B). The reaction is catalyzed by chalcone isomerase (CHI; Figure 11.10), of which there are two types in higher plants. Type I CHIs are broadly distributed and convert tetrahydroxychalcone into the flavanone, naringenin. Type II CHIs are unique to legumes, and convert the chalcone, isoliquiritigenin, into the flavanone, liquiritigenin. The type II CHIs are associated with isoflavonoid production (see below) and CHI gene expression appears to be coordinately regulated with isoflavonoid-specific genes.

The chalcones are relatively unstable compounds that will isomerize spontaneously in solution to form flavanones and they tend not to accumulate in plants for this reason. CHI, like CHS, is induced by light, ultraviolet B, and pathogens. However, because flavanones can form spontaneously at physiological pH, CHI was not thought to be able to exert much control over pathway flux. Despite this assertion, overexpression of a *Petunia CHI* gene in tomato was found to result in a 78-fold increase in fruit flavonols. This response indicates that CHI does exert some control over flux into the flavonoid pathway.

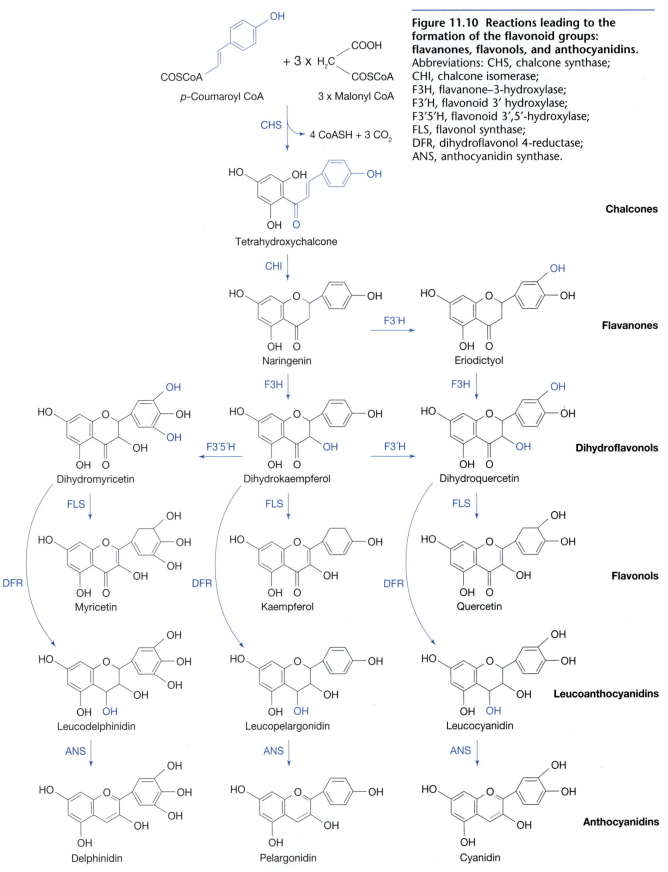

Figure 11.10 Reactions leading to the formation of the flavonoid groups: flavanones, flavonols, and anthocyanidins. Abbreviations: CHS, chalcone synthase; CHI, chalcone isomerase; F3H, flavanone–3-hydroxylase; F3'H, flavonoid 3' hydroxylase; F3'5'H, flavonoid 3',5'-hydroxylase; FLS, flavonol synthase; DFR, dihydroflavonol 4-reductase; ANS, anthocyanidin synthase.

The variety of anthocyanin structures, and consequently of the range of flower and fruit colors, can be traced back to reactions that introduce variety into the structures of the flavanones, dihydroflavonols, and flavonols (Figure 11.10). First, the flavanones are converted to dihydroflavonols by the action of flavanone-3-hydroxylase (F3H), a member of the highly conserved 2-oxo-glutarate-dependent dioxygenase family, that adds a hydroxyl group at the C-3 position within the central C-ring. However, it is the additions to the B-ring that are the primary cause of color differences among the anthocyanin pigments. B-ring modifications include the addition of hydroxyl groups to flavanones and dihydroflavonols, in reactions catalyzed by at least two different hydroxylase enzymes. Flavonoid 3′-hydroxylase (F3′H) catalyzes the hydroxylation at the 3′ position, while flavonoid 3′, 5′-hydroxylase (F3′5′H) introduces hydroxyl groups at both the 3′ and 5′ positions. Consequently, the dihydroflavonols dihydromyricetin (three hydroxyl groups within the B-ring), dihydrokaempferol (one hydroxyl group), and dihydroquercetin (two hydroxyl groups) are formed. These dihydroflavonols serve as precursors for both the flavonols (myricetin, kaempferol, and quercetin, respectively) and the leucoanthocyanidins (leucodelphinidin, leucopelargonidin, and leuco-cyanidin, respectively), each of which retains the hydroxyl groups at the same positions within the B-ring as they were in the precursor dihydroflavonol. Although colorless to humans, the flavonoids are important co-pigments in flowers, as they absorb within the ultraviolet spectrum and can form nectar guides for insect pollinators (Box 11.2).

The leucoanthocyanidins are also colorless and are converted to the colored anthocyanins by anthocyanidin synthase (also called leucoanthocyanidin dioxygenase in some texts). The resulting anthocyanidins (delphinidin, pelargonidin, and cyanidin; Figure 11.10) are unstable and do not accumulate *in vivo*. The attachment of a sugar, to form the stable anthocyanidin glycosides, or anthocyanins (Figure 11.11) both stabilizes the structure and makes it water-soluble so that these pigments can accumulate in the vacuole. The color of the resulting anthocyanin pigment is determined by the anthocyanidin group, which, in turn, depends on the leucoanthocyanidins.

Figure 11.11 The general structure of an anthocyanin molecule and examples of anthocyanin structures and enzymes that modify them. Note that anthocyanins consist of an anthocyanidin molecule attached to a sugar. The addition of hydroxyl groups to the B-ring, as well as methoxylations, complexes with metal ions and co-pigments, and attachment of different sugars (e.g. glucose, rhamnose) all contribute to the range of colors of the various anthocyanins. For example, pelargonidins are orange, pink or red; cyanidins are red or magenta, while delphinidins produce blue and deep purple colors (see Box 11.2).

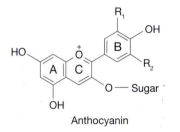

Anthocyanin

Examples of anthocyanin structures

Anthocyanin	R_1	R_2	Sugar
Cyanidin 3-O-β-D-glucoside	OH	H	Glucose
Delphinidin 3-O-β-D-glucoside	OH	OH	Glucose
Pelargonidin 3-O-β-D-glucoside	H	H	Glucose
Malvidin 3-O-β-D-glucoside	OCH$_3$	OCH$_3$	Glucose
Petunidin 3-O-β-D-glucoside	OH	OCH$_3$	Glucose

Examples of enzymes that modify anthocyanin structures

Enzyme	Example reaction	Modification
Anthocyanidin 3-glucosyl transferase	Cyanidin + UDP-glucose → Cyanidin 3-O-glucoside + UDP	Adds a glucosyl group to an anthocyanidin to form an anthocyanin
Rhamnosyltransferase (RT)	Cyanidin 3-O-glucoside + rhamnose → Cyanidin rutinoside	Adds a rhamnosyl group to an anthocyanin
O-Methyl transferase (OMT)	Petunidin 3-O-glucoside + S-adenosyl methionine → Malvidin 3-O-glucoside + S-adenosyl homocysteine	Transfers a methyl group from S-adenosyl methionine to an anthocyanin

Box 11.2 Phenolics and flower color

Flavonoid pigments produce most of the colors in flowers

Two groups of chemicals are responsible for virtually all of the different colors found in plants, the flavonoids and carotenoids.

The carotenoids are terpenoid pigments that produce yellow, orange, and red/orange coloration. Because of their hydrophobicity—carotenoids, like most terpenoids, are oily hydrocarbons—they accumulate in a specialized type of plastid, the chromoplast (see Chapter 3) rather than in the vacuole. The structure and biosynthesis of these terpenoid pigments is described further in Chapter 12.

The flavonoids, in contrast to the carotenoids, are water-soluble phenolic compounds that accumulate in the cell vacuole. Flavonoids produce a far wider variety of colors than the carotenoids, ranging across the spectrum from red, through crimson and yellow, to the blues and purples. The main flavonoid group responsible for this color range is the anthocyanins, with other flavonoids acting as accessory pigments.

Anthocyanin color is influenced by several factors, one of the most important of which is the B-ring substitution pattern. As described in this chapter, the B-ring may be hydroxylated in different positions to produce different classes of anthocyanin pigments (see Figures 11.10 and 11.11). The hydroxyl groups are introduced at an early stage in the pathway by one of two different flavonoid hydroxylases. F3′H adds a hydroxyl group at the 3′ position in the B-ring, while F3′5′H adds two hydroxyl groups, at the 3′ and 5′ positions. Further reactions (described in the text and in Figure 11.11) produce the respective anthocyanin groups: pelargonidins from dihydrokaempferol, cyanidins from dihydroquercetin, and delphinidins from dihydromyricetin. The pelargonidins provide orange, pink, and red colors (e.g. *Lupinus podophyllus*, Plate 11.2), while the cyanidins form the red and magenta colors (*Papaver rhaeas*, Plate 11.3) and delphinidins the purple and blue colors (e.g. *Delphinium occidentale*, Plate 11.4).

Further changes in color can result from the introduction of methoxyl groups to form peonidin, petunidin, and malvidin (see Figure 11.11), which provide bright red, deep purple, and burgundy red, respectively.

Anthocyanin pigments are also affected by pH. As the pH becomes more alkaline, hydrogen ions are lost from the B-ring hydroxyl groups. The electrons are less confined in the basic form and the color becomes progressively more blue. The opposite effect occurs as the pH becomes more acidic, for example, with delphinidin (Figure 1).

This pH effect has confounded many attempts to alter the color of flowers both by conventional breeding and by genetic engineering, as explained later in this box.

Anthocyanin color can also be highly influenced by the presence of metal ions, such as Mg^{2+}, Al^{3+}, Cu^{2+}, and Mn^{2+}. These metal ions form complexes with anthocyanins and, in some cases further complexes may form with flavonols and nonflavonoid co-pigments. Metal ions tend to have a blueing effect on the overall color. Metal ion–anthocyanin complexes are known to be responsible for the blue colors of a number of plants, including the Himalayan blue poppy (*Meconopsis grandis*), where ferric (Fe^{3+}) and magnesium (Mg^{2+}) ions form complexes with anthocyanins and flavonols.

Flavonol and flavone co-pigments are important, not just because they form complexes with anthocyanins, but also because they absorb ultraviolet light. The flavonols (e.g. kaempferol, quercetin) absorb at slightly longer wavelengths (about 360–380 nm) than flavones (e.g. luteolin, apigenin; 335–350 nm) but the effect is the same. These pigments can be seen by insects, such as bees, that can see in the UV range of the spectrum. Distinct patterns, invisible to us, can be produced by the distribution of flavones and flavonols across the flower to form so-called nectar guides that draw the pollinator towards the pollen and nectar. These patterns can be visualized by ultraviolet photography, as shown in Plate 11.5 (*Potentilla anserina*), and are found in a wide range of species.

Figure 1 The effect of pH on the anthocyanin pigment delphinidin.

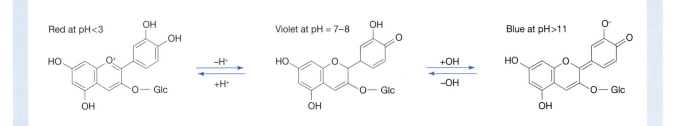

Box 11.2 Phenolics and flower color (continued)

Some plants change the color of their flowers when growing on different types of soil

The composition of soils can affect the color of flowers in some plant species. One well-known example is the hydrangea (*Hydrangea macrophylla*), a popular garden plant in Europe and North America and a native plant of Japan and China (see Plate 11.6).

When grown in alkaline soils the flowers are pink, while on acid soils the flowers are blue. Having seen the effect of pH on anthocyanin color (as explained above) this observation seems counterintuitive—alkaline solutions cause anthocyanins to turn blue. However, the effect on hydrangea color is not due to differences in the pH of the cell. It is the result of the effect of soil pH on the uptake of aluminum ions. As the soil pH becomes more acidic, aluminum becomes more soluble and is more readily taken up by plant roots. Once inside the cell, aluminum forms a complex with delphinidin 3-*O*-glucoside, the only anthocyanin pigment in hydrangea. The complex is further stabilized by the incorporation of quinic acid groups to render the pigment blue, even though the pH of hydrangea vacuoles (pH 4.0) would normally turn the pigment pink. Hence, in the absence of aluminum, the complex does not form and the delphinidin pigment turns pink within the acidic vacuole. It is possible to grow blue hydrangeas on alkaline soils by spraying the plants regularly with an aluminum sulfate solution. Similarly, adding lime to acid soils can render the aluminum insoluble and the hydrangea develops pink flowers.

Changing the color of flowers for commercial purposes—the quest for the blue rose

A true blue rose (*Rosa multiflora*) has been the target for plant breeders since 1840, when the horticultural societies of Britain and Belgium offered a prize of 500,000 Francs to the first person to produce one. It is, in fact, impossible to produce one through conventional breeding. This is because the rose species lacks the gene for the F3′5′H that produces dihydromyricetin (see above). Consequently, roses cannot produce the delphinidin anthocyanins needed to produce a true blue flower.

Recent progress has been made by scientists working within the Florigene company. They have introduced the *F3′5′H* gene from Petunia into roses, along with another Petunia gene that encodes a cytochrome *b*, that is needed for optimal F3′5′H activity. The resulting flowers were a burgundy red due to the mixture of naturally occurring rose cyanidins as well as the newly acquired delphinidins produced as a result of the Petunia genes. This breakthrough was achieved by the mid-1990s. A further step forwards was achieved recently with the use of RNAi technology to switch off the gene encoding dihydroflavonol reductase in the rose. The result of this experiment was to prevent the formation of any of the colored anthocyanins, as their synthesis is blocked at the stage that results in leucoanthcyanidin synthesis (see Figure 11.10). By introducing the *DFR* gene from *Iris* (which functions within the delphinidin pathway) into these roses, along with the *F3′5′H* gene from *Viola* the group was able to produce a rose that synthesized the blue delphinidin pigments along with just a small amount of cyanidins. The result was a mauve-colored rose. Still not a true blue, but the closest anyone has achieved so far. The remaining barrier is the pH of the vacuole. At pH 4.5, the delphinidin pigments remain mauve-colored. Hence, the true blue color will only be achieved when either the vacuolar pH can be modified, or some stabilizing co-pigments (such as those present in the Himalayan poppy described above) are also produced.

Hence, leucopelargonidins result in the formation of the pelargonidin anthocyanidins, which are orange, pink, or red, while leucocyanidins form the red/magenta cyanidins, and the leucodelphinidins lead to the purple/blue delphinidins (structures shown in Figure 11.11 and discussed further in Box 11.2). Further variations to the color of the anthocyanin pigments can be provided by the introduction of methoxyl groups (catalyzed by *O*-methyl transferases, Figure 11.11), the attachment of sugars, such as rhamnose (by rhamnosyl transferase, Figure 11.11) or by acylation with organic acids or hydroxycinnamic acids. Differences in vacuolar pH, the presence of co-pigments and metals can also add to this diversity of color (see Box 11.2).

As the dihydroflavonols can serve as substrates for both flavonol and leucoanthocyanidin formation, it is likely that there is competition between the two enzymes, flavonol synthase (for flavonol production) and dihydroflavonol 4-reductase (for leucoanthocyanidin formation; Figure 11.10). Indeed, these enzymes are targets for breeding and genetic manipulation experiments that aim to alter the balance between anthocyanin and flavonol production. Reduction in flavonol synthase expression in *Petunia* and tobacco, for example, resulted in the predicted reduction in flavonol formation and an increased accumulation of anthocyanin pigments. Further attempts to alter flower colors by modifying these, and other reactions of the flavonoid pathway, are described in Box 11.2.

As well as being precursors for anthocyanin formation, the leucoanthocyanidins and anthocyanidins are thought to be required for the production of proanthocyanidins (condensed tannins). Proanthocyanidins are polymers of the flavan-3-ols, (+)-catechin and (–)-epicatechin (Figure 11.2g). These flavan-3-ols are formed from a branch point of the central flavonoid pathway (Figure 11.9 and in detail in Figure 11.12). The precise details of the reactions that form the flavan-3-ols need further characterization. The reactions have proven difficult to resolve because the substrates are unstable and are not readily available. Furthermore, the substrates are stereoisomers and both the (+) and (–) forms are present within the polymer. The (–)-epi-flavan-3-ols, such as (–)-epicatechin, are thought to be formed from anthocyanidins via anthocyanidin reductase. Different forms of (–)-epi-flavanols form according to the type of anthocyanidin that is available (i.e. pelargonidin, cyanidin, or delphinidin). The (+)-flavan-3-ols, such as (+)-catechin, can be formed from leucoanthocyanidins, via leucoanthocyanidin reductase. The product, again, varies according to the type of leucoanthocyanidin substrate. The way in which the proanthocyanidin polymers are formed is still poorly understood. It is uncertain whether proanthocyanidin polymerization is enzyme-catalyzed, or whether it forms simply as the result of the provision of substrate to the

Figure 11.12 Possible routes for the biosynthesis of proanthocyanidins. Abbreviations: ANS, anthocyanidin synthase; LAR, leucoanthocyanidin reductase; ANR, anthocyanidin reductase.

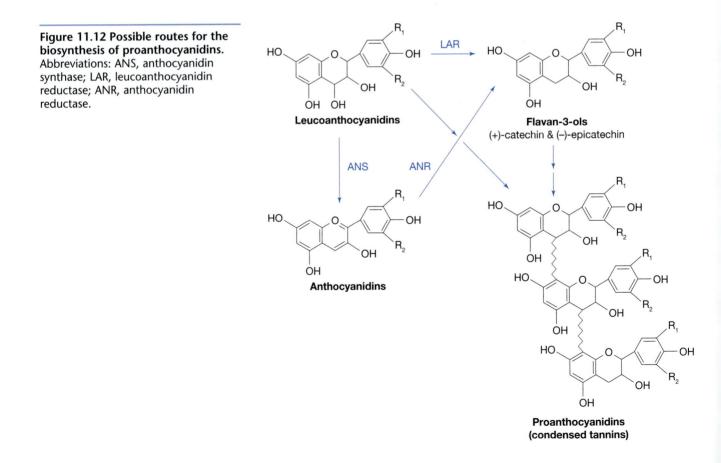

vacuole. Even the starter units for polymerization are unclear. While some species appear to use (+)-catechin as the initial template, to which is added (–)-epicatechin extension units, other species appear to use (–)-epicatechin as both the starter and extension unit. Other monomer units have been suggested by some workers, including some models showing leucoanthocyanidin as a component, while other investigators dispute this (see Further Reading).

As we have seen, the hydroxylation of flavanones is an essential step in the reactions leading to the biosynthesis of anthocyanins and proanthocyanidins. Other modifications to the flavanones provide precursors for two further flavonoid groups, the flavones, which result from the dehydration of flavanones, and the phlobaphenes, whose precursors originate from the reduction of flavanones (see Figure 11.9 for overview and Figure 11.13 for more detail).

The first step in the formation of flavones from the flavanone naringenin is a dehydration reaction catalyzed by flavone synthase (FS, also abbreviated to FNS in some texts) that forms apigenin. The reaction results in the desaturation of the C-ring; i.e. a C=C double bond is introduced during the conversion of a flavanone to a flavone (Figure 11.13). FS is restricted to a limited number of species that are capable of producing flavones, including parsley (*Petroselinum crispum*), *Chrysanthemum*, and *Gerbera*. Different forms of FS are found in different species. FS1, which is found in parsley, is a 2-oxoglutarate-dependent dioxygenase, while FS2, found in *Gerbera*, is a cytochrome P450 enzyme. Flavones are important co-pigments in flowers, where they can alter the color by forming complexes with anthocyanins (see Box 11.2).

Figure 11.13 Formation of flavones and phlobaphenes from flavanones. Abbreviations: FS1, FS2 flavone synthases 1 and 2; DFR, dihydroflavonol 4-reductase; F3′H, flavonoid 3′ hydroxylase.

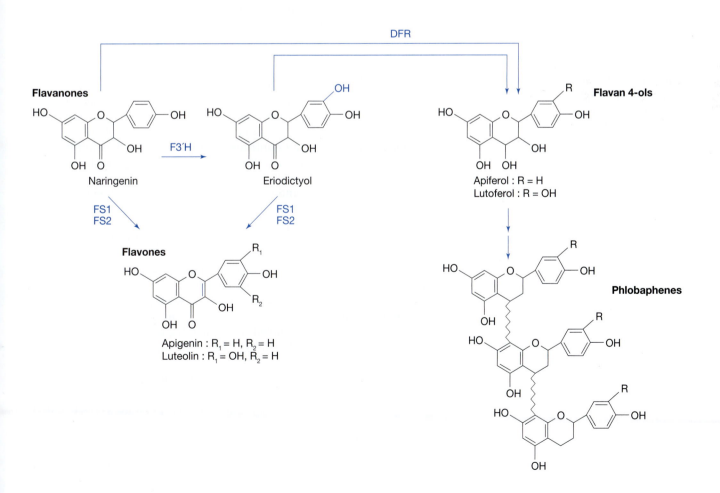

The phlobaphenes, isoflavonoids, aurones, and stilbenes

The phlobaphenes, isoflavonoids, aurones, and stilbenes occur in a limited number of species but have valuable biological and commercial properties. They are a relatively rare group of flavan 4-ol polymers that provide red pigmentation in a few species, including sorghum (*Sorghum bicolor*), maize (*Zea mays*), and gloxinia (*Sinningia cardinalis*). Their synthesis originates from a branch-point from the flavanone, naringenin, catalyzed by dihydroflavonol reductase, which carries out an NADPH-dependent reduction to form the flavan 4-ols (Figure 11.13). Dihydroflavonol 4-reductase also features later on in the flavonoid pathway, where it forms leucoanthocyanidins from dihydroflavonols, as described above (Figure 11.10).

The few species that form phlobaphenes have a unique form of dihydroflavonol 4-reductase that can carry out both reactions, i.e. it will react with dihydroflavonols to form leucoanthocyanidins (Figure 11.10), and with flavanones (naringenin, eriodictyol; Figure 11.13) to form flavan 4-ols. The balance between these two dihydroflavonol 4-reductase reactions is influenced by the availability of its two substrates, the dihydroflavonols and flavanones. As the flavanones may be converted to dihdroflavonols by F3H (Figure 11.10), one might expect this enzyme to play an important part in determining the relative flux through these competing dihydroflavonol 4-reductase reactions. This appears to be the case. In maize lines that differ in the content of anthocyanin and phlobaphenes in the grain, those with the highest phlobaphene content also had the lowest F3H activity. Hence, as F3H activity is reduced it would appear that the flavanones are diverted into phlobaphene, rather than to anthocyanin production in maize. F3H is a target for both breeding and genetic manipulation approaches to changing the quality and quantity of flavonoids (e.g. see Box 11.2 with respect to flower color, and below for a discussion of isoflavonoid manipulation).

Isoflavonoids are produced from a branch-point in the flavonoid pathway that only occurs in members of the Leguminoseae. Indeed, isoflavonoids are mostly found in a subfamily of the leguminosae, the Papilionoideae, which includes many of the commercially important legumes such as soybean (*Glycine max*), green beans (*Phaseolus vulgaris*), peas (*Pisum sativum*), and alfalfa (*Medicago sativa*). The isoflavonoids serve a wide range of functions, as antimicrobial (phytoalexin) and anti-insect compounds and as inducers of the nodulation genes in symbiotic nitrogen-fixing *Rhizobium* bacteria.

The key reaction in isoflavonoid biosynthesis is catalyzed by isoflavone synthase (IFS), a cytochrome P450 enzyme, which reacts with the flavanones liquiritigenin or naringenin to produce 2-hydroxyisoflavanone (Figure 11.14). The IFS reaction results in the migration of the B-ring of the flavanone from the C-3 to the C-2 position on the central C-ring and also adds a hydroxyl group to the C-2 to produce 2-hydroxyisoflavanone. A second enzyme, 2-hydroxyisoflavanone dehydratase, dehydrates the 2-hydroxyisoflavanone to form the isoflavones, daidzein and genistein. Other products of this pathway include medicarpin, which is the main phytoalexin in alfalfa (*Medicago sativa*), and the rotenoids produced in tropical legumes, that inhibit mitochondrial Complex I (NADH dehydrogenase; see Chapter 6).

Some isoflavones are believed to be beneficial to human health. For example, a regular dietary intake of the soy isoflavones, genistein and methylgenistein, is correlated with a reduced risk of breast and prostate cancer. Other

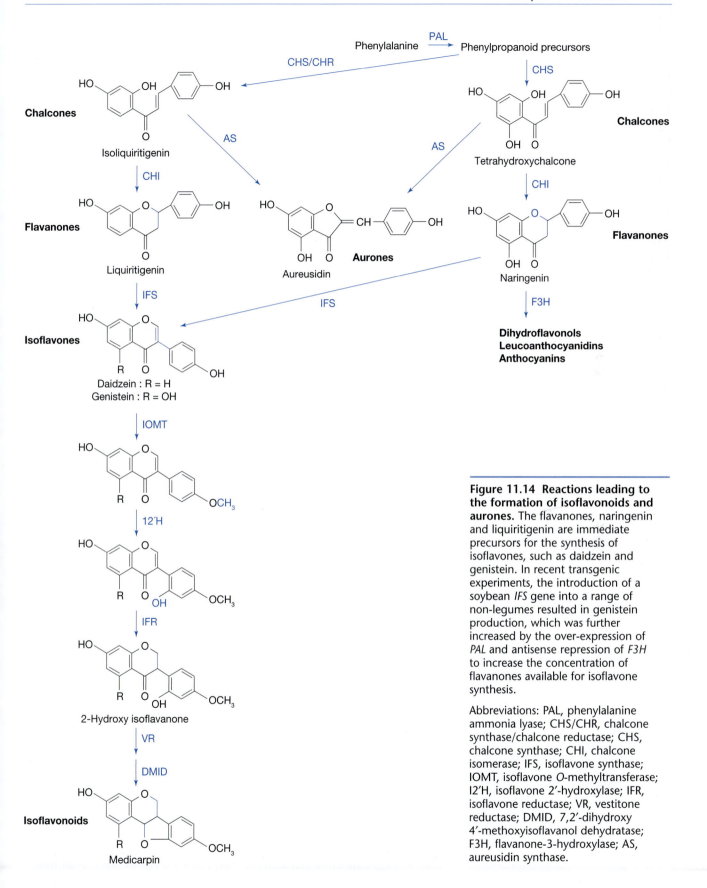

Figure 11.14 Reactions leading to the formation of isoflavonoids and aurones. The flavanones, naringenin and liquiritigenin are immediate precursors for the synthesis of isoflavones, such as daidzein and genistein. In recent transgenic experiments, the introduction of a soybean *IFS* gene into a range of non-legumes resulted in genistein production, which was further increased by the over-expression of *PAL* and antisense repression of *F3H* to increase the concentration of flavanones available for isoflavone synthesis.

Abbreviations: PAL, phenylalanine ammonia lyase; CHS/CHR, chalcone synthase/chalcone reductase; CHS, chalcone synthase; CHI, chalcone isomerase; IFS, isoflavone synthase; IOMT, isoflavone *O*-methyltransferase; I2'H, isoflavone 2'-hydroxylase; IFR, isoflavone reductase; VR, vestitone reductase; DMID, 7,2'-dihydroxy 4'-methoxyisoflavanol dehydratase; F3H, flavanone-3-hydroxylase; AS, aureusidin synthase.

isoflavones may reduce the incidence of osteoporosis and cardiovascular disease. Hence, there is considerable research and commercial interest in isoflavonoid biosynthesis. Significant effort is currently being directed towards the goal of introducing this legume-specific pathway into non-leguminous crops that can be grown in a wider range of climates than, for example, soybean. Recent success in this area of research has come from the introduction of the soybean *IFS* gene into tobacco, petunia, and lettuce. In all cases, the resulting transformed plants were able to produce genistein, albeit at relatively low concentrations. Further increases in genistein production were achieved by manipulating genes to increase the availability of precursors for genistein biosynthesis; e.g. the introduction of a *PAL* gene, to increase the rate of production of precursor phenylpropanoids, together with anti-sense suppression of *F3H* in order to divert the flavanones away from anthocyanin production (see Figure 11.14). Hence, this study provides an example of how the legume-specific isoflavonoid pathway might be successfully introduced into non-leguminous crops (see Further Reading).

Aurones (from the Latin *aurum* meaning gold) occur in only a limited number of plant species where they produce the yellow flower colors in some members of the Scrophulariaceae (e.g. *Antirrhinum majus*) and Compositae (e.g. *Cosmos, Dahlia*). The conversion of chalcones to aurones requires hydroxylation of the B-ring and/or oxidative cyclization to give the aurone structure. The enzymes involved in these reactions are still being investigated. Aureusidin synthase (Figure 11.14) has been identified in *Anthirrhinum majus* and found to react with a range of chalcones to form aurones, including aureusidin. There is evidence that aureusidin synthase actually reacts with chalcone glucosides, rather than with the free chalcone. One current view is that a cytoplasmically localized chalcone 4'-*O*-glucosyltransferase converts chalcones to chalcone glucosides that are then transported into the vacuole where aureusidin synthase converts them to aurone 6-glucosides. Evidence for this proposal has come from the creation of transgenic *Torenia hybrida* plants expressing genes, from *Antirrhinum majus*, encoding chalcone 4'-*O*-glucosyltransferase and aureusidin synthase. Wildtype *T. hybrida* flowers are blue because this species cannot normally make aurones. The introduction of the two genes from *Antirrhinum* enables the transgenic plants to produce aurone 6-*O*-glucosides, and the resulting flowers are bright yellow in color.

The stilbenes, while not members of the flavonoid group, are formed from the same precursors: *p*-coumaroyl CoA and malonyl CoA. The reaction, catalyzed by STS, is very similar to that catalyzed by CHS (see overview Figure 11.9 and detailed reaction in Figure 11.15). In both cases, decarboxylation of three malonyl CoA molecules generates three acetyl units that are sequentially condensed to the phenylpropanoid (*p*-coumaroyl CoA or cinnamoyl CoA). The reactions differ at the ring folding stage, so that CHS produces a chalcone (naringenin), while STS produces a stilbene (e.g. resveratrol from *p*-coumaroyl CoA, or pinosylvin from cinnamoyl CoA). The genes encoding STS and CHS share 70% homology at the deduced amino acid level. While CHS is widely distributed among plants, STS is limited to those plants that produce stilbenes, including grape (*Vitis vinifera*), peanut (*Arachis hypogaea*), and pine (*Pinus sylvestris*). Hence, the *STS* gene is believed to have evolved from *CHS*. Stilbenes are present in bryophytes, pteridophytes, gymnosperms, and angiosperms. More than 300 different types have been identified, yet we know very little about how these are produced, beyond the initial STS reaction. Stilbenes are important defense compounds in plants, having antifungal and antimicrobial properties. STS is induced by pathogen attack and this results in increased stilbene production as a systemic defense response. Hence, stilbenes are of interest to crop breeders with the aim of improving

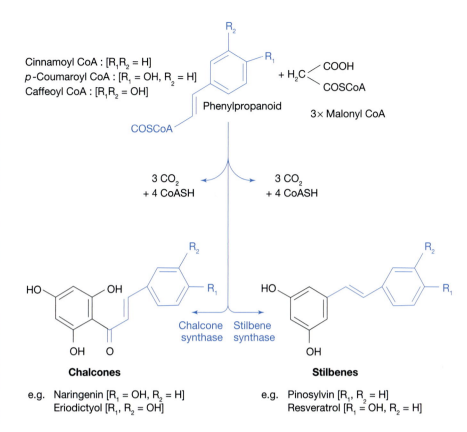

Cinnamoyl CoA : [R₁R₂ = H]
p-Coumaroyl CoA : [R₁ = OH, R₂ = H]
Caffeoyl CoA : [R₁R₂ = OH]

Phenylpropanoid

+ H₂C COOH / COSCoA

3× Malonyl CoA

3 CO₂ + 4 CoASH

3 CO₂ + 4 CoASH

Chalcone synthase   Stilbene synthase

Chalcones

e.g. Naringenin [R₁ = OH, R₂ = H]
Eriodictyol [R₁, R₂ = OH]

Stilbenes

e.g. Pinosylvin [R₁, R₂ = H]
Resveratrol [R₁ = OH, R₂ = H]

Figure 11.15 The stilbene synthase and chalcone synthase reactions. Chalcone synthase and stilbene synthase catalyze the transfer of acyl groups, from malonyl CoA, onto a phenylpropanoid. The reaction proceeds via the decarboxylation of three molecules of malonyl CoA, followed by the sequential addition of each of three acyl groups and the release of three CO_2 and 4 CoASH molecules (one from each malonyl CoA and one from the phenylpropanoid). The main reaction catalyzed by chalcone synthase is to form naringenin from p-coumaroyl CoA. However, chalcone synthases from different plant species, will also react with caffeoyl CoA to form eriodictyol, and (rarely) with cinnamoyl CoA to form 5,7-dihydroxyflavanone. Stilbene synthase will form resveratrol from p-coumaroyl CoA, and pinosylvin from cinnamoyl CoA.

resistance to fungal pathogens. In addition, stilbenes, in particular resveratrol (present in red wines), are attracting commercial interest due to their cardio-protective and anticancer properties.

Simple phenolics from the basic phenylpropanoid pathway are used in the biosynthesis of the hydrolyzable tannins

Hydrolyzable tannins are polymers of carbohydrates and the simple phenolics, gallic and ellagic acids (Figure 11.2g). Gallic acid biosynthesis is described above (see benzoic acid derivatives). The simplest form of hydrolyzable tannin is β-glucogallin (Figure 11.16) where one gallic acid molecule is esterified with one molecule of glucose. Increasing substitutions to this ester result in the formation of a glucose residue with each of its available hydroxyl groups replaced with a galloyl residue. The resulting molecule is a pentagalloylglucose (1,2,3,4,5-penta-O-galloyl-β-D-glucopyranose; Figure 11.16). Pentagalloylglucose can also be subject to oxidation reactions that form linkages between suitably orientated galloyl residues to yield a second subclass of tannins, the ellagitannins. Further galloyl residues may be added to the pentagalloylglucose structures, but now the gallic acid combines with phenolic hydroxyls on other galloyl residues, rather than with the –OH of glucose. Several enzymes, with different affinities for penta-, hexa-, and heptagalloyl-glucose substrates, are involved in this process. The reactions have been identified in extracts from oak leaves. An example of a complex hydrolyzable tannin is shown in Figure 11.2g.

Figure 11.16 Gallic acid, and gallic acid esters that form the hydrolysable tannins. Gallic acid is esterified with a glucose molecule to form the simplest form of hydrolyzable tannin, β-glucogallin. Further gallic acid molecules may form esters with the same glucose molecule, eventually forming a pentagalloylglucose (1, 2, 3, 4, 5-penta-O-galloyl-β-D-glucopyranose; where each of the hydroxyls on the glucose molecule is esterified).

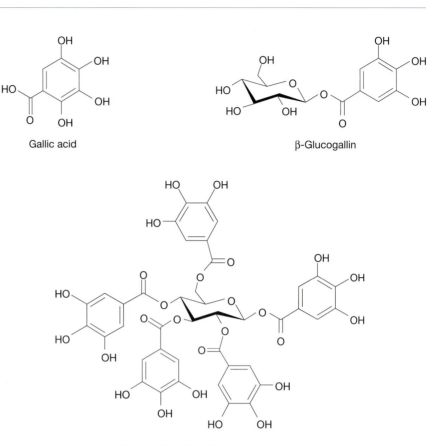

Gallic acid   β-Glucogallin

1, 2, 3, 4, 5-Penta-O-galloyl-β-D-glucopyranose

Lignin is formed from monolignol subunits in a complex series of reactions that are still being unraveled

The pathway for lignin biosynthesis is still the subject of considerable debate. It was revised extensively during the late 1980s and early 1990s and yet there are many aspects that remain unresolved.

Lignin is formed mainly from the monolignols, *p*-coumaryl, coniferyl, and sinapyl alcohols (Figure 11.17). Lignin derived from these three different units is known as *p*-hydroxyphenyl (H), guaiacyl (G), and syringyl (S) lignin, respectively, although most lignins contain a combination of the three units. Gymnosperm lignin is composed almost entirely of G, with small amounts of H; dicot angiosperm lignin contains a mixture of G and S, while monocot grasses also contain H.

Monolignols are synthesized from coumaric acid produced by the phenylpropanoid pathway

The early steps in monolignol biosynthesis are the core reactions of the phenylpropanoid pathway described above (Figure 11.5). These reactions are shown within the context of the lignin biosynthesis pathway in Figure 11.18.

To form monolignols from coumaric acid precursors, oxygen and methyl (–CH₃) groups have to be incorporated into the aromatic ring. This is achieved by aromatic hydroxylation (adding –OH) and O-methylation reactions (forming

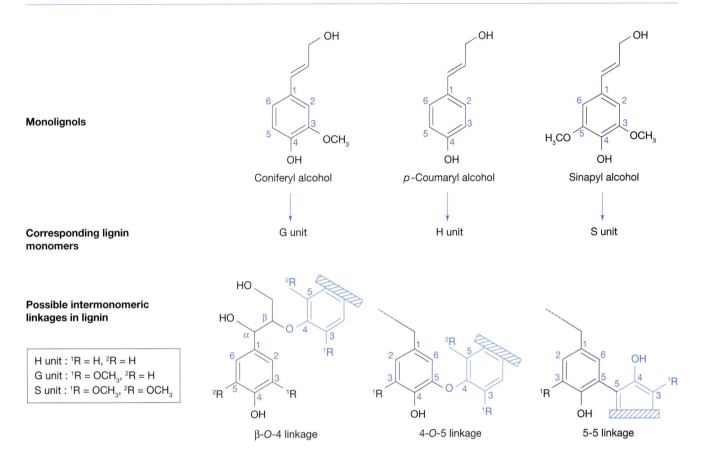

Monolignols

Coniferyl alcohol   *p*-Coumaryl alcohol   Sinapyl alcohol

Corresponding lignin monomers

G unit   H unit   S unit

Possible intermonomeric linkages in lignin

H unit : ^1R = H, ^2R = H
G unit : ^1R = OCH_3, ^2R = H
S unit : ^1R = OCH_3, ^2R = OCH_3

β-*O*-4 linkage   4-*O*-5 linkage   5-5 linkage

a methoxyl (–OCH_3 group). These reactions determine the type of monolignol formed, as these only differ in the number and positioning of methoxyl groups attached to the aromatic ring (Figure 11.17).

The fact that hydroxylation and methylation reactions are involved in monolignol formation is not disputed. It is the substrates for these reactions that remain uncertain. The original version of the pathway (until as recently as the late 1990s/early 2000s) considered that the free acids, i.e. caffeic, ferulic, 5-hydroxyferulic, were sequentially methylated by *O*-methyltransferase reactions to form sinapic acid. These intermediates were then thought to be converted, along with *p*-coumaric acid, to their corresponding aldehydes (coumaryl, caffeyl, and coniferyl aldehydes, 5-hydroxyconiferaldehyde and sinapaldehyde) and then to the monolignols (*p*-coumaryl, caffeyl, coniferyl, and sinapyl alcohols). These reactions can be followed in the metabolic grid shown in Figure 11.18. The reason for this pathway to be presented as a grid is primarily because many of the enzymes have been found to react with multiple substrates when assayed *in vitro*. For example, caffeic acid *O*-methyltransferase was initially named as a bifunctional caffeic acid/5-hydroxyferulate *O*-methyltransferase because it can use both of these substrates. Other enzymes, such as 4CL, cinnamoyl CoA reductase, and cinnamyl alcohol dehydrogenase also react with a range of substrates when assayed *in vitro*, as discussed below. Consequently, the grid provides a means of illustrating the different routes that are possible according to our current understanding of the substrate preferences of the component enzymes.

A further reason for this diversity of reactions is that several of the enzymes exist as multiple isoforms. These isoforms differ in their cellular localization and response to environmental and developmental signals, and therefore

Figure 11.17 Examples of monolignol structures. The monolignols differ in the number and positioning of methoxyl (CH_3O–) groups attached to the phenyl ring. When incorporated into the lignin polymer, these monolignols (*p*-coumaryl, coniferyl, and sinapyl alcohols) are called, respectively, the *p*-hydroxyphenyl (H), guaiacyl (G), and syringyl (S) units. The structure of the lignin polymer differs according to the relative amounts of H, G, and S units (and other minor constituents) that are present. Adjacent H, G, and S units form intermonomeric linkages within the lignin polymer. While the H and G units can form bonds from both the C_4 and C_5 of the phenyl ring, only the C_4 of S units can bond because the methoxyl group on C_5 of the sinapyl alcohol structure restricts further bonding. Consequently, H and G units can form 5-5, and 4-*O*-5 linkages as well as β-*O*-4 linkages, while S units can only form β-*O*-4 linkages. Other linkages are possible, for example from C_1 (not shown).

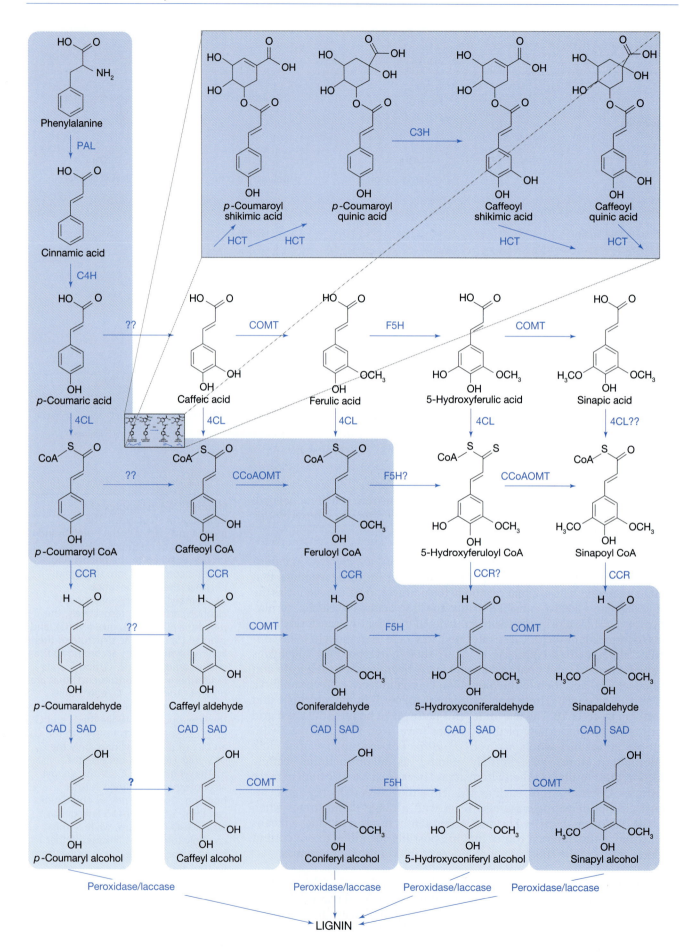

Figure 11.18 Phenylpropanoid and monolignol biosynthesis leading to lignin. The reactions highlighted with the dark background are believed to be the most likely route to monolignol production in Angiosperms. Species-specific reactions are highlighted in light blue. The reactions shown against a white background constitute the free acid route that was once favored, but is now thought not to play a significant role in monolignol biosynthesis.

Abbreviations: CAD, cinnamyl alcohol dehydrogenase; 4CL, 4-coumarate:CoA ligase; C3H, p-coumarate 3-hydroxylase;

C4H, cinnamate 4-hydroxylase; CCoAOMT, caffeoyl CoA O-methyltransferase; CCR, cinnamoyl CoA reductase; COMT, caffeic acid O-methyltransferase; HCT, p-hydroxycinnamoyl CoA: quinate shikimate p-hydroxycinnamoyltransferase; F5H, ferulate 5-hydroxylase; PAL, phenylalanine ammonia-lyase; SAD, sinapyl alcohol dehydrogenase. Reactions indicated by '?' are uncertain. ?, conversion demonstrated; ??, direct conversion not convincingly demonstrated; 4CL??, some species have 4CL activity toward sinapic acid; CCR? and F5H?, substrate not tested.

add further to the diversity of the pathway both within (i.e. in tissues and cells) and between species. For example, the methylation steps can occur at the level of the CoA derivative, rather than the free acid, with caffeoyl CoA being converted to feruloyl CoA, and 5-hydroxyferuloyl CoA to sinapoyl CoA. While caffeic acid O-methyltransferase catalyzes the reactions involving the free acids, caffeoyl CoA O-methyltransferase reacts with the CoA derivatives (Figure 11.18).

More recently, the results of radiotracer experiments and further enzymatic analysis have pointed to the preferential hydroxylation and methylation of the aldehyde and alcohol forms of the intermediates. Thus, ferulic acid 5-hydroxylase will react with coniferaldehyde to form 5-hydroxyconiferaldehyde, which is then methylated by caffeic acid O-methyltransferase. Indeed, caffeic acid O-methyltransferase will preferentially methylate the aldehyde intermediates caffeyl aldehyde, 5-hydroxyconiferaldehyde, and 5-hydroxyconiferyl alcohol but the precise reactions may be species-specific.

It is only recently that the reactions resulting in the conversion of p-coumaroyl CoA to caffeoyl CoA have been identified (see insert in Figure 11.18). p-Coumarate 3-hydroxylase was previously thought to convert p-coumaric to caffeic acid. It is now believed to react, instead, with both the shikimic and quinic acid esters of p-coumaroyl CoA. The resulting caffeoyl CoA esters are converted to caffeoyl CoA by the same enzyme, hydroxycinnamoyl CoA:shikimate/quinate hydroxycinnamoyltransferase that is believed to form the esters from p-coumaroyl CoA in the first place.

The involvement of 4CL in sinapic acid formation for S monolignol synthesis has been somewhat controversial. A long-standing problem has been that, while almost all 4CL isoforms will react with p-coumaric, caffeic, and ferulic acid, to form their respective CoA esters, the ability to react with sinapic acid seems to be confined to only one or two species. If sinapoyl CoA cannot be formed by 4CL, it is not possible to produce sinapyl alcohol for S lignin biosynthesis via the free acid route (pathway intermediates shown on a white background in Figure 11.18). This would seem to be the case for most species. However, both soybean (*Glycine max*) and arabidopsis have been found to possess a specific 4CL isoform whose capacity to react with sinapic acid appears to have arisen from the deletion of a single amino acid residue (valine in *G. max*; leucine in arabidopsis). Hence, in these species at least, there is the possibility that the free acid pathway may operate in S lignin biosynthesis.

The formation of the coniferyl, coumaryl, and sinapyl alcohols via reduction of their respective aldehydes was generally believed to occur via a multifunctional enzyme, cinnamyl alcohol dehydrogenase. However, a sinapyl alcohol dehydrogenase has been discovered in aspen (*Populus tremuloides*), where it is believed to be responsible for sinapyl alcohol formation for S lignin synthesis. This would leave cinnamyl alcohol dehydrogenase to be involved only in cinnamyl alcohol formation for G lignin synthesis. The relative contributions of cinnamyl alcohol dehydrogenase and sinapyl alcohol dehydrogenase to these final stages of monolignol formation are likely to differ according to the tissue and species. Arabidopsis, for example, appears only to use cinnamyl alcohol dehydrogenase, and not sinapyl alcohol dehydrogenase to form both sinapyl and coniferyl alcohols. Further work with transgenic plants from a range of species will help to clarify the situation. For the meantime, we have labeled this step in the monolignol pathway as cinnamyl alcohol dehydrogenase/sinapyl alcohol dehydrogenase to recognize the emerging status of these reactions (Figure 11.18).

The reactions that are currently believed to be most important in angiosperm monolignol biosynthesis are shown with the darkest shading in Figure 11.18. The lighter-shaded reactions may also be involved, depending on species and under certain conditions. It is likely that some of the substrates and the positioning of some of the component enzymes will be revised further as new information is obtained. The reactions shown against a white background, where the free acids are used, are believed to occur in only a few species.

Monolignol polymerization is a poorly understood aspect of lignin biosynthesis

The way in which lignin is formed from the monolignol precursors is still poorly understood. Lignin is a complex and heterogeneous polymer that varies among species, cell types, and individual cell wall layers (Chapter 7). Its composition is also influenced by environmental and developmental signals, such as wounding and compression or tension of the tissue. This heterogeneity of composition led to earlier hypotheses that the final stages of polymerization resulted from random couplings of monolignol units that did not require the involvement of enzymes. This opinion has changed in recent years, although our understanding of the polymerization process is still very sketchy.

Lignin forms at distinct sites within the cell wall. The monolignols are transported there from the cytoplasm, either via Golgi-derived vesicles or through specific transporters on the plasma membrane. There is evidence that the Golgi releases proteins that locate to the region of the cell wall where lignin is being formed. While these proteins may be involved in monolignol transport, there is also speculation that they might serve as scaffolds on which the lignin polymer is built. There is no firm evidence for either supposition at the moment.

The monolignols are thought to be transported in the form of glycosides, i.e. with glucose molecules attached. After transport to the cell wall, the monolignols are oxidized to form radicals that combine to produce the lignin polymer. There are several classes of enzymes present in cell walls that could catalyze monolignol oxidation, including peroxidases, laccases, and polyphenol oxidases. The fact that so many of these enzymes are present in the cell wall adds to the difficulty in identifying which, if any, is involved in the process. Arabidopsis, for example, has about 70 different peroxidase genes. Added to this multiplicity of enzymes is the likelihood of redundancy and overlapping function, and the task of unraveling this aspect of the pathway becomes even more of a challenge.

A remaining, and unresolved, controversy surrounding the mechanism of lignin formation is whether it results from the random chemical coupling of monolignol radicals, or whether it is subject to enzymatic control. Lignin-like polymers can be formed artificially by adding either peroxidase together with H_2O_2 or laccase with O_2, which serve as oxidizing agents, along with the monolignol precursors. The oxidizing agents generate the monolignol radicals and these combine to form different lignin structures, depending on the rate of supply of the monomers, the rate of radical generation, the presence of polysaccharides, and the existing structure of the growing lignin polymer. This artificial, non-enzymatic construction of lignin provided the most substantial evidence that the same process could result in lignin production *in vivo*. However, this view has been challenged with the discovery of the involvement of a dirigent protein in the synthesis of lignans.

Lignans are dimers, or sometimes higher oligomers, of the monolignol coniferyl alcohol, which have important defensive functions, as antimicrobial or antifungal agents in plants. Most lignans are optically active, with (+) and (−) forms occurring consistently within specific tissues and plants (Figure 11.19). This specific distribution indicated that the dimers were not being formed by random chemical interactions and the discovery of the dirigent protein (from *dirigere* (Latin): to guide, align) provided evidence that their synthesis was, indeed, subject to enzymatic control. The dirigent protein was discovered in *Forsthyia* where it catalyzes the coupling of two coniferyl alcohol radicals in such a way as to form only (+)-pinoresinol and not (−)-pinoresinol (see Figure 11.19 for comparison of the two structures). The dirigent protein was cloned and shown to encode a cell wall-localized protein. Following this discovery, the suggestion was made that lignins, like lignans, might also be formed in an ordered manner with similar, dirigent proteins organizing the precise structure of the growing polymer. There is still no biochemical or genetic evidence to support this idea of an organized assembly of lignin and we await the generation of suitable transgenic plants or mutants with which this can be tested.

Summary

Plant phenolics are a diverse range of chemicals with equally diverse biological functions. Some of the more simple phenolics, e.g. phenylpropanoids, coumarins, and benzoic acid derivatives are volatile and provide mobile signals that attract pollinators and deter herbivores. Other biological functions include phytoalexin and allelopathic roles that protect the plant from pathogens and from competition from neighboring plants. The more complex phenolics include the flavonoids, a large group with a common C_6–C_3–C_6 structure. Flavonoids include the widespread anthocyanin pigments that provide much of the coloration of flowers and fruits, and current research is aimed at manipulating this pathway in order to generate novel flower colors. Other flavonoids are more limited in their distribution. The isoflavonoids, for example, are restricted to the legumes where they provide signals involved in the symbiotic association with nitrogen-fixing bacteria. Some phenolic compounds provide the basis for complex polymers, such as lignin, which is vital for structural support, and the tannins that serve as feeding deterrents and antifungal agents. While we have a good understanding of the biosynthesis of some of these phenolic compounds, many reactions remain to be characterized and there is still much to learn about these pathways and their regulation.

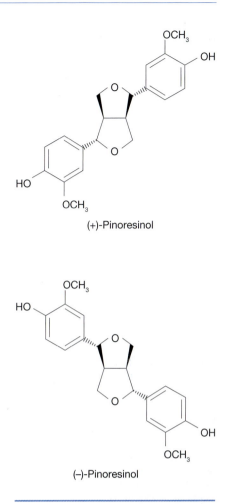

(+)-Pinoresinol

(−)-Pinoresinol

Figure 11.19 Examples of the lignans, (+) and (−) pinoresinol. Lignans are dimers, or sometimes higher oligomers, of the monolignol coniferyl alcohol. Conventional terminology defines true lignans as being formed from 8-8′ bonds between the monomers. Where other bonds are involved, the resulting dimer is classed as a neolignan. Most lignans are optically active, with (+) and (−) forms occurring consistently within specific tissues and plants. The (+) and (−) forms of pinoresinol are shown here for comparison.

Further Reading

Benzoic acids

Wildermuth MC (2006) Variations on a theme: synthesis and modification of plant benzoic acids. *Curr. Opin. Pl. Biol.* 9, 288–296.

A good, recent review of the biosynthesis of benzoic acids in plants.

Proanthocyanidins

Xie D-Y & Dixon RA (2005) Proanthocyanidin biosynthesis— still more questions than answers? *Phytochemistry* 66, 2127–2144.

A very detailed discussion of the structure and biosynthesis of the proanthocyanidins (condensed tannins) and a consideration of the gaps in our current knowledge.

Lignin

Boerjan W, Ralph J & Baucher M (2003) Lignin biosynthesis. *Annu. Rev. Pl. Biol.* 54, 519–546.

A good, detailed review article.

Baucher M, Halpin C, Petit-Conil M & Boerjan W (2003) Lignin: Genetic engineering and impact on pulping. *Crit. Rev. Biochem. Mol. Biol.* 38, 305–350.

A very comprehensive article, covering commercial applications of wood products, and the generation of transgenic plants for research and commercial purposes.

Flower color

Grotewold E (2006) The genetics and biochemistry of floral pigments. *Annu. Rev. Pl. Biol.* 57, 761–780.

A recent, detailed review of the pathways and approaches to their modification through breeding and genetic manipulation.

Tanaka Y, Katsumoto Y, Brugliera F & Mason J (2005) Genetic engineering in floriculture. *Pl. Cell Tissue Organ Culture* 80, 1–24.

This review describes current and potential uses of genetic engineering to alter the color of flowers.

Floral scents

Pichersky E & Dudareva N (2007) Scent engineering: toward the goal of controlling how flowers smell. *Trends Biotechnol.* 25(3), 105–110.

This review describes successful and not so successful attempts to use genetic engineering to alter floral scents. It also indicates possible targets for engineering reactions within the phenylpropanoid and terpenoid pathways.

Phenylpropanoids

Dixon RA, Achnine L, Kota P, Lio C-J, Reddy MSS & Wang L (2002) The phenylpropanoid pathway and plant defence—a genomics perspective. *Mol. Pl. Pathol.* 3(5), 371–390.

A fairly specialized review that provides some useful background into the defensive role of phenylpropanoids, and also identifies potential targets for genetic manipulation of the pathway.

Isoflavonoids

Liu R, Hu Y, Li J & Lin Z (2007) Production of soybean isoflavone genistein in non-legume plants via genetically modified secondary metabolism pathways. *Metab. Eng.* 9, 1–7.

This paper describes a successful strategy for introducing isoflanoid biosynthesis into non-legumes.

General interest

Manach C, Scalber A, Morand C, Rémésy C & Jiménez L (2004) Polyphenols: food sources and bioavailability. *Am. J. Clin. Nutr.* 79, 727–747.

An interesting review that discusses the dietary sources of polyphenols and includes a readily digestible introduction to the chemical structures of the major classes of phenolic compounds.

Terpenoids

12

Key concepts

- Terpenoids are a large group of oily compounds that are composed of isoprene units.

- Terpenoids have a range of functions, from poisons and feeding deterrents, to scents and pigments.

- Many terpenoids have important pharmaceutical properties and include cancer treatments and male and female contraceptives.

- A number of plant growth regulators are derived from terpenoid structures.

- Isopentenyl diphosphate is the precursor for terpenoid biosynthesis.

- There are two pathways for isopentenyl diphosphate biosynthesis in higher plants, the deoxyxylulose phosphate pathway in the plastids and the mevalonic acid pathway in the cytosol.

- Isopentenyl diphosphate subunits are joined together by prenyltransferases to form C-10, C-15, C-20, and larger precursors for mono-, sesqui-, and diterpene synthesis.

- Terpene synthases introduce considerable variety into the terpenoid structures and many terpene synthases are capable of producing multiple products from a single substrate.

Terpenoids are a diverse group of essential oils that are formed from the fusion of five-carbon isoprene units

Terpenoids represent the largest class of natural products with as many as 25,000 different forms identified in higher plants. The biological functions of terpenoids range from light harvesting in photosynthesis, electron transport, growth regulation, and as volatile attractants and toxic deterrents in animal–plant interactions (see Box 12.1). Terpenoids are not unique to higher plants. They are also synthesized in animals, bacteria, fungi, and algae, as discussed below.

The name terpenoid originates from the hydrocarbons that were first identified in extracts from turpentine oil, which is produced from the distillation of resin from various species of conifer trees. While the term terpenoid is generally used interchangeably with terpene to define this group of plant chemicals, the two names refer to different chemical classes. Terpenes are

Box 12.1 Terpenoids involved in animal–plant interactions

Note that the structures of many of the terpenoids described in this section are shown in Figure 12.4.

Many plant terpenoids serve important functions in the interactions of plants with animals, either as attractive signals for pollinators and seed-dispersers, or as deterrents against herbivores.

The low molecular weight terpenoids, the hemi-, mono-, and sesquiterpenes, are volatile and many of them serve as mobile scents to attract pollinators or to deter herbivores. Sometimes the same chemical can have both effects. The monoterpene 1,8-cineole, for example, which is found in a wide range of species, can serve as both a pollinator-attractant for bees, moths, and bats, and as a herbivore deterrent to hares and deer. Other volatile monoterpenes include linalool (a floral scent in a number of species, including *Freesia*), limonene (the major scent of citrus flowers and fruits), and geraniol (the scent of geraniums and roses). Sesquiterpene scents include α-(–)-bisabolol, in orange blossom, valencine, in grape (*Vitis vinifera*) flowers, and β-ionine in violets.

The aroma of flowers is usually composed of a complex mixture of chemicals, including a number of volatile terpenoids together with low molecular weight phenolics (see Chapter 11). There may be as many as a hundred different chemicals within the scent. These complex mixtures can serve as tracking signals, as the more volatile components will disperse over a greater distance, hence animals can follow the changing scent as they move towards the source.

The four main volatile terpenoids in conifers are the monoterpenes myrcene, limonene and α and β pinene. Insects respond to these volatiles in different ways, depending on the species. Myrcene, for example, is an attractive scent to a number of species of pine bark beetle (e.g. *Dendroctonus brevicomis*, the Western pine beetle) that sequester it from their diet and release it as a pheromone. Limonene tends to have a deterrent effect as it is fairly toxic. While α pinene, like myrcene, is frequently used as a dietary pheromone, β pinene is generally repellent due to its toxicity. The pine bark beetle can tolerate relatively high doses of β pinene, whereas the beetle (*Dendroctonus pseudotsugae*) that feeds on Douglas fir (*Pseudotsuga menziesii*) finds β pinene highly repellent. Hence, Douglas fir trees with a high β pinene/α pinene ratio and a high limonene content tend to be less susceptible to insect attack.

Although the volatility of the lower molecular weight terpenoids tends to deter many potential herbivores, some of these mono- and sesquiterpenoids are still capable of poisoning animals that ignore the warning. Pyrethrins, for example, are a class of irregular monoterpenes with strong insecticidal properties that occur in Chrysanthemum.

Among the diterpene poisons, phorbol, which is present as an ester in the latex of spurges (Euphorbiaceae) causes inflammation on contact with the skin, and can cause tumors. Rhododendron and some members of the Ericaceae contain grayanotoxins, a group of diterpene poisons. These toxins bind to sodium channels in membranes and keep the cell in a depolarized condition, affecting the central nervous system and skeletal muscles. Grayanotoxin poisoning was recorded in the fourth century BC with a written account of the poisoning of 10,000 Greek soldiers who had eaten honey produced from the nectar of Rhododendron.

The triterpenoids represent the largest class of terpenoid poisons. The main group of toxins is the saponins. These are glycosylated (i.e. attached to one or more sugars). The saponins are named after their soapy, detergent properties. The sugar moiety is hydrophilic while the steroidal group is hydrophobic. Hence, saponins can penetrate cell membranes, causing hemolysis of red blood cells, which helps the poison to spread. Saponins occur in a number of species, including ivy (*Hedera helix*) and the horse chestnut tree (*Aesculus hippocastanum*). Diosgenin is a saponin that occurs in yam (*Dioscorea*) that is used as a source of progesterone (see Box 12.3).

Another group of triterpenoid glycosides within the saponin family is the cardenolides, or cardiac glycosides (also called steroidal glycosides). These are so-called because of their effects on the heart. They inhibit the Na/K-ATPases, causing a slowing of the heartbeat that makes them useful in the treatment of heart disease. Even so, cardenolides are highly toxic to animals, including humans, and digitoxin (from the foxglove, *Digitalis purpurea*) can cause fatalities by absorption through the skin. Some insects are able to sequester cardenolides in order to deter predators. A well-documented case is of the monarch butterfly larvae (*Danaus plexippus*) that store cardenolides derived from the milkweed (*Asclepia* spp.). The cardenolides are retained in the body of the moth, which carries orange and black markings as a warning coloration. Not all predators heed the warning. Nevertheless, the blue jay (*Cyanocitta cristata*) is said to need just one experience of the vomiting that follows from eating a monarch butterfly to avoid feeding on monarchs ever again.

The triterpenoid insecticide, azadirachtin, which occurs in the neem tree (*Azadirachta indica*; native to South-east Asia) is a bitter-tasting feeding deterrent that can be effective in minute amounts (parts per billion). It also affects insect growth by interfering with the production of ecdysones, hormones that are required for molting. As such, it disrupts the normal development of insects, with lethal consequences. Azadirachtin is a component of a number of insecticides currently being marketed in North America and in India.

Another class of triterpenoids, the phytoecdysones, are close structural analogs of insect hormones, such as the juvenile hormones that regulate molting. Juvabione was the first such phytoecdyson to be recognized. Its discovery arose by accident when a Czechoslovak biologist, Karel Sláma, was trying to culture the European bug *Pyrrhocoris apteris* in the laboratory. Sláma found that the bug failed to develop past the fifth larval stage when the Petri dishes were lined with paper from US sources, whereas there was no problem with European or Japanese papers. The US paper was made largely from balsam fir

(*Abies balsamea*), which is not used elsewhere. It was subsequently discovered that *A. balsamea* contained juvabione, a close structural analog of the juvenile hormones of the European bug. Juvabione was inhibiting the action of the endogenous hormones and arresting the development of the bug. Since this discovery, in the 1960s, several other phytoecdysones have been identified, including juvocimene 2, in basil (*Ocimum basilicum*) and α and β ecdysone, present in a range of species, including the fern *Polypodium vulgare*.

---

hydrocarbons (e.g. α- and β-pinene and limonene; see Figure 12.4), while terpenoids are oxygen-containing analogs of the terpenes (e.g. α-1,8-cineole and menthol; see Figure 12.4). Terpenoids may also be referred to as isoprenoids because many of them break down to release the five-carbon gas, isoprene (Figure 12.1).

All terpenoids consist of a combination of one or more isoprene units, as illustrated in Figure 12.2. Indeed, the presence of isoprene units can be diagnostic of a terpenoid, although biochemical modifications can sometimes make this identification less than straightforward to determine, especially in the case of complex terpenoids. Isoprene is not actually a substrate

Isoprene

Figure 12.1 Isoprene.

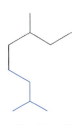

Head to middle

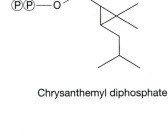

Chrysanthemyl diphosphate

Head to tail

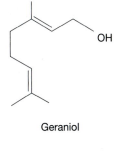

Geraniol

Head to head

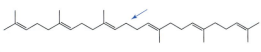

Squalene

Figure 12.2 Terpenoids are synthesized from 5-carbon units that can be joined together in different orientations. The isoprene units (from isopentenyl diphosphate; IPP) are joined together by head-to-head, head-to-tail or head-to-middle bonding to form a variety of terpenoid structures. An example of a head-to-tail conjugation is the monoterpene oil geraniol. The 30-carbon terpenoid squalene is formed from a head-to-head condensation (marked with the arrow) of two farnesyl diphosphate (FPP) molecules that are themselves formed from head-to-tail bonding between IPP and geranyl diphosphate (GPP) as explained in the text. The less-common irregular terpenoids are formed from head-to-middle bonding as in the example, chrysanthemyl diphosphate. This irregular terpenoid is a precursor for the synthesis of pyrethrin, a naturally occurring terpenoid insecticide found in Chrystanthemum (Box 12.1).

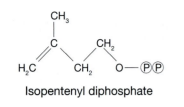

Isopentenyl diphosphate

Figure 12.3 Isopentenyl diphosphate.

for terpenoid biosynthesis; the biochemical precursor is isopentenyl diphosphate (IPP), first identified by Feodor Lynen, in 1964, and called active isoprene. The abbreviation IPP derives from the earlier name, isopentenyl pyrophosphate, that is still in use today. The structure of IPP is shown in Figure 12.3.

Given that terpenes are hydrocarbons, they tend to be hydrophobic and to have an oily consistency. Many of the terpenoids are also relatively hydrophobic, especially those with few oxygen groups. Consequently, terpenes and many terpenoids are generally found in specialized cells, such as the resin ducts of conifer wood or glandular hairs on the epidermis of leaves and petals.

Terpenoids are classified according to the number of five-carbon units from which they are formed. The terminology can be confusing because the 10-carbon terpenoids were the first ones to be identified and, as these were assumed to be the simplest forms, they were named monoterpenes (i.e. single terpenes) despite the fact that they consist of two five-carbon units. The simplest terpenoids, which consist of just one isoprene unit, are therefore termed hemiterpenes (half terpenes). It follows that the 15-carbon terpenoids are termed sesquiterpenes (i.e. one and one-half terpenes, or three five-carbon units); 20-carbon terpenoids are termed diterpenes (two terpenes, four five-carbon units), while the larger terpenoids are 30-carbon triterpenes and 40-carbon tetraterpenes. Larger polymers are classed as polyterpenes, consisting of eight or more five-carbon units. Representative structures of each class of terpenoid are given in Figure 12.4.

Note that we have sometimes used different formats for illustrating the structure of the various terpenoids and precursors in this chapter. We have used skeletal formulae where the terpenoid structures are relatively complex, or where this format provides greater clarity within multistep pathways. We have also reoriented some of the structures where we consider that this helps to illustrate the chemical reactions within the pathways. For example, we have presented some structures in a linear format (e.g. see Figure 12.7, geranyl diphosphate, GPP), while in other figures we have represented the same molecule in a more cyclized style (e.g. Figure 12.8, GPP). These particular styles are chosen as a means of illustrating the relationship between the structure of the substrate and that of its product; hence when cyclic structures are produced it is usual to illustrate the precursors in a pseudo-cyclic form (as in Figure 12.8).

Terpenoids serve a wide range of biological functions

Although terpenoids are often referred to as secondary metabolites many of them are essential components of primary metabolic pathways. Without them the plant would not be able to function. Terpenoids within this category include: the carotenoids (e.g. carotene, xanthophylls; Figure 12.4), which function in photosynthesis and also protect against photooxidation (see Chapter 4); the phytol side chain (a diterpenoid) of chlorophyll, and the electron carriers plastoquinone and ubiquinone both of which contain polyterpenes (see Chapters 4 and 6, respectively).

The plant growth regulators, abscisic acid and the cytokinins, gibberellins and brassinosteroids are all derived from terpenoid precursors. While the brassinosteroids are triterpenes (C-30) and the gibberellins (Figure 12.4) are diterpenes (C-20) the 15-carbon abscisic acid is actually produced by cleavage of a tetraterpene (C-40) carotenoid. Cytokinins, such as zeatin, are meroterpenes (meaning that terpenoids form only part of the structure) that consist of an isoprenoid (C-5) side chain attached to an adenosine moiety (see Box 12.2 for the biosynthesis of these plant growth regulators).

Hemiterpenes

Isoprene

Released as a gas from conifer trees, may help protect against heat stress

Monoterpenes

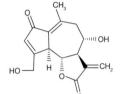

Menthol

α-Pinene   β-Pinene

* = CH₃

Linalool   Limonene   1,8-Cineole

Volatile oils that are important as attractive or deterrent scents, many have insecticidal properties

Sesquiterpenes

Lactucin

Bitter-tasting in lettuce and chicory

Gossypol

Insecticidal properties, can cause infertility due to its effects on spermatogenesis. Found in cotton seeds

β-Caryophyllene

The major scent of Scots pine
(*Pinus sylvestris*)

α-Farnesene

Volatile signal in Norway spruce
(*Picea abies*)

Germacrene B

Volatile signal in spruce (*Picea abies*)
and juniper (*Juniperus communis*)

Capsidiol

A phytoalexin

Diterpenes

Forskolin

Insecticidal and anti-feedant properties.
Treatment for glaucoma

Taxol

Phytoalexin found in *Taxus* spp. (yew).
Anticancer drug.
The blue region marks
the terpenoid moiety

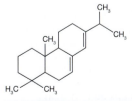

Abietadiene

Phytoalexin found in resin of grand fir (*Abies grandis*)

Grayanotoxin

Poison present in rhododendron

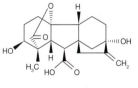

Gibberellin A₁

Plant growth regulator

continued

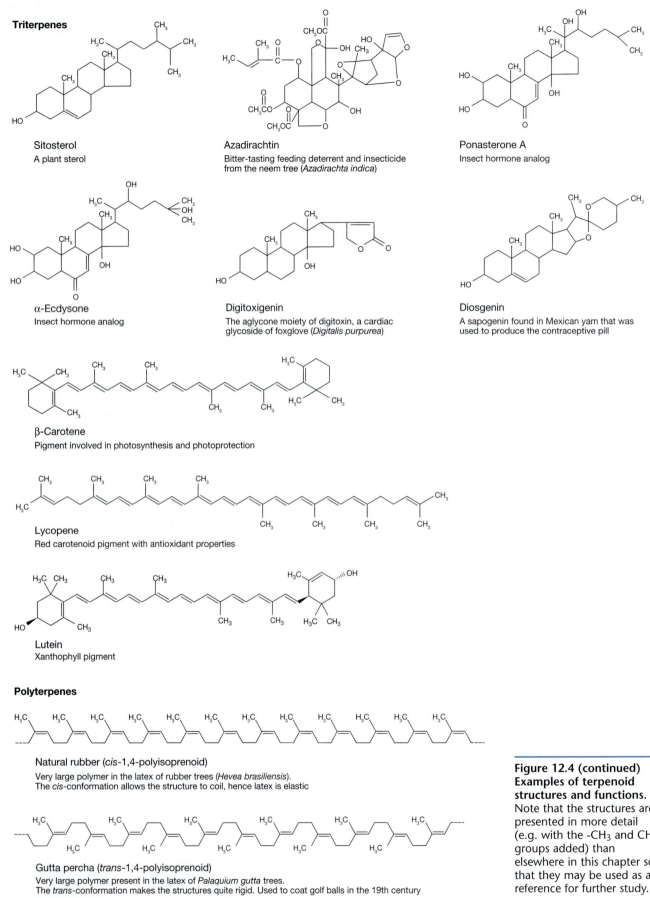

Triterpenes

Sitosterol
A plant sterol

Azadirachtin
Bitter-tasting feeding deterrent and insecticide from the neem tree (*Azadirachta indica*)

Ponasterone A
Insect hormone analog

α-Ecdysone
Insect hormone analog

Digitoxigenin
The aglycone moiety of digitoxin, a cardiac glycoside of foxglove (*Digitalis purpurea*)

Diosgenin
A sapogenin found in Mexican yam that was used to produce the contraceptive pill

β-Carotene
Pigment involved in photosynthesis and photoprotection

Lycopene
Red carotenoid pigment with antioxidant properties

Lutein
Xanthophyll pigment

Polyterpenes

Natural rubber (*cis*-1,4-polyisoprenoid)
Very large polymer in the latex of rubber trees (*Hevea brasiliensis*).
The *cis*-conformation allows the structure to coil, hence latex is elastic

Gutta percha (*trans*-1,4-polyisoprenoid)
Very large polymer present in the latex of *Palaquium gutta* trees.
The *trans*-conformation makes the structures quite rigid. Used to coat golf balls in the 19th century

Figure 12.4 (continued) Examples of terpenoid structures and functions. Note that the structures are presented in more detail (e.g. with the -CH₃ and CH₂ groups added) than elsewhere in this chapter so that they may be used as a reference for further study.

Several types of terpenoids serve as phytoalexins, i.e. chemicals that have antibacterial and/or antifungal properties. These include the diterpenes, casbene (found in castor bean, *Ricinus communis*) and taxol (in the bark of yew, *Taxus baccata*; see Plate 12.1, Box 12.3, and Figure 12.4), and the sesquiterpenes, capsidiol (found in capsicum peppers, *Capsicum annuum*) and gossypol (Figure 12.4; found in the seeds of cotton, *Gossypium hirsutum*). As well as its phytoalexin properties, gossypol has insecticidal and hormonal properties. It interferes with sperm production in mammals and, as such, it has been tested for use as a male contraceptive (Box 12.3).

Many conifers produce oleoresins in response to wounding. Oleoresins are mixtures of monoterpene olefins (e.g. turpentine oils, mixtures of mono-, di-, and sesquiterpenes) and diterpene resin acids (rosin). These oleoresins have phytoalexin properties and also provide a means of sealing the wound site to prevent infection. The seal arises from the polymerization of the nonvolatile

Box 12.2 Some important plant growth regulators are formed from terpenoid precursors

Cytokinins

Cytokinins promote plant cell division and act in concert with auxins to regulate cell differentiation. The cytokinins are adenine derivatives that are attached to an isopentenyl-based side chain.

The first step in cytokinin biosynthesis (Figure 1) is catalyzed by an isopentenyltransferase (IPT) that carries out a prenyltransferase reaction (see page 418), adding an isopentenyl group from dimethylallyl diphosphate (DMAPP) to ATP, ADP, or AMP. The respective products are isopentenyl ATP (iPTP), isopentenyl ADP (iPDP) and isopentenyl AMP (iPMP). Recent evidence indicates that the plant enzyme uses ATP or ADP and is less active with AMP, which is the major substrate for the fungal and bacterial enzymes. The zeatin-type cytokinins are formed following the addition of a hydroxyl group to the isopentenyl side chain, in a reaction catalyzed by a P450 monoxygenase, CYP735A. The precise details of the pathway remain to be characterized. Alternative routes have been proposed, for example, using prenylated tRNAs as a source of the isopentenyl side chain, but these remain to be verified.

Abscisic acid (ABA)

ABA is involved in seed dormancy and also in responses to water stress, such as stomatal closure. It is a C-15 compound produced by the breakdown of a C-40 terpenoid, 9'-*cis*-neoxanthin.

The ABA biosynthesis pathway (Figure 2) shares its early steps with the carotenoid biosynthesis pathway (see Box 12.4). Neoxanthin is cleaved by 9-*cis*-epoxy-carotenoid dioxygenase (NCED) within the thylakoids to form xanthoxal. Subsequent oxidative steps convert xanthoxal to ABA-aldehyde and then to ABA, in reactions catalyzed by

short chain dehydrogenase/reductase (SDR) and aldehyde oxidase (AO), respectively. The reactions from xanthoxal onwards are still subject to some conjecture and two alternative routes have been proposed (see Further Reading).

Gibberellic acid (GA)

The GAs stimulate shoot growth and promote seed germination. About 125 different GAs have been identified, of which 100 are exclusive to plants. Their nomenclature is somewhat misleading as they are numbered in the order in which they were discovered (i.e. GA_1 to GA_{125}) hence the numbers have nothing to do with their structure (Figure 3).

As with all of the diterpenes, the early stages of GA biosynthesis involve the formation of geranylgeranyl diphosphate (GGPP) from isopentenyl diphosphate (IPP) as described in the text (see Figure 12.8). The first committed step is the conversion of GGPP to *ent*-kaurene, catalyzed by a plastid-located terpene cyclase. *Ent*-kaurene is then converted to GA_{12} or GA_{53}, depending on species (the latter is most common), via a series of reactions within the endoplasmic reticulum. Subsequent reactions within the cytosol convert GA_{12} and GA_{53} to their respective active GA products, GA_4 and GA_1. The reactions are summarized in a much-simplified form in Figure 3. The enzymes of the GA biosynthetic pathway have been identified and many of the corresponding genes have also been isolated and characterized.

Brassinosteroids

Brassinosteroids are found in algae, ferns, and angiosperms. They stimulate stem elongation, pollen tube growth, and unrolling of grass leaves. More than 40

continued

Box 12.2 Some important plant growth regulators are formed from terpenoid precursors (continued)

types of brassinosteroid have been identified, with brassinolide being the most biologically active. The brassinosteroids, as their name suggests, are steroid structures with 27, 28, or 29 carbon residues. The plant sterols, campesterol, sitosterol, cholesterol, and isofucosterol, are considered to be precursors for the various brassinosteroid structures. Of these, campesterol has been most thoroughly studied and it serves as the precursor the C-23 brassinosteroids. Campesterol (see Box 12.3 Figure 3 for structure), like all plant sterols, is synthesized from squalene, following the head-to-head condensation of two molecules of farnesyl diphosphate (Figure 12.11a). There would seem to be two potential routes for the formation of the brassinosteroid, brassinolide, from campesterol. The first stage is common to both routes and involves the

three-step conversion of campesterol to 5α-campestanol (Figure 4). Following this, the two routes differ with respect to the stage at which a hydroxyl group is introduced at the C-6 position. In the early C-6 oxidation route the hydroxyl group is added to 5α-campestanol. A series of oxidations and reductions result in the formation of castasterone, the immediate precursor to brassinolide. The late C-6 oxidation route is less well characterized although it would appear that the hydroxyl group is introduced at the penultimate step, with the formation of 6-α-hydroxy-6-deoxocastasterone that is subsequently converted to castasterone. In the final step, common to both routes, castasterone is oxidized to brassinolide. A much simplified version of the pathway is shown in Figure 4.

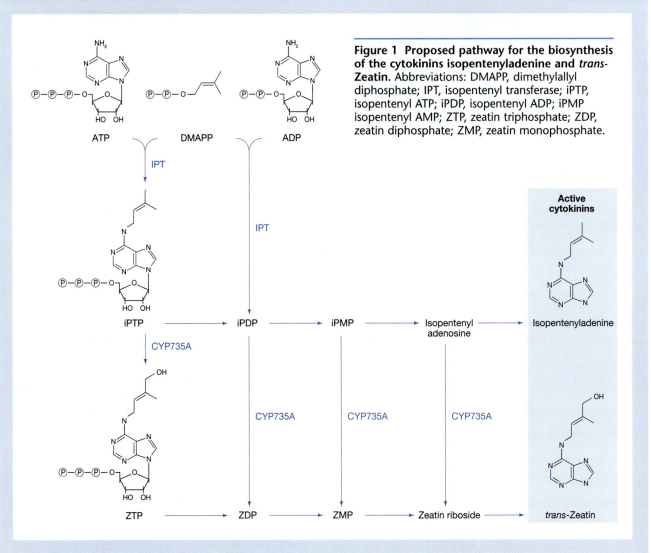

Figure 1 Proposed pathway for the biosynthesis of the cytokinins isopentenyladenine and *trans*-Zeatin. Abbreviations: DMAPP, dimethylallyl diphosphate; IPT, isopentenyl transferase; iPTP, isopentenyl ATP; iPDP, isopentenyl ADP; iPMP isopentenyl AMP; ZTP, zeatin triphosphate; ZDP, zeatin diphosphate; ZMP, zeatin monophosphate.

Box 12.2 Some important plant growth regulators are formed from terpenoid precursors (continued)

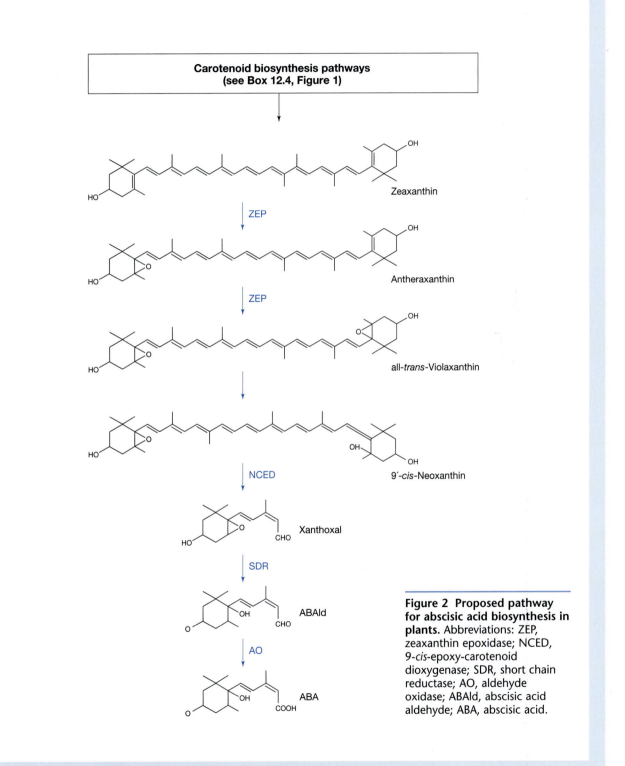

Figure 2 Proposed pathway for abscisic acid biosynthesis in plants. Abbreviations: ZEP, zeaxanthin epoxidase; NCED, 9-*cis*-epoxy-carotenoid dioxygenase; SDR, short chain reductase; AO, aldehyde oxidase; ABAld, abscisic acid aldehyde; ABA, abscisic acid.

continued

Box 12.2 Some important plant growth regulators are formed from terpenoid precursors (continued)

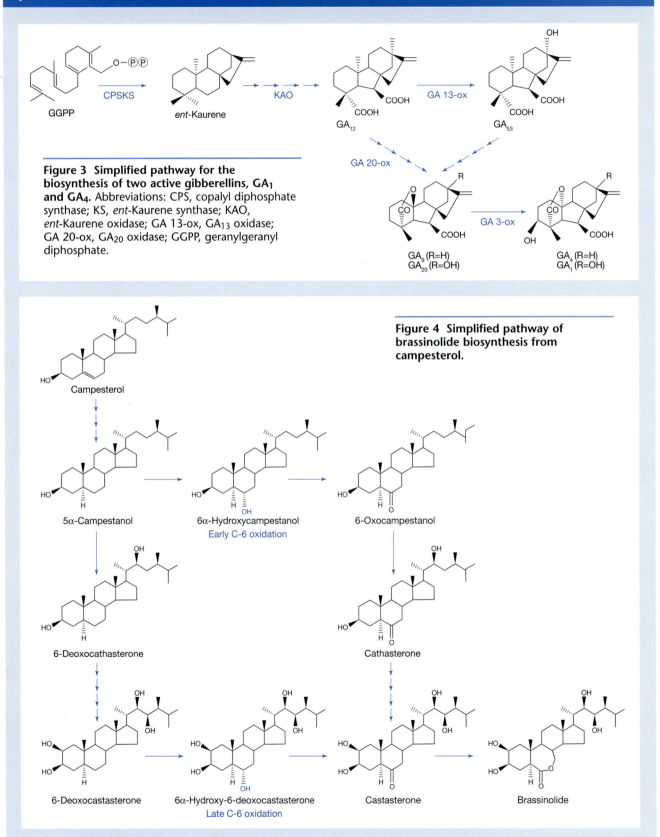

Figure 3 Simplified pathway for the biosynthesis of two active gibberellins, GA₁ and GA₄. Abbreviations: CPS, copalyl diphosphate synthase; KS, *ent*-Kaurene synthase; KAO, *ent*-Kaurene oxidase; GA 13-ox, GA₁₃ oxidase; GA 20-ox, GA₂₀ oxidase; GGPP, geranylgeranyl diphosphate.

Figure 4 Simplified pathway of brassinolide biosynthesis from campesterol.

resins to form a crystalline structure, which will eventually form into amber after millions of years of fossilization.

The hemiterpene, isoprene, appears to protect some plants from heat stress. Isoprene is volatile and is released as a gas from many plants. This release of isoprene can be quite substantial at high temperatures, resulting in the blue haze phenomenon that is commonly observed in the appropriately named Smoky Mountains (Eastern Tennessee, USA), Blue Ridge Mountains (Virginia, USA), and the Blue Mountains of New South Wales (Australia). The enzyme that produces isoprene, isoprene synthase, is induced at high temperatures and it is believed that the subsequent release of isoprene serves to cool the leaf by evaporative heat loss—equivalent to the transpiration cooling of water but without the risk of dehydration (see Further Reading). The release of isoprene from trees prompted the former US President Ronald Reagan in 1980 to claim that 80% of air pollution was caused by plants and trees. His aides later

Box 12.3 Terpenoids and human health

Many of the terpenoids are important in human health, either as essential components of our diet, or as medicines.

The carotenoids are a particularly important group of terpenoids within the human diet. β-Carotene, for example, is converted to vitamin A (retinol) by the action of a dioxygenase in the intestine. Vitamin A deficiency is a major cause of infant mortality in developing countries. The carotenoids, lutein and zeaxanthin, are the main components of the macular pigments, i.e. within the central region of the retina. A diet that is rich in these carotenoids can evidently protect the eye against age-related macular degeneration, which is the leading cause of blindness in the Western world.

The antioxidant properties of carotenoids are thought to be important in protecting cells from free radical damage that is implicated in illnesses such as cancer and cardiovascular disease. For example, there is evidence that lycopene, which is present in high concentrations in tomato, may offer some protection against prostate cancer, while diets rich in β-carotene are frequently correlated with a reduced risk of cancer and cardiovascular disease.

These health benefits of the carotenoids have been a driving force in the development of crops with enhanced carotenoid content, as discussed in Box 12.4.

Taxol (Figure 12.4a) is a diterpene that has become a widely used treatment for cancer. It was originally discovered in bark from the Pacific yew (*Taxus brevifolia*; see Plate 12.1). This discovery came about through a major research program that was launched in the 1960s in the USA. The program involved chemists and plant scientists in an extensive screening of over 100,000 plant extracts, to look for potential new drugs. The Pacific yew bark extract was the only candidate to emerge from these trials. It was

found to inhibit the growth of cancer cell cultures and, within a few years the active ingredient, taxol, was isolated. Approval for clinical trials was granted in 1984 and, by 1988 taxol was found to be effective against ovarian tumors. However, a major problem was the limited supply of Pacific yew bark. Even the best synthetic methods at the time could only produce half a gram of taxol from a mature Pacific yew tree, that could have taken 200 years to reach maturity. It would take 360,000 trees to produce enough taxol to treat all of the ovarian cancer victims in the USA. Clearly, the situation was unsustainable. Although the supply problem could be solved by chemical synthesis, the taxol structure was so complex that chemists were unable to make much progress towards this aim.

The breakthrough in taxol chemistry came with the discovery in France of a related chemical, 10-deacetylbaccatin (10-DAB), which could be used to manufacture taxol. The added bonus was that 10-DAB could be extracted in relatively large quantities from the needles of the English yew (*Taxus baccata*), without having to harvest the entire tree. Prunings and clippings from English yew trees were collected from many of the large country estates in Europe and sent to the pharmaceutical companies, in a major effort to provide enough taxol for clinical needs. As the supply improved, extensive trials could be completed and Taxol® (registered trademark of Bristol-Myers Squibb) was launched on to the market in 1993. Today, Taxol® is approved for treatment of ovarian cancer, breast and lung cancers, and Kaposi's sarcoma. Taxol® is the best-selling cancer drug ever manufactured, with annual sales peaking at $1.6 billion in 2000.

A number of terpenoid compounds are effective as contraceptives. Indeed, the contraceptive pill for women has its chemical origins in the Mexican yams (*Dioscorea*

continued

Box 12.3 Terpenoids and human health (continued)

macrostachya and *Dioscorea barabasco*). The yam tubers contain the triterpenoid saponin, diosgenin, which was used by a biochemist, Russell Marker, to synthesize progesterone in the 1940s. Marker set up a new company in Mexico to produce steroids, such as progesterone, from the Mexican yams. However, progesterone is not a good contraceptive because it has to be injected. It was another chemist, Carl Djerassi, who joined the company in 1949 and succeeded, 2 years later, in converting progesterone into norethindrone, a form that could be taken orally. This was the breakthrough that was needed, and Mexican yams were the sole source of diosgenin for steroid production up until the 1980s (Figure 1).

Although a reliable male contraceptive pill has yet to be produced, the terpenoid, gossypol (Figure 2), was an early candidate. Gossypol is a sesquiterpene (Figure 12.4) that occurs in high concentrations in cotton. It is an effective phytoalexin with antifungal activity, and its synthesis is induced by fungal elicitors. Gossypol causes male infertility

in humans and other mammals, by inhibiting spermatogenesis. It was first tested as a potential male contraceptive in the 1970s but has now been abandoned due to its toxic side-effects and the increased risk of infertility that results from prolonged use.

Plant sterols (phytosterols) are triterpenoids that serve the same function in plants as the animal steroid, cholesterol. They stabilize the phospholipid bilayer in cell membranes. Phytosterols have been used for the past 50 years to reduce cholesterol levels in the blood. The first evidence for their efficacy came from the early 1950s, and the weight of scientific evidence now supports this view. Initially, phytosterols were administered as drugs but, more recently, they have been incorporated into margarines, fruit juices, and ice creams. The major phytosterols are β-sitosterol, campesterol, and stigmasterol (Figure 3). As they share similar structural properties with cholesterol, they are thought to act by inhibiting cholesterol absorption into the blood.

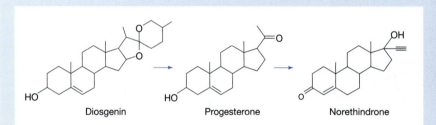

Diosgenin   Progesterone   Norethindrone

Figure 1 Norethindrone, a component of the female contraceptive pill, was first produced from progesterone that was itself synthesized from diosgenin, a triterpenoid saponin extracted from the Mexican yam.

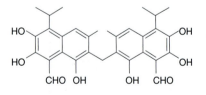

Gossypol

Figure 2 Gossypol is classed as a sesquiterpene as its basic skeleton is derived from the 15-carbon precursor farnesyl pyrophosphate.
The last step in the pathway involves the condensation of two molecules of the sesquiterpene hemigossypol to form gossypol. Hence gossypol is classed as a sesquiterpene dimer, even though it has the appearance of a triterpenoid.

Hemigossypol

Box 12.3 Terpenoids and human health (continued)

Although phytosterols are effective at reducing blood cholesterol levels, the most widespread treatment is the use of statins. These drugs inhibit HMG CoA reductase, which catalyzes an early step in the mevalonic acid pathway for terpenoid biosynthesis (see Figure 12.5). Statins are the most valuable pharmaceutical commodity in the world, with an estimated market value of $22 billion. Hence, knowledge of the terpenoid biosynthetic pathway, that led to the development of these pharmaceuticals, has brought rich returns, both in monetary value and in terms of human health.

Another inhibitor of terpenoid biosynthesis that has potential use as a pharmaceutical is fosmidomycin, which inhibits DXP reductoisomerase (the second enzyme in the DXP pathway; Figure 12.6). It is of medicinal use as an inhibitor of sterol formation in the malaria parasite, *Plasmodium falciparum*. Fosmidomycin appears to be well tolerated in both adults and children and although it does not achieve a total cure, it is highly effective in the treatment of asymptomatic carriers of malaria parasites.

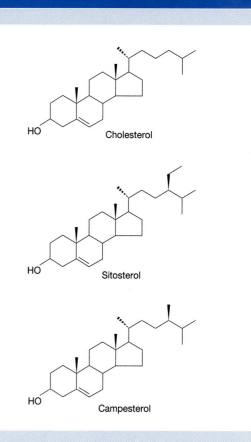

Figure 3 Plant sterols, sitosterol and campesterol, and the animal sterol cholesterol.

said that he had been misquoted as he was speaking only about hydrocarbons. In this respect, Reagan was correct. Coniferous forests can produce massive quantities of hydrocarbons in the form of isoprene. A recent twist in the story, which could never have been predicted at the time of Reagan's speech, has been recognized by scientists in Australia and in the USA. They found that isoprene can interact with nitrous oxides from car exhausts to produce ozone, a highly damaging gas that can cause harm to plants, animals, and humans. So, although trees cannot be held responsible for pollution, it would seem that the emission of isoprene and other volatile terpenoids is exacerbating the effects of other atmospheric pollutants.

While isoprene appears to serve a special physiological function in plants, other volatile terpenoids are primarily involved in animal–plant interactions (as discussed in Box 12.1).

The biosynthesis of terpenoids

Terpenoid biosynthesis can be divided into four stages:

1. Production of the basic five-carbon units, IPP and dimethylallyl diphosphate (DMAPP).

2. Condensation of IPP and DMAPP by prenyltransferases to form the terpenoid precursors such as geranyl and farnesyl diphosphate.

3. Conversion of these precursors to the parent structures of the terpenoid products, catalyzed by terpene synthases.

4. Modification of these terpenoid skeletons to introduce structural diversity.

Stage 1: the formation of isopentenyl diphosphate can occur via two distinct pathways

Common to all terpenoids, irrespective of their structure, is that they are synthesized from IPP. There are two quite distinct pathways for IPP formation, the mevalonic acid pathway (MVA pathway) and the deoxyxylulose phosphate pathway (DXP pathway; after the initial product, 1-deoxy-D-xylulose 5-phosphate (DXP)).

The DXP pathway was only discovered in the 1990s, first in bacteria, and later in plants. Up until then, the MVA pathway was believed to be the only means of producing IPP and, consequently, terpenoids.

We now know that many organisms do not depend on the MVA pathway alone. In fact, the DXP pathway is essential for terpenoid biosynthesis in virtually all eubacteria and also in the green algae (Chlorophyta), which lack the MVA pathway. In contrast, the Archaea, fungi (including yeasts), animals, and some eubacteria and algae lack the DXP pathway and rely entirely on the MVA pathway. Higher plants, mosses, liverworts, marine diatoms, and some algae (e.g. cyanobacteria) use both the DXP and MVA pathways, with the complete DXP pathway being present in plastids, and the MVA pathway being cytosolic (Table 12.1).

There are some interesting exceptions to these generalizations. Within the eubacteria, for example, a small group of actinomycetes, e.g. *Streptomyces,*

Table 12.1 Distribution of the mevalonic acid and deoxyxylulose phosphate pathways

	Mevalonic acid pathway		Deoxyxylulose phosphate pathway
Algae	✓	and/or	✓
Animals	✓		
Archaea	✓		
Eubacteria	✓	or	✓
Fungi	✓		
Mosses	✓		✓
Liverworts	✓		✓
Higher plants Plastids			✓
Cytosol	✓		
Protozoa	✓		✓

uses both pathways. Similarly, *Euglena gracilis* differs from all other photo-synthetic eukaryotes in using the MVA pathway alone. The lack of genes encoding DXP pathway enzymes in organisms that do not contain plastids indicates that higher plants acquired this pathway from the cyanobacterial ancestor of plastids. The reason why both the MVA and DXP pathways have been retained in higher plants is perhaps due to the need to produce different types of terpenoids in the plastids and cytosol, as discussed later.

The mevalonic acid pathway produces IPP from acetyl CoA within the cytosol

The MVA pathway begins with the condensation of two acetyl CoA molecules to form acetoacetyl CoA (Figure 12.5). Reaction with a further molecule of acetyl CoA results in the formation of a six-carbon product, 3-hydroxy-3-methylglutaryl CoA (HMG CoA). Until recently, it was thought that in plants a single bi-functional enzyme catalyzed both reactions, whereas in other organisms two enzymes, acetoacetyl CoA thiolase (AACT) and HMG CoA synthase (HMGS), were responsible. Indeed, many textbooks and reviews still present this as a fact. The evidence for a single enzyme, AACT/HMGS, in plants has come from studies of radish (*Raphanus sativus*) and *Catharanthus roseus* where, in each case, it proved impossible to separate AACT and HMGS activities during protein purification. However, functional complementation experiments indicate that there are two separate enzymes in some plants, as in other organisms. In these studies, a cDNA from arabidopsis that encoded HMGS was able to restore growth to a yeast mutant that lacked HMGS, but the same arabidopsis cDNA could not complement a yeast –AACT mutant. Similarly, an arabidopsis AACT cDNA could complement a yeast mutant that lacked AACT, but could not complement a yeast –HMGS mutant. Consequently, these experiments indicate that there are distinct genes, and enzymes, for the HMGS and AACT reactions in arabidopsis. While there is emerging evidence of the involvement of multigene families that encode either HMGS or AACT, there is no evidence, as yet, for the existence of a gene encoding a bi-functional AACT/HMGS enzyme.

MVA is formed from HMG CoA by an NADPH-dependent reductive reaction catalyzed by HMG CoA reductase (HMGR). This enzyme is highly regulated in animals where it is an important regulatory step in cholesterol biosynthesis (see Box 12.3). Plant HMGR is also highly regulated and appears to be particularly important in the control of sterol formation. A tobacco mutant with increased HMGR activity, for example, accumulates 10-fold more sterols than wild type. Plant HMGRs are encoded by a small family of genes, with each gene exhibiting distinct patterns of expression, from constitutive to tissue-specific to wound-induced. There is also evidence that the plant enzyme may be subject to posttranslational regulation via phosphorylation.

The remaining steps in the MVA pathway involve two consecutive phosphory-lation steps, catalyzed by mevalonate kinase and 5-phosphomevalonate kinase to produce mevalonate diphosphate. Finally, mevalonate diphosphate is decarboxylated by 5-diphosphomevalonate decarboxylase to release IPP. DMAPP can

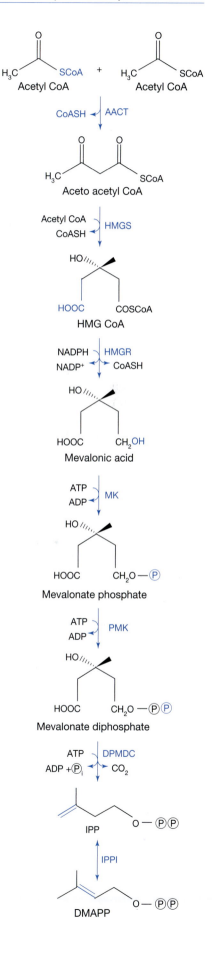

Figure 12.5 The mevalonic acid (MVA) pathway. The entire pathway takes place in the cytosol in higher plants. Abbreviations: AACT, acetoacetyl CoA thiolase; HMGS, 3-hydroxy-3-methylglutaryl CoA synthase; HMGR, 3-hydroxy-3-methylglutaryl CoA reductase; MK, mevalonate kinase; PMK, 5-phosphomevalonate kinase; DPMDC, 5-diphosphomevalonate decarboxylase; IPPI, IPP isomerase; HMG CoA, 3-hydroxy-3-methylglutaryl CoA; IPP, isopentenyl diphosphate; DMAPP, dimethylallyl diphosphate.

be formed by the isomerization of IPP (via IPP isomerase). These two isomers can be condensed together to form the 10-carbon precursor to monoterpenoids, as explained later. The pathway is shown in detail in Figure 12.5.

The deoxyxylulose phosphate pathway forms isopentenyl diphosphate in the plastid from pyruvate and glyceraldehyde 3-phosphate

The DXP pathway occurs in the plastids of higher plants. It converts two glycolytic products, pyruvate and glyceraldehyde 3-phosphate, to IPP and also forms the IPP isomer, DMAPP. The pathway was first characterized in bacteria, using classical isotope-feeding experiments (see page 416) that resulted in the identification of many of the reaction steps. More recently, a bioinformatics approach has been taken in order to identify genes and enzymes that had remained uncharacterized following these earlier investigations. Putative genes were subsequently expressed *in vitro* and subjected to further biochemical analysis in order to establish their function. All of the genes have now been isolated from bacterial and plant sources. The plant genes carry plastid-targeting sequences and are all nuclear-encoded. There remains some speculation about the reaction mechanisms, and the later stages of the pathway are in particular need of further research in higher plants. Nevertheless, the following summary and the accompanying figure (Figure 12.6) provide an account of our current knowledge of the DXP pathway in higher-plant plastids.

One confusing aspect of the DXP pathway is the terminology. The pathway itself is variously called the methylerythritol phosphate pathway (MEP pathway, after one of its intermediates, 2-*C*-methyl-D-erythritol phosphate; MEP), the DOXP pathway (DOXP is used as an alternative abbreviation for 1-deoxy-D-xylulose 5-phosphate), and the non-mevalonate pathway. Furthermore, many of the enzymes, genes, and substrates have alternative names. One of the reasons for this is that the genes were often isolated before the corresponding enzyme reactions were identified. For example, in the analysis of the bacterial DXP pathway the component genes were denoted *isp* (note that bacterial genes, by convention, are denoted in lower case) and many review articles continue to use this nomenclature, with the corresponding enzymes referred to as IspC protein, IspD protein, etc. In order to help the reader when referring to other reference material, we have therefore listed alternative names in the legend of Figure 12.6.

The DXP pathway begins with the condensation of glyceraldehyde 3-phosphate with pyruvate to produce 1-deoxy-D-xylulose 5-phosphate (DXP), a five-carbon sugar phosphate (catalyzed by DXP synthase; DXS; Figure 12.6). The reaction is a transketolase-like decarboxylation that transfers a two-carbon unit from pyruvate to glyceraldehyde 3-phosphate to form DXP. The enzyme requires thiamine pyrophosphate as well as divalent cations (Mg^{2+} or Mn^{2+}). *DXS* genes have been cloned from a range of plants, including mint (*Mentha* × *piperita*), pepper (*Capsicum annuum*), and arabidopsis. The genes are present in the nuclear genome and the resulting proteins are targeted to the plastid by N-terminal targeting sequences.

The second reaction of the pathway occurs in two steps, both of which are catalyzed by DXP reductoisomerase (DXR). DXP is first rearranged into the intermediate, 2-*C*-methyl-erythrose-4-phosphate, which is then reduced by NADPH to produce the product, 2-*C*-methyl-D-erythritol 4-phosphate (MEP). The antibiotic, fosmidomycin, is an analog of the intermediate and is an effective inhibitor of DXR. The *DXR* gene has been cloned from a number of higher plants, and plastid-targeting sequences have been predicted in all of them. Fosmidomycin is a very effective inhibitor that has proved useful in the elucidation of the pathway. It is also proving to be useful as a treatment for malaria (Box 12.3).

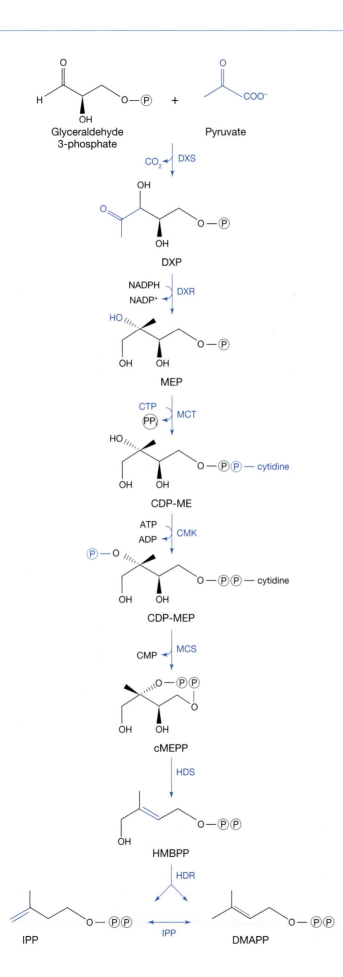

Figure 12.6 The deoxyxylulose phosphate (DXP) pathway.

The entire pathway takes place in the plastids in higher plants.

Note that many of the enzymes in the DXP pathway have more than one name. We include alternative names here to aid with further reading. Also, as the genes were discovered they were assigned the 3-letter name Isp (isoprenoid) and a letter (c to g). In some cases, the enzymes were named later, once their catalytic activities had been identified. As this renaming has only occurred recently, many publications still use the Isp terminology and we include it here for comparative purposes.

Abbreviations:

Substrates:
DXP, 1-deoxy-D-xylulose 5-phosphate;
MEP, 2-C-methyl-D-erythritol 4-phosphate;
CDP-ME, 4-diphosphocytidyl-2-C-methyl-D-erythritol;
CDP-MEP, 2-phospho-4-(cytidine 5'-diphospho)-2-C-methyl-D-erythritol;
cMEPP, 2-C-methyl-D-erythritol-2,4-cyclodiphosphate;
HMBPP, 1-hydroxy-2-methyl-2-(E)-butenyl 4-diphosphate;
IPP, isopentenyl diphosphate;
DMAPP, dimethylallyl diphosphate.

Enzymes:
DXS, DXP synthase; DXR, DXP reductoisomerase, also called MEP synthase (IspC protein); MCT, MEP cytidylyltransferase, also called CDP-MEP synthase (IspD protein); CMK, 4-diphospocytidyl-2C-methyl-D-erythritol kinase, also called CDP-ME kinase (IspE protein); MCS, cMEPP synthase (IspF protein); HDS, cMEPP reductase, also called HMBPP synthase (IspG protein); HDR, HMBPP reductase, also called Isopentenyl/dimethylallyl diphosphate synthase (IspH protein); IPPI, IPP-DMAPP isomerase.

The next three reactions result in the conversion of MEP into a cyclic 2,4-diphosphate, 2-*C*-methyl-D-erythritol-2,4-cyclodiphosphate (cMEPP). In the first of these reactions (catalyzed by MCT, 2-*C*-methyl-D-erythritol 4-phosphate cytidylyltransferase or MEP citidylyltransferase), a cytidyl phosphate residue (from cytidine triphosphate, CTP) is introduced into MEP to form CDP-ME (4-diphosphocytidyl-2*C*-methyl-D-erythritol). This reaction appears to be important in the synthesis of *ent*-kaurene, a precursor of the plant growth regulator, gibberellic acid. Supporting evidence has come from the analysis of arabidopsis plants in which expression of the *MCT* gene has been suppressed by antisense technology. The resulting plants contained reduced amounts of both *ent*-kaurene and gibberellic acid, indicating that these two terpenoid products depend on the DXP pathway for precursors. The plastid localization of both the DXP pathway and of gibberellin biosynthesis is consistent with this conclusion.

In the next reaction CDP-ME is phosphorylated via an ATP-dependent kinase reaction (catalyzed by CMK, also referred to as CDP-ME kinase, which is encoded by the *ispE* gene) that adds a phosphate group at the C-2 hydroxyl position to form CDP-MEP (2-phospho-4-(cytidine 5′-diphospho)-2-*C*-methyl-D-erythritol).

In the next step cytidine monophosphate is released and the two remaining phosphate groups associate to form a cyclic structure (cMEPP, 2-*C*-methyl-D-erythritol-2,4-cyclodiphosphate). The reaction is catalyzed by MCS (cMEPP synthase; encoded by the *ispF* gene). Evidence that this reaction occurs in plants has come from the analysis of an arabidopsis mutant that does not express the *ispF* gene. The mutant is albino because it fails to produce chlorophyll or carotenoid pigments because of the loss of this step in the terpenoid pathway.

The ring is opened in the next reaction (catalyzed by HDS (cMEPP reductase; encoded by the *ispG* gene) where cMEPP is reduced so that the phosphate groups are separated to form a linear structure (HMBPP; 1-hydroxy-2-methyl-2-(*E*)-butenyl 4-phosphate) once again. Although this reaction has yet to be fully characterized, recent research indicates that, in chloroplasts, photosynthesis generates the reductant directly, probably via ferredoxin. In the dark, the reductant appears to be provided by ferredoxin NADP$^+$ oxidoreductase and NADPH, as in the case of the provision of reductant for nitrite reduction in nonphotosynthetic plastids (see Chapter 8).

The final step in the DXP pathway results in the formation of IPP and its isomer, DMAPP. This reaction, catalyzed by HDR (HMBPP reductase), favors the formation of IPP over DMAPP (ratio of 6:1) although the balance can be adjusted by IPP isomerase, which interconverts the two. DMAPP and IPP are the end products of the DXP pathways. They are subsequently combined together to form the diverse range of terpenoid compounds that are produced in higher plant plastids, as described later.

The deoxyxylulose phosphate pathway was discovered in bacteria before it was identified in higher plants

The first experiments that led to the discovery of the DXP pathway were carried out with bacteria in the early 1990s. Before this, the MVA pathway was considered to be the only means of producing the terpenoid precursor, IPP. The first indication that this might not be the case came from feeding experiments, using the stable isotope of carbon, ^{13}C (Chapter 2, page 15). When a range of bacteria were grown on a medium containing ^{13}C-acetates, the pattern of label that was found in hopanoids (terpenoid compounds) was inconsistent with their synthesis via the MVA pathway. Further experiments with

^{13}C-glucose showed that label could enter IPP from a three-carbon triose unit and from pyruvate, with the pattern being consistent with a head-to-head condensation of glyceraldehyde 3-phosphate and pyruvate. The first product of this condensation reaction was subsequently identified as DXP. Feeding of ^{2}H-labeled DXP to *Escherichia coli* cultures then resulted in incorporation of label into the side chain of ubiquinone, a terpenoid compound. Hence, the early stages of the pathway began to be unraveled. Further labeling studies followed, using a range of species of bacteria, algae, and plants, which eventually led to the identification of many of the reactions of the DXP pathway. These early experiments can be followed by referring to the Further Reading section at the end of this chapter.

As well as the classical isotope-feeding experiments, the main strategy taken to identify the individual steps in the DXP pathway was to use a functional genomics approach. This approach uses the strategy of identifying component genes, usually by screening for mutants that are defective in terpenoid biosynthesis. In plants, mutations in the plastid DXP pathway usually result in an albino phenotype, due to an inability to produce chlorophylls (where the phytol tail is a terpenoid) or carotenoids. These mutants can then be used in order to test whether any of the candidate genes are able to restore activity when they are expressed in the mutant. In addition, the genes are expressed *in vitro* and the resulting proteins are tested for enzymatic activity. Furthermore, the generation of overexpressing or antisense lines has enabled the effects on the pathway to be tested for many of the candidate genes.

The later stages of the DXP pathway were first investigated using a bioinformatics approach. Whole genomes were compared between organisms known to use the DXP pathway with those that used the MVA pathway. As a result, putative genes were identified that occurred in all DXP-dependent organisms and not in any MVA-dependent organisms. Sequence information was then used to determine other features, such as plastid-targeting (in the case of plants) and potential substrate and cofactor-binding sites. Finally, *in vitro* expression of the proteins then enabled the enzyme to be tested for catalytic activity with potential substrates, so that the entire pathway was finally unraveled.

Transgenic plants and the manipulation of the DXP pathway

Transgenic plants have been used to unravel the reactions of the deoxyxylulose phosphate pathway and to manipulate the production of terpenoid products in commercially important plants. Mutants and/or transgenic plants have been produced for each of the genes encoding steps in the DXP pathway shown in Figure 12.6. In each case, when the expression of the gene is repressed, the plants display an albino phenotype indicative of a failure to produce terpenoid pigments (chlorophyll, carotenoids). In a recent study, for example, the technique of virus-induced gene silencing was used to block the expression of the last two enzymes in the pathway (HDS and HDR; Figure 12.6; *ispG* and *ispH* genes) in *Nicotiana benthamiana*. The resulting plants were albino. Apart from demonstrating that these two enzymes were essential for plastid terpenoid biosynthesis, this work also shows that the MVA pathway, which produces IPP in the cytosol, cannot compensate for the loss of IPP synthesis in the plastid. Further discussion of the interchange between these two pathways follows later in this chapter.

The generation of albino mutants clearly demonstrates the essential role of the DXP pathway in producing chlorophylls and carotenoids. Similar experiments have further established the requirement for this pathway for the production of precursors for tocopherols, gibberellins, and abscisic acid. For example, arabidopsis plants expressing antisense *DXS* (the gene encoding DXS, the first enzyme in the pathway; Figure 12.6) were deficient in chlorophylls, carotenoids, tocopherols, abscisic acid, and gibberellins. Similarly, arabidopsis plants that were antisensed for *MCT* (encoding MCT, the third enzyme in the DXP pathway; Figure 12.6) had a reduced capacity to produce gibberellins.

Manipulation of the DXP pathway has also demonstrated its role in the synthesis of monoterpenoids. In peppermint (*Mentha × piperita*), for example, overexpression of the *DXR* gene (encoding DXR, the second enzyme in the pathway; Figure 12.6) resulted in a 50% increase in monoterpenoid concentration in the leaves. Lavender (*Lavandula latifolia*) monoterpenoid content has also been increased, in this case by the introduction of the *DXS* gene from arabidopsis. The resulting plants had up to three times the concentration of monoterpenoids compared with wild type. These results suggest that the supply of IPP via the DXP pathway within the plastid may be limiting the rate of formation of monoterpenoids. As such, the pathway is an important target for breeding and transformation in order to increase the production of essential oils in plants such as mint and lavender (see Further Reading).

Stage 2: prenyltransferases (which combine the five-carbon IPP and DMAPP units to form a range of terpenoid precursors)

Following the synthesis of IPP and DMAPP, either in the cytosol or plastid, the next stage in terpenoid production involves the repetitive addition of these five-carbon units to form more complex terpenoid precursors. The reactions are catalyzed by a group of enzymes called prenyltransferases (Figure 12.7).

Prenyltransferases catalyze condensation reactions between IPP and DMAPP, and then add further IPP units on to the growing polymer. It is worth noting that each of the prenyl diphosphates (i.e. C-10, C-15, C-20) is formed by a specific prenyltransferase enzyme, which begins its reaction with IPP and DMAPP and continues to add C-5 units to the elongating chain until the correct chain length is achieved (Figure 12.7). The final chain length is determined by the product specificity of the individual enzyme, and is thought to be related to the amino acid sequence and three-dimensional structure of the enzyme. Hence, geranyl diphosphate synthase forms GPP, while geranylgeranyl diphosphate synthase forms geranylgeranyl diphosphate (GGPP). In each case the intermediates remain attached to the enzyme, which continues to add five-carbon units until the end-product is formed and released from the enzyme. The reaction mechanism is described in Figure 12.7.

The prenyltransferases that catalyze the formation of these linear isoprene polymers belong to a distinct class of prenyltransferases that carry out a head-to-tail (or 1′-4 condensation) of the isoprene units. Within this class there are two types of prenyltransferase, the *trans*-prenyltransferases that add the C-5-prenyl group in the *trans* configuration, and the *cis*-prenyltransferases that add the C-5 group in the *cis* configuration (Figure 12.7). In either case, the product is a linear chain, although the properties of the end-product will differ. For example, rubber prenyltransferase is a *cis*-prenyltransferase. Consequently, in natural rubber the stereochemistry is the same at each double bond, i.e. each isoprene unit is joined head-to-tail in a *cis* 1,4-linkage (Figure 12.4). This conformation allows the rubber polymer to twist and coil, making natural rubber flexible and elastic. In contrast, gutta percha, where the isoprene units are joined by *trans* 1,4-linkages, is a very rigid structure (Figure 12.4) that was used in the past as a coating for golf balls, for example.

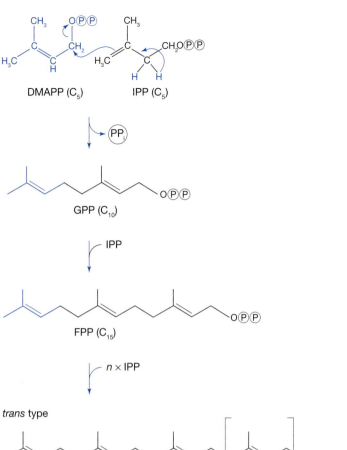

DMAPP (C₅) IPP (C₅)

GPP (C₁₀)

FPP (C₁₅)

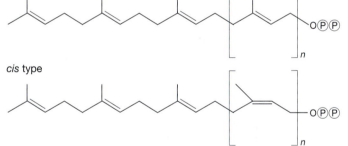

trans type

cis type

Figure 12.7 Prenyltransferase reaction mechanism. Prenyltransferases join together two prenyl diphosphates, e.g. IPP and DMAPP in the first instance. The further addition of IPP units increases the chain length of the product by 5-carbons each time.

In the example shown here, the prenyltransferase catalyzes a head-to-tail (or 1′-4 condensation) of the isoprene units. The reaction begins with the alignment of the diphosphate (OPP or head end) of the acceptor with the terminal double bond (tail end) of the donor. Removal of the pyrophosphate from the donor creates a carbocation (positively charged carbon), which then reacts with the double bond of the donor molecule to form a new carbon-carbon bond. The double bond is restored, one carbon further along, by the stereospecific removal of a proton at the C-2 position (shown in blue). The stereospecificity of this reaction depends on the type of prenyltransferase involved. A *trans*-prenyltransferase adds isoprene units in the *trans* position, while a *cis*-prenyltransferase adds them in the *cis* position, as shown. Further additions of 5-carbon isoprene units occur in the same way so that the chain length continues to increase until the final product is formed, at which point the product is released from the prenyltransferase. Abbreviations: DMAPP, dimethyallyl diphosphate; IPP, isopentenyl diphosphate; GPP, geranyl diphosphate; FPP, farnesyl diphosphate.

A second class of prenyltransferases forms head–head or head–middle condensations that produce irregular terpenoids such as chrysanthemyl diphosphate (Figure 12.3). A third class of prenyltransferases adds a prenyl group on to a non-terpenoid acceptor as, for example, in the formation of chlorophylls, tocopherols, cytokinin, and ubiquinone where terpenoid groups form only part of the final molecule; these mixed structures are classed as meroterpenes. Prenyl groups are also added to some proteins in order to anchor them into the hydrophobic environment of membranes.

The end result of the prenyltransferase stage of terpenoid biosynthesis is the formation of the terpenoid precursors. Thus, the 10-carbon GPP provides the carbon skeleton for all monoterpenes, similarly 15-carbon farnesyl diphosphate (FPP) results in sesquiterpenes, 20-carbon GGPP in diterpenes and so on (Figure 12.8). These precursors are used in the next stage to form the basic terpenoid structures, as explained in the following section.

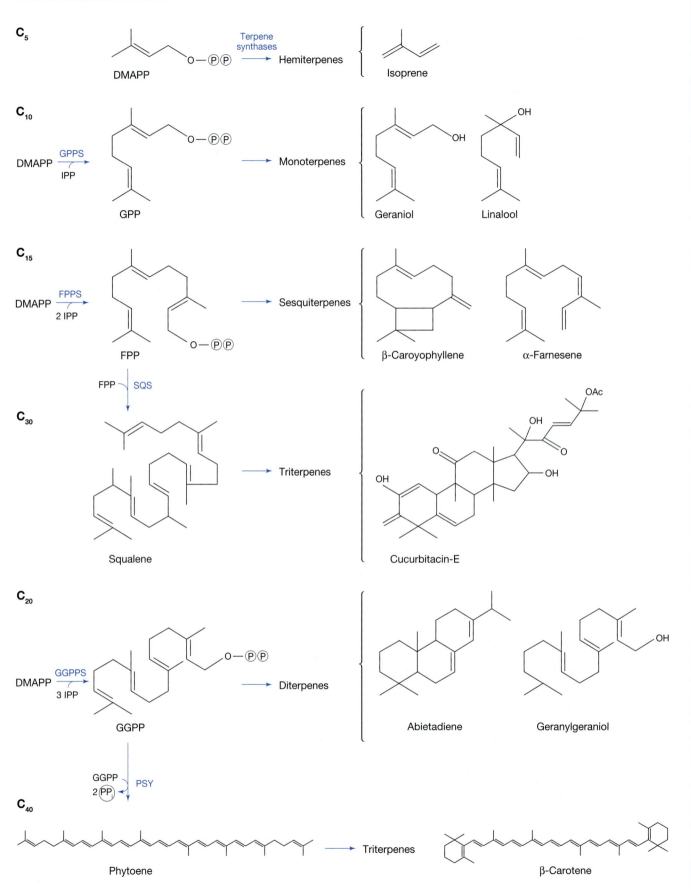

C$_5$

DMAPP

Terpene synthases → Hemiterpenes

Isoprene

C$_{10}$

DMAPP —GPPS/IPP→ GPP → Monoterpenes

Geraniol   Linalool

C$_{15}$

DMAPP —FPPS/2 IPP→ FPP → Sesquiterpenes

β-Caroyophyllene   α-Farnesene

FPP —SQS→

C$_{30}$

Squalene → Triterpenes

Cucurbitacin-E

C$_{20}$

DMAPP —GGPPS/3 IPP→ GGPP → Diterpenes

Abietadiene   Geranylgeraniol

GGPP —PSY/2 PP$_i$→

C$_{40}$

Phytoene → Triterpenes

β-Carotene

Figure 12.8 Formation of the major classes of terpenoids from isopentenyl diphosphate (IPP) and its isomer, dimethylallyl diphosphate (DMAPP). Prenyltransferases add 5-carbon units from IPP to form 10-carbon (geranyl diphosphate, GPP), 15-carbon (farnesyl diphosphate, FPP), and 20-carbon (geranylgeranyl diphosphate, GGPP) precursors to the mono-, sesqui- and diterpenes, respectively, as explained in Figure 12.7. Terpene synthases convert these precursors to specific terpenoids (see Figures 12.9 and 12.10).

Larger terpenoids may be formed by joining two smaller terpenoid precursors together. For example, triterpene (C_{30}) precursors are formed from two C_{15} farnesyl diphosphates (FPP) in a reaction catalyzed by squalene synthase (SQS). Also, tetraterpene (C_{40}) precursors are formed from two C_{20} geranylgeranyl diphosphates (GGPP) in a reaction catalyzed by phytoene synthase (PSY). These two reactions are shown in detail in Figure 12.11.

Abbreviations: GPPS, geranyl diphosphate synthase; FPPS, farnesyl diphosphate synthase; GGPPS, geranylgeranyl diphosphate synthase; SQS, squalene synthase; PSY, phytoene synthase; IPP, isopentenyl diphosphate; DMAPP, dimethylallyl diphosphate; GPP, geranyl diphosphate; FPP, farnesyl diphosphate; GGPP, geranylgeranyl diphosphate; PP_i, pyrophosphate.

Stage 3: terpene synthases

Terpene synthases convert the terpenoid precursors, GPP, FPP, and GGPP into the basic terpenoid groups, the mono-, sesqui-, di-, tri-, tetra-, and polyterpenoids. They are a large family of enzymes and are classified according to the terpenoid groups that they form, i.e. monoterpene synthases, diterpine synthases, sesquiterpene synthases, etc. They are sometimes referred to as terpene cyclases, because many of their products are cyclic. These enzymes are remarkable as regards the diversity of products that they generate and, as such, they are responsible for introducing considerable structural variation into the carbon skeletons that are produced at this stage of terpenoid biosynthesis.

The terpene synthase reaction resembles that of the prenyltransferases in that it binds the allylic diphosphate substrate and cleaves away the diphosphate group to leave behind a reactive carbocation. The carbocation remains bound to the terpene synthase, so that it is protected from being prematurely quenched or dispersed into the surrounding medium (i.e. the soluble phase of the cytosol or plastid) until the reaction is complete. While it is in this state, the carbocation undergoes a number of reactions before the end product(s) are finally released from the enzyme. The reaction mechanism is shown in detail in Figure 12.9 (for monoterpene synthases) and in Figure 12.10 (for diterpene synthases).

Many terpene synthases are capable of producing multiple products from a single substrate. For example, the pinene synthases from sage (*Salvia officinalis*) and grand fir (*Abies grandis*) can produce both α- and β-pinene using GPP as the substrate. Similarly, limonene synthase, the principal monoterpene synthase of spearmint (*Mentha spicata*) and peppermint (*Mentha* x *piperita*) produces myrcene as well as α- and β-pinene. One particularly impressive example is of the sesquiterpene synthases found in grand fir (*Abies grandis*), each of which is capable of producing in excess of 30 distinct products.

While the mono- and sesquiterpene synthases employ identical reaction mechanisms, the sesquiterpene synthases can generate a much larger array of products. The reason for this difference is that the sesquiterpene carbocation intermediate has an additional five carbon atoms with which to interact, hence a greater number of cyclizations is possible. Consequently, there are over 7000 sesquiterpenes and about 1000 different monoterpenes.

The diterpene synthases, which react with the C-20 precursor GGPP, can carry out two distinct types of reaction. The first type closely resembles that of the monoterpene and sesquiterpene synthases, where ionization of the diphosphate ester (i.e. GGPP in this case) initiates cyclization (i.e. ring closure). An

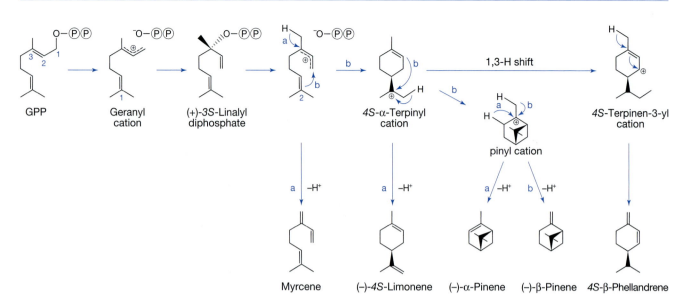

Figure 12.9 Terpene synthases produce a diverse range of products from a small number of substrates, as illustrated by the monoterpene synthase reaction mechanism.

The reaction mechanism is the same for virtually all terpene synthases, although the example shown here is for monoterpene synthases. Monoterpene synthases use geranyl diphosphate (GPP) as the substrate. The structure of GPP is such that it cannot be transformed directly into a cyclic product, because the double bond between C2 and C3 (see GPP structure) prevents rotation around C2 that would enable a cyclic structure to form (i.e. via bonding between C1 and another part of the molecule). Hence the first stage of the reaction involves the isomerization of GPP, where GPP is first ionized by interacting with a divalent metal ion (e.g. Mg^{2+}, as in the prenyltransferase reaction) to form the geranyl cation (1). This intermediate can now isomerize, by rotation about the C2-C3 bond to form either (S)- or (R)- linalyl diphosphate (formation of the R or S stereoisomers is enzyme-specific—the (S)- form is shown here). The OPP group is now split off to leave a highly reactive carbocation intermediate (2) that

remains bound to the enzyme. This intermediate may be deprotonated directly (a) to form myrcene or cyclized (b) to form the terpinyl cation ((S)- form shown here but may also be the (R)- form).

The terpinyl cation is a universal intermediate in the terpene synthase reaction and this molecule can undergo a variety of transformations, only some of which are shown here. For example, it may be deprotonated (a) to form limonene; it may undergo further cyclization (b) to form the pinyl cation, which can be deprotonated (a, b) to form α- or β-pinene. The terpinyl cation may also be transformed by hydride shifts–where a hydrogen atom migrates, along with a pair of electrons, from one carbon atom to another positively charged carbon atom. In the example shown here, the hydrogen migrates from C1 to C3 to generate the terpinen-3-yl cation that leads to the formation of the phellandrenes. A similar transformation results in the terpinen-4-yl cation that leads to the thujanes (not shown). The terpinyl cation may also be transformed to α-terpineol, which can be further cyclized to produce 1,8-cineole in some species (not shown).

example of this kind of diterpene synthase reaction is that catalyzed by cas-bene synthase (Figure 12.10a). The second type of diterpene synthase reaction differs in that protonation (i.e. the addition of a H⁺ ion) initiates cycliza-tion. In the case of abietadiene synthase (Figure 12.10b) GGPP is protonated at the C-14=C-15 double bond, and cyclization results in the formation of a two-ringed, stable intermediate, copalyl diphosphate. An ionization-induced cyclization then follows and a series of reactions (e.g. migration of a methyl group, reprotonation) results in the formation of abietadiene, one of the diterpenoid resin acids involved in wound sealing in conifers. Copalyl diphosphate is an important intermediate that is common to the reactions of all of the terpene synthases that undergo this type of protonation-initiated cyclization. Kaurene synthase, for example, uses this mechanism to produce kaurene, a precursor in gibberellin biosynthesis.

The formation of the 30-carbon triterpenoids begins with the head–head condensation of two molecules of FPP to form squalene (Figure 12.11a). The

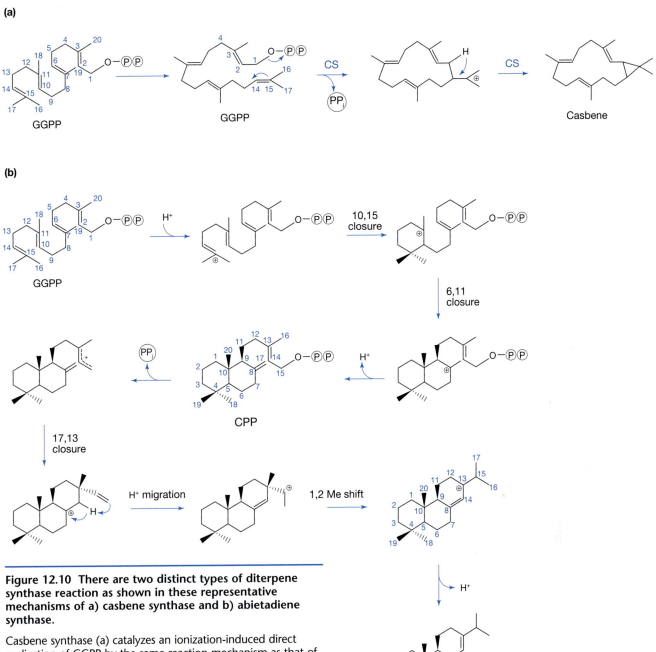

Figure 12.10 There are two distinct types of diterpene synthase reaction as shown in these representative mechanisms of a) casbene synthase and b) abietadiene synthase.

Casbene synthase (a) catalyzes an ionization-induced direct cyclization of GGPP by the same reaction mechanism as that of the monoterpene (as shown in Figure 12.9) and sesquiterpene synthases. The substrate, geranylgeranyl diphosphate (GGPP) is ionized by interacting with a divalent metal ion (Mg^{2+}), followed by cleavage of the pyrophosphate group (PPi). The carbocation intermediate is subsequently converted to casbene through the formation of a bond between C1 and C14.

Some diterpene synthases carry out a distinct type of reaction that differs from that of the other terpene synthases. The proposed reaction catalyzed by abietadiene synthase (b) is a typical example. It begins with the protonation (i.e. addition of H+) of GGPP at C14, which promotes attack by the C10-C11 double bond on C15 and this leads to the formation of a bond between C10 and C15 to form a closed ring. A similar attack of the C6-C7 double bond on C11 results in a second ring closure via the formation of a bond between C6 and C11. Deprotonation (i.e. loss of a H+) then results in the formation of copalyl diphosphate (CPP) as a stable intermediate. A

typical terpene synthase type of ionization-induced cyclization reaction follows (note the different numbering used for the CPP molecule as compared with the GGPP structure). This reaction results in closure of the third ring. In subsequent reactions a proton migrates from C17 to C15, a methyl group moves from C1 to C2 (1,2 Me shift) and finally deprotonation results in the formation of abietadiene. Note that other forms of abietadiene may be produced, depending on the site at which deprotonation occurs (not shown). As in the case of all terpene synthases, the entire reaction sequence takes place while the intermediates remain bound to the enzyme.

reaction is catalyzed by squalene synthase, which is classed as both a prenyl-transferase and a terpene synthase as it carries out both reactions. Squalene is the precursor for a number of membrane components, such as cholesterol and sitosterol. It is also used to form brassinosteroids (plant growth regulators; Box 12.2), saponins and cardenolides (see Box 12.1 and Figure 12.4). The cyclization reactions in the triterpene groups are complex and result in multiple ring structures, through several different mechanisms. In most cases squalene is initially oxidized to oxidosqualene before becoming cyclized.

The 40-carbon tetraterpenes are formed from the head–head condensation of two GGPP (20-carbon) molecules to produce phytoene (Figure 12.11b). The reaction is catalyzed by phytoene synthase (PSY), which, like squalene synthase, is a member of both the prenyltransferase and terpene synthase families. PSY and squalene synthase share similar properties. They have a similar reaction mechanism and also share three well-conserved regions within their amino acid sequence. PSY is inhibited by the herbicide norflurazon, which

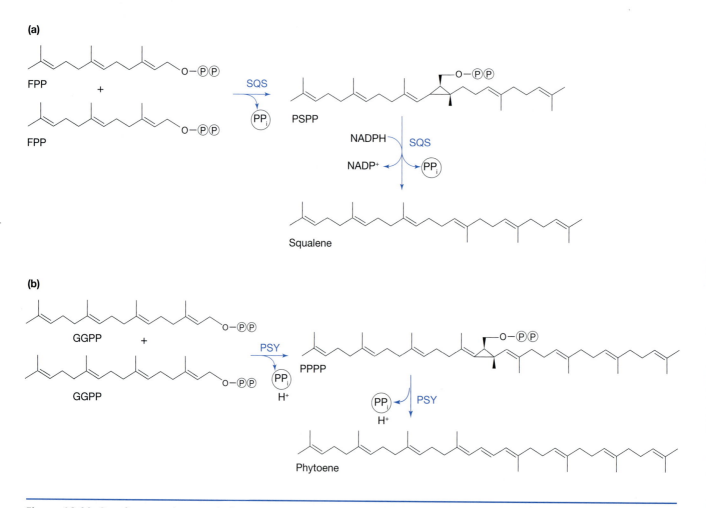

Figure 12.11 Squalene synthase and phytoene synthase reactions. (a) Squalene synthase (SQS) converts two molecules of farnesyl diphosphate (FPP) into squalene in a two-step reaction. In the first reaction, a pyrophosphate (PP_i) group is removed from the FPP donor and this is followed by a head-to-head condensation with an FPP acceptor to produce pre-squalene diphosphate (PSPP). In the second reaction, which requires NADPH, PSPP undergoes a complex rearrangement involving removal of the pyrophosphate group and the formation of a new carbon-carbon double bond.

(b) Phytoene synthase (PSY) converts two molecules of geranylgeranyl diphosphate (GGPP) into the tetraterpene, phytoene. This is a two-stage process that closely resembles the reactions catalyzed by SQS, except that PSY uses Mn^{2+} rather than Mg^{2+}, and SQS requires NADPH. The first step generates the intermediate, prephytoene diphosphate (PPPP). The mechanistic similarity is apparent when SQS is incubated with FPP and Mn^{2+} in the absence of NADPH. Under these conditions SQS produces dehydrosqualene, a 30-carbon analog of phytoene.

Box 12.4 Golden Rice: increasing the carotenoid content of rice endosperm to tackle vitamin A deficiency

Vitamin A deficiency is a major health problem throughout the developing world. As many as 500,000 children become blind each year through vitamin A deficiency and half of them die within 12 months of losing their sight. Rice (*Oryza sativa*), which lacks vitamin A in its endosperm (the edible part), is the staple food in at least 33 developing countries where malnutrition is widespread. If vitamin A could be produced within the rice endosperm, this could result in a major improvement in human health.

Vitamin A can be produced in the body from a dietary intake of β-carotene (pro-vitamin A). Hence the target for improving the nutritional quality of rice has been to introduce genes involved in carotenoid production. At present, this can only be achieved through genetic engineering because there are no rice genotypes capable of synthesizing carotenoids within the endosperm.

Golden Rice is the name that has been given to all genetically modified rice that produces β-carotene in the endosperm. The grain is golden yellow as a result of the increased carotenoid content.

The first attempt at producing Golden Rice was to introduce the phytoene synthase gene (*PSY*) from daffodil (*Narcissus pseudonarcissus*) via *Agrobacterium* transformation. Rice endosperm was known to be capable of producing geranylgeranyl diphosphate (GGPP), hence the introduced gene was expected to increase carotenoid production by increasing the supply of phytoene (Figure 1).

Figure 1 Carotenoid biosynthesis. The carotenoids are tetraterpenes, formed from the head-to-head condensation of two geranylgeranyl diphosphate (GGPP) molecules, catalyzed by phytoene synthase (PSY). In higher plants the desaturation of phytoene requires three further enzymes to form lycopene. In bacteria, a single enzyme, carotene desaturase (CRTI) converts phytoene to lycopene. Abbreviations: GGPP, geranylgeranyl diphosphate; PSY, phytoene synthase; PDS, phytoene desaturase; ZDS, ζ-carotene desaturase; CRTISO, carotenoid isomerase; LCYB, lycopene β-cyclase; LCYE, lycopene ε-cyclase; CHYB, carotenoid β-ring hydroxylase; CHYE, carotenoid ε-ring hydroxylase; ZEP, zeaxanthin epoxidase; VDE, violaxanthin de-epoxidase; CRTI, carotene desaturase (bacterial).

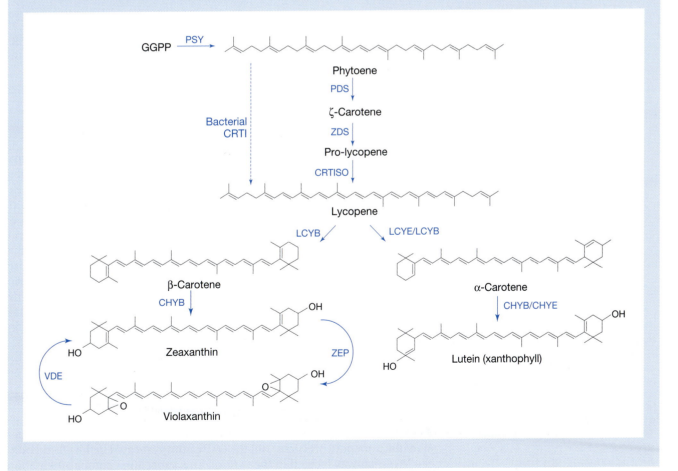

continued

Box 12.4 Golden Rice: increasing the carotenoid content of rice endosperm to tackle vitamin A deficiency (continued)

The transgenic rice did, indeed, produce more phytoene but it was not capable of producing any of the carotenoids. This was thought to be because the rice lacked the capacity to carry out the desaturation steps that convert phytoene to lycopene. Hence, a further gene was added. This time, a bacterial gene was used, CRTI, that encodes a carotene desaturase capable of converting phytoene to lycopene (Figure 1). The plants were expected to produce lycopene, as it was assumed that the endosperm lacked the cyclase enzymes (lycopene β-cyclase (LCYB) and lycopene ε-cyclase (LCYE)) needed to convert lycopene to α- and β-carotene. However, the plants *did* form α- and β- carotene, as well as variable amounts of lutein (xanthophyll) and zeaxanthin. This was a surprising result, given that the earlier study, where only *PSY* was overexpressed, seemed to indicate that rice endosperm lacked the capacity to convert phytoene to carotenoids. The explanation seems to be that rice endosperm contains all of the enzymes necessary to convert phytoene to the various carotenoids, with the exception of PSY. Consequently, introduction of the *PSY* gene provides the transgenic plants with the capacity to form

phytoene. However, this transformation is not sufficient to enable the plants to produce carotenoids, because the activity of the intermediate steps, catalyzed by phytoene desaturase (PDS) and ζ-carotene desaturase (ZDS) is too low. Therefore, the additional introduction of the *CRTI* gene enabled the plants to produce sufficient lycopene, which was then converted to carotenoids and xanthophylls by the cyclases (LCYB, LCYE) and other enzymes that are present in wild-type rice.

The most thoroughly studied Golden Rice that has been produced to date is from the agricultural company, Syngenta. These plants can produce up to 6 μg of carotenoids g^{-1} dry weight under field conditions. Despite the fact that Syngenta has been granted a license and commercial rights over these plants, the company has chosen not to use them. Instead, the Golden Rice seeds have been donated to the Golden Rice Humanitarian Board, along with the related technology and commercial rights. The objective is to enable further research to continue into the improvement of rice for developing countries.

causes bleaching of the leaves due to the loss of protective carotenoid pigments (see Chapter 4, Box 4.6). PSY is also a target for increasing the production of carotenoids and related products in plants (see Box 12.4).

Stage 4: the modification of the basic terpenoid skeletons to produce a vast array of terpenoid products

The terpenoid skeletons formed as a result of terpene synthase activity (described in the previous section) can be further modified by the action of a variety of enzymes such as cytochrome P450 hydroxylases, dehydrogenases, reductases, glycosyl transferases, and methyl transferases. These reactions variously oxidize (e.g. cytochrome P450 hydroxylase and dehydrogenase reactions), reduce (reductases), and add glucose (glycosyl transferases) and methyl groups (methyl transferases) to introduce a massive variety into the final range of terpenoid products. These modifying enzymes, like those involved in the formation of diverse alkaloid and phenolics structures, form large multifamily groups that are generally highly substrate specific. The resulting terpenoid products are consequently chemically diverse and also functionally diverse, as illustrated in the examples given in Figure 12.4 and in the earlier sections of this chapter.

Subcellular compartmentation is important in the regulation of terpenoid biosynthesis

As we have seen, higher plants possess two separate pathways for the synthesis of the terpenoid precursor IPP, the DXP pathway located in the plastids and the MVA pathway located in the cytosol. This subcellular compartmentation also seems to influence the type of terpenoids that are produced. In general, the hemiterpene isoprene, the monoterpenes, diterpenes, and tetraterpenes are produced in the plastid, while sesquiterpenes and most triterpenes (e.g. sterols) and polyterpenes (e.g. rubber) are formed in the cytosol (Figure 12.12). This general summary overlooks the apparent occurrence of some exchange of intermediates between the plastid and cytosol. Several studies have been carried out into this metabolic cross-talk between the MVA and DXP pathways. At present there is no consensus, with some experiments showing that there is no interchange, while others indicate that there is a substantial exchange of intermediates. Stable isotope feeding experiments, for example, have estimated that in *Ginkgo biloba* the diterpene ginkgolide is formed from three IPP residues derived from the MVA pathway and one residue derived from the plastid DXP pathway. Similarly, in *Catharanthus roseus* cell cultures it was estimated that phytol, which is formed in the plastid, derives 60% of its carbon from the DXP pathway and 40% from the MVA pathway. The inhibitors fosmidomycin (which inhibits the DXP pathway at DXR) and lovastatin (which inhibits the MVA pathway at HMGR) have also proved useful. While fosmidomycin inhibits chlorophyll formation completely, lovastatin only partially inhibits the formation of cytosolic sterols. These results were interpreted as showing that the plastidic pathway could export IPP to support cytosolic terpenoid synthesis, but the MVA pathway could not compensate for a lack of plastidic IPP synthesis. Experiments with mutants also support this conclusion. Arabidopsis mutants lacking either DXR, MCT, HDS, or HDR are all albino, lacking both chlorophyll and carotenoids. Hence it seems that the cytosolic MVA pathway is not able to supply IPP for pigment biosynthesis when its production is blocked within the plastid, either by inhibitors or by mutation.

If the plastidic and cytosolic compartments do, indeed, share some pathway intermediates then they have to be able to cross the plastid envelope. There is evidence that this transport occurs, although no specific transport protein

Figure 12.12 Subcellular compartmentation of terpenoid biosynthesis. In general, the hemiterpene isoprene, the monoterpenes, diterpenes and tetraterpenes are produced in the plastid, while sesquiterpenes and most triterpenes (e.g. sterols) and polyterpenes (e.g. rubber) are formed in the cytosol. Note that the mitochondrion is the site of synthesis of the electron carrier ubiquinone, from IPP produced via the cytosolic MVA pathway, while plastoquinone is produced in the plastid from IPP produced in the DXP pathway. This general model is a simplification as evidence is emerging that indicates there is some movement of precursors (e.g. IPP) between the plastid and cytosol, as discussed in the text.

Abbreviations: DXP pathway, deoxyxylulose phosphate pathway; MVA pathway, mevalonic acid pathway; IPP, isopentenyl diphosphate; DMAPP, dimethylallyl diphosphate; GPP, geranyl diphosphate; GGPP, geranylgeranyl diphosphate; FPP, farnesyl diphosphate; SPP, solanesyl diphosphate.

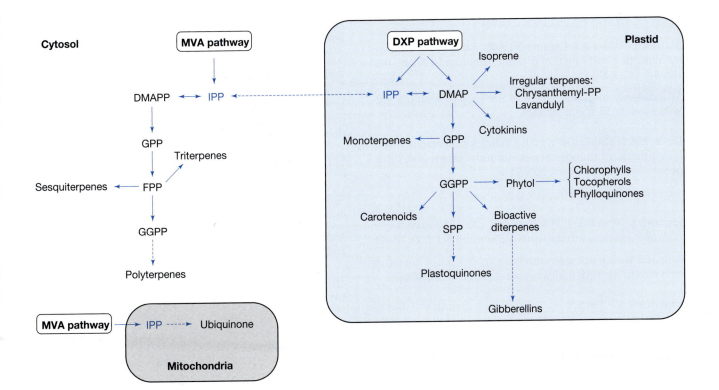

has been isolated yet. Evidence has come from experiments on isolated chloroplasts, showing that they are able to transport IPP, GPP, FPP, and DMAPP (in descending order of transport rate). GGPP is not transported. The well-characterized plastid triose phosphate/phosphate translocator (see Chapter 2) is not involved in this process, and it seems likely that there is a novel IPP/isoprenoid translocator in plastids.

Summary

The terpenoids are a diverse group of plant chemicals with an equally broad range of functions. Some are involved in animal–plant interactions, as volatile signals, poisons, and feeding deterrents. Other terpenoids serve essential functions as photosynthetic pigments, electron transport chain components, plant growth regulators, and wound-healing resins and anti-bacterial compounds. Terpenoid biosynthesis may be considered as a four-step process. The first step, leading to the synthesis of the terpenoid precursor, IPP, occurs in two separate subcellular compartments via two quite distinct pathways; the cytosolic MVA pathway (that also occurs in animals) and the plastidic DXP pathway. In the second stage, IPP and its isomer DMAPP are condensed together by prenyltranferases to form the terpenoid precursors GPP, FPP, GGPP, etc. The terpene synthases subsequently introduce considerable variety into the carbon skeleton of the various terpenoid groups. This diversity is further increased by the final stage in the biosynthetic pathway, where a range of enzymes (e.g. hydroxylases, dehydrogenases, reductases, glycosyl transferases, and methyl transferases) add a variety of groups to the terpenoid skeletons. Commercial interest has driven research into terpenoids that affect human health. These include taxol that is proving to be effective as an anticancer agent, the carotenoid precursors for vitamin A biosynthesis, as well as terpenoids (e.g. yamogenin and gossypol) with contraceptive activity.

Further Reading

Reviews of terpenoid biosynthesis

Bouvier F, Rahier A & Camara B (2005) Biogenesis, molecular regulation and function of plant isoprenoids. *Prog. Lipid Res.* 44, 357–429.

An extremely thorough review of the subject, with particularly good detail about steroids.

Davis EM & Croteau R (2000) Cyclization enzymes in the biosynthesis of monoterpenes, sesquiterpenes, and diterpenes. *Topics Curr. Chem.* 209, 53–95.

A useful review of the terpene synthases.

Eisenreich W, Rohdich F & Bacher A (2001) Deoxyxylulose phosphate pathway to terpenoids. *Trends Pl. Sci.* 6(2), 78–84.

A review with particular emphasis on the genes encoding enzymes in the DXP pathway from a range of organisms.

Eisenreich W, Bacher A, Arigoni D & Rohdich F (2004) Biosynthesis of isoprenoids via the non-mevalonate pathway. *Cell Mol. Life Sci.* 61, 1401–1426.

Shows how stable and radioactive isotopes were used to identify the DXP pathway.

Hwang I & Sakakibara H (2006) Cytokinin biosynthesis and perception. *Physiol. Plantarum* 126, 528–538.

A thorough review of the biosynthesis of cytokinins and of their role in sensing and signaling.

Lichtenthaler HK (1999) The 1-deoxy-D-xylulose-5-phosphate pathway of isoprenoid biosynthesis in plants. *Annu. Rev. Pl. Physiol. Pl. Mol. Biol.* 50, 47–65.

Reviews the early work that led to the discovery of the DXP pathway in plants.

Seo M & Koshiba T. (2002) Complex regulation of ABA biosynthesis in plants. *Trends Pl. Sci.* 7, 41–48.

A detailed review of the abscisic acid biosynthetic pathway and its regulation within a physiological context.

Tholl D (2006) Terpene synthases and the regulation, diversity and biological roles of terpene metabolism. *Curr. Opin. Pl. Biol.* 9, 297–304.

A second, more recent, review of the terpene synthases.

Comparison of DXP and MVA pathways

Lange BM, Rujan T, Martin W & Croteau R (2000) Isoprenoid biosynthesis: the evolution of two ancient and distinct pathways across genomes. *Proc. Natl Acad. Sci. USA* 97(24), 13172–13177.

This review seeks to trace the origin of the two pathways for IPP biosynthesis.

Rohmer, M. Knani M, Simonin P, Sutter B & Sahm H (1993) Isoprenoid biosynthesis in bacteria: a novel pathway for the early steps leading to isopentenyl diphosphate. *Biochem. J.* 295, 517–524.

An early paper showing how isotope-labeling experiments were used to elucidate the DXP pathway in bacteria.

Genetic engineering for terpenoid production

Aharoni A, Jongsma MA & Bouwmeester HJ (2005) Volatile science? Metabolic engineering of terpenoids in plants. *Trends Pl. Sci.* 10(12), 594–602.

A very thorough review of the strategies and targets for genetic engineering of terpenoid biosynthesis.

Al-Babili S & Beyer P (2005) Golden rice—five years on the road—five years to go? *Trends Pl. Sci.* 10(12), 565–573.

This review covers the science behind the production of golden rice as well as the political issues involved in making the rice available to developing countries.

Munoz-Bertomeu J, Arrillaga I, Ros R & Segura J (2006) Up-regulation of 1-deoxy-D-xylulose-5-phosphate synthase enhances production of essential oils in transgenic spike lavender. *Pl. Physiol.* 142, 890–900.

The production of transgenic lavender with a three-fold increase in the concentration of essential oils.

Schaub P, Al-Babili S, Drake R & Beyer P (2005) Why is golden rice golden (yellow) instead of red? *Pl. Physiol.* 138, 441-450.

This paper explains in some detail the production and analysis of plants with altered carotenoid content.

Wildung MR & Croteau RB (2005) Genetic engineering of peppermint for improved essential oil composition and yield. *Transgenic Res.* 14, 365–372.

This paper provides an overview of recent transgenic strategies for the improvement of peppermint.

Index

Note: Note: Entries which are simply page numbers refer to the main text. Other entries have the following abbreviations immediately after the page number: F, Figure; T, Table.

A

Abies, 401, 403F, 421
Abietadiene, 403F, 420F, 421, 423F
Abietadiene synthase, 420F, 421, 422
Abscisic acid (ABA), 95, 132, 322, 402, 405, 407F, 417, 418
    biosynthesis, 405
Acetoacetyl CoA thiolase (AACT), 413, 414F
Acetobacter, 227, 241
Acetohydroxy acid isomeroreductase, 277F, 278F, 279
    synthase, 277F, 278F, 279
Acetolactate synthase, 271, 278
Acetyl CoA, 343F, 347F, 358F, 359, 370, 371F, 378F, 380F, 412, 413F
    carboxylase (ACCase), 307, 308F, 309F, 310, 311F, 380F
    in amino acid synthesis, 245, 270, 278F, 279, 280F, 282F, 283F, 284F
    in lipid synthesis, 307, 309, 308F, 311F, 312, 313, 331F, 332F, 333
    in respiration, 155, 158F, 159F, 162–172, 163F, 164F, 168F, 178F
    synthetase, 307, 309F
Acetylserine, 267, 280F, 283F
Aconitase, 158F, 159F, 165, 169F, 172
Actinomycetes, 242, 412
Active oxygen species *see* Oxygen, reactive species
Acyl-ACP thioesterase, 308F, 315F
Acyl carrier protein (ACP), 307, 312, 316, 323
Acyl CoA, 312, 314, 315F, 316, 324, 326, 327, 331F, 332F
    oxidase, 331F, 332F
    reductase, 327
    synthetase, 308F, 314, 315F, 316, 331F
Adenine nucleotides, 34, 150, 161F
Adenosine 5′-phosphosulfate (APS), 237, 238, 267, 268F, 269F
    reductase, 267, 268F, 269F
Adenosylmethionine, 280, 282, 343F, 348F
ADP-glucose pyrophosphorylase (AGPase), 197, 211F, 212F, 213T, 214T, 223
Aerenchyma, 55

Aerobacter, 263
Aerobic, 151, 163F
    bacteria, 57, 65, 67 , 241, 242T
    dinoflagellates, 101
    respiration, 143, 144, 152
Aesculus, 400
Agarose, 11, 14, 15F
Agave, 133
Agmatine, 350F
Agrobacterium, 17, 373, 425
Air pollution, 409
    sulfur emissions, 266
Aizoaceae, 130
Ajmalicine, 340, 342F
Ajmaline, 340, 342F
Alanine, 112, 122, 123F, 124F, 125, 168F, 270, 271F, 274F, 275, 276, 286F, 291F, 293
    synthesis, 275
Albumins, 247, 287, 290, 292, 293
Alcohol dehydrogenase (ADH), 152F, 213T
Alder, 242
Aldolase, 16F, 98F, 106, 107, 147F, 148, 149F, 199F
Aleurone layer, 32, 217, 218
Alfalfa, 242, 246T, 264, 365, 388
Algae, 412T
Alkaloids, 2, 168F, 270, 273, 281, 322, 335–361
    biosynthesis, 168F, 336–357, 359
    diversity, 358
    in animals, 337
    medicinal properties, 340
Alkanes, 324, 325F
Allantoic acid, 246
Allantoin, 246
Allelopathy, 336, 360, 365
Allene oxide synthase (AOS), 321F, 322
Allosteric topics, 30, 31F, 150, 198, 200, 203, 212, 277
Alternanthera, 125
Alternative pathway, 177
    see also Cyanide-resistant respiration
Alternative oxidase, 144, 171, 172, 173F, 174F, 175F, 177, 178, 179F, 180–183
Amaranthaceae, 123

Amaranthus, 123
Amber, 409
Amino acid biosynthesis, 155, 168F, 237, 258F, 259, 270–286, 272F, 273F, 274F, 278F, 280F, 282F, 283F, 285F, 286F, 372F, 373
Amino acid transporters, 247
Aminobutyric acid (GABA), 286F
Aminotransferases, 237, 256, 257F, 273, 274F, 276, 277F, 278F, 279, 285F, 286F
Ammonia, 59, 60, 66, 114F, 116F, 238, 239, 242, 245, 259, 260, 275, 276, 375
Ammonium, 2, 3, 190, 237, 238, 239T, 245, 246, 247, 248F, 251, 252, 253, 255, 258, 259, 261, 262, 263, 274, 275
    assimilation, 237, 238, 245, 247, 248F, 255, 258, 259, 264, 265, 271
    formation, 258F, 259F, 265
    -proton symport, 255
Amorphophallus, 180, Plate 6.1
Amplified fragment length polymorphisms (AFLPs), 21
Amylase, 32, 216, 217, 218F, 219F, 292F
Amylopectin, 209, 210F, 211F, 213, 214T, 215, 217, 218F, 219
Amyloplasts, 63F, 198, 209, 211F, 213, 218
Amylose, 209, 210F, 211F, 213, 214T, 215, 217, 218F
Amytal, 173
Anabaena, 242, 242T
Anabasine, 353
Anacystis, 104
Anaerobic, 3, 67, 101, 104, 151, 152, 241, 242T, 248, 266F
Ananas, 133
Anaplerotic reactions, 168, 169F, 170, 171F
Anatabine, 350F, 353
Animals, 1, 41, 49, 54, 63, 173, 191, 265, 266, 270, 271, 280, 285, 287, 294, 299, 326, 335, 336, 337, 349, 352, 357, 371, 399, 400, 411, 412T, 413, 428
Anion-permeable membrane, 186
Annonaceae, 179

Anoxia, 27, 30, 151, 152, 251T
Antenna (chlorophylls etc), 69, 72, 76, 80, 81, 86, 88, 89
Antheraxanthin, 89, 90, 407F
Anthers, 51, 321
Anthocyanidin, 368, 371F, 381, 380F, 382F, 383F, 386F
Anthocyanin, 49, 366F, 368, 381, 383F, 384, 386, 387, 390, 397, Plate 11.2, Plate 11.3, Plate 11.4
Anthranilate, 285F, 342F
Antibiotics, 173, 174, 188, 360, 365, 367, 414
Antibodies, 14, 35, 217, 226, 232, 290
Anticancer agents, 340, 391
Antimicrobials, 336, 340, 344, 363, 381, 388, 390, 397
Antimycin A, 174, 176
Antioxidant, 87, 88, 280, 365, 377, 404, 409
Antiporters, 48
Antirrhinum, 17, 390
Ants, 337
Aphanizomenon, 104
Apiferol, 387F
Apigenin, 387
Apoplast, 55, 56, 197F, 205, 206F, 208F, 224, 248
APS kinase pyrophosphatase, 269
Arabidopsis, 17, 19, 20, 22, 25, 105, 117, 137, 161, 176, 178, 182, 183, 191, 222, 223, 227, 231, 232, 233F, 260, 263, 264, 265, 266, 273, 278, 296, 298, 309, 310, 314, 319, 324, 327, 332, 333, 349, 374, 377, 378, 395, 396, 413, 414, 416, 417, 418, 427
Arabinans, 232
Arabinogalctans, 232
Arabinose, 225, 229, 232
Arabinoxyloglucans (AXG), 229
Araceae, 179, 299
Archaea, 101, 109, 412T
Arachis, 390
Arbuscular mycorrhizae, 247, 248
Arctia, 337
Arginine, 112, 254F, 258, 272F, 282, 284F, 291F, 338, 339F, 349, 350F
    biosynthesis, 282, 284F
    decarboxylase (ADC), 349, 351
Aristolochiaceae , 179
Arnold, 71
Arnon, Daniel, 82
Arogenate, 285F, 373F
Arogenate dehydratase, 374, 375
Arogenate dehydrogenase, 372F, 374
Arogenic acid, 285F, 372F, 373, 374
Aromatic amino acids, 108, 137, 153, 168F, 270, 271, 273, 286, 370, 371, 373, 374, 375
    biosynthesis, 284

Aromoline, 346F
Arsenate, 161F
Arum, 3, 179F, 180, 365, Plate 6.1, Plate 6.2, Plate 6.3
Ascidians, 71
Asclepia, 337, 401
Ascorbate, 88
    peroxidase, 88
Asparaginase, 276
Asparagine, 244, 246T, 258F, 272F, 274F, 275F, 276, 288, 297, 298
    biosynthesis, 275
Asparagus, 275
Aspartate, 120, 123F, 124F, 125, 128, 161, 168F, 258, 270, 272F, 274F, 275F, 276, 277F, 279, 282, 284F, 339F
    biosynthesis, 275
Aspartate kinase, 276, 277
Aspartate semialdehyde, 276, 277F
Aspergillus, 373
Asteraceae, 354, 359
Atmospheric pollutants, 254, 266, 329, 411
Atomic force microscopy, 85
ATPase, 95F, 102, 135, 185, 187, 205
    see also ATP synthase
ATP synthase,
    chloroplast, 66, 68, 70, 83, 84F
    mitochondrial, 183–187, 188F, 189
    see also ATPase
Atractyloside, 161F
Atriplex, 120
Atropa, 340, 352, 353
Atropine, 340
Aureusidin, 390
Aureusidin synthase (AS), 389F, 390
Aurones, 388, 389F, 390
Autoradiography, 14, 15
Autotrophy, 1–3, 27, 191, 237
Auxin, 234, 285, 354
    biosynthesis, 285
Avidin, 23, Plate 2.1
Avocado, 330T
Azadirachta, 400, 403F
Azadirachtin, 400
Azide, 175, 178
Azolla, 242
Azorhizobium, 243T
Azotobacter, 242T

B

Bacillus, 189
Bacteria, 57, 59, 65, 371, 412T, 416, 425F, 426
Bacteriochlorophyll, 68F, 71
Balsam fir, 401
Bananas, 1

Barley, 1, 96, 104T, 111, 117, 136T, 139, 197, 212F, 213, 214, 220, 228, 249, 260, 262, 264, 284, 287, 289, 290, 291
Basil, 401
Bassham James, 96
Batelli, Frederic, 156
Batrachotoxins, 337, Plate 10.1
Bats, 400
Beans, 1, 388
Bees, 384, 400
Beetles, 385
Beetroot, 208
Behenic acid, 305T
Belemnitella, 139
Belladonna, 340
Bendall, Derek & Fay, 70
Benson, Andrew, 96
Benzaldehyde, 378F
Benzene, 7, 9
    tricarboxylate, 161F
Benzoic acid, 378F, 379F
    derivatives, 364, 365, 366F, 376F, 378, 391
Benzophenanthridine alkaloids, 340, 345, 348F
Berbamine, 346F
Berbamunine, 346F, 358
Berberine, 340, 344F, 345, 347, 348F
Berberis, 344, 345
Bergapten, 365
Beta, 259
Bicarbonate, 77, 97, 121, 122, 123, 125, 130, 135, 170
Bienertia, 128, 129, 138
Bifidobacteria, 222
Bifurcose, 220F, 221
Biofuels, 1, 326, 329, 330
    see also Industrial oils
Bioinformatics, 17, 22, 24, 37, 417
Biotechnology, 17, 19F, 37
Biotin, 23, 250, 308, 309F, 310, 311F, 312, Plate 2.1
    sulfoxide reductase, 250
Biotin carboxyl carrier protein (BCCP), 308, 309F, 310, 311F
Biotin carboxylase (BC), 308, 309F, 310, 311F, 312
Bisabolol, 400
1,3-Bisphosphoglycerate, 98F, 101, 109F, 125F, 147F, 148, 149F, 150, 151
Bladder, 340
Blindness, 409
Blue haze, 409
Blue jay, 400
Blue Mountains (Australia), 409
Blue Ridge Mountains (USA), 409
Boraginaceae, 341
Bordered pits, 55
Boron, 228, 232, 233F, 234
Borszczowia, 127, 128, 138

Botrytis, 365
Bovine serum albumin, 247
Boyer, Paul, 189
Bradyrhizobium, 243T
Brassica, 326, 327
Brassinolide, 406, 408F
Brassinosteroids, 402, 406, 424
    biosynthesis, 406, 408F, 424
Brazil nut, 293
Breast cancer, 388, 409, Plate 12.1
Brown seaweeds, 74
Bulbs, 63
Bundle sheath cells, 62, 121F, 122,
    123F, 124F, 125, 126F, 127F, 128,
    129, 136T, 206F, 250, 325
Butia, 327
Butiniflavan, 369
Butylmalonate, 161F
Butyric acid, 313

C

C_3 photosynthesis, 62, 134
    plants, 62, 119F
C_4 photosynthesis, 3, 62, 120, 121,
    122, 128, 134
    chloroplasts, 126, 128, 129
    evolutionary origin, 119, 126
    plants, 62, 119F, 120, 121F, 123
    variants, 125F, 126F
Cabbage, 1
Cactaceae, 130
Caffeic acid, 364, 365, 366F, 376, 393,
    394F, 395
    O-methyltransferase (COMT),
    394F
Caffeine, 19, 339F, 355, 356, 357F, 360,
    Plate 10.5
Caffeoyl CoA, 391F, 394F, 395
    O-methyltransferase (CCoAOMT),
    393, 394F, 395
Caffeyl alcohol, 393, 394F
    aldehyde, 393, 394F, 395
Calcium, 48, 49, 75, 77F, 95F, 228, 231,
    232
    channels, 95
    pump, 48
Calibration curve, 12
Calvin, Melvin, 96, 97
Calvin cycle, 60, 96, 97, 98F, 106, 108,
    109F, 113F, 122, 123F, 124F, 272,
    309F
Calystegines, 350F, 352F
Camellia, 355
Campestanol, 406, 408F
Campesterol, 406, 408F, 410, 411F
Canadine, 348F
Cancer, 156, 340, 360, 390, 409
Cannabis, 370
Canola, 1, 326, 327, 330T, 373
Capric acid, 305T, 330T

Caprylic acid, 305T, 330T
Capsaicin, 364
Capsicum peppers, 365, 405
Capsidiol, 403F, 405
Carbamate, 111F, 112F
Carbamylation, 111F, 112
Carbocation, 322, 419F, 421, 422F,
    423F
Carbohydrates, 1–4, 31, 34, 52, 53, 55,
    60, 94, 96, 106, 108, 109, 144,
    146, 151, 155, 157F, 167, 172,
    182, 195–236, 237, 239, 244, 273,
    290, 329, 333, 352F, 370, 371
    metabolism, 195, 197F
Carbon assimilation, 93
Carbon dioxide, 60, 94, 121, 123, 124F,
    126, 127F, 128, 130, 131F, 133, 135
    atmospheric concentration, 119,
    120, 139
    concentration and Rubisco
    reaction, 118
    release during photorespiration,
    113F, 114F, 115F
Carbon flow, 271
Carbon game (Calvin Cycle), 107
Carbon isotopes, 138, 139, 416
Carbon skeletons for amino acid
    synthesis, 272F, 273F, 274F
Carbonic anhydrase, 121F, 122, 123,
    135, 298
Carbonyl carbons, 197
2-Carboxy 3-ketoarabinitol
    1,5-bisphosphate, 112
Carboxylase reaction of Rubisco, 97,
    99F, 100, 104, 105, 111F, 112F,
    119
Carboxylase/oxygenase isotope ratio
    (Rubisco), 104
Carboxylation phase of the Calvin
    cycle, 97, 98F
Carboxyltransferase, (CT) 309F, 310,
    311F, 312
Cardenolide, 401
Cardiac glycosides, 401
Cardiolipin, 319F
Cardiovascular disease, 390, 409
Carnivorous plants, 55
Carotene, 72F, 74, 404F, 405, 409, 420F,
    421, 425F
    desaturase (CRTI), 426
Carotenoids, 62, 63, 69, 72F, 73, 74, 79,
    87, 90, 307, 384, 404F, 402, 409,
    417, 426
    biosynthesis, 424, 425F, 426
    deficient plants, 90, 424
Carrion, 179
Carrots, 62, 74, 232, 263, 370
Caryophyllene, 403F, 420F, 421
Casbene, 409, 423F
    synthase, 422F
Casein, 247

Casparian strip, 248
Cassava, 215, 297, 298, 299
Castasterone, 406, 408F
Castor bean, 51, 296, 328, 329, 405
Castor oil, 30, 329, 330T
Casuarina, 242
Catalase, 58F, 113F, 114F, 117, 330,
    331F
Catechin, 369, 386
Caterpillars, 337, 364, Plate 10.2
Catharanthus, 340, 341, 342, 344, 413,
    427
Cathasterone, 408F
cDNA, 22, 23, 183, 212, 222, 309, 327,
    329, 413
CDP-Choline, 318, 327, 328F
Celery, 364
Celiac disease, 291
Cell growth, 49, 54, 224, 225, 234, 355,
    357
    plate, 54
    shape, 49, 55, 225
    wall, 46, 47F, 48, 49, 52, 53, 208F,
    224, 225, 396
    wall polysaccharide, 52, 224
Cellulose, 35, 48, 55, 95F, 195, 196,
    198, 224, 225, 226
    synthase, 47, 207, 226F, 228F
Central nervous system, 340, 341, 355
Centrifugation, 36, 52, 201F
Cerebrosides, 332
CESA gene family, 227, 228
Chalcone, 366F, 368, 379, 380F, 381,
    382F, 390
Chalcone synthase (CHS), 379, 380F,
    381, 382F
Chalcone tetrahydroxychalcone, 382F
Chaperonins, 102F, 103F
Cheilanthifoline, 348F
Chemical potential, 184, 185, 186
Chemiosmotic hypothesis, 186
Chenopodiaceae, 128
Chicken, 139
Chicory, 220, 223, 403F
Chitinase, 295, 299
Chlamydomonas, 79, 104, 112
Chlorella, 96
Chlorenchyma cells, 127, 128
Chloridoideae, 123
Chlorin, 71
Chlorine, 75, 77F, 95F
Chloroform, 7, 303
Chlorophyll, 60, 71F, 402, 417, Plate 4.1
    biosynthesis, 71, 256F, 257F, 270,
    275, 419F, 427
Chlorophyll *a*, 71F, 72F, 75, 80, 256,
    257F
Chlorophyll *b*, 71 72F, 256
Chlorophyll *d*, 71
Chlorophyll protein complexes, 73
Chloroplast ATP synthase, 83, 84F, 85F

Chloroplasts, 60, 108, 131F, 164F, 169F, 190F 199F, 427
  bundle sheath, 114F, 123F, 124F, 126F, 127F, 129
  mesophyll, 123, 123F, 124F, 125, 126F, 127F, 129, 250
Cholesterol, 45, 303, 406, 410, 411F, 413, 424
Choline, 42F, 43F
Chondodendron, 340
Chorismate, 285F, 342F
  mutase, 371, 372F, 373, 374
Chorismic acid, 285, 370, 371, 372F, 373, 374, 378, 379F
Chromate, 267
Chromatium, 104
Chromatography, 5, 6–11, 26, 96, 223, 303
Chromoplasts, 62, 63
Chrysanthemum, 387, 400
Chrysanthemyl diphosphate, 402F, 427F
Cinchona, 340
Cineole, 403F
Cinnabar moth, 337, Plate 10.2
Cinnamaldehyde, 364
Cinnamate 4-hydroxylase (C4H), 375
Cinnamic acid, 258, 364, 370, 371F, 375, 376F, 378F, 394F
Cinnamon, 364
Cinnamoyl-CoA, 378, 390, 391F, 393
  reductase (CCR), 393, 394F
Cinnamyl alcohol dehydrogenase (CAD), 393, 395F, 396
Circadian rhythm (diurnal rhythm), 133, 196, 212, 219, 251T, 276
Cis-aconitate, 156, 157, 159F, 161F, 165
Citrate, 156, 157, 158F, 159F, 161, 162F, 164, 165, 169, 171F, 183, 190, 309F, 332F
  lyase, 307, 309F
  synthase, 164, 165, 172
Citric acid cycle (*see* TCA cycle), 15, 16, 26, 31, 155–157
Citrulline, 284F
Citrus flowers, 400
Clathrin-coated vesicles (CCV), 296, 297F
Climate, 222, 298
Cloning, 18, 20, 21, 222, 327
Cloves, 364
Cocaine, 340
Coccochloris, 104
Coclaurine, 358
Cocoa
  beans, 355
  butter, 330T
Coconut, 1, 329, 330T
Codeine, 340, 344F, 345, 346F, 347
Codeinone reductase (COR), 346F, 347
Coffea, 355, 356, Plate 10.5

Coffee, 355
  decaffeinated, 338, 356, 360,
Cola, 355
Colchicum, 340
Cold,
  -acclimated plants, 45
  -sensitive plants, 46
  -tolerant plants, 45
Columbamine, 347, 348F
Comfrey (*Symphytum*), 341
Compartmentation, 2, 27, 33, 45, 46, 63, 426, 427F
Complex I–V, 173, 174F, 175F, 184, 185
Confocal microscope, 51
Coniferaldehyde, 393, 394F, 395
Conococcum, 247
Coniferyl alcohol, 234, 369, 392, 393F, 394F, 395, 396, 397
Coniferyl aldehydes, 393
Contraceptives, 404F, 405, 409, 410F, 428
Convolvulaceae, 299, 352
Copalyl diphosphate (CPP), 422
Copper, 67, 75, 80, 82, 175, 246, Plate 4.1
Coptis, 340, 344, 345, 347, 348
Corms, 63, 297, 299
Corn (*Zea mays*), 1, 139, 290, 293, 294, 330T
Cosmetics, 326, 327, 329
Cosmos, 390
Cosmosoma, 337, Plate 10.3
Cotton, 330T, 373, 403F, 405, 410
Cotyledons, 58, 218, 288, 299, Plate 8.1
4-Coumarate-CoA ligase, (4CL) 376F
Coumaric acid, 364, 376F, 369, 375, 377F
Coumarins, 364, 365, 366F
p-Coumaroyl CoA, 371F, 375, 376F, 377, 379, 380F, 381, 382F, 390, 391F, 394F, 395
Coumaryl alcohol, 234, 369, 392, 393F, 394F, 395
Coupling factor, 83
Cowpea, 246T
Crassulacean Acid Metabolism (CAM), 3, 96, 130–140, 131F, 132F
  evolutionary origin, 133
Cristae, 57
Crops, 1, 3, 4, 17, 100, 106, 215, 217, 239F, 242, 246, 266, 271, 287, 293, 294, 298, 299, 326, 329, 330T, 348, 355, 390
  yield, 101, 266, 271, 329
Cryomicroscopy, 84
CSL gene families, 33
Cucumbers, 62
Cucurbitaceae, 62, 197
Cuourbiacin-E, 420F, 421
Cuphea, 330T
Cupins, 289

Cupric ion, 196
Cuprous ion, 196
Curare, 337
Curarea, 337
Cuticle, 322, 323F, 324
Cutin, 322, 323F, 324
Cutinase, 322
Cyanide-resistant respiration, 177, 178, 365
Cyanidin, 382F, 383F, 384, Plate 11.4
β Cyanoalanine hydrolase, 275
β Cyanoalanine synthase, 275
Cyanobacteria, 60, 66, 71, 104, 110, 240, 242, 263, 315, 412, 413T
Cyanocitta, 400
Cyclanthaceae, 179
Cyclic photophosphorylation, 79, 85, 86F
Cylindrotheca, 104
Cynodon, 136T
Cyperaceae, 120
Cystathione, 277, 282
Cysteine, 75, 83, 84, 110, 172, 180, 227, 228F, 238, 250, 254, 261, 265, 266F, 267, 268F, 274F, 275, 277F, 279, 280F, 281, 282, 283F
  in seeds, 288, 289, 290, 291F, 293T, 298
  synthase, 33, 267, 280, 283
Cytidine, 415
  diphosphate, 317F, 415, 416
  monophosphate, 416
  triphosphate, 317F, 416
Cytidyl phosphate, 416
Cytochrome a/a$_3$, 174F, 175
  b$_5$, 250, 314
  b$_5$, reductase 250, 314
  b$_{559}$, 76
  b$_6$ complex, 69, 78, 79F, 80, 89
  b$_6$f, 78, 79F, 85, 174F
  c, 250
  c$_1$, 174F
  c-553, 88
  f 78, 79F
  oxidase, 174, 175, 244
  pathway, 173F, 177, 178, 179, 180, 181,182, 183
  P450, 250, 357, 387, 388
  P450, monoxygenase 358, 375, 376
  P450, oxidase 426
  see also cytochrome oxidase
Cytokinesis, 224
Cytokinins, 402, 406F
  biosynthesis, 406F, 419, 427F
Cytoplasm, 1, 33, 39, 46, 47, 48, 51, 52, 54, 56, 57, 58, 128, 134, 185, 199F, 200, 206, 208F, 227, 234, 235, 245, 248, 249, 253, 255, 266, 292, 295,296, 297, 309F, 310, 314, 315, 316, 341, 396, 412, Plate 8.1

Cytoplasm to surface area ratio, 49
Cytoskeleton, 48

D

D1 protein, 75
D-3-Hydroxybutyryl-ACP, 313
Daffodil, 62, 425
Dahlia, 220, 390
Daidzein, 365, 366F, 388, 389F
Danaus, 337, 400
D-apiose, 232
D-arabinose (Ara), 225, 229, 232
Datura, 341, 349, 352
10-Deacetylbaccatin (DAB), 409
Deacetylvindoline, 343F
   -4-hydroxylase (D4H), 342, 343F
   -4-*O*-acetyltransferase (DAT), 341, 343F
Decaffeinated coffee, 19, 338, 356, 360
Deer, 400
Dehydroascorbate, 31
Dehydroquinic acid, 372F, 373
Dehydroshikimic acid, 372F, 373
Delphinidin, 382F, 383F, 384, Plate 11.4
Delphinium, 384, Plate 11.4
Demethylsuberosin, 377F
Dendrobates, 337, Plate 10.2
Dendroctonus, 400
3-Deoxy-D-*arabino*heptulosonate 7-phosphate (DAHP) synthase, 372F, 373, 375
Deoxyhypusine synthase (DHS), 355
Deoxyxylulose phosphate (DXP), 412, 417
   pathway 411–413, 413T, 415F, 416, 417, 426, 427F
   phosphate synthase (DXS), 418
   reductoisomerase (DXR), 416
Derris, 173
Desiccation, 94, 222, 289
   tolerant plants, 134, 222, 290
Desmotubule, 56
Detergents, 326, 329
Development, 304, 320, 321, 321F, 322, 324, 325, 327
Dextrans, 56
D-galactose (Gal), 41, 42, 43, 196, 225, 229, 231
D-galacturonic acid (GalA), 225, 231, 232, 233
D-glucuronic acid (GlcA), 225
Diacylglycerol (DAG), 304F, 306, 317F, 318F, 319F, 320F, 327, 328F
Diatoms, 74
Dibromothymoquinone, 90
Dicarboxylate translocators, 168, 170
Dicarboxylate-tricarboxylate translocator, 161F, 162F

Dictyosome, 52
Diesel engines, 329
Diffenbachia, 49
Digalactosyl diacylglycerol, 306, 317F, 318F
Digitalis, 400, 404F
Digitaria, 136T
Digitoxigenin, 404F
Digitoxin, 400, 404F
Dihydrodipicolinate synthase, 276, 277F
Dihydroflavonols, 380F, 381, 382F, 383, 385, 388, 389F
Dihydrokaempferol, 382F, 383, 384
Dihydrolipoyl acetyltransferase, 155, 163F
   dehydrogenase, 155, 163F
Dihydromyricetin, 382F, 383, 384
Dihydroquercetin, 382F, 383, 384
Dihydroxy acid dehydratase, 277F, 278F, 279
Dihydroxy-3-isovalerate, 280
Dihydroxyacetone, 196F
   phosphate (DHAP), 16, 98F, 101, 106, 148, 149F, 198,199F, 203, 204F, 205, 315, 316F
   phosphate reductase, 315
Dimethyl sulfoxide (DMSO), 250
Dimethyl sulfoxide reductase, 250
Dimethylallyl diphosphate (DMAPP), 411, 413, 413F, 415F, 416, 418, 419F, 420F, 421, 427F
Dinitrogenase, 240F, 241F
Dinitrophenol (DNP), 187F
Dioscorea, 400, 409, 410
Dioscorin, 298
Diosgenin, 400, 404F, 410F
Dioxygenase, 321
Diphosphatidylglycerol, 319F
Dirigent protein, 397
Disaccharide, 41, 196, 197, 222
Diterpenes, 402, 403F, 420F, 422, 427
   synthase, 420F, 421, 422F
Dithiol-disulphide interconversion, 31
Divinyl esters, 321F
Divinyl ether synthase (DES), 321F, 322
Djerassi, Carl, 410
D-malate, 166
D-mannose (Man), 196, 225, 229
DNA microarrays, 23, Plate 2.1
Docosanoic acid, 305T
Dodecanoic acid, 305T
Dopamine, 345, 346F, 347
Dough, 290
Douglas Fir, 400
Drugs, 1, 335, 341, 344, 347, 409–411
Duboisia, 349, 352
D-xylose (Xyl), 225, 232
Dyes, 1, 13, 15

E

Ecdysones, 400, 404F
Ectomycorrhizae, 247
Eicosenoic acid, 305T
Elastic, 189, 291, 404
Elasticity coefficient, 28
Electrical potential, 83, 184, 185, 186
Electrochemical gradient, 80, 95F, 174F, 178, 183–189
Electron transport chain, 3, 31, 89, 94, 101, 115, 118, 122, 125F, 129, 130F, 138, 143, 144, 155, 158F, 166, 171, 244, 250, 251, 253F, 267, 315, 428
   chloroplast, 69–87, 70F, 77F, 79F, 81F, 83F, 86F
   mitochondrial, 3, 172–191, 173F, 174F, 175F, 190F
Electron microscope, 40, 61
Electron volts, 70
Electrophoresis, 6F, 11
Eleostearic acid, 305T
Ellagic acid, 370, 391
Ellagitannins, 370
Elongase, 313, 314, 323, 324, 327
Embryo, 208, 216, 217, 356
Emerson, Ralph, 71
Emetine, 340
Endocytosis, 47, 53
Endocytotic vesicles, 47
Endo-inulinases, 221
Endomembrane system, 49
Endoplasmic reticulum (ER), 49, 50F, 293, 295, 297F, 375
   lipid biosynthesis, 316, 317F, 325F
Endosperm, 211F, 212, 213T, 216, 217, 291, 292, 294, 295, 300, 425, 426
Endosymbiotic bacteria, 59
Enolase, 147F, 149F
5-*Enol*pyruvylshikimate-3-phosphate (EPSP) synthase, 372F, 373, 374, 375
Enoyl-ACP reductase, 308F, 312, 313, 323F
Environmental conditions, 1, 25, 26, 248, 251, 260, 270, 321F, 344
   stress, 1, 27, 271, 321, 411
Enzyme assay, 7
   rate determination, 27, 32
   reaction mechanisms, 16
   substrate interactions, 27
Epicatechin, 369, 386
Epidermis, 51, 56, 95, 131, 264, 322, 401
Epimerization, 225
Epiphytic plants, 132
Epoxidase, 89, 90
Epoxyhydroxy fatty acids, 321F
ER lumen, 49, 50F
   rough, 49, 50F, 294
   smooth, 49, 50F

Eriodictyol, 382F, 387F, 388, 391F
Erucic acid, 305T, 326, 327, 330T
Erusic acid, 366, 370
Erythrose 196, 414
Erythrose 4-phosphate, 106, 286F, 371F, 372F, 373
Erythroxylon, 340
Escherichia coli, 102, 105, 188, 212, 223, 277, 417
Eschscholzia, 340, 344, 345, 347, 348
Escutelin, 365
Ethanoic acid, 305T
Ethanol, 152
Ethanolamine, 43F
Ethylene (Ethene), 281F
    biosynthesis, 281F
Ethylmethane sulfonate (EMS), 117
Etioplast, 61F
Eucalyptus, 247, 248
Eugenol, 364
Euglena, 104, 413
Eukaryotic, 144, 306, 314, 315F, 316F, 317F, 333, 355, 373
    algae, 96, 101, 104, 110
    cells, 40, 57, 60, 68
    plants, 81
Eupatorium, 337
Euphorbiaceae, 130
Evening primrose, 330T
Evolution, 1, 2, 40, 45, 54, 58, 60, 65, 66, 67, 68F, 69, 71, 75, 80, 81, 104, 105, 117, 119, 128, 133, 134, 139, 229, 247, 287, 289, 292, 299, 335, 354, 356, 357, 359, 360
Expansins, 234
Extensins, 234

F

F_0F_1 ATP synthase, 184, 188
Fabaceae, 20, 25
Faraday's constant, 186
Farnesene, 420F, 421
Farnesyl diphosphate (FDP), 424F, 427F
Fatty acids, 4, 41, 304–318, 329, 332F
    oxidation, 329, 331F, 332F
    biosynthesis, 31, 307–318, 323F
Fatty acid synthase (Type II complex), 308, 312, 313, 315, 324
Fatty seeds, 326
Fermentation, 151, 152, 232, 240, 241
Ferredoxin (Fd), 31, 81F, 82, 83, 240, 248, 253F, 254, 255F, 263, 267, 268F
Ferredoxin-dependent glutamine-oxoglutarate aminotransferase (Fd-GOGAT), 116–119
    see also glutamate synthase
Ferredoxin-NADP⁺-oxidoreductase (FNR), 82, 250, 253, 255F, 263

Ferredoxin-thioredoxin couple, 31, 110F
Ferric ions, 66, 75
Ferricyanide, 250
Ferrous ions, 66, 75, 87
Ferulic acid, 364, 365, 366F, 369, 376, 377F, 393, 394F, 395
    5-hydroxylase (F5H), 394F
Festuca, 220, 310
Fire ants, 337
Flavan-3-ol, 368, 369, 386F
    -4-ol, 380F, 387F, 388
Flavanones, 366F, 380F, 381, 382F, 383, 386, 388, 389F
Flaveria, 125
Flavin adenine dinucleotide (FAD), 165
    oxidoreductases, 250
Flavone, 366F, 368, 369, 384, 387
    synthase, 387F
Flavonoids, 243, 269F, 271, 308, 322, 364, 365, 366F, 367, 368, 369, 370, 371F, 374, 375, 376F, 387, 379, 381, 384, 388, 397, Plate 11.5
    skeleton, 379F, 381
Flavonols, 366F, 368, 369, 380F, 381, 382F, 383, 384
Flavoprotein, 82
Flavors, 364
Flax, 17
Flies, 3, 179, 180
Flipases, 42
Flour, 290
Flower colors, 364, 383, 384, 385, 386, 390
Fluid mosaic model, 41F, 45
Fluorescent labeling, 52
Flux control coefficient, 27, 28, 29F, 311
Forskolin, 403F
Forsythia, 397
Fosmidomycin, 411, 414, 427
Fossil fuel, 329
Fossil record, 120
Fracture faces, membranes, 44F
Frankia, 242T
Freesia, 400
Freeze fixation, 47F
Fructans, 146F, 148, 195, 196, 197F, 220F, 221F, 222
Fructan exohydrolase, 221
Fructokinase, 146F
Fructose, 16, 144, 148, 196, 197F, 198, 205F, 208F, 209, 220, 221F, 226F
    1,6- bisphosphatase, (F1,6BPase) 98F, 106, 109F, 110, 111, 198, 199F, 203, 204F
    1,6-bisphosphate (F1,6BP), 16, 98F, 106, 107, 109F, 146F, 147F, 148, 149F, 151, 198, 199F, 203, 204F
    2,6-bisphosphatase (F2,6BPase), 203, 204F

    2,6-bisphosphate (F2,6BP), 32, 150, 197, 200, 203, 204F
    6-phosphate, 34, 98F, 106, 107, 109F, 145, 146F, 149F, 153, 154F, 155, 198, 203f, 204F
    6-phosphate 1-phosphotransferase (PFP), 31, 149F
    6-phosphate 2 kinase (F6P2PK), 203F, 204F
Fruit dispersal, 364
Fucogalactoxyloglucans, 229
Fucose, 225, 229, 231, 232, 233, 299
Fucosyltransferase, 230, 231
Fucoxanthin, 74
Fumarase, 158F, 160F, 166
Fumarate, 156, 157, 158F, 160F, 166, 168F, 173, 174F, 276, 284F, 332F
Fungus, 371, 412T
Furanocoumarins, 365, 366F, Plate 11.1
Furanose, 225

G

Galactans, 232
Galactolipids, 41, 304T, 306
Galactomannans, 230
Galactose, 41–43, 225, 229, 231–233
Galacturonans, 229, 231
Galacturonic acid, 225, 231–233
Gallic acid, 366F, 369, 370, 371F, 391, 392F
Gallotannins, 370
Gas liquid chromatography, 8F
Gated channels, 48
Gel filtration, 8, 9F
Genes, 17
    expression, 23
    mapping, 21
    regulation, 2
    silencing, 17
Genistein, 365, 388, 389F, 390
Genomic sequence, 20
Geraniol, 338, 341, 342F, 400, 401F, 402F, 420F
Geranium, 400
Geranyl diphosphate (GPP), 402, 418, 419, 422F, 427F
    synthase (GPPS), 418
Geranylgeranyl, 420F, 421
Geranylgeranyl diphosphate (GGPP), 405, 408F, 418, 419, 420F, 423F, 424F, 425F, 427F
    synthase (GGPPS), 418
Gerbera, 387
Germacrene B, 403F
Germin, 289
Germination, 253, 254, 258, 290, 296, 297
    inhibitors, 336
Giant hogweed, 364, 365, Plate 11.1
Giardia, 173

Gibberellins (Gibberellic acid), 402, 403F, 408F, 416, 417
biosynthesis, 408F, 416, 422, 427F
Gibbs Free Energy, 186
Ginger, 364
Gingerols, 364
Ginkgo, 427
Ginkgolide, 427
Glaucoma, 403F
Gliadins, 290, 291F
Globulins, 287, 288F, 289, 290F, 291, 295
Gloxinia, 388
Glucanase, 299
Glucan-water dikinase, 219F
Glucogallin, 391, 392F
Glucomannans, 229
Gluconeogenesis, 167, 273, 331, 332F, 333
Gluconeogenic tissues, 198
Glucose, 94, 196
6-phosphate (G6P), 109F, 145, 146F, 148, 149F, 153, 154F, 155, 198F, 201F, 202, 211F, 219F, 223F, 255F
dehydrogenase(G6PDH), 31, 109F, 110, 153, 154F, 155, 255F
Glucosinolates, 269, 270F
Glucuronic acid, 225
Glucuronoarabinoxylans, (GAX), 229
Glutamate (Glu), 161F, 167, 246, 256, 257F, 258F, 259F, 261, 264, 265, 272F, 273, 274F, 275F, 276, 282, 284F, 285F, 286F, 291
in photosynthesis, 114, 116, 117, 123F, 124F, 125
Glutamate synthase (GOGAT), 115, 168, 169F, 246, 247, 248F, 258, 259F
ferredoxin dependent GOGAT (Fd-GOGAT), 259, 263
function, localization, 264, 265
NADH-dependent GOGAT (NADH-GOGAT), 259, 263
function, localization, 264, 265
structure and synthesis, 263
Glutamate-glyxoylate aminotransferase, 113
aminotransferase reaction, 113, 114F, 169F
Glutamate-malate translocator, 116
Glutamine (Gln), 76, 114F, 116, 117, 168, 169, 238, 244, 246T, 251T, 252F, 258, 259F, 260, 261, 262, 264, 265F, 272F, 273, 274F, 275F, 284, 285F, 291–294, 339F
biosynthesis, 274F, 282
Glutamine synthase (GS), 115F, 116, 169F, 237, 246, 247, 248F, 259F, 271
regulation, 261, 262
structure, 259
synthesis, 260, 262
Glutathione, 31, 88, 118, 265, 268F, 280, 282T, 283, 294T
reductase, 88, 282
Glutelins, 287
Gluten, 290, 291, 292
Glutenins, 290, 291F
Glycans, 50, 53, 230, 299
Glyceraldehyde, 196F
Glyceraldehyde 3-phosphate (GA3P), 16, 198, 199F, 342F, 351, 414, 415F, 417
dehydrogenase, 101, 148
in glycolysis, 145, 147F, 148, 149F, 150–153, 154F, 155
in photorespiration, 98F, 106, 107F, 101, 109F, 110, 125F
Glycerides, 306
synthesis, 314, 315, 316F, 317F
Glycerol, 4, 41, 42F, 43F, 303, 304F, 306, 315, 316F, 317, 325F, 326, 328, 329, 332F, 333
3-phosphate, 315, 316F, 317, 318, 326, 328F, 333
Glycerolipids, 304T, 306, 313, 314, 315F, 316F, 317, 325F, 327, 333
biosynthesis, 303, 314, 315F, 316F, 317F, 333
Glycine, 161, 164F, 167, 169, 172, 190F, 242, 258F, 262, 272F, 274F, 276, 277F, 291–293, 339F
decarboxylase, (GDC) 115, 116F, 126, 127F, 190F
in photosynthesis, 109, 113, 114F, 115, 116F, 117–119, 126, 127F
translocator, 115
Glycine max, 242, 243T, 246T, 289, 388, 395
Glycine/serine exchange, 115
Glycolate, 113, 114F, 115–117
Glycolate/glycerate translocator, 113, 114F, 117
Glycolipids, 317, 318F, 328
Glycolysis, 3, 30, 31, 144, 145, 146F, 147F, 148–153, 154F, 155, 164F, 167, 168F, 170, 171F, 177, 190, 191, 245, 285, 309F, 371F
Glycoproteins, 52, 53, 234
Glycosidic bonds, 197, 210F
Glycosylation, 50, 52, 291, 359
Glycosyltransferases, 227, 359
Glyoxylate, 168, 169F, 332
cycle, 165, 331, 332F
in photorespiration, 113, 114F, 115, 117
Glyoxysomes, 58 165, 329, 330, 331F, 332
Glyphosate, 373
Golden rice, 425
Golf balls, 418
Golgi bodies/apparatus, 40, 46–54, 224, 230, 232, 295, 296F, 377, 396
Gondoic acid, 305T
Gonyaulax, 101
Gossypium, 405
Gossypol, 403F, 405, 410F, 428
Gout, 340
Gramineae, 120, 287, 307, 377
Grana, 61, 62F
Granule bound starch synthase (GBSS), 211F, 213, 214T, 217
Grape, 365, 390
Grapefruit, 63
Grayanotoxin, 400, 403F
Green algae, 96, 102, 106, 412
Green fluorescent protein (GFP), 50F
Green sulphur bacteria, 66, 68F
Griffithsia, 106
Groudnut oil, 330T
Guaiacyl monolignols, 234, 369, 392
Guaraná, 355
Guard cells, 55, 56, 94, 95F, 137
Gunning, Brian, 40
Gutta percha, 404F
Gymnosperms, 55

H

H^+ gradient, 79, 95F, 135, 173, 185
H^+ sucrose translocator, 205
Hares, 400
Hatch, Hal, 120
Heartbeat , 340, 401
Hebeloma, 247, 248
Hechtian strands, 48
Hedera, 400
Heme, 71, 79
biosynthesis, 256
Hemoprotein kinase, 245
Hemicellulose, 55, 229
Hemiterpenes, 402, 403F, 420F, 421
Heptose, 196
Heracleum, 364, 365
Herbicides, 90, 270, 271, 310, 424
Herbivores, 2, 49, 298, 335, 336, 341, 342, 353, 354, 355, 356, 365
Heroin, 345
Heteromannans, 229
Heteroxylans, 229
Hevea, 404F
Hexacosanoic acid, 305T
Hexadecanoic acid, 305T
Hexadecenoic acid, 305T
Hexose, 147F, 148, 151, 196, 200, 204, 205, 207, 208F, 209, 213, 332F
phosphate isomerase (HPI), 146F, 149F, 198
High pressure rapid freezing, 47F
Hill, Robin, 70, 71

Histidine, 76, 81, 178, 240, 241F, 261, 270, 272F, 273F, 293
  biosynthesis, 285F
Histones, 57, 59
HIV, 352
HMG CoA synthase (HMGS), 413
Hodgkin's disease, 340
Hogweed, 364, 365, Plate 11.1
Homeopathic medicines, 340
Homeostasis, 26, 283
Homocitrate, 240, 241F
Homogalacturonans, 231, 232, 233
Homoserine, 277F, 286
Homoserine dehydrogenase, 277F, 276
Homospermidine, 339F, 354F, 355
  synthase, (HSS) 354F, 355
Honey, 400
Hooke, Robert, 39
Hopanoids, 416
Hornworts, 229
Horse 341, Plate 10.4
Horse chestnut, 400
HPLC, 6F, 9, 10F, 11, 15, 22
Humans, 1, 41, 152, 270, 271, 280, 287, 289, 290, 293, 294, 295, 298 299, 337, 340, 341, 383, 389, 410, 411
Hydrangea, 385, Plate 11.6
Hydrogen peroxide, 31, 331F
  superoxide, 182
Hydrogen sulfide, 66
Hydrolysable tannins, 391, 392F
Hydroperoxide lyase (HPL), 321F, 322
Hydrostatic pressure, 46
Hydroxybenzoate, 369
D-3-Hydroxybutyryl-ACP, 313
Hydroxy fatty acids, 321F
3-Hydroxy-3-methyl glutaryl (HMG CoA), 412, 413F
Hydroxyphenyl, 234, 369, 392, 393F
4-Hydroxyphenylacetaldehyde, 345, 346F
Hydroxyphenyl-propionyl CoA, 378F
Hydroxypyruvate, 114F, 117
  reductase, 115F, 117
Hyoscyamine, 339F, 341, 349, 350F, 351, 352, 353F, 360
Hyoscyamine 6β-hydroxylase (H6H), 350F, 351, 352, 353F
Hyoscyamus, 349, 352, 353

I

Ice cream, 410
Idioblasts, 342
Ilex, 355
Imidazole-alkaloids, 336
  ring, 240, 285
Immunolabelling, 35, 84, 295
Industrial oils, 326, 329, 330T
Infection thread, legume nodules, 243

Infertility (animal), 364, 403F, 410
Inositol, 318, 319
  IP3, 320F
Insecticides, 355, 400, 403F, 409
In situ hybridization, 35
Intracellular pathways, 31
Inulin, 220
Inulin neoseries, 220, 221F
Invertase, 146F, 195, 197F, 205F, 206F, 207F, 207T, 208F, 209
Invertebrates, 222, 227, 336
Ionine, 400
Ion channels, 48, 95
Ion pumps, 48, 58
Ipomoea, 298
Iris, 385
Iron, 67, 71, 75, 82, 166, 174, 178, 182, 240, 241F, 250, 253, 265, 321
  bands, 66
  sulfur clusters, 182, 265
  sulfur proteins, 78, 79, 80, 82, 174, 182
Isoamylases, 219
Isochorismic acid, 378, 379F
Isocitrate, 156, 157, 158, 159F, 160F, 161, 162F, 165, 168F, 169F, 171F, 332F, 332
  dehydrogenase, (IDH) 158F, 159F, 165, 169F, 175F
  lyase, 58, 332
Isocoumarin, 370, 371
Isoelectric focusing, 12F, 13F
Isoenzymes, 2, 4, 258, 259, 260, 261
Isoetes, 133, 134
Isoflavones, 366F, 368, 369, 380F, 388, 389F, 390
Isoflavonoids, 365, 388, 389F, 390
Isolation of organelles, 34
Isoleucine, 167, 258, 272F, 276, 277F, 278, 279, 280, 286, 310, 339F, 354F
  biosynthesis, 276, 277F
Isoliquiritigenin, 366F, 380F, 381, 389F
Isopentenyldiphosphate (IPP), 338, 412, 413F, 415F, 416F, 417, 418, 419F, 420F, 421, 426, 427F
Isopentenyl pyrophosphate, 402F
Isoprene, 401F, 402F, 411, 420F, 421, 426
Isoprenoid translocator, 427
Isopropylmalate dehydratase, 278F, 279
  dehydrogenase, 278F, 279
  synthase, 278F, 279
Isoquinoline alkaloids, 339F, 340, 344F, 345, 346F, 347, 348F
Isotopes, 14, 15, 104, 138, 139

J

Jasmonate, 269F, 270, 320, 321F, 344, 354
Jasmonic acid, 320, 322

biosynthesis, 321F, 322
Jojoba oil, 327
Joliot, Pierre, 75
Juncus, 55
Juniperus, 403F
Juvabione, 401
Juvocimene, 401

K

Kacser, 27, 28
Kaempferol, 365, 366F, 382F, 383, 384
Kalanchöe, 132
Kaposi's sarcoma, 409
Kaurene, 405, 408F, 416, 422
  synthase, 408F, 416, 422
Kennedy pathway, 326
Kestose, 220F, 221F
3-Ketoacyl-ACP synthase (KAS), 312, 313, 323F, 368
3-Ketoacyl-ACP reductase, 308F 312, 313, 323F
Keto fatty acids, 321F
Klebsiella, 263
Knoop, Franz, 156
Kortschak, 120
Kranz leaf anatomy (C_4), 62, 121F, 136T
Krebs Cycle (*see* TCA cycle), 57, 155, 156
Krebs, Hans, 155, 156

L

Laccaria, 247, 248
Laccase, 234, 394F, 397
L-aceric acid, 232
Lactate dehydrogenase, 152F
Lactic acid (lactate), 152F, 157F, 161F
Lactonase, 153
Lactone, 153
Lactucin, 403F
Lamiales, 229
Larix, 247
Lathyrus, 275
Laticifers, 342
Lauric acid, 305T, 329
Lavandula, 418
Lavender, 418
Lavendulyl, 427F
Lea, 263
Leather, 370
Lectin, 299
Leghemoglobin, 244, 245
Legumain, 297
Legumes, 1, 3, 173, 242T, 243T, 246, 249, 259, 263, 265, 272, 275, 286, 292, 295
Legumins, 287, 288F, 289, 290, 297
Lemons, 63
Lens culinaris, 242

Lentil, 1, 242
Lettuce, 1, 220, 390, 403F
Leucine, 167, 271F, 278F, 279, 280, 293
   293, 294, 310, 339F, 354, 395
   synthesis, 279
Leucine/isoleucine aminotransferase,
   278F, 279, 280
Leucoanthocyanidin, 381, 382F, 383,
   385, 386F, 387
   reductase (LAR), 386F
Leucocyanidin, 381, 382F, 383
Leucodelphinidin, 382F, 383
Leucopelargonidin, 382F, 383
Levan, 220, 221
   neoseries, 220, 221
L-fucose (Fuc, 6-deoxy-L-galactose),
   225, 232
L-galactose, 225, 232
Lhca, 73
Lhcb, 73
LHCI, 73F, 81
LHCII, 74F, 76, 81, 88
L-hyoscyamine, 339F, 341, 349, 350F,
   352, 353, 360
Light absorption spectrum, 71, 72F
Light harvesting chlorophyll protein
   (LHCP) complexes, 69, 73F, 74,
   81
Light microscope, 39, 40, 224
Light reaction, 65, 66, 67, 71, 75, 78
Lignans, 397F
Lignin, 55, 234, 364, 365, 367F, 369,
   370, 375, 392, 393F, 394F, 395–397
Lignoceric acid, 305T
Lilies, 229
Limnanthes, 327
Limonene, 400, 403F, 421, 422F
Linalool, 400, 403F, 420F, 421
Linoleic acid, 41, 304F, 305T, 321F,
   322, 324, 326, 330T
Linolenic acid, 41, 304F, 305T, 321F,
   322, 324, 326, 330T
Linseed oil, 330T
Lipase, 321F, 328, 329, 332F
Lipid bilayers, 40, 41F, 42, 43, 44, 67
Lipidomics, 304
Lipids, 4, 40–46, 58, 73, 76, 144, 167,
   168, 182, 238, 303–333
Lipochitin oligosaccharides, 243
Lipoic acid, 159F, 160F, 163F, 165, 265
   biosynthesis, 307
Lipo-oligosaccharides, 243
Liposomes, 41
Lipoxygenase (Lox), 321F, 322
Liquiritigenin, 366, 381, 388, 389F
Liverworts, 174, 229, 412T
L-malate (*see* Malate), 160F, 166
Loganin, 342
Lolium, 220
Lonchocarpus, 173
Lotus, 263

Lovastatin, 427
L-rhamnose (Rha, 6-deoxy-L-
   mannose, *see* Rhamnose), 225
L-scopolamine, 349
Lubicrants, 326, 327, 329
Lucerne, 24
Lung cancer, 409
Lupin (*Lupinus*), 242, 244, 275, 384,
   Plate 11.2,
Lutein, 76, 404F, 425F
Luteolin, 387F
Lutoferol, 387F
Lutonarin, 365
Lycopenes, 63, 404F, 425F
Lynen, Feodor, 401
Lysine, 19, 82, 111F, 112F, 167, 272F,
   275, 276, 277F, 286, 293, 294,
   338, 351, 353
Lysine-ketoglutarate
   reductase/saccharopine
   dehydrogenase, 167
Lysophosphatidic acid (LPA), 316F,
   326, 328F
Lytic vacuoles (LV), 295, 296, 297F

M

Macular degeneration, 409
Magnesium, 71, 109F, 110, 111, 240,
   256, 257F, 280
Maize, 17, 104T, 120, 122, 136T,
   137–139, 152, 157, 212, 213T,
   214, 215T, 217, 250, 260, 264,
   287–295, 299, 388
   Quality Protein Maize, 294, 299
Malaria, 411
Malate, 31, 49, 95F, 115, 116, 149, 150,
   177, 184T, 190, 190F, 245, 249,
   332F
   dehydrogenase (MDH), 31, 115,
      125, 131F, 132F, 149, 150F, 158F,
      160F 166, 170, 191F, 245
   in C4 photosynthesis, 120, 122,
      123F, 124F, 128
   in CAM photosynthesis, 131F,
      132F, 133, 135, 137
   in TCA cycle, 156, 157, 158F,
      160F–162F, 166, 167, 168F, 169,
      170, 171F
   translocator, 116, 245
   synthase, 332
Malic enzyme, 122, 123, 124F, 130,
   131F, 136T, 138
   in respiration, 149, 150F, 158F, 170,
      171F, 172, 177
Malonate, 157, 161F, 174
Malonic acid, 370, 371F, 379, 380F
Malonyl CoA, 307, 308F, 309, 311F,
   312, 313, 314, 324, 325F, 327,
   333, 371F, 375, 376F, 379, 380F,
   381, 382F, 390, 391F

Malonyl-thioesters, 311
Maltooligosaccharides (MOSs), 219F
Maltose, 218, 219F, 298
Mammalian PDC, 160, 162
Mandrake (*Mandragora*), 341, 349
Manganese, 75, 76, 77F
Manioc, 1, 299
Mannans, 227, 229, 230
Mannose, 196, 225, 229, 230, 298, 299
Marker, Russell, 410
Marmesin, 377F
Marrows, 62
Martin, 96
Martius, Carl, 156
Matrix, cell wall, 55, 225, 228, 230,
   232, 234
   mitochondrial, 57, 115, 155, 161F,
      164F, 165, 166, 170, 173, 174F,
      176, 178, 184, 188, 190
Meconopsis, 384
Medicago, 242, 264, 388
Medicarpin, 365, 388, 389F
Meers, 263
Melons, 62
Membranes, 40
   membrane fracture faces, 44F, 61
   membrane particles (proteins), 40,
      43F, 61
   membrane rafts, 45
   membrane recycling, 47
   membrane structure, 40, 41F
Ménière's disease, 340
Mentha, 414, 418, 421
Menthol, 403F
MEP, 415F, 416
Meristematic cells, 48, 60, 61, 232, 264
Meroterpenes, 405, 419
Mesembryanthemum, 132
Mesophyll cell, 34, 62, 121F, 122, 123F,
   124F, 125, 126F, 127F, 128, 129,
   135, 136T, 206F, 250, 260, 262,
   264, 265F, 365
Metabolic control, 27, 28, 201
   flux, 27, 32
   pathways 2, 5–37, 116, 144, 151,
      273, 276, 355, 359, 360, 402
Metabolomics, 2, 6F, 20, 25, 26
Metabolons, 32, 33, 191
Methionine, 81, 258, 265, 266F, 267,
   270, 272F, 276, 277F, 280F, 281,
   282, 283F, 286, 289, 290, 292,
   293
6-Methoxymellein, 370, 371
Methoxypsoralen, 365, 366F
Methyl jasmonate, 320, 345, 354
Methyl transferases, 385, 426, 428
Methylamine, 116F
Methylcoclaurine, 346F
Methylerythritol phosphate (MEP), 414
Methylesterase, 232
Methylgenistein, 388

Methylxanthine, 356, 357F
Mevalonate diphosphate (MVAPP), 413F
Mevalonate kinase (MK), 413F
Mevalonic acid pathway, (MVA) 411, 412, 413F, 414, 426, 427F
Mice, 298, 348
Microbodies, 34, 40, 58F
Microfibrils, 47, 55, 95F, 224–228, 230, 231, 234
Microsatellites, 21
Microtubules, 48
Middle lamella, 54, 234
Miflin, 263
Milkweed, 400
Mint, 414, 418
Miocene period, 120
Mitchell, 185, 187
Mitochondria, 51, 57, 113, 114F, 155, 157, 164F, 169F, 171, 172, 176, 183, 184, 190F, 283, 286F, 309F, 376, 427F
    inner membrane transport 161F, 162F
Mitochondrial 2-oxoglutarate translocator (OGT), 161F, 162F
Mitochondrial Carrier Family (MCF), 161
Mitochondrial electron transport chain, 3, 172, 173F, 174F, 175F, 190F
Mollugo, 125
Molybdate, 267
Molybdenum, 240, 241F, 250F, 252
Monarch butterfly, 337, 400
Monodehydroascorbate reductase, 88
Monogalactosyl diacylglycerol (MGDG), 306, 317F, 318F
Monogastric animals, 271
Monolignol, 392, 393F, 394F, 395–397
Monoterpenes, 402, 403F, 420F, 421, 426, 427F
    synthase, 421, 422F
Montanic acid, 305T
Moricandia, 125, 127
Morphinan alkaloids, 344, 345F, 346F, 347, 348, 359
Morphine, 336, 340, 344F, 345
Mosses, 110, 222, 229, 289, 412T
Moths, 337, 400
Motor cells, 55
Movement proteins, 56
Multienzyme complex, 165, 167, 280, 283, 314, 324
Muscle, 155, 156, 340
Mutant plants, 6, 16–22, 58, 214, 217, 233, 261, 262, 263, 264, 265, 285, 294, 314, 326, 413,416, 417
Mycorrhiza, 246, 247, 248
Mycorrhizal fungi, 237, 246, 247
    roots, 246, 247

Myo-inositol, 318, 319
Myrcene, 400, 421, 422F
Myricetin, 382F, 382F, 383
Myristicin, 364
Myrothamnus, 222
Myxothiazol, 174

N

N-acetylglucosamine, 298
NADH, 58, 115, 151
    dehydrogenase, 173, 174F, 175F, 176, 183, 184, 190F, 388
    dehydrogenase complex, 173
NAD-isocitrate dehydrogenase (NAD-IDH), 169F
    -malic enzyme, 122, 136T, 167
NADP, 65, 67, 69F, 70F, 82
NADP$^+$, 79, 82
NADP-dependent glyceraldehyde 3-phosphate dehydrogenase (NADP-GAPDH), 145, 149F, 150
NADPH, 60, 66, 67, 79, 85, 153
NAD(P)H dehydrogenase, 174F, 176, 177, 178, 190, 191
NADP-malate dehydrogenase, 31, 131F, 132F
    -malic enzyme, 122
Narcissus, 62, 425
Naringenin, 381, 382F, 387F, 389F, 390, 391F
Nastic movements, 55
Necic acid, 354F
Necine base, 339F, 354F,
Nectar guides, 383, 384, Plate 11.5
Neem tree, 400, 404F
Negative free energy, 27, 149, 198
Negative water potential, 54, 224
Negelein, 175
Neokestose, 220F, 221
Neoxanthine, 407F
Nerve gas, 340
Neurospora, 373
Nicolson, 45
Nicotiana, 349, 352, 353, 417
Nicotinamide, 196
Nicotine, 340, 341, 349, 350F, 351, 353, 354, 360
Nicotinic acid, 350F, 351–353,
Nitrate, 3, 60, 115, 116, 137, 190F, 237, 238, 239F, 243, 247, 248F, 249, 263, 272, 276
Nitrate assimilation, 3, 60, 238, 248, 249
Nitrate reductase (NR), 30, 32, 115, 247, 248F, 250F, 252, 254
    distribution, 250
    regulation, 251T, 252F, 253F
    inhibitor protein, (NIP) 252F, 253F
Nitrate transporters, 249, 250
    uptake, 248, 249F

Nitric oxide (NO), 253, 254, 337
    synthase, 250, 254
Nitrite, 82, 115, 190F, 237, 238, 248, 250, 251, 252F, 253, 254, 255, 258, 262, 416
Nitrite reductase (NiR), 237, 247, 248F, 253F, 255F, 256, 269
    synthesis and structure, 253, 254, 255F, 269
Nitrogen, 3, 8, 15, 59, 100, 135, 200, 238, 272, 274F
Nitrogen assimimilation, 3, 153, 171F, 190, 273
    cycle, 239F
Nitrogen dioxide, 254
Nitrogen fertilizers, 238, 239T, 248
Nitrogen fixation, 238, 239F, 264
    -fixing bacteria, 238, 242T, 397
Nitrogenase, 68, 240F, 241, 244, 245
N-linked glycans, 50, 52
N-methylputrescine, 349, 351
Nod factors, 243, 365, 388
Nodules, 59, 242–246, 243T, 243F, 246T, 263, 264, 267T, 275, 388
    cortex, 244
    infection thread, 243F
Noodles, 217
Nonphotochemical quenching (NPQ), 73, 74, 88, 89, 90
Norcoclaurine, 339F, 345, 346F, 357
    synthase (NCS), 345, 347
Norethindrone, 410F
Norflurazan, 424
Nuclear envelope, 51, 57
    pores, 57
Nuclei, 15, 39, 40F, 51, 56, 57, 102, 103, 105, 224, 288, 293, Plate 8.1
Nucleic acids, 14, 15F, 23, 56, 57, 145, 153, 168F, 237, 238
Nucleoside phosphates, 238
Nucleosomes, 57
Nucleotide biosynthesis, 108
Nutmeg, 364
Nutrient, 4, 6, 27, 49, 180, 183, 237, 238, 261, 266, 287, 299
Nymphaceae, 179

O

O-acetylserine, 267, 280F, 283F
    (thiol) lyase (OASTL), 268, 280F, 283F
Oak, 391
Oats, 96, 197, 220, 287, 290
Obamegine, 346F
Ocimum, 401
Octanoic acid, 305T
Octadecanoic acid, 305T
Octulose, 196
Oenothera, 330T
Ogren, 117

Oils, edible and industrial, 326, 327, 329, 330T
  synthesis, 328F
Oilseed rape, 1, 326, 327, 330T, 373
Okadaic acid, 201F, 202
Olefins, 409
Oleic acids, 304F, 305T, 306, 324, 326, 327, 329, 330T
Oleoresins, 405
Oleoyl ACP, 308F, 316
O-linked glycans, 53
Olisthodiscus, 104
Olives, 1, 330T
Onion, 220
Opium, 340, 345
O-phosphohomoserine, 277, 277F, 282, 283F
Oranges, 63
  blossom, 400
Orchidaceae, 354
Orchids, 229
Organelle isolation, 34
Ornithine, 156, 282, 284F, 339, 349, 350F, 351
  decarboxylase (ODC), 349, 350F, 351
Oryza sativa, 425
Osmoprotector, 202
Osmotic potential, 46, 54, 95
Osteoporosis, 390
Ovarian tumors, 409
Oxalate, 49
Oxaloacetate (OAA), 190F, 245, 270, 272F, 274, 275, 276, 332F, 333
  in photosynthesis, 95F, 115, 124F, 125, 131F, 137
  in TCA cycle, 150F, 156, 157, 158F, 159F, 160F, 161F, 162F, 166, 168F, 171F
  translocator, 161F, 162F
Oxalosuccinate, 159F, 165
β-oxidation pathway, 331F, 333, 378
Oxidative pentose phosphate pathway (OPPP), 3, 106, 109F, 110F, 144, 153, 154F, 253, 255F, 267, 270, 284, 314, 315, 371F
Oxidosqualene, 424
2-Oxoglutarate, 34, 248F, 258F, 259F, 270, 272F, 273, 274F, 285F, 286F, 343F, 351, 353F, 372F, 383
  aminotransferase, 116, 169F, 259
  dehydrogenase, 158, 160, 165, 167, 171, 286
  in photosynthesis, 114F, 116, 117, 124F, 125
  in TCA cycle, 156, 157, 158F, 159F, 160F, 161F, 162F, 165, 167, 168F, 169F, 170, 171F
Oxygen, 1, 58, 65–69, 71, 74, 75, 79, 88, 90, 93, 144, 151, 152, 156, 173, 175F, 177, 178, 181, 196, 225, 238, 240, 241, 242, 244, 245,

251T, 254, 256, 321, 322, 340, 343F, 347F, 353F, 359, 376, 392, 401, 402
  diffusion, 245
  electrode, 173
  evolving center, 67, 75, 76F, 77F, 80
  radicals, 86, 87, 90
  ratio, 184
  reactive (active) species, 74, 76, 88, 89, 90, 182, 280
Oxygenase reaction (Rubisco), 97–120
Oxylipins, 303, 320, 321F
Ozone, 67, 411

P

P680, 73–77, 78, 80F, 83F, 86, 89
  reaction center, 75
P700, 73, 80F, 81, 82, 83F, 85, 86F, 89, Plate 4.1
Paints, 329
Palaquium, 404F
Palmae, 330T
Palmitic acids, 304F, 305T, 326
Palmitoleic acid, 305T
Palmitoyl-ACP, 313, 316
Palm oil, 329, 330T
Panicoideae, 123, 294
Panicum, 125, 136T
Pantetheine, 312
Papain-type proteases, 297
Papaver, 340, 344, 345, 347, 384, Plate 11.3
Paper production, 369
Paraquat, 90
Pasta, 217
Patatin, 298
Pathogen attack, 183, 225, 227, 281, 321F, 322, 344, 365
Pathogens, 299, 322, 324, 335, 344
Paulliania, 355
Paxillus, 247
p-coumaryl-CoA, 390
PCR amplification, 17, 18F, 20–23, Plate 2.1
Peanuts, 1, 244, 390
Pea seed globulins, 288, 289, 297
Peas, 1, 41, 96, 104, 105, 117, 214, 218, 219, 242, 243T, 246T, 260, 263, 264, 289, 296, 314, 388
Pectins, 55, 225, 229, 231, 240
Pee Dee Belemnite rock (stable isotope analysis), 139
Pelargonidin, 382F, 383F, 384, 385 Plate 11.2
Pennisetum glaucum, 123
Pentagalloylglucose, 391, 392F
Pentasaccharide, 196
Pentose phosphate pathway, *see* oxidative pentose phosphate

PEP carboxylase (Phospho*enol*pyruvate carboxylase),
  in C$_4$ photosynthesis, 120–122, 123F, 124F, 125–128, 129F
  in CAM photosynthesis, 133, 135, 136T, 137, 138, 140
  in respiration, 149, 150F, 151, 170, 171F
  PEPC kinase, 128, 129F, 130, 131F, 132F,133
Peppermint, 418, 421
Peppers, 63, 414
Perfumes, 1 *see also* Scents
Peroxidases, 234, 394F, 396
Peroxides, 31, 58, 87, 88, 90, 113, 182
Peroxisomal glutamate:glyoxylate aminotransferase (GGT), 169F
Peroxisomes, 2, 58F, 59, 113, 114F, 115, 165, 166, 169F, 309F, 329, 331
Peroxygenase, 321F, 324
Peroxynitrite, 254
Pesticides, 1
  see also Insecticides
Petroselinum, 387
Petunia, 381, 386, 390
pH, 13, 30
  pigment color, 384
  soil, 385
  vacuoles, 385
Pharmaceuticals, 335, 338, 360, 411
Phaseolin, 295, 297
Phaseolus, 259, 295, 297, 388
Phellandrene, 422F
Phenol, 364F
Phenolics, 25, 123, 137, 324, 335, 363–369
Phenylalanine, 273F, 285F, 291F, 371F, 372F, 374, 375, 376F
  ammonia lyase (PAL), 375, 390
Phenylpropanoids, 33, 155, 234, 258F, 271, 282T, 363–394, 366F, 371F, 372F, 377F, 379F, 389F, 394F
  lactones, 365
  metabolism, 155
Pheophytin, 75, 76, 78
Pheromone, 400
Phlobaphene, 380F, 387F, 388
Phloem, 121F, 204, 205, 206F, 207F, 208F, 244, 262, 267T, 272, 276
  parenchyma, 56
Phoma, 365
Phorbol, 400
3'-Phospho-5'-adenylsulfate (PAPS), 269F, 270F
Phosphatidic acid (PA), 303, 306, 316F, 317F, 326, 327, 328F, 333
Phosphatidylcholine (PC), 306, 317F, 318F, 319F, 327, 328F, 329
Phosphatidylethanolamine (PE), 306, 317F, 318F, 319F

Phosphatidylglycerol (PG),
  73, 306, 317F, 318F, 319F
Phosphatidylinositol (PI), 306, 318F,
  319F, 320F
Phosphatidylserine (PS), 306, 318, 319F
Phospho*enol*pyruvate (PEP), 34F, 95F,
  150F, 286F, 332F, 371F, 372F, 373
  carboxykinase, 132
  carboxylase (PEP carboxylase),
    119, 122, 125, 130, 131F, 132,
    136T, 150F, 170, 245
Phosphofructokinase (PFK), 31, 32,
  145, 146F, 149F, 200
Phosphoglucan-water dikinase, 219F
Phosphoglucomutase, 146F, 211F
6-Phosphogluconate, 109F, 153
  dehydrogenase, 110, 153, 154F, 155
3-Phosphoglycerate (3-PGA), 34, 164,
  168F, 198, 203, 204, 212, 272,
  274F, 277, 307
  in glycolysis, 145, 147F, 148, 149F,
    151
  in photosynthesis, 96, 98F, 101,
    109F, 103F, 104F, 119, 124
Phosphoglycerate kinase (PGA
  kinase), 101, 145, 147F
Phosphoglyceromutase (PGM), 147F,
  148, 149F
Phosphoglycolate, 99F, 100, 113F, 117,
  127F
Phosphoinositides, 318
Phospholipase C, 318, 320F
Phospholipid, 41, 42F, 43F, 73, 304F,
  304T, 306, 328, 333, 410
Phosphoprotein phosphatase, 201F
Phospho-pyruvate dehydrogenase,
  kinase (PDP) 157, 163
Phosphoribosyl, 285F
Phosphoribulokinase, 97, 98F, 108,
  109F, 110, 111
Phosphorylation-dephosphorylation,
  31
Phosphotransferase, (PFP) 149F, 204F
Photoassimilates, 34, 197, 205, 206F
Photochemical reactions, 65–75, 80,
  89, 239
Photoinhibition, 86, 118, 138
Photons, 67, 72
Photooxidative damage, 86, 87
Photophosphorylation, 79, 85, 86F,
  212
Photorespiration, 2, 3, 99F, 100,
  113–119, 113F, 114F, 116F, 187,
  258F, 260, 262, 264, 265, 272,
  276, 277F
Photosensitive reaction, 364
Photosynthesis, 3, 65–90, 93, 187, 190
Photosynthetic carbon assimilation,
  93–140
  electron flow, 31, 69–87, 70F, 77F,
    79F, 81F, 83F, 86F, 253

Photosystems, 67
  PSI, 62, 69F, 70, 71, 73, 80, 81F, 82,
    Plate 4.1
  PSII, 62, 69F, 71, 73, 75, 76F, 82
Phototoxic, 364, Plate 11.1
Phragmoplast, 54
Phthalonate, 161F
Phyllobates, 337
Phylloquinone, 80, 81, 427F
Phytoalexins, 365, 388, 403F, 405
Phytochelatins, 266, 283
Phytochrome, 103, 132, 256, 260, 263,
  275
Phytochromobilin, 256, 257F
Phytoecdysones, 401
Phytoene, 420F, 421, 424F, 425F, 426
  synthase (PSY), 420F, 421, 424F,
    425F, 426
Phytoglycogen, 209, 213
Phytol, 71F, 76, 256, 402, 417, 427F
Phytosterols, 410, 411
Picea, 247, 403F
Piericidin A, 173
Pigments, 4, 49, 60, 62, 63, 68–74, 86,
  90, 237, 304, 307, 363, 364, 367,
  381, 383, 384, 385, 386, 387, 397,
  Plate 11.5
Pineapple, 136T
Pine bark beetle, 400
Pinene, 400, 401, 403F, 421, 422F
Pinoresinol, 397F
Pinosylvin, 367F, 390, 391F
Pinus, 390, 400, 403F
Piperidine alkaloids, 336, 337, 350F,
  351, 353
Plant biotechnology, 17, 19F, 37
Plant growth regulators, *see* auxins,
  brassinosteroids, cytokinins,
  ethylene, gibberrelins,
  jasmonic acid
Plant nutrition, 4, 6, 27, 49, 180, 183,
  238, 261, 266, 287
Plant pigments, *see* flavonoids,
  chlorophylls, anthocyanins,
  carotenoids
Plant transformation, 15, 17, 53, 105,
  222
Plasma membrane, 40, 44, 45, 46, 48,
  49, 51, 53–58, 95F, 188, 205–208,
  224, 227, 228, 230, 234, 234, 243,
  247, 249, 255, 267
  recycling, 47, 54
  transport, 46
Plasmodesmata, 50F, 52, 55, 56F, 121,
  123, 205, 206F, 208F
Plasmodiophora, 223
Plasmodium, 340, 411
Plastics, 329
Plastidic dicarboxylate translocators,
  168, 170
Plastids, 51, 59, 145, 198, 209, 213,

  235, 253, 259, 260, 263, 267, 269,
  283, 412T, 416, 418, 426, 427
  fatty acid biosynthesis, 310, 314,
    315F, 316F, 317F
  transformation, 105
  triose phosphate/phosphate,
    transporter, 34F, 97, 101, 124
Plastocyanin, 67, 69, 70, 78, 79, 80,
  81F, 82, 83F, 85, 89, 89, Plate 4.1
Plastoquinol, 69, 76, 78, 79F, 82, 86F, 90
Plastoquinone, 69, 75, 76, 78, 79F, 86F,
  90, 402, 427F
Plastosemiquinone, 78, 79F
Poaceae, 120, 310
Poison dart frogs, 337, Plate 10.1
Poisons, 363, 364, 367, 401
Polar residues, 223
Pollen, 180, 233, 303, 321F, 329, 384,
  405
Pollinators, 178,179, 180, 191, 335,
  336, 364, 383, 397, 399, 400,
  Plate 11.5
Polyacrylamide gel, 11, 12Fs
Polyamines, 280, 282, 349, 351, 354
Polymerase chain reaction (PCR), 17,
  18F, 20–23, Plate 2.1
Polypodium, 401
Polysaccharides, 47, 52, 195, 196,
  209, 224, 225, 228, 230, 231,
  232, 234
Polyterpenes, 404F
Ponasterone A, 404F, 426
Popcorn, 293
Poppy, 340, 344, 345, 347, 348,
  Plate 11.3
Populus, 262, 369, 395
Porins, 57, 113, 115, 169F
Porphyra, 104
Porphyridium, 104
Porphyrin, 71, 72, 256F
Potassium, 46, 48, 94, 95F, 246
  cyanide, 173
Potatoes, 1, 177, 183, 190, 198, 205,
  208, 209, 212, 215, 217, 298, 299
Potentilla, 384, Plate 11.5
Prenyl diphosphates, 418, 419F
Prenyltransferase, 418, 419F, 424
Prephenate, 285F
  aminotransferase, 373F, 374
Prescribed drugs, 335
Primary cell walls, 54, 55, 224
Printing ink, 329
Proanthocyandins, 365, 368F, 369,
  370, 371F, 380F, 386F, 387
Progesterone, 400, 410F
Prolamella bodies, 61F
Prolamins, 287, 290, 291, 291F, 292F,
  293
Proline, 112, 262, 272F, 282, 284F, 290,
  291F, 292F, 293
Proplastids, 60

Prostate cancer, 388, 409
Proteases, 49, 247, 296, 297
Proteins, 12, 40
    14-3-3 proteins, 200, 203
    storage, 51, 238, 265, 286–299
Protein bodies, 288F, 291, 292, 294, 295, 296F, Plate 8.1
Protein kinase C, 318, 320F
Protein purification, 7
Protein storage vacuoles (PSV), 294, 295F, 296F, Plate 8.1
Proteoglycans, 53
Proteolytic cleavage, 297
Proteomic analysis, 6F, 12, 13, 14, 24, 25, 26, 36
Protochlorophyll, 61
Protons, 68, 75, 78, 80, 83
    electrochemical gradient, 79F, 90, 184, 186
    gradient, 68, 83, 90, 185, 249, 266
    motive force, 83, 184, 186, 249
    pump, 78, 79, 249
    sulfate cotransporter, 267
Protopine alkaloids, 344F, 346F, 348F
Protozoa, 412T
Pseudotsuga, 400
Psoralen, 365, 366F, 377F
Pteridium, 365
Pullulanases, 219F
Pulse-chase experiments, 15
Purine alkaloids, 339, 355
Purple photosynthetic bacteria, 66, 68F, 104
Putrescine 3, 39F, 349, 350F, 351, 354F, 355
    N-methyltransferase (PMT), 349, 350F, 351, 352, 353
Pyranose, 225
Pyrethrins, 400, 402F
Pyridine, 351F, 353
Pyridine alkaloids, 336
Pyridoxal 5′-phosphate, 274
Pyrophosphatase, 123F, 124F, 129, 213T, 269
Pyrophosphate F6P-1
    phosphotransferase (PFP), 203, 204F
Pyrrhocoris, 401
Pyrroles, 256, 341
Pyrrolizidine alkaloids, 336–338, 339F, 341, 349, 350F, 351, 354F, 357, 359, Plate 10.2, Plate 10.3, Plate 10.4
    biosynthesis, 354F, 357, 360
Pyruvate, 16, 58, 95, 99, 121F, 122, 123F, 124F, 125, 129, 131F, 144, 179, 180, 181, 190, 271, 272F, 279, 307, 309F, 315
    amino acid biosynthesis, 270, 271, 274F, 276, 279, 280F, 283F, 286F,
    decarboxylase, 152

dehydrogenase (PDH), 31, 155, 157, 158F, 159F, 163F, 307, 309F, 313
dehydrogenase complex (PDC), 31, 155, 157, 160, 162, 163F, 164F, 315
dehydrogenase kinase (PDK), 157
in TCA cycle, 156, 157, 158F, 159F, 161, 163F, 164F, 168F, 171F, 172
in glycolysis, 146, 147F, 148, 149F, 150F, 151, 152, 155
kinase, 30, 147F, 149F, 150, 151, 171F
oxidation, 162, 164,183
phosphate dikinase (PPDK), 129F, 134, 137, 147F
translocator, 155, 161F, 162F

Q

Q cycle, 69F, 78, 79F, 85
Q ubiquinone pool, 173, 174F, 175F, 177, 181
Quality Protein Maize, 294, 299
Quanta, 66, 72
Quercetin, 365, 382F, 383
Quinazoline alkaloids, 336
Quinine, 340, 342F
Quinoline alkaloids, 336
Quinolinic acid, 350F
Quinolizidine alkaloids, 336
Quinone, 79

R

Rabinowitch, Eugene, 70
Radioautography, 15
Radioisotopes, 6F, 14, 96, 120
Radish, 413
Raffinose, 197
Ragwort, 341, Plate 10.2, Plate 10.4
Rainforest, 180
Rape seed oil, 326, 329, 330T
Raphanus, 413
Rat poison, 340
Raucaffricine, 358
Rauvolfia, 340
Red algae, 71
Red blood cells, 400
Redox couples, 70
Redox potential, 70F, 75, 76, 83, 88, 153, 240
Regan, Ronald, 409, 411
Reporter genes, 35
Respiration, 3, 57, 58, 67, 113, 132, 133, 137, 143–193, 173F, 187, 196, 205, 207, 209, 240F, 241, 244, 256, 315, 365
Resurrection plants, 196, 222
Resveratrol, 365, 367F, 369, 391F
Reticuline, 339F, 345, 346F, 347, 348

Retinol, 409
Retronecine, 354F
Rhamnogalacturonan I, 231, 232
    II, 231, 232, 233F
Rhamnose, 225, 232, 352F, 359, 383F, 385
Rhizobia, 240, 241, 242, 243F, 244, 245
Rhizobium, 238, 240, 241, 242T, 243T, 260, 264, 365, 388
Rhizomes, 297
Rhizosphere, 243
Rhodobacter, 241
Rhododendron, 400, 403F
Rhodophyta, 104
Rhodopseudomonas, 104
Rhodospirillium, 104
Ribose 5-phosphate, 98F, 106, 153, 154F, 155, 271, 273F
    isomerase, 98F, 153
Ribosomes, 40, 42, 43, 49, 50, 51, 52F, 57F, 59, 60F, 102, 253
Ribulose 1, 5-bisphosphate (RuBP), 96, 97, 98F, 99F, 106, 108
Ribulose 1,5-bisphosphate
    carboxylase oxygenase see Rubisco
Ribulose 5-phosphate, 98F, 108, 109F, 145, 153, 154F
    epimerase, 98F, 108
Rice, 1, 3, 20, 96, 104T, 136T, 137, 138, 203, 212, 213, 214, 215, 222, 228, 260, 262, 263, 264, 265, 287, 288, 290, 330T, 425, 426
    endosperm, 425, 426
    Golden Rice, 425, 426
Ricinoleic acid, 328, 329, 330T
Ricinus, 405
Rieske proteins, 75, 78, 79, 83F
RING-finger region, 228
RNA, 17, 49
    dsRNA, 17, 19
    mRNA, 57
    microRNA, 19
    siRNA, 18
RNA interference, 18
Root hair cells, 56, 243F
Root nodules, 59F, 242–245, 246T, 264, 267, 275
ROS see Oxygen, reactive species
Rosaceae, 197, 385
Roses, 400
Rosettes, cellulose synthase, 226
Rosin, 405
Rotenone, 173
Roundup™, 373
Rubber, 298, 404F, 418, 427
Rubisco, 29F, 97–119, 98F, 99F, 102F, 127F, 136T, 138, 139, Plate 5.1
    activase, 111F, 112F
    assembly, 103F

improvements, 105
oxygenase, 113F, 114F, 115, 118
regulation, 111
specificity, 104, 110
Rye, 218, 290, 291

S

Saccharomyces, 173, 373
Saccharum, 120, 122
Sachs, 71
S-adenosylmethionine
    280, 282, 343F, 348F
Safflower, 330T
Salicylhydroxamic acid (SHAM), 173F,
    181
Salicylic acid, 365, 366F, 370, 378,
    379F
Saliva, 340
Salutaridinol, 346F, 358
Salvia, 421
Sanguinaria, 344, 345
Sanguinarine, 344F, 348F
Sapogenin, 403F
Saponifiable lipids, 303, 304T, 306
Saponins, 400, 424
Scarlet-bodied wasp moth,
    Plate 10.3
Scenedesmus, 96, 104
Scents, 3, 179, 191, 335, 363, 364, 399,
    400, 403F
Scopolamine, 340, 341, 349, 350F, 351,
    352, 353F
Scopoletin, 365, 377F
Scoulerine, 345, 346F, 347
Scutellum, 216
SDS, 11, 14
    -PAGE electrophoresis, 11, 14, 201
Seaweeds, 74
Secologanin, 338, 341, 342F, 343F
Secondary cell wall, 54, 55
Secondary metabolites, 25, 151, 168,
    308, 322, 335, 402
Secretory proteins, 49–52
Secretory vesicles, 44, 47, 48, 52F, 53,
    54, 226, 296
Sections, microscopy, 39
Sedoheptulose, 196
    1,7-bisphosphatase, 106
    1,7-bisphosphate, 106, 196
Seeds, 53, 58, 286, 297
    germination, 297
    oil, 325, 327, 328F, 329, 330T, 333
    proteins, 265F, 286, 294, 295
Selaginella, 222
Selenate, 267
Senecio, 337, 341, Plate 10.2,
    Plate 10.4
Senecionine, 339F, 341, 354F
Senescing leaves, 272
Senna, 369

Serine, 31, 43, 162, 168F, 190F, 202,
    203, 250, 252F, 253F, 258F, 262,
    270, 274F, 276, 277F,, 280F, 283F
    312, 318
    acetyltransferase (SAT), 267, 280F,
    283F
    biosynthesis, 277
    glyoxylate aminotransferase
        reaction, 114F, 117
    hydroxymethyltransferase (SHMT),
        114F, 115, 116F
    in photosynthesis, 114, 115, 116F,
        117, 118, 127F, 128
    phosphorylation sites, 160
    threonine kinase, 128
Sesame oil, 330T
Sesquiterpene synthases, 421
Sesquiterpenes, 402, 403F, 405, 426
Shade leaves, 62, 87
Sheep, 341, 364
Shikimate, 285F
    phosphate, (S3P) 372F
Shikimic acid, 285, 370, 371F, 372F,
    373, 374, 375
    pathway, 284, 34F, 371, 372F, 373,
    374, 375
Sieve elements, 205, 206F
    tubes, 56, 205, 206F, 245
Signal peptides, 33
Silica, 9
Simmondsia, 325, 327
Simple sequence length
    polymorphisms, 21
Sinapaldehyde, 394F
Sinapic acid, 394F, 395
Sinapyl alcohols, 234, 392, 393F, 394F,
    395
    dehydrogenase (SAD), 394F, 396
Singer, 45
Sink tissue, 4, 197, 260, 275
Sinningia, 388
Siroheme, 254, 256, 257F
    biosynthesis, 256, 257F
    center, 254, 255F, 269
Site directed mutagenesis, 76, 105, 327
Sitosterol, 45, 226, 404F, 410, 411F, 424
    cellodextrin (SCD), 226
    β-glucoside (SG), 226
    lipid, 226
Skeletal muscles, 400
Slack, 120
Sláma, Karel, 401
Smoky Mountains (USA), 409
Solanales, 229, 298
Solaneesyl diphosphate (SPP), 427F
Solanum, 298, 349
Solar radiation, 49, 86
Solenopsin A, 337
Solenopsis, 337
Solutes, 8, 46, 47, 49, 54, 55, 56, 224
Somerville, Christopher, 117

Sorbitol, 197
Sorghum, 1, 121F, 275, 388
Source tissue, 205
Soybean, 1, 197, 246T, 373, 395
    oil, 326, 329, 330T
Spadix, 179F, 180
Spartina, 120, 123
Spearmint, 421
Specific antibodies, 14
Sperm, 403F, 405, 410
Spermidine, 349, 354F
    synthase, 349, 350F, 351
Spermine, 349
Sphingolipid, 313
Sphondin, 365
Spiders, 337
Spinach, 1, 34, 41, 84, 85, 104, 111,
    112, 145, 197, 200, 201, 202, 212,
    220, 252F, 279, 311, 314
Sporamin, 298, 299
Spurges, 400
Squalene, 401F, 402F, 406, 420F, 422, 424
    synthase (SQS), 422, 424F
Stachyose, 197
Starch, 4, 7, 11, 32, 34F, 40, 60F, 63F,
    94, 95F, 98, 122, 126F, 131F, 132F,
    146F, 148, 151, 164F, 179, 196,
    197F, 198, 200, 204, 205, 209,
    210F, 211, 212, 214T, 215T, 291,
    294, 300, Plate 8.1
    biosynthetic pathway, 211F, 214
    branching enzyme (SBE), 214
    commercial, 216T, 217
    degradation, 216, 217, 218F, 219F
    formation, 213
    mobilisation, 218
    novel starches, 215
    storage, 209
    structure, 210F
    synthesis, 209, 211F
State 2 transition, 88
Statins, 411
Stearic acids, 304F, 305T, 306, 326,
    330T
Steel, 67
Stern, Lina, 156
Steroidal alkaloids, 336, 337
Steroidal glycosides, 401
    see also cardiac glycosides
Sterols, 45, 322, 406, 410, 411, 413,
    426, 427F
Stigmasterol, 410
Stilbene, 365, 367F, 369, 375, 376F,
    388, 390, 391F
    synthase (STS), 391F
Stomata, 55, 86, 93, 95F, 100, 106, 118,
    121F, 130, 131F, 132F, 133,
    136,137, 140, 254
Storage lipids, 306, 325, 329, 332F, 333
Storage proteins, 51, 265F, 266,
    286–299, Plate 8.1

Storage tubers, 287, 297–299
Streptomyces, 173, 413
Stress protectant, 222, 223
Strictosidine, 338, 339F, 341, 342F, 343F, 344, 345, 357
Stroma, 30, 31, 59, 60, 67, 69, 71, 75, 76, 78, 79, 80F, 81, 82, 83, 84, 86, 88, 100, 102, 110, 111, 113, 114, 115, 164F, 191F, 200, 211F, 219, 307, 311, 314, 315F, 316
    lamellae, 60, 61, 62F
Strychnine, 340, 342F
Strychnos, 337, 340
Stylopine, 348F
Suberin, 55, 324
Substrates, 33
Substrate concentrations, 28, 30F
Succinate, 173, 174F, 175F, 184, 286F, 331F, 332
    dehydrogenase (SDH), 155, 157, 158F, 160F, 166, 173F, 174F
    in TCA cycle, 156, 157, 158F, 160F, 161F, 162F, 165, 166, 168F, 172
Succinyl-CoA, 160
    synthetase, 165
Sucrose, 4, 34, 36, 94, 98F, 101, 146F, 194–220, 221F, 226F, 245, 251T, 252F, 260, 263, 275, 298, 315
    6-phosphate, 198, 199F
    phosphatase, 199F, 200
    phosphate synthase (SPS), 32, 197, 198, 199F, 200–204
        regulation, 32, 201, 202
    synthase (SuSy), 146F, 205F, 207T, 226, 245
Sugar beet, 1, 219, 259
Sugar cane, 1, 120, 122
Sulfate, 83, 161, 237, 238, 265, 266F, 267T, 268F, 269, 280, 283, 284
    assimilation, 237, 238, 267, 268F, 269F, 282
    transport, 237, 266, 267T
Sulfation, 269F
Sulfite, 238, 250, 254, 256, 268F, 269
    oxidase, 250
    reductase, (SiR) 269
Sulfolipid, 304T
Sulfoquinovose, 317
Sulfoquinovosyl diacylglycerol (SQDG), 306, 317F, 318F
Sulfotransferase, 269F
Sulfur, 66, 266F, 282, 299
    biogeochemical cycle, 266F
    deficiency, 266, 282
Sunflower, 1, 51, 293, 330T
Superoxide dismutase, 88
Sweet pea, 275
Sweet potato, 1, 215, 297, 298
Syagrus, 327
Symbiodinium, 101
Symbiosomes, 243

Symbiotic nitrogen fixation, 3, 242, 243T
Symphytum, 341
Symplast, 56, 206F, 248
Symporters, 48
Synechococcus, 104, 105
Synge, 96
Syngenta, 426
Syringic acid, 365
Syringyl monolignols, 234, 392, 393F
Szent-Györgi, Albert, 156

T
Tabersonine, 343F
Tannins, 364, 368F, 370, 371F, 391
Tapetum, 51F
Tapioca, 215, 299
Tarins, 299
Taro, 298, 299
Taxol, 403F, 405, 409, 428, Plate 12.1
Taxus, 403F, 405, Plate 12.1
TCA cycle, 2, 3, 15, 16, 26, 31, 57, 143–172, 157F, 158F, 159F, 160F, 168F, 169F, 309F
    biosynthetic function, 167
    regulation, 170, 171, 190
T-DNA, 17, 22
Telophase, 51, 57
Temperature-dependent phase transitions (Tm), 41, 45
Tempest, 263
Tephrosia, 173
Tequila, 133
Terpenes, 399
Terpene synthase, 421, 422F, 423F, 424
Terpenoids, 335, 399–428
    biosynthesis, 168F, 341, 342F, 343F, 399, 401, 402, 411, 426, 427F
    human health, 409
    indole-alkaloids, 339F, 340
Tetrahydrocannabinol, 370
Tetrapyrrole biosynthesis, 254, 256, 257F, 275
Tetrasaccharide, 196
Tetraterpenes, 402, 426
Theobroma, 355
Theobromine, 355, 356, 357F
    synthase, 357F
Thermococcus, 101
Thermogenesis, 178, 179
Thiamine pyrophosphate, 270, 279
Thioredoxin, 31, 32F, 109–113, 129F, 172, 180, 181, 183, 212, 218, 223, 268
    regulation of Calvin cycle, 109, 110F
    regulation of, OPPP 109F
Thioredoxin/NADPH redox system, 172
Thornapple, 341
Threonine, 31, 128, 129, 258, 272F, 276, 277F, 279, 286

Thunberg, Thorsten, 156
Thylakoids, 61, 62, 74, 75, 122, 126F, 405
    lumen , 61, 89, 110
Tiger moth, 337
Timber products, 234, 369, 369
Titan arum, 180, Plate 6.1, Plate 6.2, Plate 6.3
Tobacco, 197, 265, 310, 322, 341, 365, 386, 413
Tocopherols, 417, 419, 427F
Tomato, 17, 63, 104T, 166, 222, 281, 298, 349, 381, 409
Tonoplast, 40, 48, 53, 170, 267T,
Torenia, 390
Trachelanthamidine, 354F
Transamination, 246, 276, 277, 279, 284
trans-cinnamic acid, 375, 376, 378
Transcriptome, 6F, 20, 22, 23, 26
Transfer cells, 56
Transferases, 53
Transformation, 15, 17, 53, 105, 222
Transgenic plants, 16, 19, 137, 183, 190, 261, 264, 265, 281, 283, 348, 356, 390, 417, 425, 426
Transitional vesicles, 52, 53
Transitory starch, 209, 218, 219F
Transketolase, 98F, 106, 107, 153, 154F
    reaction, 97
Translation initiation factor, 355
Translocators, metabolite, 3, 34, 101, 113, 115–118, 124, 130, 155, 161F, 162F, 168, 169, 170, 190, 197F, 205, 245, 267, 428
Transmembrane proteins, 43F, 44, 57, 113
    TM sequences, 43F
Transporte, r 33, 34, 47, 161
Transposon, 17, 20,
    tagging, 14, 18F, 22
Tree frog, 336, 337, Plate 10.1
Tree trunks, 55, 234
Trefoil, 246T
Trehalose, 222, 223F
    6-phosphate phosphatase (TPP), 223F
    6-phosphate synthase (TPS), 223F
Triacylglycerols (TAGs), 304T, 304F, 306, 307, 325F, 327, 328F, 329, 332F, 333
Tricarboxylate translocator, 161F, 162F
Tricarboxylic acid cycle *see* TCA cycle
Trichomes, 51
Triose, 196
    phosphate, 94, 95F
    isomerase, 98F, 101, 147F, 148, 199F
    translocator, 34, 101, 124F, 125, 197F, 267, 428

Triplet electrons, 72
Trisaccharide, 196, 221F
Triterpenes, 402, 403F, 404F, 424, 426
Triterpenoids, 400, 401, 410, 422
Triticum, 291
Tropane alkaloids, 339F, 340, 349, 350T, 351, 352
Tropinone, 352
Trypanosoma, 177
Trypsin inhibitors, 293, 298
Tryptamine, 338, 341, 342F, 343F
Tryptophan, 82, 273F, 284, 285, 293, 294, 338, 339F, 341, 342F, 343F, 344, 372F, 374, Plate 4.1
  biosynthesis, 285, 286F
  essential amino acid, 294, 295
Tryptophan decarboxylase (TDC), 342F
Tubers, 40, 63, 198, 205, 208, 209, 212, 286, 287, 298, 299
Tubocurarine, 340
Tulips, 220
Turgor, 54, 95, 207, 224
Turpentine oil, 399, 405
Tyramine, 346F, 358
Tyria, Plate 10.2
Tyrosine, 31, 75, 76, 81, 88, 254, 271, 273F, 278, 284, 285F, 294, 338, 339F, 344, 345, 346F, 358F, 359, 371, 372F, 374
  biosynthesis, 284, 285F

U

Ubiquinol (QH_2), 166, 174
Ubiquinone, 166, 173, 174F, 175F, 179, 402, 417, 419, 427F
UDP-glucose pyrophosphorylase, 208
Umbelliferone, 365, 366F, 377F
Uncoupling protein (UCP), 183
Uragoga, 340
Ureides, 59, 244, 246, 258
Uric acid, 59
Uricase, 59
Urochloa texana, 123
Uroporphyrinogen, 256, 257F
Uterus, 340
UV, 15, 50F, 66, 321F, 384
UVB, 365, 381
UVC, 67
UV-protectants, 365

V

Vacuoles, 35, 40F, 47F, 48, 49, 53, 95F, 136, 197, 221, 255, 295, 296F, 297F, 299, 341, 385
Valencine, 400
Validamycin A, 223
Valine, 167, 270, 271F, 278F, 279, 293, 294, 395
Valine aminotransferase, 278F, 279
van Niel equation, 94
Vanilla, 364
Vanilla orchids, 364
Vanillin, 364, 366F, 378
Varnishes, 329
Vegetable oils, 329
Vegetation canopy, 55, 86
Vegetative storage proteins, 298, 299
Vernolic acid, 328
Vertebrates, 336
Vesicles, 44, 47, 48, 52, 53, 54, 224, 226, 230, 247, 296F
Vicia, 289, 295F, 297, Plate 8.1
Vicilins, 287, 288, 289
Vinblastine, 340, 341, 342F
Vincristine, 340, 341, 342F
Vindoline, 343F
Violaxanthin, 89, 90, 407F, 425F
Violets, 385, 400
Viruses, 18, 56
Vitamins, 1, 88, 156, 265, 273, 280, 303, 428
Vitamin A, 303, 409, 425
Vitis, 390, 400
Vitrification, 223
Volatile attractants, 399
Vomilenine, 358

W

Walker, John, 189
Warburg, Otto, 71, 156, 175
Water oxidation, 80, 85
  potential, 93
Water relations of cells, 46, 49, 54
Water splitting manganese ion cluster, 77
Water-use-efficiency (WUE), 94, 100, 135
Wax, 51, 313, 322, 323F, 324, 325F
  wax esters, 326
Western pine beetle, 400

Wheat, 1, 34, 96, 104T, 136T, 139, 197, 212, 213, 214, 215, 217, 220, 262, 272, 287, 290, 291, 293, 294
Wood, 225, 234, 369, 370, 401

X

Xanthine, 246
  oxidase, 250
Xanthophyll, 63, 73, 74, 76, 89, 404F, 425F, 426
  cycle, 73, 89, 90
Xanthosine, 339F, 355, 357F
Xanthotoxin, 365
Xanthoxal, 407
X-ray crystallography, 76
Xylans, 229
Xylem, 55, 56, 244, 246, 248, 249, 267, 272, 275, 276
Xyloglucan endohydrolase (XEH), 230
  endotransglycosylases (XETs), 230
Xyloglucans, 229, 230F
Xylose, 225, 229, 230, 231, 232, 298
Xylosyltransferases, 230
Xylulose 5-phosphate, 98F, 99F, 106, 107, 108, 154, 412, 414, 415F

Y

Yamogenin, 428
Yams, 297, 298, 404F
Yeast, 20, 151, 172, 173, 188, 191, 207, 228, 310, 412, 413
Yew, 403F, 405, 409, Plate 12.1
Yucca, 299

Z

Zea (see also maize), 122, 126F, 293, 327, 388
Zeatin, 402, 405, 406F
Zeaxanthin, 74, 89, 90, 407F, 409, 425F, 426
Zein, 292, 293, 294
Z-scheme, 70F, 82, 83F, 85, 86, 88
Zucchini, 62